NEUROBIOLOGY OF ESSENTIAL FATTY ACIDS

ADVANCES IN EXPERIMENTAL MEDICINE AND BIOLOGY

Recent Volumes in this Series

NEUROBIOLOGY OF ESSENTIAL FATTY ACIDS

Edited by

Nicolas G. Bazan

Louisiana State University Eye Center and
. Neuroscience Center
Louisiana State University School of Medicine
New Orleans, Louisiana

Mary G. Murphy

Dalhousie University
Halifax, Nova Scotia, Canada

and

Gino Toffano

FIDIA Research Laboratories
Abano Terme, Italy

SPRINGER SCIENCE+BUSINESS MEDIA, LLC

Library of Congress Cataloging-in-Publication Data

Neurobiology of essential fatty acids / edited by Nicolas G. Bazan,
 Mary G. Murphy, and Gino Toffano.
       p.   cm. -- (Advances in experimental medicine and biology ; v.
 318)
    "Proceedings of a symposium on the neurobiology of essential fatty
 acids, held July 10-12, 1991, in Palm Cove, Far North Queensland,
 Australia"--T.p. verso.
    Includes bibliographical references and index.
    ISBN 978-0-306-44233-9   ISBN 978-1-4615-3426-6 (eBook)
    DOI 10.1007/978-1-4615-3426-6
    1. Essential fatty acids--Physiological effect--Congresses.
 2. Neurochemistry--Congresses. 3. Second messengers (Biochemistry)-
 -Congresses. 4. Cellular signal transduction--Congresses.
 I. Bazán, Nicolás G. II. Murphy, Mary G. III. Toffano, G.
 IV. Series.
    [DNLM: 1. Fatty Acids, Essential--physiology--congresses.
 2. Second Messenger Systems--congresses.   W1 AD559 v.318 / QU 90
 N494 1991]
 QP752.E84N48   1992
 612.8'042--dc20
 DNLM/DLC
 for Library of Congress   92-16989
                                                              CIP

Proceedings of a symposium on the Neurobiology of Essential Fatty Acids,
held July 10-12, 1991, in Palm Cove, Far North Queensland, Australia

ISBN 978-0-306-44233-9

© 1992 Springer Science+Business Media New York
Originally published by Plenum Press, New York in 1992

PREFACE

It is increasingly evident that polyunsaturated fatty acids (PUFA), which in the past were often believed to be mere components of cellular membranes of neural tissue, are actually major determinants of the functional properties of neural cells and are intimately involved in brain disease processes. The two families of PUFA, which are derived from the essential polyenes, linoleic ($18:2\omega6$) and α-linolenic ($18:3\omega3$) acids, constitute a major proportion (~30%) of the total fatty acids esterified to brain phospholipids. Each family is distributed in membranes in a highly specific manner with respect to brain region, cell type, and position within each of the phosphoglycerides. Unlike their behavior in other organs, the PUFA components of brain turn over slowly and are relatively resistant to dietary modification. Their unique distribution and stability suggest that they play an important role in determining the functional properties of neural cells. However, until recently very little was known regarding the precise nature of their involvement.

A great deal of excitement has been generated by the increasing realization that PUFA and phospholipids play integral roles in transmission of signals across neural membranes. PUFA, and particularly arachidonic acid (AA), can act as second messengers *per se*, or as substrates for synthesis of eicosanoids (e.g., prostaglandins, 12-lipoxygenase products) that fulfill this function. Release of PUFA from parent phospholipids, such as phosphatidylcholine and phosphatidylinositol, involves the concerted actions of various lipases, including phospholipase A_2 and C, and diacylglycerol lipase. Each of these appears to be distinct with respect to its substrate requirements and sensitivity to inhibitors.

A variety of cell-surface receptors are now known to use PUFA or their derivatives as second messengers. In response to inhibitory transmitters such as FRMFa or dopamine, AA is released from *Aplysia* sensory neurons and 12-HPETE, and related bioactive products are produced. This triggers increased opening of K^+ channels, hyperpolarization of sensory neurons, and a consequent reduction in activity in the postsynaptic motor cell. By comparison, stimulation of NMDA-type glutamate receptors results in release of free AA (and perhaps other PUFA), which appears to be associated directly with inhibition of glutamate uptake into neurons and astrocytes. Depolarization of hippocampal mossy fibers also involves release of AA which is related to stimulation of the glutamate receptors that play a central role in the development of long-term potentiation. It is reasonable to assume that PUFA are intimately involved in this phenomenon.

Independent of their roles as specific messengers for signal transduction, fatty acids and phospholipids modulate cell functions through interactions with functional proteins. Considerable excitement has been generated from observations that unsaturated fatty acids act synergistically with diacylglycerol (DAG) to activate species of protein kinase C (PKC), even at basal levels of Ca^{2+}, and may thereby provide a mechanism for sustained cross-talk between receptors. Other studies are using molecular biological approaches to identify specific regions of lipid-PKC interactions.

It has been known for some time that acute pathological events, such as ischemia, epileptic seizures, hypoglycemic coma, and trauma, trigger phospholipase activation and accumulation of free PUFA (particularly arachidonic and docosahexaenoic acids) and other potential second messengers (e.g., DAG, IP_3, eicosanoids). Possible consequences of these events include uncontrolled receptor and/or enzyme activation, loss of ionic (e.g., Ca^{2+}) homeostasis, and lipid peroxidation. Another product of phospholipid remodeling, and a relative newcomer in lipid neuropathology, is the alkyl-acetyl phospholipid, platelet-activating factor (PAF). PAF produced in response to ischemic injury appears to exert its deleterious effects through actions at synaptic membranes, with possible involvement of excitatory amino acids. However, it is now evident that PAF also acts at intracellular receptors where it stimulates expression of immediate-early genes, including c-*fos* and c-*jun*. These rapid but brief genomic responses can be linked with long-term transcriptional changes and, in this manner, can provide a mechanism for the neuronal plasticity that is required for repair processes. Such a mechanism may also be involved in the phenomenon of long-term potentiation.

This book summarizes the results of a Satellite Symposium, "The Neurobiology of Essential Fatty Acids," that was held in Palm Cove, Far North Queensland, Australia, in July, 1991. The conference was generously supported by the Fidia Research Foundation. The purpose of the symposium was to discuss exciting recent developments concerning the dynamic involvement of fatty acids and phospholipids both in nervous system function and in disease processes. The topics spanned a variety of research interests including the molecular biology of enzyme and gene regulation, physiological and pharmacological aspects of lipid metabolism, mechanisms of biological signal transduction (and possible behavioral relationships), as well as the molecular mechanisms of cerebral ischemia, convulsive disorders, and inborn errors of fatty-acid catabolism. We were particularly pleased to have some newcomers in the group. It is largely through this interdisciplinary approach that we are able to gain new perspectives and to fully appreciate the many challenges that lie ahead. This is certainly one of the most exciting scientific research fields of our time.

We are grateful to the Fidia Research Foundation for providing the financial support that made the symposium possible and to Ms. Oriana Casadei, the meeting's manager. Many thanks also to Ms. Paula Gebhardt for her expert and patient editorial assistance in the production of this book. And, of course, we are grateful to the speakers and other participants who contributed their efforts to make this symposium so valuable.

Nicolas G. Bazan
Mary G. Murphy
Gino Toffano

CONTENTS

METABOLITES OF ESSENTIAL FATTY ACIDS AND OTHER
MEMBRANE-DERIVED SECOND MESSENGERS IN CELL SIGNALING

DIETARY SUPPLY OF ESSENTIAL FATTY ACIDS,
SYNAPTOGENESIS, AND PHOTORECEPTOR MEMBRANE BIOGENESIS

THE INDUCTION OF CELLULAR GROUP II PHOSPHOLIPASE A$_2$ BY

CYTOKINES AND ITS PREVENTION BY DEXAMETHASONE

H. van den Bosch,[1] C. Schalkwijk,[1]
J. Pfeilschifter,[2] and F. Märki[2]

[1]Centre for Biomembranes and Lipid Enzymology
Padualaan 8, 3584 CH Utrecht
The Netherlands

[2]Research Department, Pharmaceuticals Division
Ciba-Geigy Ltd.
CH-4002 Basel
Switzerland

ABSTRACT

Treatment of rat glomerular mesangial cells with interleukin-1ß, tumor necrosis factor or forskolin resulted in the secretion of phospholipase A$_2$ activity into the culture medium. Essentially all of this secreted phospholipase A$_2$ activity was recognized by monoclonal antibodies elicited against rat liver mitochondrial 14 kDa group II phospholipase A$_2$. Immunoblot analysis and gel filtration confirmed the presence of only 14 kDa phospholipase A$_2$ in the culture supernatant. This enzyme could hardly be detected in unstimulated mesangial cells and after a lag period of 6 to 8 hours becomes detectable in both cells and culture medium. The results indicate that the increased phospholipase A$_2$ activity upon treatment of the cells with cytokines is not due to activation of an existing cellular pool of enzyme but is caused by induced synthesis of group II phospholipase A$_2$. Pretreatment of the cells with dexamethasone, a known inhibitor of prostaglandin synthesis, dose-dependently inhibits cytokine-induced phospholipase A$_2$ activity. Western immunoblot analysis of cells and culture medium demonstrates that this is not due to inhibition of existing phospholipase A$_2$ but because dexamethasone prevents the cytokine-induced synthesis of phospholipase A$_2$ protein.

INTRODUCTION

Phospholipase A$_2$ is believed to play an important role in the formation of important lipid mediators such as eicosanoid and platelet activating factor through regulation of the release of precursors, i.e., arachidonate and lyso-platelet activating factor, from membrane phosphoglycerides (van den Bosch, 1980). It has become clear in recent years that cellular phospholipase A$_2$ can be divided into low molecular weight, 14 kDa, and high-molecular-weight, 70-110 kDa, Ca^{2+}-dependent enzymes. The low molecular weight enzymes in turn can be divided into two groups based on their amino acid sequences. Group I phospholipases A_2 are characterized by the presence of Cys 11 and within mammals are expressed mainly in the pancreas, but also in spleen, gastric mucosa and lung. A homologous non-pancreatic group II phospholipase A_2 lacking Cys 11 has been reported for rat (Hayakawa et al., 1988), rabbit (Mizushima et al., 1989) and human (Kramer et al., 1989) platelets, rat liver (Aarsman et al., 1989) and spleen (Ono et al., 1988), and human placenta (Lai and Wada, 1988) and spleen (Kanda et al., 1989). This type of phospholipase A_2 is also found in soluble form at inflammatory sites such as peritoneal exudates (Forst et al., 1986) and human rheumatoid arthritis (Hara et al., 1989; Kramer et al., 1989; Seilhamer et al., 1989), suggesting its involvement in inflammatory processes. In line with this suggestion enzymes purified from human synovial fluid (Vadas et al., 1985) or rat platelets (Murakami et al., 1990) proved pro-inflammatory.

Interleukin-1 and tumor necrosis factor, potent pro-inflammatory cytokines, have been shown to enhance phospholipase A_2 activity and/or prostaglandin formation in chondrocytes (Suffys et al., 1988; Kerr et al., 1989; Lyons-Giordano et al., 1989; Campbell et al., 1990), human synovial cells (Godfrey et al., 1987), rat (Pfeilschifter et al., 1989 a,b; Pfeilschifter and Mühl, 1990) and human (Topley et al., 1989) mesangial cells, rat vascular smooth muscle cells (Nakano et al., 1990a), and rat and human hepatoma cells (Crowl et al., 1991). Treatment of rabbit chondrocytes (Kerr et al., 1989; Lyons-Giordano et al., 1989), rat smooth muscle cells (Nakano et al., 1990b), and human HepG2 cells (Crowl et al., 1991) with cytokines was shown to increase the levels of mRNA for group II phospholipase A_2.

In previous studies with cultured rat mesangial cells it was observed that treatment with interleukin-1ß and tumor necrosis factor α induced the release of a phospholipase A_2 activity into the culture medium in parallel with a highly enhanced prostaglandin E_2 synthesis (Pfeilschifter et al., 1989a,b; Pfeilschifter and Mühl, 1990). However, the type of phospholipase A_2 responsible for the secreted phospholipase A_2 activity remained to be characterized. When we embarked on these experiments the only type of phospholipase A_2 reported to be present in mesangial cells was a hormonally regulated high-molecular-weight phospholipase A_2 that showed a molecular weight of 60-70 kDa upon gel filtration (Gronich et al., 1988) and later was shown to migrate as a 110 kDa protein in SDS-PAGE (Gronich et al., 1990). This enzyme is translocated from the cytosol to membranes in a Ca^{2+}-dependent manner and becomes fully activated over the range of 0.1 to 1 μM calcium. It shares these properties with a number of high-molecular-weight phospholipases A_2 that show a preference for arachidonate and that have quite recently been purified from the human monocytic cell line U937 (Clark et al., 1990; Kramer et al., 1991) and from bovine (Kim et al., 1991b) and rabbit (Kim et al., 1991a) platelets.

In this manuscript we demonstrate that this high-molecular-weight phospholipase A_2 is not secreted from rat mesangial cells. The phospholipase A_2 activity that becomes elevated in cells and culture supernatants upon treatment of the cells with inter-leukin-1ß, tumor necrosis factor, or forskolin is caused by induced synthesis and secretion of an additional 14 kDa group II phospholipase A_2 that has hitherto not been identified for mesangial cells (Schalkwijk et al., 1991). In addition, we demonstrate that the cytokine-induced synthesis of this enzyme can be prevented by pretreatment of the cells with dexamethasone.

RESULTS

Characterization of Secreted Phospholipase A_2

Figure 1 shows that the little phospholipase A_2 activity is secreted from control mesangial cells and that release into the medium is highly stimulated by interleukin-1ß (IL-1ß), tumor necrosis factor α (TNF), and IL-1ß plus forskolin, whereas forskolin alone is less effective. This figure also shows that in all cases essentially the total phospholipase A_2 activity is bound by monoclonal antibody (McAb)-Sepharose using McAbs against rat liver phospholipase A_2. These McAbs have previously been shown to recognize rat group II phospholipase A_2 and to have no cross-reactivity with the rat group I enzyme (de Jong et al., 1987; Aarsman et al., 1989). The nearly complete binding of the secreted phospholipase A_2 activity by the McAb-Sepharose raised the question whether these antibodies show cross-reactivity with the high-molecular-weight phospholipase A_2 known to be present in mesangial cells (Gronich et al., 1988; 1990) or whether these cells contain an additional 14 kDa phospholipase A_2. This was resolved by the Western immunoblot depicted in Figure 2. In all cases except the supernatant of control cells, only a band corresponding to a 14 kDa phospholipase A_2 was detected. Moreover, the intensities of these bands correlated favorably with the activities measured in Figure 1, i.e., they increased in the order: forskolin, TNFα, IL-1ß and IL-1ß + forskolin. Since all activity bound to McAb-Sepharose and no cross-reactivity with a high-molecular-weight phospholipase A_2 was observed, these data indicate that only 14 kDa enzyme is secreted. This was confirmed in an experiment in which the cells were stimulated with IL-1ß + forskolin in the presence of a cocktail of six protease inhibitors to exclude the possibility that a high-molecular-weight phospholipase A_2 was in fact secreted and then became proteolytically converted to a 14 kDa enzyme. The culture supernatant obtained in this experiment was subjected to gel filtration and found to contain a single phospholipase A_2 activity peak corresponding to a molecular mass of 14 kDa (data not shown) (Schalkwijk et al., 1991).

Induction and Secretion of Phospholipase A_2

We then addressed the question whether the enhanced levels of phospholipase A_2 activity and phospholipase A_2 protein in the culture supernatants of stimulated cells

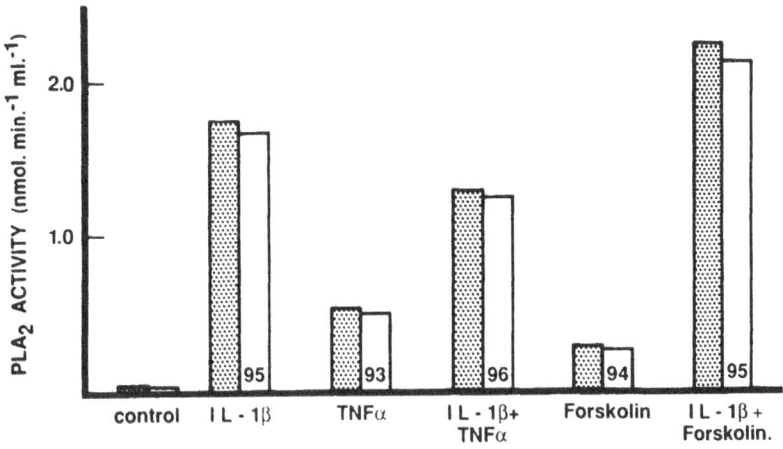

Figure 1. Phospholipase A_2 activity in culture medium of rat mesangial cells. Cells were stimulated for 48 hours by the indicated agents at concentrations given in parentheses below. Cell culture, measurement of phospholipase A_2 activity, and immunoprecipitation were done as previously described (Pfeilschifter et al., 1989a; Schalkwijk et al., 1991). Abbreviations: IL-1ß, interleukin-1ß (1 nM); TNF, tumor necrosis factor (1 nM); forskolin (10 μM). Stippled bars: phospholipase A_2 activity in culture medium; open bars: phospholipase A_2 activity bound by McAb-Sepharose. Numbers in bars indicate percentage of total activity recovered from McAb-Sepharose precipitates.

3

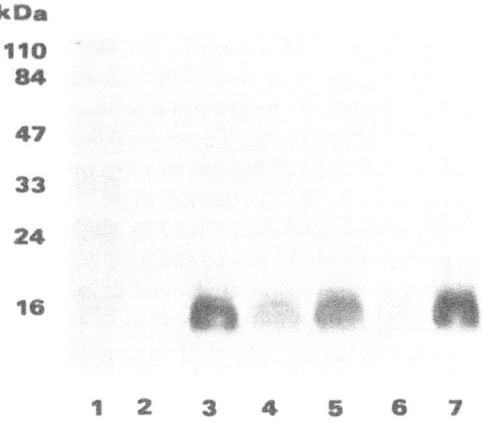

Figure 2. Immuno-detection of phospholipase A$_2$ secreted from rat mesangial cells. Lanes: 1, molecular weight markers; 2 to 7, culture medium from control cells and cells stimulated with: 3, IL-1ß; 4, TNFα; 5, IL-1ß + TNFα; 6, forskolin; and 7, IL-1ß + forskolin. Reproduced with permission (Schalkwijk et al., 1991).

were caused by enhanced secretion of enzyme from a pre-existing cellular pool (which then would become depleted upon stimulation) and whether all enzyme became secreted. This was done by repeating the Western blot experiment with the cells rather than the culture medium. The results clearly showed that the enzyme could not be detected in unstimulated cells (data not shown). Thus, there is no preexisting cellular store of the secreted enzyme. Rather, the enzyme was newly synthesized upon stimulation. The time-dependency of phospholipase A$_2$ synthesis and secretion as induced by IL-1ß + forskolin was then followed by activity measurements and Western blot analysis. As can be seen in Figure 3 at zero time (control cells) the enzyme can be detected neither in the cells nor in the medium. After a lag period of 6 to 8 hrs, phospholipase A$_2$ activity rapidly increased in the stimulated cells in conjunction with increased enzyme protein levels. No increases in enzyme protein levels were observed in control cells and only blots of stimulated cells are shown. The experiments with

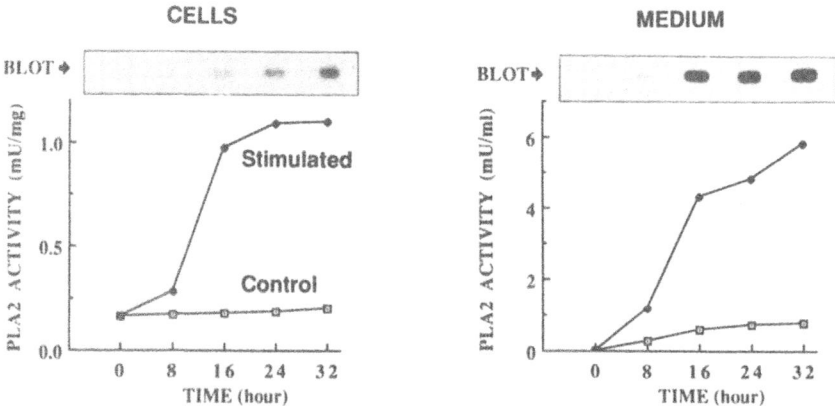

Figure 3. Induction of phospholipase A$_2$ activity and enzyme protein in rat mesangial cells and its secretion into the culture medium as a function of time after stimulation with interleukin-1ß + forskolin. Reproduced with permission (Schalkwijk et al., 1991).

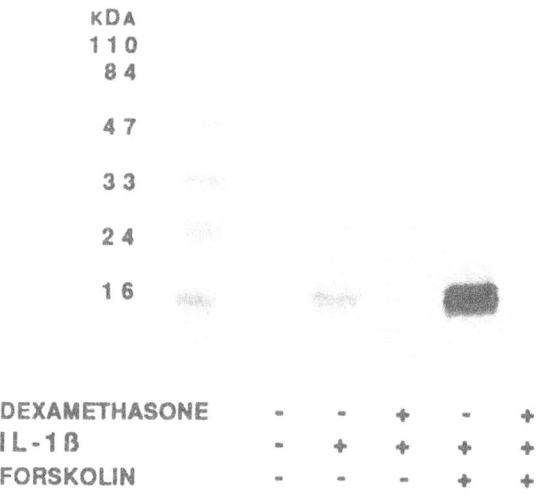

KDA
110
84

47

33

24

16

DEXAMETHASONE   -   -   +   -   +
IL-1ß   -   +   +   +   +
FORSKOLIN   -   -   -   +   +

Figure 4. Pretreatment of mesangial cells with dexamethasone prevents IL-1ß- and IL-1ß + forskolin-induced synthesis and secretion of phospholipase A_2.

stimulated cells demonstrate that the increased phospholipase A_2 activity is not due to activation of existing enzyme (or relief of enzyme inhibition), but is caused by induced synthesis of new enzyme. The time-courses of enzyme activity and protein secretion were similar. A comparison of total phospholipase A_2 activity in cells and medium demonstrated that at all time points after the lag period 85 to 90% of the activity was present in the medium indicating secretion of the majority of the newly synthesized enzyme.

Effect of Dexamethasone on Induced Synthesis of Phospholipase A_2

It is well known that glucocorticosteroids inhibit prostaglandin formation in many cells and animal models of inflammation without affecting cyclooxygenase. This inhibition has been ascribed to glucocorticoid-induced synthesis of phospholipase A_2 inhibitory proteins, termed *lipocortins* (Flower, 1988; Davidson and Dennis, 1989). In view of the fact that the cytokine-stimulated formation of prostaglandin E_2 in rat mesangial cells could also be prevented by dexamethasone (Pfeilschifter et al., 1989b) and considering the induction of phospholipase A_2 in these cells by cytokines as described above, we investigated the effect of dexamethasone on the cytokine-induced synthesis of phospholipase A_2. Figure 4 confirms again that no phospholipase A_2 can be detected by immunostaining of Western blots of the culture medium from unstimulated cells and that IL-1ß and especially IL-1ß + forskolin induce the secretion of phospholipase A_2. In both cases the induced secretion can be completely prevented by pretreatment of the cells with dexamethasone. The dose-dependency of this process, with respect both to phospholipase A_2 activity and enzyme protein levels is shown in Figure 5. Phospholipase A_2 activity in the culture medium increased sharply upon stimulation with either IL-1ß alone or IL-1ß + forskolin. When the cells were pretreated with increasing concentrations of dexamethasone for 6 hours prior to addition of the stimulators, a gradual decrease in the amount of phospholipase A_2 activity secreted into the medium was observed. This lower activity was not caused by

5

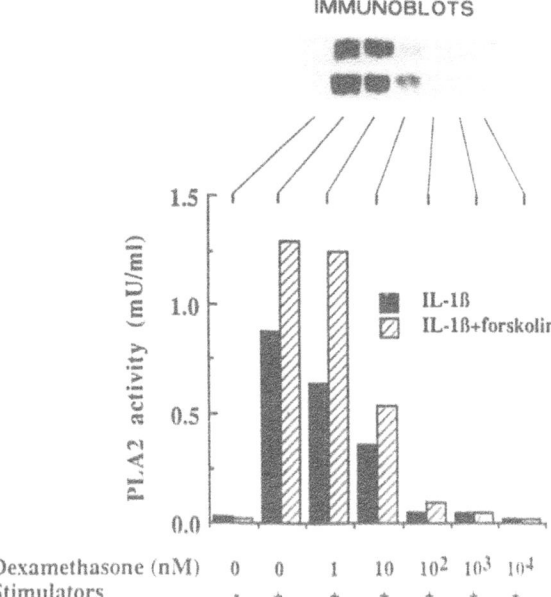

IMMUNOBLOTS

Figure 5. Dose-dependency of the inhibition by dexamethasone of cytokine- and forskolin-induced synthesis and secretion of phospholipase A_2. Cells were pretreated for 6 hr with the indicated concentrations of dexamethasone and then stimulated for 24 hr with either IL-1ß or IL-1ß + forskolin in concentrations as given in the legend of Figure 1.

inhibition of secreted phospholipase A_2 but, as shown in the blots, was due to the fact that gradually less enzyme protein was secreted. As observed for the culture medium (Fig. 5), the patterns of activity and enzyme protein levels in the cells were similar (data not shown). These results convincingly demonstrate that dexamethasone dose-dependently prevents the cytokine- and forskolin-induced synthesis and subsequent secretion of group II phospholipase A_2 from mesangial cells.

DISCUSSION

Interleukins and tumor necrosis factor have been shown to induce prostaglandin E_2 synthesis in parallel to phospholipase A_2 release into the medium of cultured rat glomerular mesangial cells (Pfeilschifter et al., 1989a). The stimulation of phospholipase A_2 secretion by forskolin has also been reported for rat mesangial cells (Pfeilschifter et al., 1991). Although a high-molecular-weight form of phospholipase A_2 had been described for mesangial cells (Gronich et al., 1988, 1990), we now demonstrate that this is not the enzyme secreted under these conditions. Rather, an additional 14 kDa group II phospholipase A_2, identified by gel filtration and immunochemical experiments, appears to be responsible for the phospholipase A_2 activity secreted upon cytokine- and forskolin stimulation. In retrospect it is not surprising that this enzyme was not previously detected in mesangial cells. As shown in Figure 3 it cannot be detected in unstimulated cells. The cytokine-induced synthesis and secretion of this phospholipase A_2 in parallel to enhanced prostaglandin synthesis suggests a role for this enzyme in the release of arachidonate for prostaglandin formation. Although part of the enhanced

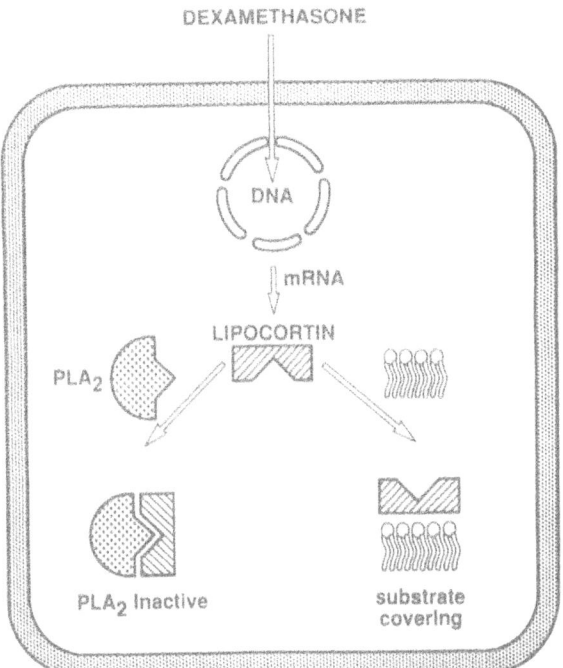

Figure 6. Previous models for the inhibition of eicosanoid formation by glucocorticosteroids through induced synthesis of the phospholipase A_2 inhibitory lipocortin.

synthesis may be explained by interleukin-increased cyclooxygenase activity (Raz et al., 1988; Coyne et al., 1990; Pfeilschifter and Mühl, 1990), it is generally believed that the release of arachidonate, rather than cyclooxygenase activity, is the rate-limiting step in eicosanoid formation.

An additional correlation between phospholipase A_2 activity and prostaglandin E_2 formation in mesangial cells was observed in the inhibition of both by the anti-inflammatory glucocorticoid dexamethasone. The inhibitory effect of dexamethasone has previously been ascribed to the induced synthesis of the phospholipase A_2 inhibitory protein lipocortin (Flower, 1988; Davidson and Dennis, 1989). Originally, it was thought that lipocortin inhibited phospholipase A_2 by association with the enzyme to give an inactive complex (Hirata, 1981; Fig. 6). Subsequently, Davidson et al. (1987) showed that inhibition of pancreatic group I phospholipase A_2 by lipocortin could only be observed at low substrate concentrations, and we confirmed this phenomenon (Aarsman et al., 1987) for cellular group II phospholipases A_2. Thus, at best, lipocortin inhibition of phospholipase A_2 involved binding of lipocortin to phospholipid substrate to make this substrate unavailable to the enzyme (Fig. 6). Even this explanation became unlikely as the sole mechanism for glucocorticoid inhibition of prostaglandin formation, when it was reported that glucocorticoids, while still reducing phospholipase A_2 activity and/or eicosanoid production, did not affect the cellular levels of lipocortin (Wallner et al., 1986; Bienkowski et al., 1989; Hullin et al., 1989; Nakano et al., 1990b). In another report (Piltch, 1989), dexamethasone increased the cellular levels of lipocortin but the decrease in phospholipase A_2 activity was noted even in lipocortin-depleted cellular fractions suggesting the presence of a lipocortin-independent mechanism of reducing phospholipase A_2 activity. The results reported in Figures 4 and 5 provide such an alternative mechanism for the anti-inflammatory action of glucocorticoids in that dexamethasone completely blocks the cytokine- and forskolin-induced expression of phospholipase A_2 protein. These observations in rat mesangial cells confirm a recent report of Nakano et al. (1990a) who demonstrated the suppression of TNF- and forskolin-induced synthesis of group II phospholipase A_2 by dexamethasone in rat smooth muscle cells. In these cells, low concentrations of dexamethasone affected forskolin-induced phospholipase A_2 synthesis by inhibition of mRNA synthesis, whereas 10-fold higher concentrations were required to completely prevent TNF-induced enzyme expression at, presumably, a post-transcriptional level (Nakano et al., 1990a). Whether IL-1ß and forskolin influence phospholipase A_2 gene expression in mesangial cells also by different mechanisms remains to be seen.

REFERENCES

Aarsman AJ, de Jong JGM, Arnoldussen E, Neys FW, van Wassenaar PD, van den Bosch H (1989) Immunoaffinity purification, partial sequence, and subcellular localization of rat liver phospholipase A_2. J Biol Chem 264:10008.

Aarsman AJ, Mijnbeek G, van den Bosch H, Rothhut B, Prieur B, Comera C, Jordan L, Russo-Marie F (1987) Lipocortin inhibition of extracellular and intracellular phospholipases A_2 is substrate concentration dependent. FEBS Lett 219:176.

Bienkowski MJ, Petro MA, Robinson LJ (1989) Inhibition of thromboxane A synthesis in U937 cells by glucocorticoids. Lack of evidence for lipocortin 1 as the second messenger. J Biol Chem 264:6536.

Campbell IK, Piccoli DS, Hamilton JA (1990) Stimulation of human chondrocyte prostaglandin E_2 production by recombinant human interleukin-1 and tumour necrosis factor. Biochim Biophys Acta 1051:310.

Clark JD, Milona N, Knopf JL (1990) Purification of a 110-kilodalton cytosolic phospholipase A_2 from the human monocytic cell line U937. Proc Natl Acad Sci USA 87:7708.

Coyne DW and Morrison AR (1990) Effect of the tyrosine kinase inhibitor, genistein, on interleukin-1 stimulated PGE_2 production in mesangial cells. Biochem Biophys Res Commun 173:718.

Crowl RM, Stoller TK, Conroy RR, Stouer CR (1991) Induction of phospholipase A_2 gene expression in human hepatoma cells by mediators of the acute phase response. J Biol Chem 266:2647.

Davidson FF and Dennis EA (1989) Biological relevance of lipocortins and related proteins as inhibitors of phospholipase A_2. Biochem Pharmacol 38:3645.

Davidson FF, Dennis EA, Powell M, Glenney J (1987) Inhibition of phospholipase A_2 by lipocortins and calpactins: an effect of binding to substrate phospholipids. J Biol Chem 262:1698

de Jong JGN, Amesz H, Aarsman AJ, Lenting HBM, van den Bosch H (1987) Monoclonal antibodies against an intracellular phospholipase A_2 from rat liver and their cross-reactivity with other phospholipases A_2. Eur J Biochem 164:129.

Flower RJ (1988) Lipocortin and the mechanism of action of the glucocorticoids. Br J Pharmacol 94:987.

Forst S, Weiss J, Elsbach P, Maraganore JM, Reardon I, Heinrikson RL (1986) Structural and functional properties of a phospholipase A_2 purified from an inflammatory exudate. Biochemistry 25:8381.

Godfrey RW, Johnson WK, Hoffstein ST (1987) Recombinant tumor necrosis factor and interleukin-1 both stimulate human synovial cell arachidonic acid release and phospholipid metabolism. Biochem Biophys Res Commun 142:235.

Gronich JH, Bonventre JV, Nemenoff RA (1988) Identification and characterization of a hormonally regulated form of phospholipase A_2 in rat renal mesangial cells. J Biol Chem 263:16645.

Gronich JH, Bonventre JV, Nemenoff RA (1990) Purification of a high-molecular-mass form of phospholipase A_2 from rat kidney activated at physiological calcium concentrations. Biochem J 271:37.

Hara S, Kudo I, Chang HW, Matsuta K, Miyamoto Y, Inoue K (1989) Purification and characterization of extracellular phospholipase A_2 from human synovial fluid in rheumatoid arthritis. J Biochem 105:395.

Hayakawa M, Kudo I, Tomita M, Nojima S, Inoue K (1988) The primary structure of rat platelet phospholipase A_2. J Biochem 104:767.

Hirata F (1981) The regulation of lipomodulin, a phospholipase inhibitory protein, in rabbit neutrophils by phosphorylation. J Biol Chem 256:7730.

Hullin F, Raynal P, Ragab-Thomas JMF, Fauvel J, Chap H (1989) Effect of dexamethasone on prostaglandin synthesis and on lipocortin status in human endothelial cells. J Biol Chem 264:3506.

Kanda A, Ono T, Yoshida N, Tojo H, Okamoto M (1989) The primary structure of a membrane-associated phospholipase A_2 from human spleen. Biochem Biophys Res Commun 163:42.

Kerr JS, Stevens TM, Davis GL, McLaughlin JA, Harris RR (1989) Effects of recombinant interleukin-1 beta on phospholipase A_2 activity, phospholipase A_2 mRNA levels, and eicosanoid formation in rabbit chondrocytes. Biochem Biophys Res Commun 165:1079.

Kim DK, Kudo I, Inoue K (1991a) Purification and characterization of rabbit platelet cytosolic phospholipase A_2. Biochim Biophys Acta 1083:80.

Kim DK, Sub PG, Rya SH (1991b) Purification and some properties of a phospholipase A_2 from bovine platelets. Biochem Biophys Res Commun 174:189.

Kramer RM, Hession C, Johansen B, Hayes G, McGray P, Pinchang Chow E, Tizard R, Pepinsky RB (1989) Structure and properties of a human non-pancreatic phospholipase A_2. J Biol Chem 264:5760.

Kramer RM, Roberts EF, Manetta J, Putnam JE (1991) The Ca^{++}-sensitive cytosolic phospholipase A_2 is a 100-kDa protein in human monoblast U937 cells. J Biol Chem 266:5268.

Lai C-Y and Wada K (1988) Phospholipase A_2 from human synovial fluid: purification and structural homology to the placental enzyme. Biochem Biophys Res Commun 157:488.

Lyons-Giordano B, Davis GL, Galbraith W, Pratta NA, Arner EC (1989) Interleukin-1ß stimulates phospholipase A_2 mRNA synthesis in rabbit articular chondrocytes. Biochem Biophys Res Commun 164:488.

Mizushima H, Kudo I, Horigome K, Murakami M, Hayakawa M, Kim DK, Kondo E, Tomita M, Inoue K (1989) Purification of rabbit platelet secretory phospholipase A_2 and its characteristics. J Biochem 105:520.

Murakami M, Kudo I, Nakamura H, Yokoyama Y, Mori H, Inoue K (1990) Exacerbation of rat adjuvant arthritis by intradermal injection of purified mammalian 14-kDa group II phospholipase A_2, FEBS Lett 268:113.

Nakano T, Ohara O, Teraoka H, Arita H (1990a) Group II phospholipase A_2 mRNA synthesis is stimulated by two distinct mechanisms in rat vascular smooth muscle cells. FEBS Lett 261:171.

Nakano T, Ohara O, Teraoka H, Arita H (1990b) Glucocorticoids suppress group II phospholipase A_2 production by blocking mRNA synthesis and post-transcriptional expression. J Biol Chem 265:12745.

Ono T, Tojo H, Kuramitsu S, Kagamiyama H, Okamoto M (1988) Purification and characterization of a membrane-associated phospholipase A_2 from rat spleen. J Biol Chem 263:5732.

Pfeilschifter J and Mühl H (1990) Interleukin-1 and tumor necrosis factor potentiate angiotensin II- and calcium ionophore-stimulated prostaglandin E_2 synthesis in rat renal mesangial cells. Biochem Biophys Res Commun 169:585.

Pfeilschifter J, Leighton J, Pignat W, Marki F, Vosbeck K (1991) Cyclic AMP mimics, but does not mediate, interleukin-1 and tumour necrosis-factorstimulated phospholipase A_2 secretion from rat renal mesangial cells. Biochem J 273:199.

Pfeilschifter J, Pignat W, Vosbeck K, Marki F (1989a) Interleukin-1 and tumor necrosis factor synergistically stimulate prostaglandin synthesis and phospholipase A_2 release from rat renal mesangial cells. Biochem Biophys Res Commun 159:385.

Pfeilschifter J, Pignat W, Vosbeck K, Marki F, Wiesenberg I (1989b) Susceptibility of interleukin 1- and tumor necrosis factor-induced prostaglandin E_2 and phospholipase A_2 release from rat renal mesangial cells to different drugs. Biochem Soc Trans 17:916.

Piltch A, Sun L, Fava RA, Hayashi J (1989) Lipocortin-independent effect of dexamethasone on phospholipase activity in a thymic epithelial cell line. Biochem J 261:395,

Raz A, Wyche A, Siegel N, Needleman P (1988) Regulation of fibroblast cyclooxygenase synthesis by interleukin-1. J Biol Chem 263:3022.

Schalkwijk C, Pfeilschifter J, Marki F, van den Bosch H (1991) Interleukin-1ß, tumor necrosis factor and forskolin stimulate the synthesis and secretion of group II phospholipase A_2 in rat mesangial cells. Biochem Biophys Res Commun 174:268.

Seilhamer JJ, Pruzanski W, Vadas P, Plant S, Miller JA, Kloss J, Johnson LK (1989) Cloning and recombinant expression of phospholipase A_2 present in rheumatoid arthritic synovial fluid. J Biol Chem 264:5335.

Suffys P, van Roy F, Piers W (1988) Tumor necrosis factor and interleukin 1 activate phospholipase in rat chondrocyte. FEBS Lett 232:24.

Topley N, Floege J, Wessel K, Hass R, Radeke HH, Kaever V, Resch K (1989) Prostaglandin E_2 production is synergystically increased in cultured human glomerular mesangial cells by combinations of IL-1 and tumor necrosis factor-α. J Immunol 143:1989.

Vadas P, Stefanski E, Pruzanski W (1985) Characterization of extracellular phospholipase A_2 in rheumatoid synovial fluid. Life Sci 36:579.

van den Bosch H (1980) Intracellular phospholipases A. Biochim Biophys Acta 604:191.

Wallner BP, Mattaliano RJ, Hession C, Cate RL, Tizard R, Sinclair LJ, Foeller C, Chow EP, Browning JL, Ramachandran KL, Pepinsky RB (1986) Cloning and expression of human lipocortin, a phospholipase A_2 inhibitor with potential anti-inflammatory activity. Nature 320:77.

BRAIN PHOSPHOLIPASES AND THEIR ROLE IN SIGNAL TRANSDUCTION

Akhlaq A. Farooqui, Yutaka Hirashima,
and Lloyd A. Horrocks

Department of Medical Biochemistry
1645 Neil Ave., Rm 471
The Ohio State University
Columbus, OH 43210

In response to stimuli, neural membranes release arachidonate which is then metabolized to biologically active eicosanoids (Chang et al., 1987). In neural membranes arachidonate is located exclusively in the sn-2 position of membrane phospholipids and is released through various direct and indirect enzymic pathways. The direct pathway involves the stimulation of phospholipase A_2. One indirect pathway requires the activation of phospholipase C followed by diacylglycerol and monoacylglycerol lipases. A second indirect pathway releases arachidonate by utilizing a lysophospholipase preceded by phospholipase A_1. The hydrolysis of phospholipids by phospholipase D produces phosphatidic acid, which can then be hydrolyzed by a phosphatidic acid-specific phospholipase A_2. The action of a phosphatase on phosphatidic acid generates diacylglycerol, which can be hydrolyzed by diacylglycerol and monoacylglycerol lipases.

Although the relative contributions of these pathways to the release of arachidonate are still obscure, the importance of the direct deacylation of phospholipids by phospholipase A_2 and the action of di- and monoacylglycerol lipases preceded by phospholipase C has been clearly established (Chang et al., 1987). Indeed, phospholipases have emerged as a family of ubiquitous enzymes that modulate cell function by regulating the levels of lipid metabolites with second messenger properties (Dennis et al., 1991). The purpose of this review is to describe the properties and regulation of brain phospholipases and to evaluate the metabolic interactions between phospholipases A_2, C, and D at the molecular level.

PHOSPHOLIPASE A_2

Phospholipases A_2 are ubiquitous enzymes that catalyze the hydrolysis of phospholipids, specifically at the sn-2 ester bond (Chang et al., 1987; Dennis et al., 1991). Bovine brain microsomes and cytosol contain a phospholipase A_2 that acts on phosphatidylinositol (Gray and Strickland, 1982a,b). Microsomal phospholipase A_2 solubilized with 0.05% Triton X-100 and a multiple column chromatographic procedure results in

1600-fold purification, compared with bovine brain (grey matter) homogenate. The activity of the purified enzyme is greatest with phosphatidylinositol (100%), compared with phosphatidylcholine (62%), phosphatidic acid (33%), phosphatidylethanolamine (25%), and phosphatidylserine (21%) (Gray and Strickland, 1982a,b). The purified enzyme shows a single band on SDS gel electrophoresis and has a molecular mass of 18.3 kDa (Gray and Strickland, 1982a). Amino acid analysis data gives a molecular mass value of 18.5 kDa. The enzyme contains large amounts of aspartate, glutamate, and glycine. Single residues of cysteine, tyrosine, and arginine are also present. This bovine brain microsomal phospholipase A_2 is stimulated with a high concentration of calcium ions (5 mM). Other metal ions, including Mg^{2+}, Mn^{2+}, and Zn^{2+}, inhibit the activity of the purified enzyme. It has been suggested that this phospholipase A_2 may be a glycoprotein (Gray and Strickland, 1982a). Phospholipase A_2 activity has also been reported in rat brain (Witter and Kanfer, 1985). The rat brain enzyme is stimulated by Ca^{2+} and mellitin and releases monounsaturated fatty acids from endogenous phospholipids. Purification and properties of this enzyme have not been reported.

Recently two forms of phospholipase A_2 activity (PLA_2-H and PLA_2-L) have been detected in rat brain cytosol (Yoshihara and Watanabe, 1990). PLA_2-H is eluted as a broad peak of 200-500 kDa, and its activity is partially inhibited by the addition of 4 mM Ca^{2+}. In contrast, PLA_2-L has a molecular mass of 100 kDa and shows an absolute requirement for Ca^{2+} (Yoshihara and Watanabe, 1990). The activity of PLA_2-L changes biphasically with increasing concentrations of Ca^{2+}. That is, the first stimulation is observed at 1 μM Ca^{2+} and the second at 10 mM Ca^{2+}. In neurons, the concentration of cytosolic Ca^{2+} is maintained at 100-300 nM in the resting state and increases to 1-2 μM during excitation (Balow et al., 1986). Another phospholipase A_2, located on the outer surface of cultured cells of neuronal and glial origin (Edgar et al., 1980), is not affected by Ca^{2+}. This enzyme has not been characterized in detail. However, it was proposed (Edgar et al., 1980) that during receptor stimulation, this ectophospholipase A_2 may be stimulated first, resulting in the opening of calcium channels and the entry of calcium into the neural cell, which could cause the stimulation of other phospholipases. A similar ectophospholipase A_2 is also found in rabbit neutrophils (Kennedy and Becker, 1987).

Using pyrenesulfonyl-labeled ethanolamine glycerophospholipids, we detected substantial amounts of phospholipase A_2 activity in bovine brain cytosol (Hirashima et al., 1990a/b; in press). One substrate for this enzyme is the acid-stable ethanolamine glycerophospholipid, the mixture of 1,2-diacylglycero-3-phosphoethanolamine [PtdEtn(Pyr)] and 1-alkyl-2-acylglycero-3-phosphoethanolamine [PakEtn(Pyr)]. The other substrate is pure 1-alk-1′-enyl-2-acylglycero-3-phosphoethanolamine [plasmenylethanolamine, PlsEtn(Pyr)]. Peak I is active with only the acid-stable ethanolamine glycerophospholipids, whereas Peak II is active with both that lipid mixture and PlsEtn(Pyr). The enzyme is precipitated by the addition of solid ammonium sulfate (40%). Ultrogel AcA54 column chromatography of ammonium sulfate fraction gives a single peak in the void volume when chromatographed without 1 M KCl. However, when the column is developed in the presence of 1 M KCl, two active peaks are obtained. One peak (Peak I) is eluted in the void volume whereas the other peak (Peak II) is washed out with an apparent molecular mass of 39 kDa (Fig. 1). The Peak I phospholipase A_2 protein has a molecular mass of 110 kDa. Partial purification of Peaks I and II by plasmenylethanolamine-Affigel 10 chromatography and HPLC on a MA7Q column indicates the existence of isozymes with different substrate specificity (Hirashima et al., 1990a/b; in press). Peaks I and II show optimal activity at pH 7.4 and 8.0 respectively and are not activated by Ca^{2+} ions. With the mixture of PtdEtn(Pyr) and PakEtn(Pyr) used as substrate, the apparent K_m and V_{max} values for Peak I are 29 μM and 13 nmol/min/mg, respectively, whereas for Peak II, the corresponding values are 70 μM and 36 nmol/min/mg (Table 1).

Table 1. Kinetic properties of bovine brain cytosolic phospholipases A_2

Enzyme	Property	Value
Peak I	pH optimum	7.4
	Apparent K_m value (μM)	29.0
	V_{max} (nmol/min/mg)	13.0
Peak II	pH optimum	8.0
	Apparent K_m value (μM)	70.0
	V_{max} (nmol/min/mg)	36.0

Summarized from Hirashima et al., in press.

The occurrence of high-molecular-weight, calcium-independent phospholipases A_2 has been recently reported in several tissues (Clark et al., 1990; Gronich et al., 1990; Kim et al., 1991; Kramer et al., 1991). Some high-molecular-weight phospholipases A_2, similarly to the bovine brain enzyme, do not respond to calcium ions (Hazen et al., 1990; Kim et al., 1991; Yoshihara and Watanabe, 1990) whereas others do require calcium ions for their activity (Yoshihara and Watanabe, 1990). A high-molecular-weight (85.2 kDa) cytoplasmic phospholipase A_2 has been recently cloned from human monocytic cell line U937 (Clark et al., 1991). Similarly to rat brain phospholipase A_2 (Yoshihara and Watanabe, 1990), this enzyme is translocated from cytosol to membrane in response to changes in free Ca^{2+}. It is reported that the amino terminal end of phos-

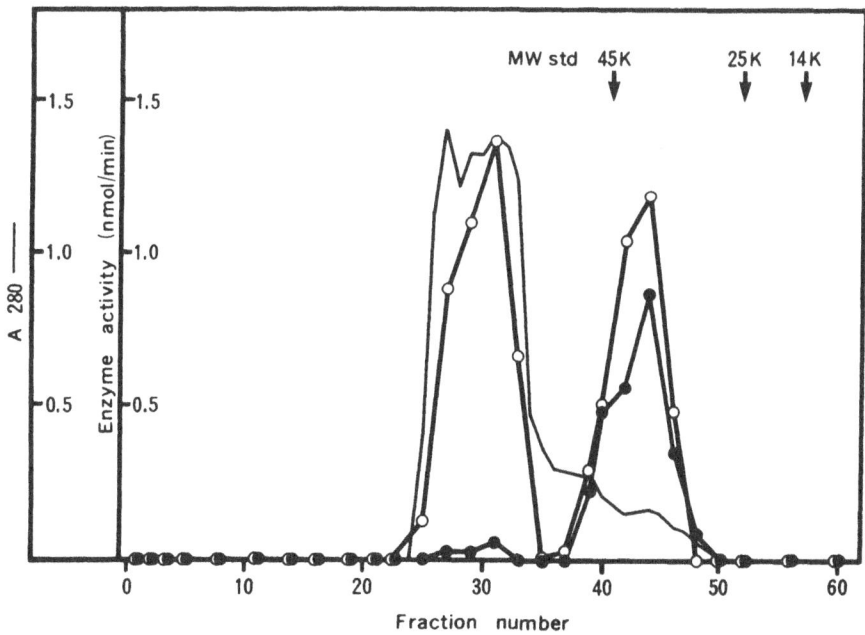

Figure 1. Elution profile of bovine brain cytosolic phospholipase A_2 through an Ultrogel AcA54 column. For enzyme activity, PtdEtn(Pyr) + PakEtn(Pyr) (open circles) and PlsEtn(Pyr) (filled circles) were used as substrate. The solid line without circles indicates absorbance at 280 nm. From Hirashima et al., in press.

pholipase A_2 that binds to membrane contains a 45-amino-acid region (Ca^{2+}-dependent phospholipid-binding domain) which has sequence homology with protein kinase C, synaptic vesicle protein p65, GTPase activating protein, and phospholipase C (Clark et al., 1991). The existence of a Ca^{2+}-dependent phospholipid binding domain in phospholipases and kinases suggests that this property may be a general mechanism which translocates and activates cytosolic enzymes (Clark et al., 1991). The relationship between low-molecular-weight phospholipases A_2 and high-molecular-weight phospholipases A_2 is not understood. However, some of high-molecular-weight phospholipases A_2 may represent aggregated forms of low-molecular-weight enzymes or the low-molecular-weight forms associated with regulatory elements.

PHOSPHOLIPASE C/DIACYLGLYCEROL LIPASE PATHWAY

Phospholipase C cleaves phospholipids at the phospho moiety, forming the polar head group and diacylglycerol (Carter et al., 1990). The most thoroughly investigated group of phospholipases C is the phosphatidylinositol-hydrolyzing phospholipase C. This enzyme liberates two second messengers, 1,2-diacylglycerol and inositol 1,4,5-trisphosphate. Diacylglycerol activates protein kinase C, stimulating the phosphorylation of a number of intracellular proteins. Inositol 1,4,5-trisphosphate releases calcium from the endoplasmic reticulum, which initiates other Ca^{2+}-dependent processes. Multiple forms of phospholipase C are known to occur in mammalian brain (Carter et al., 1990). Five immunologically distinct phospholipase C (PLC) activities have been reported to occur in the cytosolic fraction of tissue homogenates. They are referred to as PLC-α, PLC-ß, PLC-γ, PLC-δ, and PLC-ξ (Table 2). Expression of PLC-ß, PLC-γ, and PLC-δ was also studied in primary cell cultures of rat brain (Mizuguchi et al., 1991). Immunoreactivity of cultured neurons, astrocytes, and oligodendrocytes was shown for all three isozymes by immunocytochemical staining and immunoblotting, with some differences in reaction intensity. Immunoblotting studies indicate that the level of expression is neurons > oligodendrocytes > astrocytes for PLC-ß and PLC-γ and astrocytes > oligodendrocytes > neurons for PLC-δ (Mizuguchi et al., 1991). Using antibodies raised against various phospholipases C, it has been shown that some isozymes are present in brain cytosolic fractions (PLC-γ) while others (PLC-ß and PLC-α) are found equally distributed between the cytosolic and particulate fractions (Carter et al., 1990). Several isozymes of phospholipase C have been purified from bovine and rabbit brain by means of multiple column chromatographic procedures (Carter et al., 1990; Dennis et al., 1991). The major form (PLC-ßm) from rabbit brain is obtained in a homogeneous state. It has a molecular mass of 155 kDa. Although the isozymes of phospholipase C differ from one another in physiochemical properties, the kinetic properties are very similar (Carter et al., 1990). Purified isozymes of phospholipase C show an absolute requirement for Ca^{2+} and are very sensitive to changes in Ca^{2+} concentration in the micromolar range (Carter et al., 1990; Fain, 1990). The isozymes are also stimulated by sodium fluoride. Additional isozymes are being reported.

DIACYLGLYCEROL LIPASE

Bovine brain contains two diacylglycerol lipases. One is localized in microsomes and the other is found in a crude plasma membrane fraction. The kinetic parameters and heat inactivation profiles of the two diacylglycerol lipases are quite similar (Farooqui et al., 1985; 1986a). Both lipases are markedly inhibited by RHC 80267, a potent inhibitor of diacylglycerol lipase (Sutherland and Amin, 1982; Farooqui et al., 1986b).

Table 2. Properties of mammalian brain phospholipases

Source	Molecular Localization	Mass (kDa)	Substrate Specificity	Reference
Bovine brain				
PLA$_2$ (Peak I)	Microsomes	18.3	PtdIns	(1)
	Cytosol	110	PtdEtn(Pyr)	(2)
			+ PakEtn(Pyr)	(2)
PLA$_2$ (Peak II)	Cytosol	39	PlsEtn(Pyr)	(2)
Rat brain				
PLA$_2$-H	Cytosol	200-500	PtdCho	(3)
PLA$_2$-L	Cytosol	100	PtdCho	(3)
Rabbit brain				
PLC				
α	Membrane, cytosol	62-65	PtdIns	(4)
ß	Membrane, cytosol	150-154	PtdIns	(4)
γ	Membrane, cytosol	145-148	PtdIns	(4)
δ	Membrane, cytosol	85-88	PtdIns	(4)
ξ	Membrane, cytosol	85	PtdIns	(4)
Rat brain				
PLD	Synaptosomal plasma membrane	–	PtdCho	(5)

References: (1) Gray and Strickland, 1982a; (2) Hirashima et al., 1990a/b; in press; (3) Yoshihara and Watanabe, 1990; (4) Carter et al., 1990; (5) Taki and Kanfer, 1979. Abbreviations: PtdIns, phosphatidylinositol; PtdEtn(Pyr), 1,2-diacylglycero-3-phosphoethanolamine; PakEtn(Pyr), 1-alkyl-2-acylglycero-3-phosphoethanolamine; PlsEtn(Pyr), 1-alkenyl-2-acylglycero-3-phosphoethanolamine; PtdCho, phosphatidylcholine.

Heparin, a highly sulfated glycosaminoglycan with anticoagulant and antilipemic properties, markedly inhibits the diacylglycerol lipases but has no effect on monoacylglycerol lipase. Mono- and diacylglycerol lipases are separated by heparin-Sepharose affinity chromatography. Microsomal diacylglycerol lipase is completely retained on a heparin-Sepharose column and is eluted with 0.5 M NaCl or 2-5 mg/ml heparin, whereas monoacylglycerol lipase is not retained on the column (Farooqui et al., 1984). Adenosine phosphates markedly affect the diacylglycerol lipases in a concentration dependent manner. ATP is the most potent inhibitor followed by ADP. AMP has no effect and cAMP slightly stimulates the diacylglycerol lipase (Farooqui et al., 1984).

A multiple column chromatographic procedure was used to purify diacylglycerol lipase 433-fold from bovine brain microsomes, with an overall recovery of 34%. Application of a similar procedure to plasma membranes resulted in 476-fold purification and 33% yield of diacylglycerol lipase activity (Farooqui et al., 1989). The purified enzyme preparations are homogeneous as judged by SDS-gel electrophoresis and by amino acid sequencing. We have partially sequenced microsomal diacylglycerol lipase (amino acids 1-38). The N-terminal end of this enzyme contains a large proportion of hydrophobic amino acids.

Microsomal diacylglycerol lipase is strongly inhibited by free fatty acids. Palmitate is the most potent inhibitor followed by arachidonate and linoleate. The addition of fatty acid-free bovine serum albumin to the reaction mixture results in reversal of this inhibition (Farooqui et al., 1989). Diacylglycerol lipases are strongly inhibited by C-MT peptide (Farooqui et al., 1989). Lipocortin I has no effect on the activities of diacylglycerol lipases. Quinacrine, a well-known inhibitor of phospholipase A_2, also has no effect on diacylglycerol lipase activity.

Bovine brain diacylglycerol lipases require no metal ions for their activity. Purified preparations of microsomal and plasma membrane diacylglycerol lipases are not affected by $CaCl_2$. The calcium binding protein, calmodulin, stimulates the activities of some phospholipases (Moskowitz et al., 1983). Diacylglycerol lipases are not affected by calmodulin in the presence or absence of $CaCl_2$. Sodium chloride and sodium azide (up to 100 mM) have no effect, but sodium fluoride at 75 mM produces 50% inhibition of microsomal and plasma membrane diacylglycerol lipases.

The retention of microsomal and plasma membrane diacylglycerol lipases on a concanavalin-A Sepharose column and their elution with methyl α-D-mannoside indicate that the binding of these enzymes to concanavalin-A Sepharose is through the carbohydrate moiety (Farooqui et al., 1986b). Determination of galactose by the orcinol procedure in purified diacylglycerol lipase indicates that these enzymes contain 10% carbohydrate. Furthermore, the acid hydrolysate of microsomal diacylglycerol lipase contains traces of hexosamine. All these properties strongly suggest that bovine brain diacylglycerol lipases are glycoproteins.

PHOSPHOLIPASE D

Phospholipase D catalyzes the hydrolysis of phospholipids, producing phosphatidic acid and the free polar head group (Billah and Anthes, 1990). Subcellular distribution studies indicate that the highest specific activity of phospholipase D is present in synaptosomal plasma membranes (Hattori and Kanfer, 1985). In addition to hydrolysis of phospholipids, rat brain phospholipase D catalyzes a transphosphatidylation reaction (Kanfer, 1980). In the presence of ethanol this phospholipase D-catalyzed reaction results in generation of phosphatidylethanol (Gustavsson and Alling, 1987). Phospholipase D was partially purified from rat brain (Taki and Kanfer, 1979; Kobayashi and Kanfer, 1987). The partially purified preparation is free from base exchange activity, but it did contain transphosphatidylation activity suggesting that phospholipase D and transphosphatidylation activity may be localized in a single protein (Taki and Kanfer, 1979; Chalifour and Kanfer, 1982). Phospholipase D is active in the absence of Ca^{2+}, is markedly stimulated by oleic acid (Chalifour and Kanfer, 1982), and in contrast to phospholipase C is blocked by fluoride. Rat and dog synaptic membrane phospholipase D is stimulated by Mg^{2+} and Triton X-100 (Chalifa et al., 1990; Qian and Drewes, 1989). Sphingosine and psychosine produce a marked stimulation of phospholipase D activity in the NG 108-15 cell line (Lavie and Liscovitch, 1990). Canine brain microsomal phospholipase D is greatly inhibited by zinc. Cholera toxin, aluminum fluoride, and GTPγS stimulate phospholipase D activity, indicating that a G protein may be involved in its regulation (Qian et al., 1979). In vascular endothelium and NIH 3T3 cells, phospholipase D is localized in the membrane fraction. Two forms of phospholipase D are known to occur in neutrophils. One form is present in the cytosol, has a neutral pH optimum and requires Ca^{2+} (Billah and Anthes, 1990). The other phospholipase D is found in the azurophilic granules, has an acidic pH optimum and does not require Ca^{2+}. Homogeneous preparations of phospholipase D are prepared from blood plasma (Huang et al., 1990).

Phospholipase A$_2$

The mechanism of regulation of receptor-mediated phospholipase A$_2$ is not well understood (Chang et al., 1987). However, it has been proposed that calcium ions, lipocortins, protein kinase C, neuropeptides, and GTP-binding proteins (G proteins) may be involved.

Several studies have indicated that Ca^{2+} is involved in the regulation of some phospholipases A$_2$ (Chang et al., 1987; Nozawa et al., 1991). These enzymes have a calcium binding site. The precise mechanism of calcium-induced stimulation remains to be established (Table 3) and may vary from cell to cell. The level of the Ca^{2+} concentration necessary to stimulate phospholipase A$_2$ in many assay systems is significantly higher than that achieved with any stimuli (Chang et al., 1987). Thus factor(s) other than Ca^{2+} may be involved in agonist-induced stimulation of phospholipase A$_2$. At least three pathways are dependent on Ca^{2+}. The non-specific phospholipase A$_2$ is activated by Ca^{2+} (Billah et al., 1989). However, platelet phospholipase acting on choline glycerophospholipid is stimulated 5-fold by Ca^{2+} (Kramer et al., 1988). Ca^{2+} not only may activate phospholipase A$_2$, but also may simultaneously inhibit the acyltransferase which recycles the lysophospholipids (Kroner et al., 1981). This latter enzyme is very active and thus serves to remove free arachidonic acid by adding it back to the lysophospholipids. By inhibiting this reacylation step, Ca^{2+} may promote accumulation of arachidonic acid. Recent studies with rat brain synaptosomal membranes have indicated that purified rat brain phospholipase A$_2$ associates with membrane in a Ca^{2+}-dependent manner. In the absence of Ca^{2+} the enzyme does not

Table 3. Effect of calcium ions on brain phospholipases

Enzyme	Effect of Calcium	Reference
Bovine brain microsomal phospholipase A$_2$	Stimulated	(1)
Bovine brain cytosolic phospholipase A$_2$		
Peak I	No effect	(2)
Peak II	No effect	(2)
Rat brain cytosolic phospholipase A$_2$		
(PLA$_2$-H)	No effect	(3)
(PLA$_2$-L)	Stimulated	(3)
Rabbit brain phospholipase C		
α	Stimulated	(4)
ß	Stimulated	(4)
γ	Stimulated	(4)
δ	Stimulated	(4)
ξ	Stimulated	(4)
Rat brain phospholipase D	No effect	(5)

References: (1) Gray and Strickland, 1982a; (2) Hirashima et al., 1990a/b; in press; (3) Yoshihara and Watanabe, 1990; (4) Carter et al., 1990; Dennis et al., 1991; (5) Qian et al., 1979.

bind to synaptosomal membranes (Yoshihara and Watanabe, 1990), indicating that Ca^{2+} ions modulate phospholipase A_2 activity in the nervous system. Phospholipase A_2 activity may also be regulated by free fatty acids and diacylglycerols (Ballou et al., 1986; Farooqui et al., 1989).

Another way in which activity of phospholipase A_2 may be regulated in the cell is with regulatory proteins. In mammalian tissues anti-inflammatory steroids induce the synthesis of a phospholipase inhibitory protein called lipocortin (Hirata, 1985; Flower, 1988). The cDNA for four lipocortins has been cloned (Pepinsky et al., 1988). The concentrations of these proteins are rather low in brain (Strijbos et al., 1990). During cell activation, lipocortins are phosphorylated by protein kinase C or protein tyrosine kinase and then lose their inhibitory activity (Khanna et al., 1986; Clark and Dunlop, 1991). The mechanism of phospholipase A_2 regulation by lipocortin is not known. Initially, a direct regulation by protein-protein interaction was proposed (Hirata, 1985). The recent recognition that lipocortins are able to bind to phospholipids (Schlaepfer and Haigler, 1987) suggests alternative modes of actions of these proteins. Hydrolysis of membrane phospholipids by extracellular and intracellular phospholipases A_2 is inhibited by lipocortin in a substrate-dependent manner (Davidson et al., 1987; Aarsman et al., 1987). This suggests that lipocortins regulate phospholipase A_2 by coating the phospholipid and thereby blocking interaction of the enzyme with its substrate. Further studies are necessary to evaluate the regulatory role of lipocortin in mammalian metabolism.

Regulation of phospholipase A_2 by a G protein has been proposed (Burch et al., 1986; Fain et al., 1988). The most direct evidence for regulation of phospholipase A_2 by a G protein is that purified ßγ subunits of transducin stimulate phospholipase A_2 activity of membranes of photoreceptors and other cells (Jelsema and Axelrod, 1987; Jelsema, 1987; Axelrod et al., 1988). The free α subunit is much less active and may inhibit the ßγ subunits. These data suggest that the dissociation of G protein can activate phospholipase A_2, whereas recombination of free α and ßγ to form the αßγ heterotrimer can turn off phospholipase A_2 activation. The occurrence of phospholipase A_2 stimulatory proteins has been recently reported in *Aplysia* neurons and rat cerebral cortex (Calignano et al., 1991). The partially purified phospholipase A_2 stimulatory protein is phosphorylated by protein kinase C. It has been proposed that phosphorylation of this stimulatory protein by protein kinase C regulates phospholipase A_2 in neurons (Calignano et al., 1991).

At least in 3T3 fibroblasts and neuroblastoma cell cultures, phospholipase A_2 activity is stimulated by bradykinin (Burch et al., 1986). This stimulation may be due to a transient increase in intracellular calcium ions.

Phospholipase C/Diacylglycerol Lipase

It is not clear at present how different isozymes of phospholipase C are regulated. PLC-γ has been sequenced. It contains regions that are similar to the tyrosine kinase regulatory domain of cytoplasmic tyrosine kinase-related oncogenes (Harafuji and Ogawa, 1980). EGF or PDGF stimulation of A431 or 3T3 fibroblasts has been shown to result in phosphorylation of PLC-γ on tyrosine residues (Taylor and Exton, 1987; Rock and Jackowski, 1987). This phosphorylation of PLC-γ was also observed in *in vitro* experiments (Kim et al., 1990). Additionally, treatment of HER 14 cells with EGF or PDGF induces a translocation of phospholipase C-γ from cytosol to membranes. The growth factors promote both the physical association of PLC-γ with their receptors and the subsequent phosphorylation of the enzyme directly by the membrane-bound receptor tyrosine kinases (Kim et al., 1990). This suggests that a possible mechanism of regulation of phospholipase C may be the phosphorylation/dephosphorylation reaction accompanied by phospholipase C translocation.

An alternative mode of receptor regulation of phospholipase C may be through G proteins (Fain et al., 1988). The stable analogs of GTP promote activation of phospholipase C in cell-free preparations from canine cerebral cortex (Qian et al., 1979) and guanine nucleotides are necessary for receptor-mediated activation of the enzyme (Chang et al., 1987). The identity of the G protein involved in the receptor-mediated activation of phospholipase C is yet to be established. However, at least in FRTL5 thyroid cells, stimulation of the α_1-adrenergic receptor activates phospholipase C and phospholipase A_2 by distinct G proteins. The G protein coupled to phospholipase A_2 is pertussis toxin-sensitive whereas that coupled to phospholipase C is pertussis toxin-insensitive (Burch et al., 1986). Like phospholipase A_2, phospholipase C activity is also stimulated by bradykinin (Burch et al., 1986).

Diacylglycerol Lipase

Little is known about the regulation of diacylglycerol lipase in brain. Treatment of neuron-enriched primary cultures with bradykinin results in a 4-fold increase in the specific activity of diacylglycerol lipase and this stimulation can be blocked with a bradykinin antagonist (Farooqui et al., 1990). The mechanism of the stimulation process is not understood. However, the following possibilities should be carefully considered: i) Covalent modification (phosphorylation/dephosphorylation) of diacylglycerol lipase or its regulatory protein; ii) regulation of enzymic activity by G protein; and finally iii) the fact that bradykinin induces a transient increase in intracellular calcium (Reiser and Hamprecht, 1985) which may stimulate diacylglycerol lipase in the membrane-bound state. Diacylglycerol lipase activity is strongly inhibited by free fatty acids (Farooqui et al., 1986a) and it may be that endogenous free fatty acid is involved in the regulation of this enzyme.

Phospholipase D

Numerous studies have indicated that phospholipase D is activated in response to cell stimulation and this enzyme may be involved in the signal transduction process (Billah and Anthes, 1990; Lavie and Liscovitch, 1990). GTPγS stimulates phospholipase D activity in rat hepatocyte membranes (Bocckino et al., 1987), in granulocyte homogenates (Anthes et al., 1989), and in saponin-permeabilized endothelial cells (Martin and Michaelis, 1989). This suggests that a G protein may be involved in the regulation of phospholipase D activity, but the identity of this G protein remains to be established. Furthermore, phorbol ester stimulates phospholipase D activity (Liscovitch, 1989) indicating the involvement of protein kinase C. There is convincing evidence that phosphatidylcholine-derived diacylglycerol may cause the sustained activation of protein kinase C to initiate and maintain long-term responses. This suggests that although the precise mechanism of phospholipase D stimulation is unclear, protein kinase C and/or G protein may regulate the activity of this enzyme.

CROSS-TALK BETWEEN RECEPTOR-REGULATED PHOSPHOLIPASES A$_2$, C, AND D

Numerous studies have indicated that phospholipases A_2, C, and D play an important role in signal transduction processes by generating lipid second messengers such as arachidonic acid, eicosanoids, diacylglycerol, inositol 1,4,5-trisphosphate, and phosphatidic acid (Nahorski et al., 1986; Dennis et al., 1991; Nozawa et al., 1991). Several of these second messengers are common to the catalytic action of various phospholipases (Fig. 2). For example, the action of phospholipase C on polyphospho-

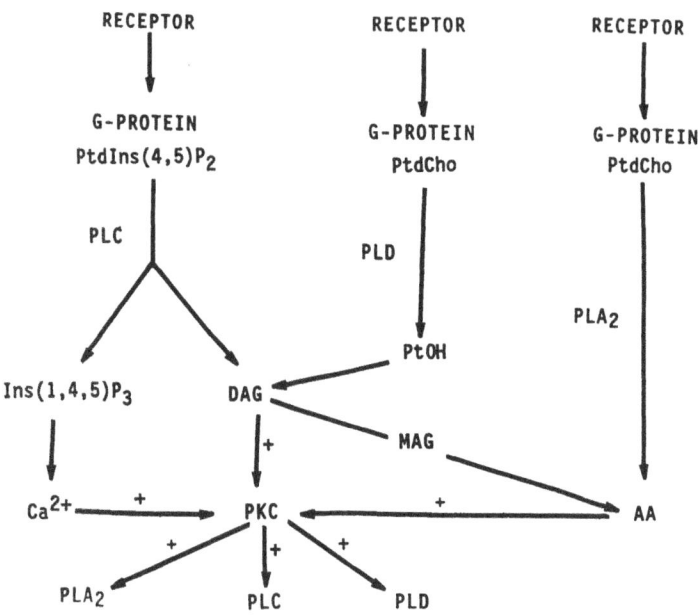

Figure 2. Proposed mechanism of activation and regulation of phospholipases A_2, C, and D in brain.

inositides generates diacylglycerol, which is phosphorylated to phosphatidic acid by a diacylglycerol kinase (Nahorski et al., 1986). The action of phospholipase D on phosphatidylcholine produces phosphatidic acid, which can be converted to diacylglycerol by phosphatidic acid phosphohydrolase (Billah and Anthes, 1990; Dennis et al., 1991). Does phospholipase C regulate the activity of phospholipase D, or is the activity of phospholipase D regulated by phospholipase C? Little information is available to answer this question. It seems possible that diacylglycerol and inositol 1,4,5-trisphosphate, generated during the activation of receptor-linked phospholipase C, regulate the activity of phospholipase D in two ways. First, diacylglycerol may activate protein kinase C, which may be involved in stimulation of phospholipase D. Second, the inositol 1,4,5-trisphosphate-induced rise in cytosolic Ca^{2+} may cause the stimulation of phospholipase D (Billah et al., 1989). In contrast, however, recent studies have indicated that phospholipase D positively modulates the activity of phospholipase C by changing the membrane microenvironment in which phospholipase C and inositol phospholipids reside (Zian and Drewes, 1991). It was shown that an active phosphatidic acid phosphatase is located in the synaptic membrane. This enzyme can rapidly convert the accumulated phosphatidic acid to diacylglycerol (Zian and Drewes, 1991). Thus phosphatidic acid phosphatase acts as a switch-off mechanism for the modulation of phospholipase C by phospholipase D (Zian and Drewes, 1991).

The relationship between phospholipase A_2 and phospholipase C is quite complex. These enzymes may mutually regulate their activity through the generation of second messengers. Activation of protein kinase C is a major signal produced by the stimulation of phospholipase C (Fig. 2). This protein kinase modulates the activity of phospholipase A_2 either by the direct phosphorylation of this enzyme or by the phosphorylation of its regulatory proteins (Zeitler et al., 1991; Bonventre, 1990). Conversely, arachidonic acid generated during the activation of phospholipase A_2 stimulates phospholipase C (Zeitler et al., 1991). Phospholipases A_2 and C may interact in a mutually regulatory manner to bring about homeostatic regulation of the plasma membranes's physicochemical state (Scott, 1984). Several parameters may be interactively modified by these two enzymes, including membrane surface charge, lamellar volume, and phospholipid composition (Scott, 1984). These observations suggest that mammalian phospholipases are part of a complex signal transduction network and that cross-talk between receptor-regulated effector systems through the generation of second messengers is essential for maintaining normal cellular function.

ROLE OF PHOSPHOLIPASES AND LIPASES IN BRAIN

The roles of phospholipases in brain are not fully understood. Under normal conditions, phospholipases may maintain cellular function by providing a variety of second messengers (Dennis et al., 1991; Nozawa et al., 1991). Phospholipase A_2 and C may be involved in detoxification of phospholipid peroxides. If lipid peroxides are not removed they can break down to free radicals, which can initiate a chain reaction to produce more lipid peroxides (Scott, 1984; van Kuijk et al., 1987). Lipid peroxides can also break down into other reactive products, such as aldehydes, which can greatly disturb cellular metabolism (van Kuijk et al., 1987). Thus, phospholipases A_2 and C may play an important role in maintaining membrane integrity (Scott, 1984; van Kuijk et al., 1987).

Phospholipase D may perform several roles in brain. It may provide a long-term increase in diacylglycerol, which may modulate the activity of numerous enzymes (Farooqui et al., 1988). Phospholipase D may provide phosphatidic acid which may act as a calcium ionophore (Billah and Anthes, 1990; Dennis et al., 1991). The second product of the phospholipase D reaction, i.e., choline, is used for the synthesis of the neurotransmitter acetylcholine (Wurtman et al., 1985).

The low levels and rapid turnover of acylglycerols in neural membranes and the relatively high activity of lipases in brain suggest that the mono- and diacylglycerol lipases may be involved in maintaining low levels of acylglycerols (Farooqui et al., 1989).

Under pathological conditions such as ischemia, spinal cord injury, and Alzheimer's disease, phospholipases and lipases may be involved in massive release of free fatty acids, diacylglycerols, and eicosanoids, which may cause serious cell damage (Farooqui and Horrocks, 1991). It remains to be seen whether increased activities of phospholipases and lipases are involved in degeneration of neurons in the above pathological situations or whether enhanced activities of lipases and phospholipases in the surrounding neurons and glial cells represent a mechanism of cell survival.

ACKNOWLEDGMENTS

Supported in part by NIH grants NS-10165, NS-29441 and grants from the Ohio Department of Aging, State of Ohio.

REFERENCES

Aarsman AJ, Mynbeek G, van den Bosch H, Rothhut B, Prieur B, Comera C, Jordan L, Russo-Marie F (1987) Lipocortin inhibition of extracellular and intracellular phospholipases A_2 is substrate concentration dependent. FEBS Lett 219:176-180.

Anthes JC, Eckel S, Siegel MI, Egan RW, Billah MM (1989) Phospholipase D in homogenates from HL-60 granulocytes: Implications of calcium and G protein control. Biochem Biophys Res Commun 163:657-664.

Axelrod J, Burch RM, Jelsema CL (1988) Receptor-mediated activation of phospholipase A_2 via GTP-binding proteins: Arachidonic acid and its metabolites as second messengers. Trends Neurosci 11:117-123.

Ballou LR, DeWitt LM, Cheung WY (1986) Substrate-specific forms of human platelet phospholipase A_2. J Biol Chem 261:3107-3111.

Balow R-M, Tomkinson B, Ragnarsson U, Zetterqvist O (1986) Purification, substrate specificity, and classification of tripeptidyl peptidase II. J Biochem 261:2409-2417.

Billah MM and Anthes JC (1990) The regulation and cellular functions of phosphatidylcholine hydrolysis. Biochem J 269:281-291.

Billah MM, Pai JK, Mullmann TJ, Egan RW, Siegel MI (1989) Regulation of phospholipase D in HL-60 granulocytes. Activation by phorbol esters, diglyceride, and calcium ionophore via protein kinase-independent mechanisms. J Biol Chem 264:9069-9076.

Bocckino SB, Blackmore PF, Wilson PB, Exton JH (1987) Phosphatidate accumulation in hormone-treated hepatocytes via a phospholipase D mechanism. J Biol Chem 262:15309-15315.

Bonventre JV (1990) Calcium in renal cells. Modulation of calcium-dependent activation of phospholipase A_2. Environ Health Perspect 84:155-162.

Burch RM, Luini A, Axelrod J (1986) Phospholipase A_2 and phospholipase C are activated by distinct GTP-binding proteins in response to alpha 1-adrenergic stimulation in FRTL5 thyroid cells. Proc Natl Acad Sci USA 83:7201-7205.

Calignano A, Piomelli D, Sacktor TC, Schwartz JH (1991) A phospholipase A_2-stimulating protein regulated by protein kinase C in *Aplysia* neurons. Mol Brain Res 9:347-351.

Carter HR, Wallace MA, Fain JN (1990) Purification and characterization of PLC-ßm, a muscarinic cholinergic regulated phospholipase C from rabbit brain membrane. Biochim Biophys Acta 1054:119-128.

Chalifa V, Mohn H, Liscovitch M (1990) A neutral phospholipase D activity from rat brain synaptic plasma membranes. J Biol Chem 265:17512-17519.

Chalifour R and Kanfer JN (1982) Fatty acid activation and temperature perturbation of rat brain microsomal phospholipase D. J Neurochem 39:299-305.

Chang J, Musser JH, McGregor H (1987) Phospholipase A_2: Function and pharmacological regulation. Biochem Pharmacol 36:2429-2436.

Clark S and Dunlop M (1991) Modulation of phospholipase A_2 activity by epidermal growth factor (EGF) in CHO cells transfected with human EGF receptor. Biochem J 274:715-721.

Clark JD, Lin L-L, Kriz RW, Ramesha CS, Sultzman LA, Lin AY, Milona N, Knopf JL (1991) A novel arachidonic acid-selective cytosolic PLA_2 contains a Ca^{2+}-dependent translocation domain with homology to PKC and GAP. Cell 65:1043-1051.

Clark JD, Milona N, Knopf JL (1990) Purification of a 110-kilodalton cytosolic phospholipase A_2 from the human monocytic cell line U937. Proc Natl Acad Sci USA 87:7708-7712.

Davidson FF, Dennis EA, Powell M, Glenney JR Jr (1987) Inhibition of phospholipase A_2 by "lipocortins" and calpactins. J Biol Chem 262:1698-1705.

Dennis EA, Rhee SG, Billah MM, Hannun YA (1991) Role of phospholipases in generating lipid second messengers in signal transduction. FASEB J 5:2068-2077.

Edgar AD, Freysz L, Mandel P, Horrocks LA (1980) Phospholipid metabolism in low and high density C6 cells. Trans Am Soc Neurochem 11:100.

Fain JN (1990) Regulation of phosphoinositide-specific phospholipase C. Biochim Biophys Acta 1053:81-88.

Fain JN, Wallace MA, Wojcikiewicz RJH (1988) Evidence for involvement of guanine nucleotide-binding regulatory proteins in the activation of phospholipases by hormones. FASEB J 2:2569-2574.

Farooqui AA and Horrocks LA (1991) Excitatory amino acid receptors, neural membrane phospholipid metabolism and neurological disorders. Brain Res Rev 16:171-191.

Farooqui AA, Anderson DK, Flynn CJ, Bradel E, Means ED, Horrocks LA (1990) Stimulation of mono- and diacylglycerol lipase activities by bradykinin in neuronal-enriched cultures from mouse spinal cord. Biochem Biophys Res Commun 166:1001-1009.

Farooqui AA, Farooqui T, Yates AJ, Horrocks LA (1988) Regulation of protein kinase C activity by various lipids. Neurochem Res 13:499-511.

Farooqui AA, Pendley CE II, Taylor WA, Horrocks LA (1985) Studies on diacyl-glycerol lipases and lysophospholipases of bovine brain. In: Phospholipids in the nervous system, Vol II: Physiological role (Horrocks LA, Kanfer JN, Porcellati G, eds) pp 179-192. New York: Raven Press.

Farooqui AA, Rammohan KW, Horrocks LA (1989) Isolation, characterization and regulation of diacylglycerol lipases from bovine brain. Ann NY Acad Sci 559:25-36.

Farooqui AA, Taylor WA, Horrocks LA (1984) Separation of bovine brain mono- and diacylglycerol lipases by heparin-Sepharose affinity chromatography. Biochem Biophys Res Commun 122:1241-1246.

Farooqui AA, Taylor WA, Horrocks LA (1986a) Membrane bound diacylglycerol lipases of bovine brain. In: Phospholipid research and the nervous system. Biochemical and molecular pharmacology, Fidia Research Series (Horrocks LA, Freysz L, Toffano G, eds) pp 181-190. Italy: Liviana Press.

Farooqui AA, Taylor WA, Horrocks LA (1986b) Characterization and solubilization of membrane bound diacylglycerol lipases from bovine brain. Int J Biochem 18:991-997.

Flower RJ (1988) Lipocortin and the mechanism of action of the glucocorticoids. Br J Pharmacol 94:987-1015.

Gray NCC and Strickland KP (1982a) The purification and characterization of a phospholipase A2 activity from the 106000 × g pellet (microsomal fraction) of bovine brain acting on phosphatidylinositol. Can J Biochem 60:108-117.

Gray NCC and Strickland KP (1982b) On the specificity of a phospholipase A_2 from the 106000 × g pellet of bovine brain. Lipids 17:91-96.

Gronich JH, Bonventre JV, Nemenoff RA (1990) Purification of a high-molecular-mass form of phospholipase A_2 from rat kidney activated at physiological calcium concentrations. Biochem J 271:37-43.

Gustavsson L and Alling C (1987) Formation of phosphatidylethanol in rat brain by phospholipase D. Biochem Biophys Res Commun 142:958-963.

Harafuji H and Ogawa Y (1980) Re-examination of the apparent binding constant of ethylene glycol bis(ß-aminoethyl ether)-N,N,N',N'-tetraacetic acid with calcium around neutral pH. J Biochem (Tokyo) 87:1305-1312.

Hattori H and Kanfer JN (1985) Synaptosomal phospholipase D potential role in providing choline for acetylcholine synthesis. J Neurochem 45:1578-1584.

Hazen SL, Stuppy RJ, Gross RW (1990) Purification and characterization of canine myocardial cytosolic phospholipase A_2. J Biol Chem 265:10622-10630.

Hirashima Y, Farooqui AA, Mills JS, Horrocks LA (in press) Purification and characterization of bovine brain cytosol phospholipase A_2. J Neurochem.

Hirashima Y, Mills J, Farooqui AA, Horrocks LA (1990a) Purification and characterization of bovine brain cytosol phospholipase A_2. Soc Neurosci Abs 16:538.

Hirashima Y, Mills JS, Yates AJ, Horrocks LA (1990b) Phospholipase A_2 activities with a plasmalogen substrate in brain and in neural tumor cells: A sensitive and specific assay using pyrenesulfonyl-labeled plasmenylethanolamine. Biochim Biophys Acta 1074:35-40.

Hirata F (1985) Receptor mediated cascade of phospholipid metabolism In: Phospholipids in the nervous system, Vol. 2 (Horrocks LA, Kanfer JN, Porcellati G, eds) pp 99-105. New York: Raven Press.

Huang K-S, Li S, Fung W-JC, Hulmes JD, Reik L, Pan Y-CE, Low MG (1990) Purification and characterization of glycosyl-phosphatidylinositol-specific phospholipase D. J Biol Chem 265:17738-17745.

Jelsema CL (1987) Light activation of phospholipase A_2 in rod outer segments of bovine retina and its modulation by GTP-binding proteins. J Biol Chem 262:163-168.

Jelsema CL and Axelrod J (1987) Stimulation of phospholipase A_2 activity in bovine rod outer segments by the beta gamma subunits of transducin and its inhibition by the alpha subunit. Proc Natl Acad Sci USA 84:3623-3627.

Kanfer JN (1980) The base exchange enzymes and phospholipase D of mammalian tissues. Can J Biochem 58:1370-1380.

Kennedy SP and Becker EL (1987) Ectophospholipase A_2 activity of the rabbit peritoneal neutrophil. Int Arch Allergy Appl Immunol 83:238-246.

Khanna NC, Tokuda M, Waisman DM (1986) Phosphorylation of lipocortins in vitro by protein kinase C. Biochem Biophys Res Commun 141:547-554.

Kim DK, Suh PG, Ryu SH (1991) Purification and some properties of a phospholipase A_2 from bovine platelets. Biochem Biophys Res Commun 174:189-196.

Kim U-H, Kim H-S, Rhee SG (1990) Epidermal growth factor and platelet-derived growth factor promote translocation of phospholipase C-γ from cytosol to membrane. FEBS Lett 270:33-36.

Kobayashi M, Kanfer JN (1987) Phosphatidylethanol formation via transphosphatidylation by rat brain synaptosomal phospholipase D. J Neurochem 48:1597-1603.

Kramer RM, Jakubowski JA, Deykin D (1988) Hydrolysis of 1-alkyl-2-arachidonoyl-sn-glycero-3-phosphocholine, a common precursor of platelet-activating factor and eicosanoids, by human platelet phospholipase A_2. Biochim Biophys Acta 959:269-279.

Kramer RM, Roberts EF, Manetta J, Putnam JE (1991) The Ca^{2+}-sensitive cytosolic phospholipase A_2 is a 100-kDa protein in human monoblast U937 cells. J Biol Chem 266:5268-5272.

Kroner EE, Peskar BA, Fischer H, Ferber E (1981) Control of arachidonic acid accumulation in bone marrow-derived macrophages by acyltransferases. J Biol Chem 256:3690-3697.

Lavie Y and Liscovitch M (1990) Activation of phospholipase D by sphingoid bases in NG108-15 neural-derived cells. J Biol Chem 265:3868-3872.

Liscovitch M (1989) Phosphatidylethanol biosynthesis in ethanol-exposed NG108-15 neuroblastoma X glioma hybrid cells. J Biol Chem 264:1450-1456.

Martin TW and Michaelis K (1989) P_2-Purinergic agonists stimulate phosphodiesteratic cleavage of phosphatidylcholine in endothelial cells. J Biol Chem 264:8847-8856.

Mizuguchi M, Yamada M, Kim SU, Rhee SG (1991) Phospholipase C isozymes in neurons and glial cells in culture: An immunocytochemical and immunochemical study. Brain Res 548:35-40.

Moskowitz N, Puszkin S, Schook W (1983) Characterization of brain synaptic vesicle phospholipase A_2 activity and its modulation by calmodulin, prostaglandin E_2, prostaglandin $F_{2\alpha}$, cyclic AMP and ATP. J Neurochem 41:1576-1586.

Nahorski SR, Kendall DA, Batty I (1986) Receptors and phosphoinositide metabolism in the central nervous system. Biochem Pharmacol 35:2447-2453.

Nozawa Y, Nakashima S, Nagata K-I (1991) Phospholipid-mediated signaling in receptor activation of human platelets. Biochim Biophys Acta 1082:219-238.

Pepinsky RB, Tizard R, Mattaliano RJ, Sinclair LK, Miller GT, Browning JL, Chow EP, Burne C, Huang K-S, Pratt D, Wachter L, Hession C, Frey AZ, Wallner BP (1988) Five distinct calcium and phospholipid binding proteins share homology with lipocortin I. J Biol Chem 263:10799-10811.

Qian Z and Drewes LR (1989) Muscarinic acetylcholine receptor regulates phosphatidylcholine phospholipase D in canine brain. J Biol Chem 264:21720-21724.

Qian Z, Reddy PV, Drewes LR (1979) Guanine nucleotide-binding protein regulation of microsomal phospholipase D activity of canine cerebral cortex. J Biol Chem 254:9761-9765.

Reiser G and Hamprecht B (1985) Bradykinin causes a transient rise of intracellular Ca^{2+}-activity in cultured neural cells. Pflugers Arch 405:260-264.

Rock CO and Jackowski S (1987) Thrombin- and nucleotide-activated phosphatidylinositol 4,5-bisphosphate phospholipase C in human platelet membranes. J Biol Chem 262:5492-5498.

Schlaepfer DD and Haigler HT (1987) Characterization of Ca^{2+} dependent phospholipid binding and phosphorylation of lipocortin-I. J Biol Chem 262:6931-6937.

Scott JA (1984) Phospholipase activity and plasma membrane homeostasis. J Theor Biol 111:659-665.

Strijbos P, Tilders F, Carey F, Forder R, Rothwell N (1990) Localization of lipocortin-1 in normal rat brain. Biochem Soc Trans 18:1234-1235.

Sutherland CA and Amin D (1982) Relative activities of cat and dog platelet phospholipase A_2 and diglyceride lipase — Selective inhibition of diglyceride lipase by RHC 80267. J Biol Chem 257:14006-14010.

Taki T and Kanfer JN (1979) Partial purification and properties of a rat brain phospholipase D. J Biol Chem 254:9761-9765.

Taylor SJ and Exton JH (1987) Guanine-nucleotide and hormone regulation of polyphosphoinositide phospholipase C activity of rat liver plasma membranes. Biochem J 248:791-799.

van Kuijk FJGM, Sevanian A, Handelman GJ, Dratz EA (1987) A new role for phospholipase A_2: Protection of membranes from lipid peroxidation damage. Trends Biochem Sci 12:31-34.

Witter B and Kanfer JN (1985) Hydrolysis of endogenous phospholipids by rat brain microsomes. J Neurochem 44:155-162.

Wurtman RJ, Blusztajn JK, Maire JC (1985) "Autocannibalism" of choline-containing membrane phospholipids in the pathogenesis of Alzheimer's disease — A hypothesis. Neurochem Int 7:369-372.

Yoshihara Y and Watanabe Y (1990) Translocation of phospholipase A_2 from cytosol to membranes in rat brain induced by calcium ions. Biochem Biophys Res Commun 170:484-490.

Zeitler P, Wu YQ, Handwerger S (1991) Melittin stimulates phosphoinositide hydrolysis and placental lactogen release: Arachidonic acid as a link between phospholipase A_2 and phospholipase C signal-transduction pathways. Life Sci 48:2089-2095.

Zian Z and Drewes LR (1991) Cross-talk between receptor-regulated phospholipase D and phospholipase D in brain. FASEB J 5:315-319.

CHARACTERISTICS AND POSSIBLE FUNCTIONS OF

MAST CELL PHOSPHOLIPASES A$_2$

Makoto Murakami, Ichiro Kudo, and Keizo Inoue

Faculty of Pharmaceutical Sciences
University of Tokyo
Hongo, Bunkyo-ku
Tokyo 113, Japan

ABSTRACT

Phospholipase A$_2$ activity in lysates of mast cells and their related cells [mouse bone marrow-derived IL-3 dependent mast cells (BMMC), rat connective tissue mast cells (CTMC), and rat mastocytoma RBL-2H3 cells] was measured using phosphatidylethanolamine (PE), phosphatidylserine (PS), and phosphatidylcholine (PC) as exogenous substrates. Both BMMC and RBL cells showed rather high phospholipase A$_2$ activity, whereas CTMC showed only weak activity. These cells contained at least three types of phospholipase A$_2$. Type 1 enzyme showed no appreciable affinity to heparin, and preferentially hydrolyzed either PC or PE, both of which have an arachidonic acid at the sn-2 position. The activity was absorbed by monoclonal antibody against rabbit platelet cytosolic 85-kDa phospholipase A$_2$. Type 2 enzyme had an affinity to heparin, and was completely inhibited by anti-rat platelet 14-kDa secretory phospholipase A$_2$. This enzyme could be expressed as an "ecto-type" enzyme on the cell surface and might be secreted from cells when mast cells are activated. Type 3 enzyme also had an affinity to heparin, but was separated from type 2 enzyme on reverse-phase HPLC. This enzyme did not interact with anti-14-kDa secretory enzyme antibody. Purified type 3 enzyme (30-kDa) specifically hydrolyzed PS. p-Bromophenacylbromide inhibited all types of phospholipase A$_2$, whereas mepacrine inhibited type 2 and type 3 enzymes, but not type 1 enzyme. Type 2 enzyme was also inhibited by the specific antibody, complement degradation product, and a small-molecular-weight inhibitor. Histamine release was inhibited by all these inhibitors, whereas PGD$_2$ production was inhibited only by p-bromophenacylbromide. Possible roles for these phospholipases A$_2$ in mast cell function are proposed.

INTRODUCTION

Mast cells sensitized with IgE immunoglobulins, which are bound to high affinity IgE receptors, are activated by specific multivalent antigens. Such activation causes degranulation and the release of histamine, proteoglycans, and proteases, as well as ela-

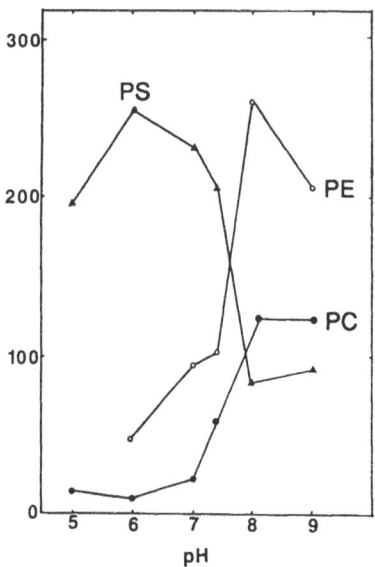

Figure 1. pH profiles of phospholipases A_2 from mouse BMMC

boration of newly synthesized lipid mediators such as prostaglandins, leukotrienes, and platelet-activating factor (PAF).

Production of such lipid mediators is induced by activation of cellular phospholipase A_2, which liberates free fatty acid at the *sn*-2 position of membrane glycerophospholipids. Lysophosphatidylserine (lysoPS), a potent stimulator of connective tissue mast cells (CTMC), may also be produced from PS via phospholipase A_2 action. Limited information has been available so far on the biochemical properties of phospholipase A_2 associated with mast cells. One of the reasons may be difficulty in preparing sufficient amounts of mast cells. In the 1980s, however, a useful method for preparing greater numbers of mast cells was developed. When mouse bone marrow cells are cultured *in vitro* in the presence of interleukin-3 (IL-3), a homogeneous population of mast cells is obtained (Razin et al., 1981). These bone marrow-derived mast cells (BMMC) share many common phenotypic characteristics with mucosal mast cells (MMC).

In the present study, we characterized the phospholipases A_2 of BMMC and of rat basophilic leukemia RBL-2H3 cells (cultured counterpart of rat MMC). Rat CTMC were also used to study the functions of phospholipase A_2 during cell activation. Three types of phospholipase A_2 were found in these cells.

CHARACTERIZATION OF MAST CELL PHOSPHOLIPASES A_2

Phospholipase A_2 activity in mast cell lysates was measured using PE, PS, and PC as exogenous substrates (Murakami et al., 1992). Both BMMC and RBL cells showed rather high phospholipase A_2 activities, whereas CTMC showed only weak activities. The PS-hydrolyzing activity in these cells was detected in the acidic to neutral pH range and showed an optimum at pH 5.5-7.4, while the activity that hydrolyzed either PE or PC was detected in the basic pH range and showed an optimum at pH 8.0-9.0 (Fig. 1). These findings indicate that MMC may contain at least two phospholipases A_2, one for PS and another for PE and PC. This possibility was further supported by the different

Table 1. Characteristics of mast cell phospholipases A_2

Characteristic	Type 1	Type 2	Type 3
Molecular mass	~85 kDa	14 kDa (group II)	30 kDa?
Substrate specificity	Arachidonoyl PE, PC	PE, PS>PC	PS
Ca^{2+} requirement	10^{-6} M	10^{-3} M	10^{-7} M
Optimum pH	8.0-9.0	7.0-9.0	5.5-7.4
Affinity for heparin	Low	High	High
Distribution	Cytosol	Secretory granules	?

elution profiles of these phospholipase A_2 activities on heparin-Sepharose column chromatography. The PC- and PE-hydrolyzing activity was mostly recovered from the flow-through fractions, although part of the activity was absorbed and eluted with a relatively low concentration of NaCl (0.3 M). In contrast, PS-hydrolyzing activity was tightly adsorbed to the column and eluted with the buffer containing approximately 1 M NaCl. Either PE or PC with arachidonate at the sn-2 position of glycerol was hydrolyzed easily by the phospholipase A_2 in MMC homogenates at pH 9.0, whereas PE and PC that contained linoleate at sn-2 were hydrolyzed only slightly under the same conditions. The anti-rabbit platelet 85-kDa phospholipase A_2 monoclonal antibody absorbed this enzyme activity efficiently, indicating that the PE-, PC-hydrolyzing enzyme shared an immunochemically common structure with the platelet cytosolic 85-kDa phospholipase A_2 (Kim et al., 1991). It can be concluded that rat RBL-2H3 cells as well as mouse BMMC contain type 1 (high-molecular-weight type) phospholipase A_2, which shows no appreciable affinity for heparin and preferentially hydrolyzes either PC or PE having arachidonate at the sn-2 position. The PS-hydrolyzing activity showed no appreciable reactivity with this antibody.

PS-hydrolyzing activity of RBL cells adsorbed to heparin-Sepharose was further divided into two fractions by reverse-phase HPLC on a Capsell Pak C8 column. One (type 2 enzyme) might correspond to 14-Kda group II phospholipase A_2 (Kudo et al., 1989), since the elution profile was similar to that of the mammalian group II enzyme. In fact, examination of the reactivity of both fractions with anti-rat group II phospholipase A_2 antibody R377 (Murakami et al., 1990) indicated that one was attributable to 14-kDa group II phospholipase A_2. This conclusion was also supported by the finding that SDS-PAGE showed a major protein band with an approximate molecular weight of 14,000. The enzyme activity of post-butyl-Toyopearl fractions of mouse BMMC was partially neutralized with anti-rat group II phospholipase A_2 antibody R377, suggesting that mouse BMMC may also express the 14-kDa group II phospholipase A_2.

To characterize activity in the other fraction, we next attempted to purify the enzyme from RBL cells transplanted and expanded in the peritoneal cavity of rats. The group II phospholipase A_2 was eliminated from the 100,000 × g supernatant of the homogenates by passing it through the anti-rat group II phospholipase A_2 antibody MD7.1-Sepharose column (Murakami et al., 1988; 1990). The enzyme was further purified by sequential use of heparin-Sepharose, butyl-Toyopearl, reverse-phase HPLC and Superose 12 gel filtration.

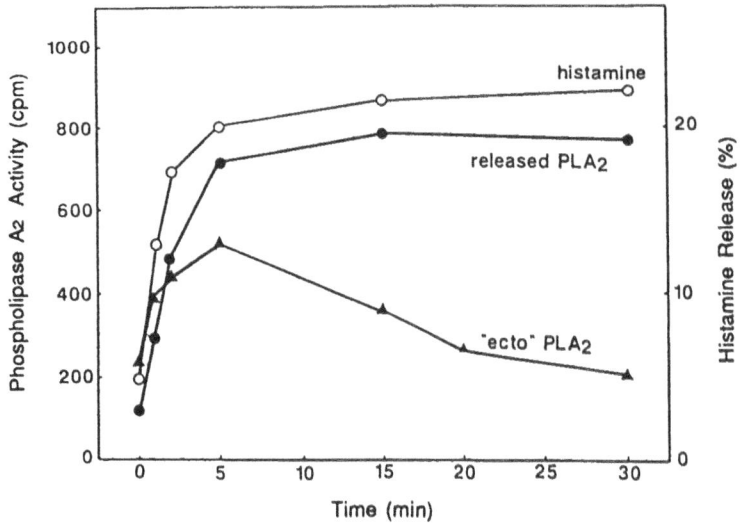

Figure 2. Release of phospholipase A_2 from mast cells stimulated with anti-IgE

The final enzyme preparation gave a single protein band with a molecular weight of approximately 30,000. This type 3 phospholipase A_2 hydrolyzed PS most efficiently at optimum pH 5.5-7.4. PE was a poor substrate, and no hydrolysis was observed with either PC or phosphatidylinositol (PI). The phospholipase A_2 activity was detected in the presence of 10^{-7} M Ca^{2+}.

It is concluded that mast cells contain at least three types of phospholipase A_2 (Table 1). Type 1 enzyme shows no appreciable affinity for heparin, and preferentially hydrolyzes either PC or PE, which has an arachidonic acid at the sn-2 position. The enzyme may be similar to rabbit platelet cytosolic 85-kDa phospholipase A_2. Type 2 enzyme has an affinity for heparin, and belongs to 14-kDa group II phospholipase A_2. Type 3 enzyme has also an affinity for heparin, but was separated from type 2 enzyme on reverse-phase HPLC. This enzyme does not interact with anti-14-kDa group II enzyme antibody. Purified type 3 enzyme selectively hydrolyzes PS. p-Bromophenacyl-bromide inhibited all types of phospholipase A_2, whereas mepacrine inhibited type 2 and type 3 enzymes, but not type 1 enzyme (Murakami et al., 1991).

STIMULUS-INDUCED EXPRESSION OF 'ECTO-TYPE' ENZYME ACTIVITY ON OUTER SURFACE OF MAST CELLS AND SECRETION OF THE ACTIVITY FROM CELLS

When mouse BMMC were sensitized with IgE and challenged with goat anti-mouse IgE antiserum, appreciable amounts of phospholipase A_2 activity were released. The release of phospholipase A_2 activity almost paralleled the release of histamine (Fig. 2).

A major part of the phospholipase A_2 activity secreted in extracellular medium was absorbed by anti-rat group II phospholipase A_2 antibody, indicating that the phospholipase A_2 released was the group II enzyme.

It is noteworthy here that a transient appearance of phospholipase A_2 activity was observed on the cell surface. Since no cell lysis was observed during incubation, the

Table 2. Effect of various compounds on group II phospholipase A_2 activity, histamine release, and PGD_2 production

Inhibitors	Inhibition of group II phospholipase A_2 (%)	Inhibition of histamine release (%)	Inhibition of PGD_2 production (%)
p-Bromophenacyl-bromide (10^{-5} M)[*1]	77.8	94.4	98.9
Mepacrine (10^{-5} M)[*2]	73.3	72.2	10.4
Antibody R377 (50 μg/ml)[*3]	80.6	55.7	0
Complement degrading product (1 μg/ml)[*3]	80.0	80.0	0
Thielocin A1 (1 μg/ml)[*3]	95.5	68.0	0
Antibody MD 7.1 (100 μg/ml)[*4]	0	31.7	0

Rat peritoneal mast cells were sensitized with IgE and then challenged with antigen in the presence of various compounds as shown in the Table. After stimulation for 10 min at 37° C, the supernatant was collected and histamine or PGD_2 was measured.

[*1]: p-Bromophenacylbromide inhibits all three types of phospholipase A_2.
[*2]: Mepacrine inhibits group II phospholipase A_2.
[*3]: Specific inhibitors of group II phospholipase A_2.
[*4]: The antibody inhibits binding of the group II phospholipase A_2 to heparin, but does not affect the enzymatic activity.

phospholipase A_2 should be expressed on the cell surface as an ecto-enzyme. The appearance of phospholipase A_2 activity on the cell surface seemed to precede that in the medium; a rapid increase was detected within 2 min (80% of maximum value) and reached maximum within 5 min. Thereafter the activity decreased gradually to the level of cells in the resting state. This ecto-phospholipase A_2 activity was also detected by binding of anti-group II phospholipase A_2 antibody on the cell surface. Thus immuno-chemical analysis revealed that the phospholipase A_2 expressed on the cell surface was also group II enzyme. Group II phospholipase A_2 would be exocytosed from activated mast cells in both the soluble and membrane-bound form.

ROLE OF PHOSPHOLIPASES A_2 IN MAST CELL FUNCTION

When rat peritoneal mast cells were sensitized with IgE and challenged with antigen in the presence of group II phospholipase A_2 inhibitors such as mepacrine, specific anti-

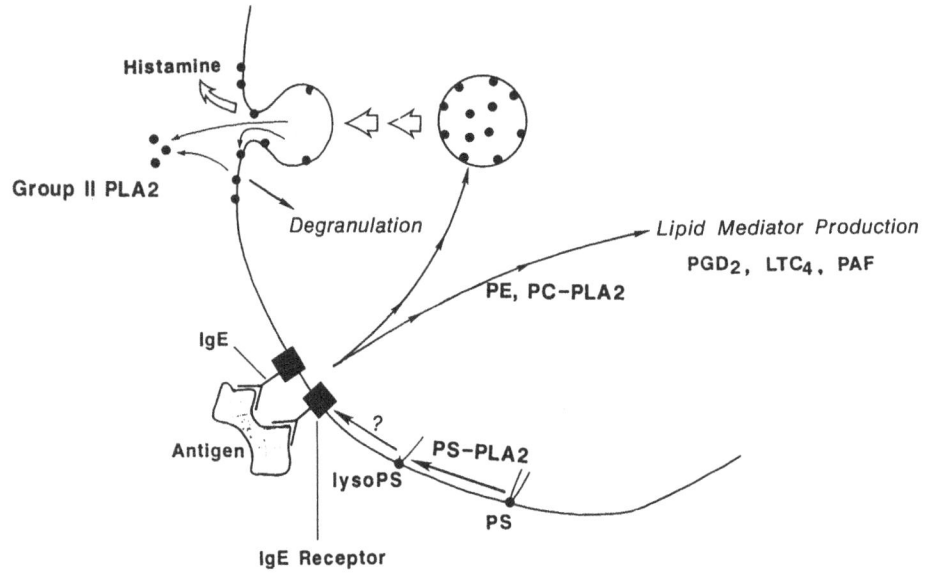

Figure 3. Possible functions of phospholipases A_2 in mast cells

body R377, complement-degrading product (Suwa et al., 1990), or thielocin A1, a low-molecular-weight compound derived from microorganisms (Yoshida et al., 1991), histamine release was inhibited whereas prostaglandin D_2 (PGD_2) generation was not affected (Table 2). PGD_2 generation was affected only when cells were treated with p-bromophenacylbromide. Type 1 phospholipase A_2 (high-molecular-weight enzyme), which was sensitive to p-bromophenacylbromide, may participate in liberating arachidonate upon stimulation of mast cells (Murakami et al., 1991).

Inhibition of histamine release was also observed with rat CTMC which were stimulated with non-immunochemical stimuli, such as Ca^{2+} ionophore A23187, compound 48/80, or substance P. The mechanisms by which each compound activates mast cells are thought to be different. Thus group II phospholipase A_2 may play an important role in common steps of the process of degranulation induced by various stimuli. A similar inhibitory effect on histamine release was observed with mouse BMMC. No inhibitory effect on histamine release was observed, however, when mast cells were pretreated with these inhibitors, washed, and then stimulated in the absence of inhibitors. Antibody cannot permeate into the intracellular space, indicating that group II phospholipase A_2 exposed on the cell surface should be involved in the process of histamine release.

Mouse BMMC were pre-labeled with radiolabeled linoleate, and the alteration of phospholipid composition upon stimulation was examined. When linoleate-labeled cells were activated, an appreciable decrease in radioactive PE was observed. This change was not observed in the cells activated in the presence of group II phospholipase A_2 inhibitors. Thus, group II phospholipase A_2 may have hydrolyzed PE in membranes of activated mast cells.

Karli et al. (1990) reported an attractive model for membrane fusion, in which treatment of plasma membrane prepared from chromaffin cells with pancreatic phospholipase A_2 triggered fusion with chromaffin granules in the presence of Ca^{2+}; the authors proposed that liberated fatty acids are metabolized to fusiogenic compound(s), rendering the plasma membrane for fusion. The group II phospholipase A_2 derived from mast cells might act on plasma membrane in an autocrine manner to generate such fusiogenic compound(s), leading to progression or enhancement of the fusion in activated mast cells between plasma membrane and secretory granules. It is interesting that anti-rat group II phospholipase A_2 monoclonal antibody MD7.1, which prevented the binding of group II enzyme to heparin, partially inhibited degranulation (Table 2). Thus the interaction of group II phospholipase A_2 with heparin might have an important role in association and/or dissociation of the enzyme with plasma membrane and, therefore, in the regulation of degranulation.

In conclusion, mast cells may be regulated by at least two distinct phospholipases A_2 independently: degranulation by type 2 enzyme (group II phospholipase A_2) and eicosanoid production by type 1 enzyme (high-molecular-weight phospholipase A_2). Another PS-hydrolyzing phospholipase A_2 (type 3 enzyme) may function as a modulator of activation by regulating the endogenous supply of lysoPS, which is an essential co-factor for antigen-induced CTMC activation (Tamori-Natori et al., 1986, Chang et al., 1988; Inoue et al., 1989). The proposed roles of the phospholipases A_2 are shown schematically in Figure 3.

REFERENCES

Chang HW, Inoue K, Bruni A, Boarato E, Toffano G (1988) Stereoselective effects of lysophosphatidylserine in rodents. Br J Pharmacol 93:647.

Karli UO, Schafer Y, Burger MM (1990) Fusion of neurotransmitter vesicles with target membrane is calcium independent in a cell-free system. Proc Natl Acad Sci USA 87:5912.

Kim DK, Kudo I, Inoue K (1991) Purification and characterization of rabbit platelet cytosolic phospholipase A_2. Biochim Biophys Acta 1083:80-88.

Kudo I, Chang HW, Hara S, Murakami M, Inoue K (1989) Characteristics and pathophysiological roles of extracellular phospholipase A_2 in inflamed sites. Dermatologica 179:72.

Inoue K, Kobayashi T, Kudo I (1989) Function and metabolism of lysophosphatidylserine in rat mast cell activation. In: Phospholipids in the nervous system: Biochemical and molecular pathology (Bazan NG, Horrocks LA, Toffano G, eds), p 225, Padova: Liviana Press.

Murakami M, Kobayashi T, Umeda M, Kudo I, Inoue K (1988) Monoclonal antibodies against rat platelet phospholipase A_2. J Biochem 104:884.

Murakami M, Kudo I, Fujimori Y, Suga H, Inoue K (1991) Group II phospholipase A_2 inhibitors suppressed lysophosphatidylserine-dependent degranulation of rat peritoneal mast cells. Biochem Biophys Res Commun 181:714.

Murakami M, Kudo I, Inoue K (1990) Immunochemical detection of "platelet type" phospholipase A_2 in the rat. Biochim Biophys Acta 1043:34.

Murakami M, Kudo I, Umeda M, Matsuzawa A, Takeda M, Komada M, Fujimori Y, Takahashi K, Inoue K (1992) Detection of three distinct phospholipases A_2 in cultured mast cells. J Biochem 111:175.

Razin E, Ihle JN, Seldin D, Mencia-Huerta JM, Katz HR, LeBlanc PA, Hein A, Caulfield JP, Austen KF, Stevens RL (1984) Interleukin 3: A differentiation and growth factor for the mouse mast cell that contains chondroitin sulfate E proteoglycan. J Immunol 132:1479.

Suwa Y, Kudo I, Igarashi A, Okada M, Kamimura T, Suzuki Y, Chang HW, Hara S, Inoue K (1990) Proteinaceous inhibitors of phospholipase A_2 purified from inflammatory sites in rats. Proc Natl Acad Sci USA 87:2395-2399.

Tamori-Natori Y, Horigome K, Inoue K, Nojima S (1986) Metabolism of lysophosphatidylserine, a potentiator of histamine release in rat mast cells. J Biochem 100:1099.

Yoshida T, Nakamoto S, Sakazaki R, Matsumoto K, Inoue K, Kudo I (1991) Thielocin A1α and A1ß, novel phospholipase A_2 inhibitors from Ascomycetes. J Antibiotics 44:1467.

EXTRACELLULAR PHOSPHOLIPASE A$_2$

Edward A. Dennis, Raymond A. Deems, and Lin Yu

Department of Chemistry
University of California, San Diego
La Jolla, California 92093

PHOSPHOLIPASE A$_2$ AND ARACHIDONIC ACID RELEASE

Various lipids have been implicated in signal transduction and as precursors of lipid mediators, among them arachidonic acid (Dennis et al., 1991). Production of the prostaglandins, thromboxanes, and prostacyclins via the cyclooxygenase pathway, as well as the production of the leukotrienes and lipoxins via the lipoxygenase pathway, all depend on the availability of free arachidonic acids. As illustrated in Figure 1, control of the production of free arachidonic acid is believed to involve an as yet undefined membrane receptor event (Dennis, 1989). Presumably this event involves the activation of a phospholipase, either directly or through mediators. Presumably this phospholipase is membrane associated or becomes so after translocation (Channon and Leslie, 1990). Because the bulk of cellular arachidonic acid is found esterified in the *sn*-2 position of membrane phospholipids, a phospholipase A$_2$ is the simplest and most logical enzyme

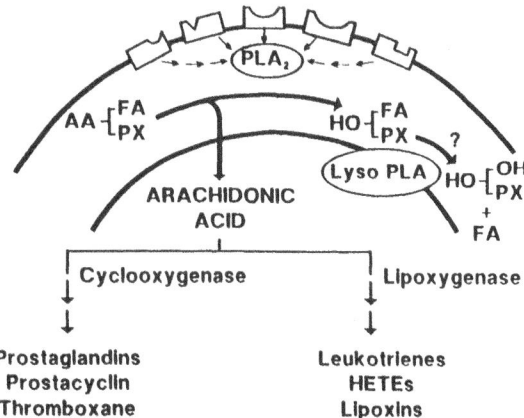

Figure 1. Role of phospholipase A$_2$ (PLA$_2$) in arachidonic acid (AA) release. (Adapted with permission from Dennis, 1989).

to fill this role. When phospholipase A_2 releases arachidonic acid, lysophospholipid is also produced. Lysophospholipids are biological detergents that must be rapidly degraded by a lysophospholipase or be reacylated. In some cases, the lysophospholipid product may lead to the production of platelet-activating factor (PAF).

The blocking of either the cyclooxygenase or the lipoxygenase pathways offers potentially important pharmacological tools for treating numerous disorders. The mode of action of aspirin and other nonsteroidal anti-inflammatory drugs as inhibitors of the cyclooxygenase enzyme is well known. There has also been a great deal of interest in inhibiting the lipoxygenase enzyme. Another potential approach would be to block the cyclooxygenase, lipoxygenase, and PAF pathways by inhibiting the phospholipase A_2.

Until recently, investigators studying the mechanisms of action and inhibition of phospholipase A_2 have focused almost exclusively on extracellular enzymes from venom or pancreas (Dennis, 1983). These enzymes were readily available and relatively easy to purify, and were the only known phospholipases A_2. It is, however, becoming clear that there are a large number of intracellular phospholipases that may be involved in many important cellular functions, a primary one being the production of prostaglandins. These enzymes are present in cells only in very small amounts, and they are difficult to purify. It is also becoming apparent that in many ways they differ dramatically from the extracellular enzymes. We are investigating these phospholipases in the macrophage (Dennis et al., 1989).

EVOLUTIONARY COMPARISONS

It is interesting to compare the sequences of the extracellular phospholipases A_2, including those from cobra venom, bovine and porcine pancreas, rattlesnake, and human synovial fluid. Each of these enzymes is made up of a single polypeptide chain of 120-130 amino acids and contains about 50% α-helical structure and a small amount of pleated sheet. All of these enzymes have an absolute requirement for calcium and a conserved calcium binding site. In addition, a histidine (His-48) and an aspartic acid (Asp-99) are involved in the catalytic mechanism. The other notable feature is the abundance of cysteines. Of the approximately 120 amino acids, there are 14 cysteines forming seven disulfide bonds.

Recently, we (Davidson and Dennis, 1990) compared all known sequences of phospholipases A_2 using the latest computer techniques and developed an evolutionary map. The sequences are clustered into three groups. Type I enzymes are obtained from cobras, kraits, other Old World snakes, and mammalian pancreas. Type II enzymes are found in rattlesnakes, other vipers, other New World snakes, and human synovial fluid. It is particularly interesting that of the two human enzymes that had been sequenced at that time, the pancreatic enzyme is Type I and the synovial fluid enzyme is Type II. These studies led us to suggest that there was a gene duplication event some 200-300 million years ago before mammals and reptiles diverged. Interestingly, the Old World snakes conscripted a Type I enzyme for their venom, while the New World snakes conscripted the Type II form for their venom. Mammals have retained both types. The Type I enzyme is a digestive enzyme produced in the pancreas. The Type II enzyme is found in many other tissues, but its precise function is not clear. One suggestion has been that it is produced by the liver as an acute phase protein (Crowl et al., 1991). Recent evidence that a lizard contains a Type III enzyme suggests that there may have been an additional gene duplication event before reptiles and insects diverged a couple hundred million years earlier. However, to date, there have been no reports of a Type III enzyme in humans.

We can make a few general statements about the characteristics of these three types of phospholipase A_2. The catalytic mechanism of all of these enzymes is virtually identical, and the geometry of the active site residues is also similar. Their substrate specificities do differ, indicating that while their structures may be very similar, they are not identical. This is also reflected in the fact that their interaction with inhibitors is also different.

All these enzymes are extracellular and contain a large number of disulfide bonds. We (Davidson and Dennis, 1990) have pointed out that a cytoplasmic enzyme should not contain many disulfide bonds. Without the constraints of seven disulfide bonds, evolutionary pressures would probably lead to an entirely different sequence for the cytoplasmic enzyme. Indeed, many groups have identified cytoplasmic enzymes that differ from the Type I, II, and III extracellular enzymes. Of particular note, Clark et al. (1990) and Kramer et al. (1991) have identified and sequenced an 85 kDa cytoplasmic phospholipase A_2 from human U937 cells. Its sequence is very different from the extracellular enzymes, and it has a micromolar requirement for Ca^{2+} instead of the millimolar requirement exhibited by the extracellular enzymes.

PHOSPHOLIPASE A_2 STRUCTURE AND MECHANISM

We (Fremont et al., 1990) have recently determined the x-ray crystal structure of the *Naja naja naja* cobra venom enzyme. When we compared the α-chain backbone of the cobra venom enzyme with that of the porcine and bovine pancreatic enzymes and the *Crotalus* enzyme, we found that they could be superimposed in many areas, especially the N-terminal α-helix and the two main helices that form the catalytic site. All of these extracellular enzymes appear to be virtually the same in this region. There are small differences between these enzymes, particularly in the C-terminus and the calcium binding site region, the beta wing and the pancreatic loop. The calcium, His-48, and Asp-99 are virtually superimposable in all these extracellular enzymes.

We (Yu et al., 1990) have synthesized a number of inhibitors of the enzyme; the most potent contained an amide in the *sn*-2 position and a thioether at the *sn*-1 position. Both the phosphatidylcholine (PC) and phosphatidylethanolamine (PE) forms of this inhibitor were potent inhibitors of thio PC hydrolysis. The ethanolamine derivative was the most potent with an IC-50 of about 0.5 μM, making it as potent an inhibitor as been reported for phospholipase A_2.

Table 1 is a partial list of the large number of closely related inhibitors that we have studied. From these studies we have identified four factors that affect inhibitor binding. i) As mentioned above, *sn*-2 amide analogues bind tighter than oxy-ester analogues. ii) PE inhibitors bind tighter than equivalent PC inhibitors. iii) Increasing the hydrophobicity of the moiety in the *sn*-1 position also increases the binding affinity. iv) The α-methylene group of the *sn*-2 chain is also crucial for binding.

With these facts in mind, we attempted to understand why the amide was such a potent inhibitor. The generally accepted mechanism for the action of this enzyme, shown in Figure 2, is that His-48, assisted by Asp-99, pulls a proton off a water molecule which

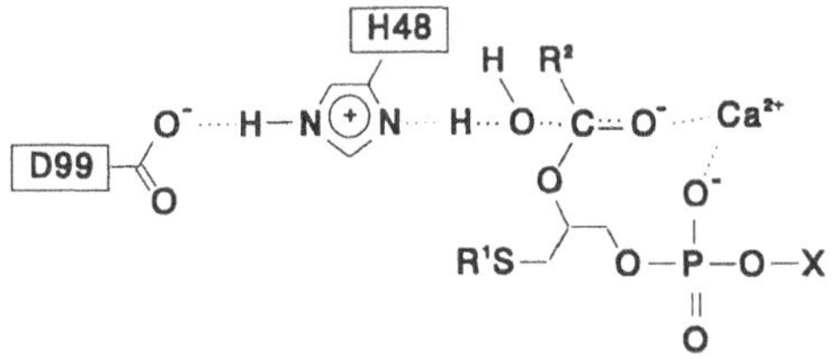

Figure 2. Postulated transition state of the phospholipid hydrolysis catalyzed by phospholipase A_2.

Table 1. Inhibition of PLA$_2$ by phospholipid analogues*

	IC$_{50}$ (µM)		IC$_{50}$ (µM)
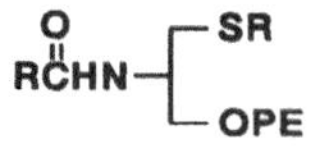 RCHN (O) —[SR / OPE]	0.00045		
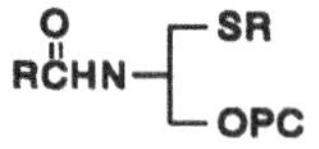 RCHN (O) —[SR / OPC]	0.002	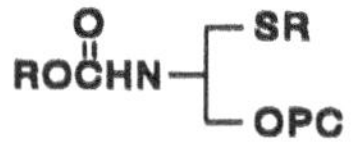 ROCHN (O) —[SR / OPC]	N.D.
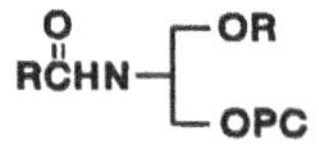 RCHN (O) —[OR / OPC]	0.04		
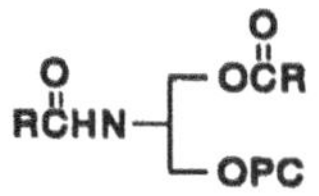 RCHN (O) —[OCR (O) / OPC]	0.16	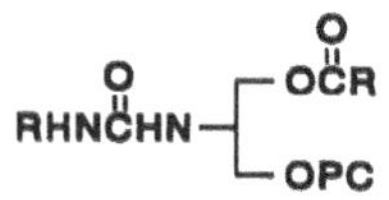 RHNCHN (O) —[OCR (O) / OPC]	N.D.

*Adapted with permission from Yu et al., 1990.

then acts as a general base to attack the carbonyl at the *sn*-2 position of the phospholipid. This carbonyl may be coordinated to the catalytic Ca^{2+}. The Ca^{2+} is also held in place by Asp-49 and is coordinated to a phosphate group on the phospholipid. We (Yu and Dennis, 1991) reasoned that if this is the mechanism and if the ester is replaced by an amide, then the amide NH must form a hydrogen bond to the His-48, and the energy obtained from this binding is what makes the amide such a potent inhibitor. This hypothesis has been borne out by recent x-ray crystallography of an amide-phospholipase A$_2$ complex (Thunnissen et al., 1990).

CONCLUSION

It should be possible to use the wealth of structural and mechanistic information now available regarding the extracellular phospholipase A$_2$ as a basis to explore the intracellular, cytosolic phospholipases A$_2$ in similar detail. The important regulatory questions about both the extracellular and intracellular enzymes are of great interest and will now be approached. It is crucial to investigate the underlying mechanisms that control arachidonic acid production in biological systems.

ACKNOWLEDGMENTS

This work was supported by NSF grant DMB 88-17392 and NIH grants GM 20,501 and HD 26171.

REFERENCES

Channon JY and Leslie CC (1990) A calcium-dependent mechanism for associating a soluble arachidonoyl-hydrolyzing phospholipase A$_2$ with membrane in the macrophage cell line RAW 264.7. J Biol Chem 265:5409.

Clark JD, Milona N, Knopf JL (1990) Purification of a 110-kilodalton cytosolic phospholipase A_2 from the human monocytic cell line U937. Proc Natl Acad Sci USA 87:7708.

Crowl RM, Stoller TJ, Conroy RR, Stoner CR (1991) Induction of phospholipase A_2 gene expression in human hepatoma cells by mediators of the acute phase response. J Biol Chem 266:2647.

Davidson FF and Dennis EA (1990) Evolutionary relationships and implications for the regulation of phospholipase A_2: From snake venom to human secreted forms. J Mol Evol 31:228.

Dennis EA (1983) Phospholipases. In: The enzymes, third edition, Vol 16 (Borer P, ed) p 307. New York: Academic Press.

Dennis EA (1989) The regulation of eicosanoid production: role of phospholipases and inhibitors. Bio/Technology 5:1294.

Dennis EA, Lister MD, Deems RA, Utevitch RJ (1989) Phospholipase A_2 from a macrophage-like cell line. In: Cell activation and signal initiation: receptor and phospholipase control of inositol phosphate, PAF and eicosanoid production. UCLA symposia on molecular and cellular biology-New Series, Vol 106 (Dennis EA, Hunter A, Berridge M, eds) p 369. New York: Alan R. Liss.

Dennis EA, Rhee SG, Billah MM, Hannun YA (1991) Role of phospholipases in generating lipid second messengers in signal transduction. FASEB J 5:2068.

Fremont DH, Anderson D, Wilson IA, Xuong N-H, Dennis EA (1990) Crystal structure of phospholipase A_2 from Indian cobra. American Crystallographic Association, Series 2 18:55.

Kramer RM, Roberts EF, Manetta J, Putnam JE (1991) The Ca^{2+}-sensitive cytosolic phospholipase A_2 is a 100-kDa protein in human monoblast U937 Cells. J Biol Chem 266:5268.

Thunnissen MM, Ab E, Kalk KH, Drenth J, Dijkstra BW, Kuipers OP, Dijkman R, de Haas GH, Verheij HM (1990) X-ray structure of phospholipase A_2 complexed with a substrate-derived inhibitor. J Biol Chem 347:689.

Yu L, Deems RA, Hajdu J, Dennis EA (1990) The interaction of phospholipase A_2 with phospholipid analogues and inhibitors. J Biol Chem 265:2657.

Yu L, Dennis EA (1991) Critical role of a hydrogen bond in the interaction of phospholipase A_2 with transition-state and substrate analogues. Proc Natl Acad Sci USA 88:9325.

ISCHEMIC BRAIN DAMAGE: FOCUS ON LIPIDS AND LIPID MEDIATORS

Bo K. Siesjö and Kenichiro Katsura

The Laboratory for Experimental Brain Research
Department of Neurobiology
Experimental Research Center
University Hospital S-221
85, Lund, Sweden

INTRODUCTION

Two decades ago, Bazan (Bazan, 1970; 1971) demonstrated that ischemia and epileptic seizures are accompanied by lipolysis with accumulation of free fatty acids (FFAs), particularly stearic acid (18:0), arachidonic acid (20:4), and docosahexaenoic acid (22:6). These results were subsequently confirmed and extended by workers interested in ischemia (e.g., DeMedio et al., 1980; Yoshida et al., 1980; 1983; 1986; Rehncrona et al., 1982; Shiu et al., 1983; Abe et al., 1987; Ikeda et al., 1987) or epileptic seizures (Siesjö et al., 1982; Rodríguez de Turco et al., 1983; Reddy and Bazan, 1987). It was also demonstrated that extensive lipolysis with accumulation of FFAs occurs in hypoglycemia (Agardh et al., 1981; Wieloch et al., 1984) and in hypoxia (Gardiner et al., 1981).

These results raised the question of the triggering mechanisms. Bazan (1976) speculated that accumulation of FFAs was related both to enhanced lipolysis, probably reflecting increased phospholipase A_2 (PLA_2) activity, and to reduced reacylation in the energy-compromised tissue. A coupling between accumulation of FFAs and reduction in cellular phosphorylation state seemed likely since energy failure is common to ischemia, hypoxia, and hypoglycemia; additionally, accumulation of FFAs was only observed if tissue hypoxia was sufficiently severe to reduce the phosphorylation state (Gardiner et al., 1981). However, this does not mean that gross energy failure is a prerequisite for lipolysis. Thus, animals subjected to status epilepticus under optimal conditions for tissue oxygenation have a near-normal cerebral energy state (Meldrum and Nilsson, 1976); yet, they show extensive accumulation of FFAs (Siesjö et al., 1982; 1989).

We may explain this obvious disparity by assuming that two factors contribute to the enhanced lipolysis: energy failure and loss of calcium homeostasis. Ischemia/hypoxia, hypoglycemia, and status epilepticus are known to give rise to neuronal damage. Since all three conditions are accompanied by loss of cellular calcium homeostasis, it is tempting to assume that the neuronal necrosis is calcium mediated (Siesjö, 1981). However, since they are also accompanied by lipolysis and accumulation of FFAs the results can be interpreted in two ways. One is to assume that loss of cellular calcium homeostasis is responsible for triggering both enhanced lipolysis and cell death, the enhanced lipolysis having very little to do with the cell death. The other interpretation is to consider cell death as a result of the enhanced lipolysis. In discussing these two alternatives, we will focus attention on lipolysis and cellular calcium metabolism in disease, on calcium-triggered events, and on the role played by changes in lipids and lipid mediators in causing cell dysfunction and cell death in the brain.

LIPOLYSIS IN BRAIN DISEASE

At any given moment, the levels of FFAs represent the balance between hydrolysis of phospholipids (and triglycerides) and their resynthesis or reacylation via pathways which depend on energy (e.g. Bazan, 1976; Sun et al., 1979). This balance is illustrated by the cyclical reactions depicted in Figure 1 (modified from Wieloch and Siesjö, 1982). Hydrolysis of phosphoinositides is exemplified by the breakdown of phosphatidylinositol bisphosphate (PIP_2) into a diglyceride (DG) and inositol trisphosphate (IP_3), a reaction catalyzed by phospholipase C (PLC). DG is then reconverted into PIP_2 by a series of reactions which are energetically driven by ATP and CTP. In this sequence, FFAs are produced when DGs are hydrolyzed by di- and monoglyceride lipases. Since polyphosphoinositides are enriched in stearic and arachidonic acids, these FFAs accumulate as a result of such lipase activity.

Other phospholipids are broken down by phospholipases of the A_2 type, yielding FFAs and lysophospholipids. Since brain tissues contain active lysophospholipases the lyso compounds are degraded with the production of FFAs. Reacylation and resynthesis of the parent compounds require energy in the form of ATP.

Clearly, energy failure will open the phospholipid cycles and lead to accumulation of DGs and FFAs. Hydrolysis of ATP and accumulation of AMP can also trigger the production of DGs from choline phosphoglycerides. This is because the two cycles are connected via a base exchange mechanism and because accumulation of AMP shifts the balance in favor of production of diacylglycerols (DAGs) (Goracci et al., 1981; Horrocks et al., 1982). If energy failure is less than complete, one would expect to see a more moderate release of FFAs, or a quasi-steady state in which the FFA concentrations are increased, albeit not increasing. In such situations, the FFA levels attained depend on two factors: the availability of high energy phosphates and the degree of activation of PLC and PLA_2. In dense ischemia both factors must operate in concert but, in other conditions such as epileptic seizures, the predominant factor is activation of lipases. It is now known that receptors exist which, upon stimulation, lead to activation of both PLA_2 and PLC (e.g. Burgoyne et al., 1987). However, to simplify the discussion, we will assume that PLC activation is mainly caused by stimulation of surface receptors, coupled to PLC, and that the major factor causing PLA_2 activation is a rise in cytosolic calcium concentration (Ca^{2+}_i). The agonists, which stimulate surface receptors and thus activate PLC, include glutamate (see Nicoletti et al., 1986; Sladeczek et al., 1988).

An additional pathway of phospholipid breakdown by PLA_2 has come into focus during recent years (see Braquet et al., 1987; Kumar et al., 1988; Snyder, 1989; Bazan, 1990; Bazan et al., in press; Lindsberg et al., in press). This is the pathway leading to the formation of platelet-activating factor (PAF) from the precursor 1-alkyl-2-acyl-*sn*-glycerophosphocholine, a minor component of the phosphatidylcholine fraction of many cell membranes. The primary products are lyso-PAF, a biologically inactive compound, and arachidonic acid. PAF is formed by the addition of an acetyl group to lyso-PAF in a reaction catalyzed by an acetyltransferase. When formed, PAF can activate PLA_2 thereby catalyzing hydrolysis of other phospholipids with additional formation of arachidonic acid, the precursor of prostaglandins, thromboxane A_2, and leukotrienes. PAF itself is proinflammatory and vasoactive, and it promotes hemostasis by causing platelet aggregation and adherence of leukocytes to endothelial cells. The physiological functions of PAF are not accurately known, but the phospholipid may act in signal transduction sequences which could well encompass gene expression (see Bazan, 1990; Bazan et al., in press). Pathologically, PAF may be a major cause of microcirculatory problems in stroke and other forms of ischemia (see below).

Ischemia

The massive accumulation of FFAs during ischemia has been well documented (see above). The major points for discussion encompass the origin of the FFAs released, the triggering mechanisms, and the secondary consequences of the lipolytic events.

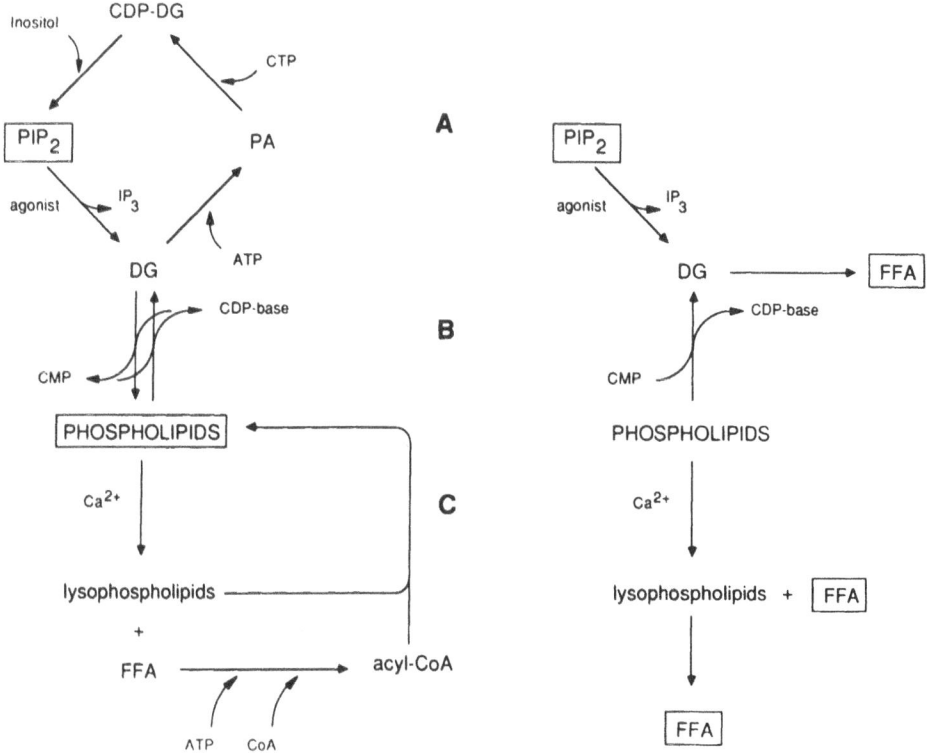

Figure 1. The main phospholipid turnover pathways in the brain under normal (left) and ischemic (right) conditions. Left panel: Top left (A) shows the agonist-stimulated breakdown of phosphatidylinositols, exemplified by the breakdown of phosphatidylinositol-4'-5'-bisphosphate (PIP_2), and the energy-dependent resynthesis by phosphorylation to phosphatidic acid (PA) and further to CDP-diglyceride (CDP-DG). Breakdown of PIP_2 gives rise to inositol triphosphate (IP_3) and diglyceride (DG). Left middle (B) illustrates the base exchange reaction. The direction of this reaction is highly dependent on the concentrations of the reactants CMP and CDP-bases. Bottom left (C) illustrates Ca^{2+} triggered breakdown of phospholipids to lysophospholipids and free fatty acid (FFA), and the ATP-dependent reacylation of lysophospholipids. Right panel: Under ischemic conditions, the anabolic pathways are inhibited. The release of transmitters triggers PIP_2 breakdown to DG and FFA, and Ca^{2+} influx into cells causes phospholipid degradation to lysophospholipids and FFAs. In addition, the increased CMP levels drive the base exchange reaction with phospholipid breakdown to DG and further to FFAs. Reproduced with permission from Siesjö et al., 1989b.

There is little doubt that part of the accumulation of FFAs during ischemia is due to enhanced PLA_2 activity (e.g., Edgar et al., 1982). However, substantial evidence exists that the FFAs released during the first few minutes are mainly derived from phosphoinositides. The first indication that this is so was based on findings that diacylglycerides enriched in 20:4 and 18:0 are rapidly produced during ischemia (Aveldaño and Bazan, 1975; see also Banschbach and Geison, 1974). Breakdown of polyphosphoinositides during ischemia, originally reported by Dawson and Eichberg (1965), was subsequently documented in detail (Ikeda et al., 1986; 1987; Abe et al., 1987). The results reported by Abe et al. (1987) showed that PIP_2, PIP, and PI were hydrolyzed during the first minute of ischemia in gerbils; in fact, the polyphosphoinositides were the only phospholipids which were reduced in concentration during the 15 min period of ischemia studied. However, this result should not be interpreted to show that the FFAs are derived from the polyphosphoinositides only. Thus, since the tissue concentrations of, for instance, the ethanolamine and the choline phosphoglycerides are so much higher, enhanced lipolysis of an extent which will raise the FFA concentrations may not be detected as a fall in the total phospholipid content. The best interpretation of available data is that, initially, the FFAs are mainly derived from phosphoinositides; at later stages a substantial contribution from other phospholipids is likely.

Since ischemia leads to presynaptic release of transmitters, to energy failure, and to influx of calcium with a rise in Ca^{2+}_i, lipolysis should be maximally stimulated. The question arises whether these events can explain the early rise in FFA concentrations during ischemia. Free fatty acids are known to accumulate within the first 30-60 sec of ischemia (Aveldaño and Bazan, 1975; Pediconi and Rodríguez de Turco, 1984; Abe et al., 1987). Figure 2 clearly illustrates this fact. Recent data in rats provide additional information on early triggering events. Thus, although a K^+ conductance seems to be activated within seconds following induction of ischemia, depolarization with influx of calcium from extracellular fluid does not occur until after 60-70 sec (Hansen, 1985; Siesjö, 1988c). In the gerbil, ischemia of brief duration (30-60 sec) was accompanied by a substantial breakdown of ATP (Abe et al., 1987). Unfortunately, no data exist which demonstrate the coupling among energy failure, loss of ion homeostasis, and accumulation of FFAs.

Collaborative work between our group and that headed by Dr. Bazan is designed to show whether accumulation of FFAs in the cerebral cortex of rats commences within the first 30-45 sec of ischemia (Katsura et al., in preparation). This is at a time when the total tissue ATP concentration exceeds 80% of control, and before calcium enters cells from extracellular fluids (see Hansen, 1985; Siesjö, 1988a,b,c). What is the trigger for an early rise in FFA concentration, if present? One possibility is that a *local* fall in ATP concentration, or in the ATP/ADP ratio, reduces reacylation/resynthesis of phospholipids to such an extent that FFAs start to accumulate. However, it is also conceivable that calcium is released from intracellular stores such as mitochondria or endoplasmic reticulum (ER). Release from the ER could reflect stimulation of surface receptors leading to activation of PLC and to production of IP_3. What speaks in favor of an intracellular release of calcium is that phosphorylase *b* to *a* conversion, a calcium-triggered event, occurs within 15 sec of ischemia (Folbergrova et al., 1990). Additional evidence that Ca^{2+} rises early during ischemia was provided by Silver and Erecinska (1990), who measured Ca^{2+}_i with microelectrodes. It remains to be shown, however, what mechanisms could cause an early release of Ca^{2+} from organelles or stimulation of surface receptors coupled to PLC (or PLA_2) at a time when massive release of transmitters has not yet occurred.

The third point concerns the secondary consequences of accumulation of FFAs. Termination of the anoxic transient leads to the normalization of the FFA contents; however, this takes time and accumulation of FFAs persists for 15-30 min during recirculation/reoxygenation (Rehncrona et al., 1980; 1982; Yoshida et al., 1980; 1984).

Thus, although arachidonic acid is metabolized more quickly than other FFAs (Yoshida et al., 1980) conditions are nevertheless at hand for an accelerated rate of metabolism of the acid along the cyclooxygenase and the lipoxygenase pathways. It has also been clearly documented that recirculation leads to a spurt of production of prostaglandins (Gaudet et al., 1980) and of leukotrienes (Moskowitz et al., 1984), and it is now becoming increasingly clear that PAF accumulates in the tissue as well (see Lindsberg et al., 1990; Gilboe et al., 1991). We will discuss below whether production of arachidonic acid metabolites and of PAF contributes to the damage incurred.

Hypoglycemia

Hypoglycemic coma, defined as hypoglycemia of sufficient severity to cause cessation of spontaneous and evoked electrical activity, leads to accelerated lipolysis and accumulation of FFAs (Agardh et al., 1981; Wieloch et al., 1984). The lipolytic events are tightly coupled to depolarization, loss of energy homeostasis, and cellular energy failure (Harris et al., 1984; Wieloch et al., 1984). As Figure 3 shows, tissue FFA concentrations are normal before, and raised just after, the first depolarization wave, which heralds hydrolysis of high energy phosphate compounds and loss of ion homeostasis. The events can be interpreted as follows (Siesjö, 1988b): Local depolarization in the energy-compromised tissue causes release of excitatory transmitters and activation of ion conductances; as a result, the energy stores are wasted, and calcium homeostasis is lost. Both events must contribute to the lipolysis and accumulation of FFAs.

Changes in lipid metabolism during hypoglycemic coma differ from those observed during ischemia in two major respects. First, breakdown of phospholipids is more extensive since the size of the phospholipid pool is reduced by about 10%, and the reduction encompasses both phosphoinositides (Ikeda et al., 1987) and other phospholipids (Agardh et al., 1981; Wieloch et al., 1984). For unknown reasons, changes in total phospholipid phosphorus and in FFA concentrations are established during the first 5 min of coma, with no obvious further changes thereafter. Second, there is a disparity between the amounts of phospholipid disappearing and the amounts of FFAs

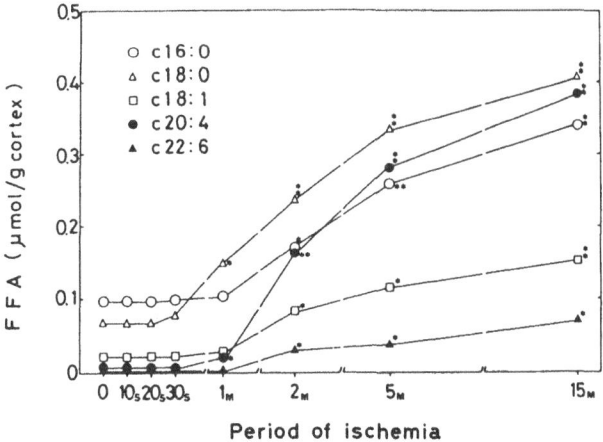

Figure 2. Changes in the contents of free fatty acids (FFA) during ischemia in gerbil cerebral cortex. Periods of ischemia were 0 (sham control), 10 s, 20 s, 30 s, 1 min, 2 min, 5 min, and 15 min. Amounts of palmitic acid, stearic acid, oleic acid, arachidonic acid, and docosahexaenoic acid increased significantly during ischemia (*p<0.05, **p<0.01). Note early rise in 18:0 and 20:4. Reproduced with permission from Abe et al., 1987.

accumulated, suggesting that FFAs are either oxidized or lost to the circulation. During the insult, oxidation would be favored by the continued supply of oxygen and loss to the circulation by the increased blood flow.

Status Epilepticus

Although seizures are accompanied by a marked rise in FFA concentrations, particularly of arachidonic acid, lipolytic events differ from those observed in ischemia and hypoglycemia (see Siesjö et al., 1982; 1989; Reddy and Bazan, 1987). First, changes in FFA concentrations during seizures are more moderate, reflecting the near-normal cellular energy state. Second, the changes mainly involve 20:4 and 18:0. This fact, as well as the rapid reduction in PIP_2 concentration, suggests that the FFAs are derived

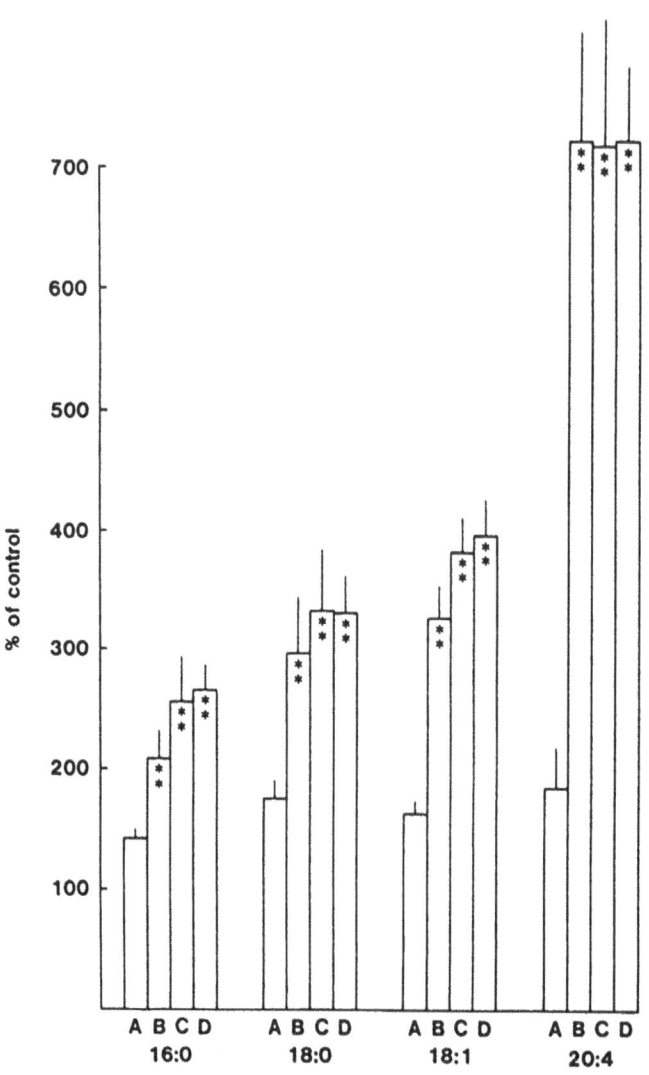

Figure 3. Cortical levels of palmitic (16:0), stearic (18:0), oleic (18:1), and arachidonic (10:4) acids during hypoglycemia as percent of control values. A: Predepolarization. B: First depolarization. C: Second depolarization. D: After 10 min of isoelectric EEG. Reproduced with permission from Wieloch et al., 1984.

mainly from poly-phosphoinositides, probably as a result of stimulation of membrane receptors coupled to PLC. Third, the total FFA concentrations increase rapidly during the first 5 min of status epilepticus, remain elevated for 20-30 min, but then decrease to levels of about 150% of control (see Siesjö et al., 1989b). This adjustment, which is not paralleled by any reduction in metabolic rate, could reflect downregulation of receptors or of PLC.

CALCIUM, LIPOLYSIS, AND BRAIN DAMAGE

Calcium serves as a first and second messenger in the stimulus-response coupling at neuronal synapses and has a key role in modulating cellular metabolism. However, by using excitatory amino acids (EAAs) to trigger the coupling and calcium to mediate the response, cells carry the seeds of their own destruction (see Choi, 1988; Siesjö, 1988c; 1990; Siesjö et al., 1991). Thus, when calcium transients become pathologically enhanced, calcium-activated reactions may run out of control, disrupting cytoskeletal and membrane structure and adversely altering the function of receptors and ion channels. Under such circumstances, calcium may also activate endonucleases, triggering fragmentation of DNA (Orrenius et al., 1988).

The calcium hypothesis of cell death postulates that a series of potentially devastating reactions are set into motion whenever Ca^{2+} values rise unduly, either because calcium influx/release is excessive or because extrusion mechanisms fail. The key events involved are summarized in Figure 4. The reactions outlined encompass enhanced lipolysis, changes in protein phosphorylation, and proteolysis/disaggregation of microtubules. The figure does not show the possible role of calcium transients in triggering changes in RNA and protein metabolism, and in activating a hypothetical suicide program in some cells (see Siesjö, in press).

As discussed elsewhere (Siesjö, in press) proteolysis and disaggregation of microtubules, with disruption of the cytoskeleton, may be key events in the cascade of reactions leading to calcium-related cell damage. In the present discussion, however, the focus will be on changes in lipid metabolism. Excessive phospholipid hydrolysis with accumulation of DAGs and FFAs is potentially harmful in three ways. One is related to the loss of membrane-bound phospholipids and to the toxicity of the primary degradation products. The latter encompass FFAs, lysophospholipids, and PAF. FFAs and lysophospholipids have a detergent-like effect on membranes and can act as ionophores, causing uncoupling of oxidative phosphorylation and changes in plasma membrane permeability. As will be discussed below, PAF is proaggregatory and hemostatic and, since it is a powerful chemotactic agent, it may cause adhesion and activation of leukocytes and, hence, an inflammatory reaction at the blood-endothelial cell interface. The second way in which accumulation of FFAs can be harmful is by triggering an uncontrolled production of metabolites for which arachidonic acid serves as a substrate, i.e., prostaglandins, thromboxanes, and leukotrienes. In one way or the other this cascade of events sets the stage for an increased production of free radicals and, hence, for lipid peroxidation and oxidative damage to proteins. The third way, finally, is one in which a pathological calcium transient leads to a sustained activation of PKC and to its translocation to membranes.

Changes Secondary to Phospholipid Hydrolysis

It is often stated that phospholipid hydrolysis *per se* must jeopardize cell function and survival. This is because phospholipids form integral and functionally important parts of membranes, influencing receptor function and ion conductances, and because the primary degradation products, i.e., lysophospholipids, DAGs, and FFAs, have pro-

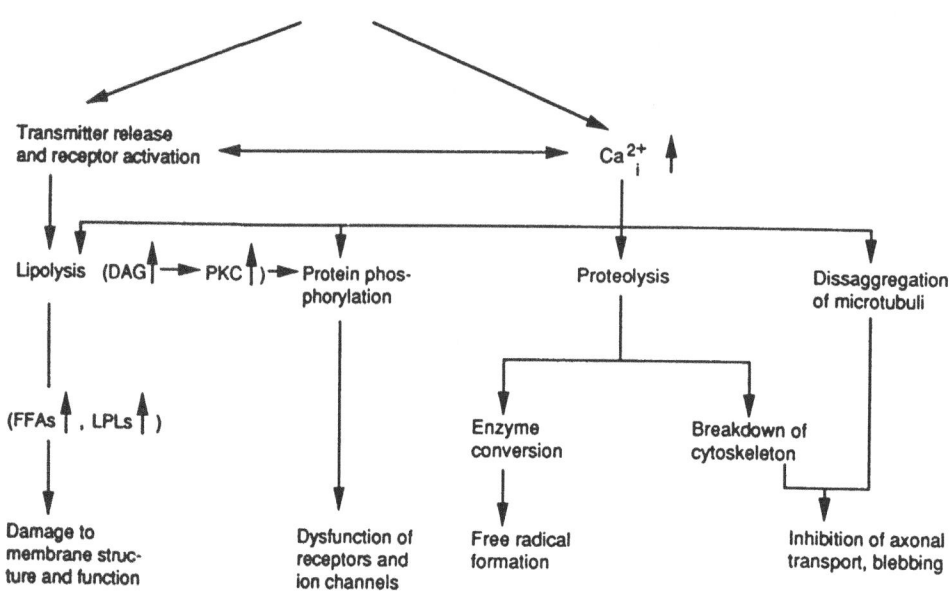

Figure 4. Schematic diagram illustrating cascade of events elicited by energy failure/depolarization and by the ensuing increase in Ca^{2+}_i. For further details, see text. Reproduced with permission from Siesjö (in press).

nounced effects on membrane functions. For example, it is conceivable that the increased membrane conductance of depolarized cells is, at least in part, caused by the presence of such components. It may also be speculated that changes in lipids are responsible for an uncoupling of oxidative phosphorylation, e.g., during hypoglycemic coma (see Siesjö, 1988b). The question arises whether such changes persist in the post-insult period, *viz.* when an adequate oxygen or glucose supply has been restored. For example, would a change in membrane structure due to phospholipid hydrolysis give rise to a sustained perturbation of membrane function? Some experiments suggest that this is not the case. For example, although hypoglycemic coma gives rise to very extensive phospholipid hydrolysis, termination of the hypoglycemic insult leads to rapid restoration of plasma membrane and mitochondrial function (Bengtsson, Katsura, and Siesjö, unpublished observations). Also following ischemia of moderate duration, such functions seem to be promptly restored (e.g., Hillered et al., 1985).

Evidence has accumulated that PAF, produced during ischemia, adversely affects recovery following recirculation. In the gerbil, PAF antagonists improved recovery of function and mitochondrial respiration, and ameliorated both tissue edema and accumulation of FFAs following short-duration transient ischemia (Spinnewyn et al., 1987; Birkle et al., 1988). Experiments with transient forebrain ischemia in rats seemed to give similar results since CA1 damage in the hippocampus was ameliorated by Gingkolide B (Oberpichler et al., 1990). Finally, Gilboe et al. (1991), working with the isolated perfused dog's brain, reported that two PAF antagonists ameliorated the deterioration of the EEG and of the cerebral energy state following 14 min of circulatory arrest.

Although these results appear consistent, they should probably not be construed to show that PAF plays a major role in determining the outcome of brief periods of cerebral ischemia. Our own group has failed to confirm that an established PAF antagonist (BN52021) ameliorates secondary hypoperfusion or histologically assessed brain damage following 15 min of forebrain ischemia in rats (M.-L. Smith and B.K. Siesjö, unpublished results). Although this result is at variance with those reported by Oberpichler et al. (1990), it is tempting to conclude that PAF antagonists only ameliorate ischemic brain damage when recirculation leads to microvascular problems. This is probably the case in gerbils, which reportedly have a low post-ischemic blood pressure (see Spinnewyn et al., 1987), and in the isolated perfused dog's brain. We thus hypothesize that although PAF plays a role in aggravating ischemic damage, this role only becomes manifest if ischemic or post-ischemic blood flow is slow enough to allow activation and adhesion of leukocytes, thereby setting the stage for an inflammatory reaction at the blood-endothelial cell interface.

This hypothesis is consistent with results demonstrating a salutary effect of PAF antagonists in more long-lasting ischemia due to middle cerebral artery occlusion (Bielenberg and Wagner, 1989), to laser-induced microvascular obstruction (Frerichs et al., 1990), or to spinal cord ischemia (Lindsberg et al., 1990). However, it remains to be shown whether there are species differences in the metabolism of PAF and in its effect on microcirculatory events. It also remains to be shown whether PAF also has a detrimental effect on extravascular tissues, such as neurons.

Changes Secondary to Metabolism of Arachidonic Acid

Production of prostaglandins, thromboxanes, and leukotrienes occurs whenever the concentration of arachidonic acid is raised and sufficient oxygen is present to allow its metabolism via the cyclooxygenase and the lipoxygenase pathways (see Wolfe, 1982; Samuelsson 1983; Walker and Pickard, 1985). The products formed by cyclooxygenase, i.e. prostaglandins, prostacyclin, and thromboxane A_2, and those formed by various lipoxygenases, mainly HETEs and leukotrienes B_4, C_4 and D_4, probably fulfill important roles in the modulation of membrane function and transduction of signals, as well as in the control of microcirculation. It is less clear what roles these compounds may play in modulating ischemic brain damage.

A thromboxane A_2 antagonist (1-benzylimidazole) has been reported to ameliorate the secondary hypoperfusion which follows upon brief transient ischemia (Pettigrew et al., 1989). However, since it has not been clearly established that the secondary hypoperfusion (observed at adequate perfusion pressures) aggravates the final brain damage, it is not known at present if thromboxane inhibitors are anti-ischemic. Predictably, such inhibitors could ameliorate damage due to prolonged ischemia, such as in stroke, simply because platelet aggregation and an inflammatory response at the platelet-endothelial interface are likely events when flow is reduced over a long period.

More definitive evidence exists that prostaglandin inhibitors and prostacyclin ameliorate ischemic damage. Hallenbeck and collaborators (see Kochanek et al., 1987, 1988) showed that a combination of heparin, prostacyclin, and indomethacin improved reflow following 60 min of ischemia in the dog. This combination likely acts by preventing a secondary compromise of circulation in a tissue which is metabolically grossly perturbed. However, a subsequent study showed that two cyclooxygenase inhibitors (piroxicam and flurbiprofen) ameliorated CA1 damage in gerbils subjected to 5 min of transient forebrain ischemia (Nakagomi et al., 1989). Since this type of damage is probably not "vascular" we have to consider the possibility that cyclooxygenase products contribute to the slowly maturing CA1 damage, e.g., by causing an upregulation of receptors and ion channels involved in membrane calcium cycling (see Siesjö et al., 1990). It seems highly justified that cyclooxygenase inhibitors be tested in experimental stroke due to middle cerebral artery occlusion.

In tissues other than the brain, leukotrienes are known to increase capillary membrane permeability, thereby predisposing to extravasation of proteins and edema (Samuelsson, 1983). Similar effects have been assumed for brain tissues (Black and Hoff, 1985; Minamisawa et al., 1988). However, there are at present no definitive data establishing a pathogenetic role for HETEs or for leukotrienes in ischemic brain. Further experiments are clearly warranted.

Production of Free Radicals

It has been assumed for almost 15 years that ischemia, particularly if followed by reperfusion, leads to increased production of free radicals and to free radical-induced damage (Demopoulos et al., 1977; for recent reviews, see Halliwell, 1989; Siesjö et al., 1989a). Such production (or damage) has been difficult to prove, however, and it has remained unknown under what conditions a deleterious production of free radicals occurs, and what are the targets. It now seems established that free radical damage does occur under certain conditions (e.g. Patt et al., 1988; Martz et al., 1989). Based on these findings, it has been postulated that free radicals become determinants of the ischemic damage when the ischemia is sustained, as in stroke, and also that a major target of the free radical damage is the microvessel compartment; possibly, enhanced free radical damage is also responsible for the exaggeration of ischemic damage seen in hyperglycemic and hyperthermic animals (Siesjö et al., 1989a).

Free radical damage is related to lipids/FFAs in at least two ways. First, free radicals are generated when arachidonic acid is metabolized by cyclooxygenase and lipoxygenase. Second, lipids are the targets of free radical attack, since they are peroxidized in a cascade of autocatalytic reactions when an allylic hydrogen atom is abstracted by a free radical such as $^{\bullet}OH$ (see Demopoulos et al., 1977; Del Maestro et al., 1980; Siesjö, 1981). Although much emphasis has been placed on lipid peroxidation and damage to the phospholipid backbone of cell membranes, damage to protein constituents or modification of their activity is now receiving increasing attention (e.g., Halliwell, 1989; Floyd, 1990). Such oxidative alteration of proteins may encompass inactivation of enzymes such as glutamine synthetase (see Floyd, 1990), or activation of others such as endothelial cell Na^+/K^+-ATPase, an alteration which has been assumed to predispose to edema (Koide et al., 1985; 1986).

Since free radicals can cause damage to proteins, lipids, and nucleic acids in membranes, cytosol, and nuclei, cell dysfunction can have a variety of origins, and be expressed in many different ways (see Freeman and Crapo, 1982). However, most interest has been focused on the ability of free radicals to induce edema and vascular dysfunction. Chan and Fishman (1980; 1984) established the link between administration of polyunsaturated fatty acids, free radical formation, and edema of both the cellular and vasogenic type. Their studies and those conducted by others (Abe et al., 1988) suggest that free radicals, partly emanating from the metabolism of arachidonic acid, or arising as a consequence of the membrane effects of polyunsaturated fatty acids, are important mediators of vasogenic brain edema. The simultaneous work by Kontos (Kontos et al., 1980; Kontos 1985) provided a link between impact trauma to nervous tissue and vascular dysfunction, in which the mediator was established to be a free radical, probably $^{\bullet}O_2^-$. Since the production of $^{\bullet}O_2^-$, and the damage, could be prevented by indomethacin, the origin of the free radical production seemed to be the cyclooxygenase reaction.

At present, the role of free radicals has not been accurately defined, nor do we know the mechanisms that link free radical production with the metabolism of phospholipids and fatty acids. However, since established free radical scavengers ameliorate damage due to long-term ischemia, it seems justified that this link be further pursued.

The cascade of reactions elicited as a result of influx/release of calcium encompasses activation of protein kinases, including PKC. It is not unlikely that *sustained* activation and membrane translocation of PKC, following a nonphysiological rise in Ca^{2+}, cause changes in excitable membranes which lead to cell death even though recirculation/reoxygenation has been achieved. It has long been known that transient global or forebrain ischemia gives rise to delayed neuronal death in some brain areas, including the CA1 sector of the hippocampus (e.g., Kirino, 1982; Pulsinelli et al., 1982). It has been assumed that the transient ischemia leads to sustained calcium cycling across the metabolically perturbed cell membranes and to delayed, calcium-related cell death (Deshpande et al., 1987; Siesjö et al., 1990). A likely molecular mechanism for such upregulation of ion channels/receptors responsible for calcium influx was suggested when Connor et al. (1988) reported that glutamate exposure, particularly following a priming stimulus, could give rise to standing Ca^{2+}_i gradients in CA1 pyramidal cell dendrites. It was subsequently proposed that glutamate-related cell death in cerebellar granular cell cultures is related, not to the calcium influx *per se*, but to the activation/translocation of PKC (Manev et al., 1989; 1990). This has led to intense exploration of PKC-related metabolic events in ischemic tissue and to therapeutic trials with PKC inhibitors (Hara et al., 1990).

SUMMARY AND CONCLUSIONS

The last two decades of research have produced detailed information not only on how ischemia causes degradation of phospholipids and accumulation of potentially cytotoxic breakdown products of such lipids, but also on reactions elicited by the subsequent conversion of these products into a series of lipids, mediating an array of cellular and intercellular reactions. It now seems clear that PAF, as well as several of the cyclooxygenase and lipoxygenase products of arachidonic acid, can induce changes, particularly in the microvasculature, which jeopardize cell survival in reperfused tissue. It is equally clear that, at least following long periods of ischemia, free radicals generated in reactions that are interacting with those producing eicosanoids and PAF play a similar role. A somewhat more speculative mechanism links sustained activation and membrane translocation of PKC to delayed neuronal death following transient ischemia. All of these interactions underscore the importance of lipolytic events for cell damage in ischemia and other conditions with a compromised cellular energy metabolism.

REFERENCES

Abe K, Kogure K, Yamamoto H, Imazawa M, Miyamoto K (1987) Mechanism of arachidonic acid liberation during ischemia in gerbil cerebral cortex. J Neurochem 48:503-509.

Abe K, Yuki S, Kogure K (1988) Strong attenuation of ischemic and postischemic brain edema in rats by a novel free radical scavenger. Stroke 19:480-485.

Agardh C-D, Chapman AG, Nilsson B, Siesjö BK (1981) Endogenous substrates utilized by rat brain in severe insulin-induced hypoglycemia. J Neurochem 36: 490-500.

Aveldaño M and Bazan N (1975) Rapid production of diacylglycerols enriched in arachidonate and stearate during early brain ischemia. J Neurochem 25:919-920.

Banschbach MW and Geison RL (1974) Post-mortem increase in rat cerebral hemisphere diglyceride pool size. J Neurochem 23:875-877.

Bazan NG (1970) Effects of ischemia and electroconvulsive shock on free fatty acid pool in the brain. Biochim Biophys Acta 218:1-10.

Bazan NG (1971) Changes in free fatty acids of brain by drug-induced convulsions, electroshock and anesthesia. J Neurochem 18:1379-1385.

Bazan NG (1976) Free arachidonic acid and other lipids in the nervous system during early ischemia and after electroshock. Adv Exp Med Biol 72:317-335.

Bazan NG (1990) Neuronal cell signal transduction and second messengers in cerebral ischemia. In: Pharmacology of cerebral ischemia (Krieglstein J, Oberpichler H, eds) pp 391-396. Stuttgart: Wissenschaftliche Verlagsgesellschaft.

Bazan NG, Squinto SP, Braquet P, Panetta T, and Marcheselli VL (in press) Platelet activating factor and polyunsaturated fatty acids in cerebral ischemia or convulsions: Intracellular PAF-binding sites and activation of a FOS/JUN/AP-1 transcriptional signaling system. Lipids 26.

Bielenberg GW and Wagner G (1989) PAF antagonists reduce infarct size in focal ischemia in the rat brain. In: Pharmacology of cerebral ischemia 1988. (Krieglstein J, ed) pp 281-284. Stuttgart: Wissenschaftliche Verlagsgesellschaft.

Birkle DL, Kurian P, Braquet P, Bazan NG (1988) Platelet-activating factor antagonist BN52021 decreases accumulation of free polyunsaturated fatty acid in mouse brain during ischemia and electroconvulsive shock. J Neurochem 51:1900-1905.

Black KL and Hoff JT (1985) Leukotrienes increase blood-brain barrier permeability following intraparenchymal injections in rats. Ann Neurol 18:349-351.

Braquet P, Touqui L, Shen TS, Vargaftig BB (1987) Perspectives in platelet activating factor research. Pharmacol Rev 39:97-145.

Burgoyne R, Cheek TR, Sullivan AJ (1987) Receptor-activation of phospholipase A_2 in cellular signalling. Trends Biochem Sci 12:332-333.

Chan P and Fishman R (1980) Transient formation of superoxide radicals in polyunsaturated fatty acid-induced brain swelling. J Neurochem 35:1004-1007.

Chan P and Fishman R (1984) The role of arachidonic acid in vasogenic brain edema. Fed Proc 43:210-213.

Choi DW (1988) Glutamate neurotoxicity and diseases of the nervous system. Neuron 1:623-634.

Connor JA, Connor JA, Wadman WJ, Hochberger PE, Wong RKS (1988) Sustained dendritic gradients of Ca^{2+} induced by excitatory amino acids in CA1 hippocampal neurons. Science 240:649-653.

Dawson R and Eichberg J (1965) Diphosphoinositide and triphosphoinositide in animal tissues: Extraction, estimation and changes post mortem. Biochem J 96:634-643.

Del Maestro RF, Thaw HH, Bjork J, Planker M, Arfors K-E (1980) Free radicals as mediators of tissue injury. Acta Physiol Scand Suppl 492:43-57.

DeMedio G, Goracci G, Horrocks L, Lazarewicz J, Mazzari S, Porcellati G, Strosznajder J, Trovarelli G (1980) The effect of transient ischemia on fatty acid and lipid metabolism in the gerbil brain. Ital J Biochem 29:412-432.

Demopoulos HB, Flamm ES, Seligman ML, Power R, Pietronigro DD, Ransohoff J (1977) Molecular pathology of lipids in CNS membranes. In: Oxygen and physiological function (Jobsis FF, ed) pp 491-508. Dallas: Professional Information Library.

Deshpande JK, Siesjö BK, Wieloch T (1987) Calcium accumulation and neuronal damage in the rat hippocampus following cerebral ischemia. J Cereb Blood Flow Metab 7:89-95.

Edgar A, Strosznajder J, Horrocks L (1982) Activation of ethanolamine phospholipase A_2 in brain during ischemia. J Neurochem 39:1111-1116.

Floyd R (1990) Role of oxygen free radicals in carcinogenesis and brain ischemia. FASEB J 4:2587-2597.

Folbergrová J, Minamisawa H, Ekholm A, Siesjö BK (1990) Phosphorylase a and labile metabolites during anoxia: Correlation to membrane fluxes of K^+ and Ca^{2+}. J Neurochem 55:1690-1696.

Freeman BA and Crapo JD (1982) Biology of disease. Free radicals and tissue injury. Lab Invest 47:412-422.

Frerichs KU, Lindsberg PJ, Hallenbeck JM, Feuerstein G (1990) Platelet activating factor antagonist protective in progressive focal brain damage. J Neurosurg 73:223-233.

Gardiner M, Nilsson B, Rehncrona S, Siesjö BK (1981) Free fatty acids in the rat brain in moderate and severe hypoxia. J Neurochem 36:1500-1505.

Gaudet R, Alam I, Levine L (1980) Accumulation of cyclooxygenase products of arachidonic acid metabolism in gerbil brain during reperfusion after bilateral common carotid artery occlusion. J Neurochem 35:653-658.

Gilboe DD, Kintner D, Fitzpatrick JH, Emoto SE, Esanu A, Braquet PG, Bazan NG (1991) Recovery of postischemic brain metabolism and function following treatment with a free radical scavenger and platelet-activating factor antagonists. J Neurochem 56:311-319.

Goracci G, Francescangeli E, Horrocks LA, Porcellati G (1981) The reverse reaction of cholinephosphotransferase in rat brain microsomes, a new pathway for degradation of phosphatidylcholine. Biochim Biophys Acta 664:373-379.

Halliwell B (1989) Superoxide, iron, vascular endothelium and reperfusion injury. Free Radic Res Commun 5:315-318.

Hansen AJ (1985) Effects of anoxia on ion distribution in the brain. Physiol Rev 65(1):101-148.

Hara H, Onodera H, Yoshidomi M, Matsuda Y, Kogure K (1990) Staurosporine, a novel protein kinase C inhibitor, prevents postischemic neuronal damage in the gerbil and rat. J Cereb Blood Flow Metab 10:646-653.

Harris R, Wieloch T, Symon L, Siesjö BK (1984) Cerebral extracellular calcium activity in severe hypoglycemia: Relation to extracellular potassium activity and energy state. J Cereb Blood Flow Metabol 4:187-193.

Hillered L, Smith M-L, Siesjö BK (1985) Lactic acidosis and recovery of mitochondrial function following forebrain ischemia in the rat. J Cereb Blood Flow Metab 5:259-266.

Horrocks L, Dorman R, Dabrowiecki Z, Goracci G, Porcellati G (1982) CDPcholine and CDPethanolamine prevent the release of free fatty acids during brain ischemia. Prog Lipid Res 20:531-534.

Ikeda M, Yoshida S, Busto R, Santiso M, Ginsberg MD (1986) Polyphosphoinositides as a probable source of brain free fatty acids accumulated at the onset of ischemia. J Neurochem 47:123-132.

Ikeda M, Yoshida S, Busto R, Santiso M, Martinez E, Ginsberg M (1987) Cerebral phosphoinositide and energy metabolism during and after insulin-induced hypoglycemia. J Neurochem 49:100-106.

Kirino T (1982) Delayed neuronal death in the gerbil hippocampus following transient ischemia. Brain Res 239:57-69.

Kochanek PM, Dutka AJ, Hallenbeck JM (1987) Indomethacin, prostacyclin, and heparin improve postischemic cerebral blood flow without affecting early postischemic granulocyte accumulation. Stroke 18:634-637.

Kochanek PM, Dutka AJ, Kumaroo KK, Hallenbeck JM (1988) Effects of prostacyclin, indomethacin, and heparin on cerebral blood flow and platelet adhesion after multifocal ischemia of canine brain. Stroke 19:693-699.

Koide T, Asano T, Matsushita H, Takakura K (1986) Enhancement of ATPase activity by a lipid peroxide of arachidonic acid in rat brain microvessels. J Neurochem 46:235-242.

Koide T, Gotoh O, Asano T, Takakura K (1985) Alterations of the eicosanoid synthetic capacity of rat brain microvessels following ischemia: Relevance to ischemic brain edema. J Neurochem 44:85-93.

Kontos H (1985) Oxygen radicals in cerebral vascular injury. Circ Res 57:508-16.

Kontos HA, Wei EP, Povlishock JT, Dietrich WL, Majiera CM, Ellis EF (1980) Cerebral arteriolar damage by arachidonic acid and prostaglandin G_2. Science 209:1242-1245.

Kumar R, Harvey SAK, Kester MK, Hanahan DJ, Olson MS (1988) Production and effects of platelet-activating factor in the rat brain. Biochim Biophys Acta 963:375-383.

Lindsberg PJ, Hallenbeck JM, Feuerstein G (in press) Platelet-activating factor in stroke and brain injury. Ann Neurol.

Lindsberg PJ, Yue T-L, Frerichs KU, Hallenbeck JM, Feuerstein G (1990) Evidence for platelet-activating factor as a novel mediator in experimental stroke in rabbits. Stroke 21:1452-1457.

Manev H, Costa E, Wroblewski J, Guidotti A (1990) Abusive stimulation of excitatory amino acid receptors: a strategy to limit neurotoxicity. FASEB J 4:2789-2797.

Manev H, Favaron M, Guidotti A, Costa E (1989) Delayed increase of Ca^{2+} influx elicited by glutamate: Role in neuronal death. Mol Pharmacol 36:106-112.

Martz D, Rayos G, Schielke GP, Betz AL (1989) Allopurinol and dimethylthiourea reduce brain infarction following middle cerebral artery occlusion in the rat. Stroke 20:488-494.

Meldrum BS and Nilsson B (1976) Cerebral blood flow and metabolic rate early and late in prolonged epileptic seizures induced in rats by bicuculline. Brain 99:523-542.

Minamisawa H, Terashi A, Katayama Y, Kanda Y, Shimizu J, Shiratori T, Inamura K, Kaseki H, Yoshino Y (1988) Brain eicosanoid levels in spontaneously hypertensive rats after ischemia with reperfusion: Leukotriene C_4 as a possible cause of cerebral edema. Stroke 19:372-377.

Moskowitz M, Kiwak K, Heikimian K, Levine L (1984) Synthesis of compounds with properties of leukotrienes C_4 and D_4 in gerbil brains after ischemia and reperfusion. Science 224:886-889.

Nakagomi T, Sasaki T, Kirino T, Tamura A, Noguchi M, Saito I, Takakura K (1989) Effect of cyclooxygenase and lipoxygenase inhibitors on delayed neuronal death in the gerbil hippocampus. Stroke 20:925-929.

Nicoletti F, Meek JL, Iadarola MJ, Chuang DM, Roth BL, Costa E (1986) Coupling of inositol phospholipid metabolism with excitatory amino acid recognition sites in rat hippocampus. J Neurochem 46:40-46.

Oberpichler H, Sauer D, Rossberg C, Mennel H-D, Krieglstein J (1990) PAF antagonist Gingkolide B reduces postischemic neuronal damage in rat brain hippocampus. J Cereb Blood Flow Metab 10:133-135.

Orrenius S, McConkey D, Jones D, Nicotera P (1988) Ca^{2+}-activated mechanisms in toxicity and programmed cell death. ISI Atlas Sci: Pharmacology 2:319-324.

Patt A, Harken AH, Burton LK, Rodell TC, Piermattei D, Schorr WJ, Parker NB, Berger EM, Horesh IR, Terada LS, Linas SL, Cheronis JC, Repine JE (1988) Xanthine oxidase-derived hydrogen peroxide contributes to ischemia reperfusion-induced edema in gerbil brains. J Clin Invest 81:1556-1562.

Pediconi M and Rodriguez de Turco E (1984) Free fatty acid content and release kinetics as manifestations of cerebral lateralization in mouse brain. J Neurochem 43:1-7.

Pettigrew LC, Grotta JC, Rhoades HM, Wu KK (1989) Effect of thromboxane synthase inhibition on eicosanoid levels and blood flow in ischemic rat brain. Stroke 20:627-632.

Pulsinelli WA, Brierley JB, Plum F (1982) Temporal profile of neuronal damage in a model of transient forebrain ischemia. Ann Neurol 11:491-498.

Reddy TS and Bazan NG (1987) Arachidonic acid, stearic acid, and diacylglycerol accumulation correlates with the loss of phosphatidylinositol 4,5-bisphosphate in cerebrum 2 seconds after electroconvulsive shock: Complete reversion of changes 5 minutes after stimulation. Neurosci Res 18:449-455.

Rehncrona S, Siesjö BK, Smith DS (1980) Reversible ischemia of the brain: Biochemical factors influencing restitution. Acta Physiol Scand Suppl 492:135-140.

Rehncrona S, Westerberg E, Åkesson B, Siesjö BK (1982) Brain cortical fatty acids and phospholipids during and following complete and severe incomplete ischemia. J Neurochem 38:84-93.

Rodríguez de Turco EB, Morelli de Liberti S, Bazan NG (1983) Stimulation of free fatty acid and diacylglycerol accumulation in cerebrum and cerebellum during bicuculline-induced status epilepticus. Effect of pretreatment with α-methyl-p-tyrosine and p-chlorophenylalanine. J Neurochem 40:252-259.

Samuelsson B (1983) Leukotrienes: Mediators of immediate hypersensitivity reactions and inflammation. Science 220:568-575.

Shiu G, Nemmer P, Nemoto E (1983) Reassessment of brain fatty acid liberation during global ischemia and its attenuation by barbiturate anesthesia. J Neurochem 40:880-884.

Siesjö BK (1981) Cell damage in the brain: A speculative synthesis. J Cereb Blood Flow Metab 1:155-185.

Siesjö BK (1988a) Historical overview. Calcium, ischemia, and death of brain cells. Ann NY Acad Sci 522:638-661.

Siesjö BK (1988b) Hypoglycemia, brain metabolism, and brain damage. Diabetes Metab Rev 4(2):113-144.

Siesjö BK (1988c) Mechanisms of ischemic brain damage. Crit Care Med 16:954-963.

Siesjö BK (1990) Calcium, excitotoxins, and brain damage. News in Physiological Science 5:120-125.

Siesjö BK (in press) The role of calcium in cell death. In: Neurodegenerative disorders: mechanisms and prospects for therapy (Price DL, Aguayo AJ, Thoenen H, eds). New York: John Wiley & Sons Ltd.

Siesjö BK, Agardh C-D, Bengtsson F (1989a) Free radicals and brain damage. Cerebrovasc Brain Metab Rev 1:165-211.

Siesjö BK, Agardh C-D, Bengtsson F, Smith M-L (1989b) Arachidonic acid metabolism in seizures. Ann NY Acad Sci 559:323-339.

Siesjö BK, Ekholm A, Asplund B (1991) Transmitter release, ion and water fluxes, and ischemic brain damage. In: Volume transmission in the brain: novel mechanisms for neural transmission (Fuxe K, Agnati L, eds) pp 539-547. New York: Raven Press.

Siesjö BK, Ekholm A, Katsura K, Memezawa H, Ohta S, Smith M-L (1990) The type of ischemia determines the pathophysiology of brain lesions and the therapeutic response to calcium channel blockade. In: Pharmacology of Cerebral Ischemia (Krieglstein J, Oberpichler H, eds) pp 79-88. Stuttgart: Wissenschaftliche Verlagsgesellschaft.

Siesjö BK, Ingvar M, Westerberg E (1982) The influence of bicuculline-induced seizures on free fatty acid concentrations in cerebral cortex, hippocampus and cerebellum. J Neurochem 39:796-802.

Silver I and Erecinska (1990) Intracellular and extracellular changes of $[Ca^{2+}]$ in hypoxia and ischemia in rat brain in vivo. J Gen Physiol 95:837-866.

Sladeczek F, Recasens M, Bockaert J (1988) A new mechanism for glutamate receptor action: phosphoinositide hydrolysis. Trends Neurosci 11:545-549.

Snyder F (1989) Biochemistry of platelet-activating factor: a unique class of biologically active phospholipids. Proc Soc Exp Biol Med 190:125-135.

Spinnewyn B, Blavet N, Clostre F, Bazan N, Braquet P (1987) Involvement of platelet-activating factor (PAF) in cerebral post-ischemic phase in mongolian gerbils. Prostaglandins 34:337-349.

Sun G, Su K, Der O, Tang W (1979) Enzymic regulation of arachidonate metabolism in brain membrane phosphoglycerides. Lipids 14:229-235.

Walker V and Pickard JD (1985) Prostaglandins, thromboxane, leukotrienes and the cerebral circulation in health and disease. Neurosurgery 12:3-90.

Wieloch T and Siesjö BK (1982) Ischemic brain injury: The importance of calcium lipolytic activities, and free fatty acids. Pathol Biol 5:269-277.

Wieloch T, Harris RJ, Symon L, Siesjö BK (1984) Influence of severe hypoglycemia on brain extracellular calcium and potassium activities, energy charge and phospholipid metabolism. J Neurochem 43:160-168.

Wolfe L (1982) Prostaglandins:thromboxanes, leukotrienes and other derivatives of carbon-20 unsaturated fatty acids. J Neurochem 38:1-3.

Yoshida S, Harik SK, Busto R, Santiso M, Martinez E, Ginsberg MD (1984) Free fatty acids and energy metabolites in ischemic cerebral cortex with noradrenaline depletion. J Neurochem 42:711-717.

Yoshida S, Ikeda M, Busto R, Santiso M, Martinez E, Ginsberg M (1986) Cerebral phosphoinositide, triacylglycerol, and energy metabolism in reversible ischemia: origin and fate of free fatty acids. J Neurochem 47:744-757.

Yoshida S, Inoh S, Asano T, Kubota M, Shimasaki H, Uera N (1980) Effect of transient ischemia on free fatty acids and phospholipids in the gerbil brain. J Neurosurg 53:323-331.

Yoshida S, Inoh S, Asano T, Sano K, Shimasaki H, Ueta N (1983) Brain free fatty acids, edema and mortality in gerbils subjected to transient, bilateral ischemia, and effect of barbiturate anesthesia. J Neurochem 40:1278-1286.

RECIPROCAL REGULATION OF FATTY ACID RELEASE IN THE BRAIN BY GABA AND GLUTAMATE

Dale L. Birkle

Department of Pharmacology and Toxicology
West Virginia University
Health Science Center North
Morgantown, WV

INTRODUCTION

Free fatty acids (FFA) and their metabolites have many effects on neurochemical processes, including altering receptor-effector coupling, modulating the activity of protein kinase C, and changing ion channel conductance in the cell membrane. However, the neurotransmitters and other factors that control the release of FFA in neurons in normal or pathological states are not well defined. The following studies investigate the regulation of FFA release in intact brain, synaptosomes, and isolated neurons in culture in response to drugs that interact at γ-aminobutyric acid (GABA) and glutamate receptors. The results suggest that neuronal excitation via stimulation of glutamate receptors or blockade of GABA receptors causes the activation of FFA release, and that phosphatidylcholine (PC) is a major source of FFA. Conversely, inhibition of neuronal activity reduces FFA release. FFA release occurs via activation of phospholipase A_2 and possibly via activation of a PC-specific phospholipase C, followed by diacylglycerol (DG) lipase. The common pathway for these effects may be alterations in intracellular calcium.

FFA are normally maintained at extremely low levels in nervous tissue (Bazan et al., 1986). Accumulation can occur by activating phospholipase A_2, which removes the fatty acid esterified at the 2-position of phospholipids (Lands and Crawford, 1976). FFA are re-esterified into lipids by the energy-requiring conversion to acyl-CoA derivatives and subsequent incorporation into lysophospholipids or mono- and diacylglycerols (via acyltransferases). DG accumulate via the activation of phospholipase C, which removes the phosphoryl-base group of phospholipids. DG are converted to phosphatidic acid (PA) via DG kinase, and to other phospholipids via cytidylphosphotransferases. Alternatively, DG can be catabolized by DG lipase to produce FFA (see Horrocks, this volume).

Calcium appears to be an important regulator of phospholipase A_2 (Ho and Klein, 1987; Moskowitz et al., 1984; Felder et al., 1990; Sanfeliu et al., 1990; Dumuis et al., 1988; Lazarewicz et al., 1990). Calcium binds to the active site of phospholipase A_2 and enhances the substrate-enzyme interaction. Intracellular calcium also regulates the translocation of phospholipase A_2 from cytosol to membrane, a step that may be critical to activation (Channon and Leslie, 1990). Activation of arachidonic acid (20:4) release in cultured neurons in response to N-methyl-D-aspartate (NMDA) (Sanfeliu et al., 1990; Dumuis et al., 1988; Dumuis et al., 1990; Lazarewicz et al., 1990), serotonin ($5HT_2$) agonists (Felder et al., 1990), and quisqualate/AMPA (Dumuis et al., 1990) is dependent on receptor-gated calcium influx. Voltage-gated calcium influx via K^+ depolarization is not sufficient in some preparations (Sanfeliu et al., 1990; Dumuis et al., 1988) but, in the retina (Birkle and Bazan, 1984) and in synaptosomes (Birkle and Bazan, 1987; Dorman, 1991), 20:4 release is stimulated by K^+ depolarization. Obviously the regulation of phospholipase A_2 by calcium is complex and may be influenced by compartmentalization within the intracellular space.

The studies described below characterize the activation of phospholipase A_2, the loss of fatty acids from particular glycerolipid classes, and the release and accumulation of FFA and DG in neurons. Studies have been conducted in the intact brain and in various model systems, including synaptosomes, primary neuronal cultures, and cultures of a neuroblastoma cell line. The advantages and disadvantages of the various experimental paradigms are discussed. The focus of the work is on the effects of inhibition and stimulation of the GABA and glutamate systems, to determine the possible reciprocal regulation of phospholipase A_2 via these neurotransmitters.

FFA AND DG ACCUMULATION IN THE RAT BRAIN *IN VIVO*

In these types of studies, animals are treated with a drug and killed by head- focused microwave irradiation. This method of sacrifice heats the brain tissue very rapidly, reaching 90°C in less than 3 seconds. Therefore enzymes are denatured rapidly, and postmortem activation of phospholipase A_2 and other lipases does not occur. The measurements of FFA and DG represent levels at a single time point, and reflect the balance between FFA release and reacylation and DG production and conversion to phospholipids, triacylglycerols (TG), or monoacylglycerols (MG), and FFA.

Increased neuronal activity causes rapid (within seconds) changes in membrane lipids in brain. A striking example of this phenomenon is the accumulation of FFA during the earliest periods of seizures (Bazan, 1970). Seizures induced by the GABA antagonist, bicuculline, cause a prompt and dramatic increase in FFA (Rodriguez de Turco et al., 1983). This occurs in whole brain and in hippocampus and cerebral cortex (Siesjo et al., 1982), and can be measured in synaptosomes isolated from bicuculline-treated rats (Birkle and Bazan, 1987). Whether these effects are mediated via direct interaction with the GABA receptor complex or via other neurotransmitter systems is not known.

There is substantial evidence that ischemia-induced elevation in FFA is due in large part to the activation of phospholipase A_2, while elevation in DG is due in large part to activation of phosphoinositide (PI)-specific phospholipase C (see Sun, this volume). Drugs that change FFA levels may be affecting phospholipase A_2 directly through their receptors or via a drug-induced alteration in another neurotransmitter system. Isolation of particular subcellular fractions from microwave-fixed brains is not possible, but specificity in the responses to pharmacological treatments can be assessed to a limited degree by studying discrete brain regions. Changes in response to a generalized stimulus from one brain region to another reflect the neurotransmitters and receptors that are present in that region.

Effects of Diazepam on FFA in the Rat Brain

Diazepam acts at the $GABA_A$ receptor complex to increase the affinity of GABA binding sites for GABA, thus enhancing inhibitory tone in the brain. These experiments

(Flynn et al., 1986) were done to test the hypothesis that positive modulation of the GABA system would affect fatty acid metabolism in a manner opposite to negative modulation, i.e., since $GABA_A$ blockade by convulsants (bicuculline, picrotoxin, pentylenetetrazol) increases FFA in brain (Bazan et al., 1986), then increasing $GABA_A$ activity may decrease FFA levels.

Rats were treated with diazepam (2 or 4 mg/kg, i.p.) or drug vehicle and were killed 1 hr later by head-focused microwave irradiation. Whole brain less cerebellum and brain stem was removed and homogenized in hexane:2-propanol (3:2) to extract lipids. FFA were isolated by thin layer chromatography (TLC) (Birkle et al., 1988b), derivatized to form methyl esters (Morrison and Smith, 1964), and quantified by gas-liquid chromatography (GLC) using an internal standard method (Birkle et al., 1988b). There was no significant effect of diazepam on the levels of saturated fatty acids (palmitic, 16:0 or stearic, 18:0) and no effect on oleic (18:1) or linoleic (18:2) acids (Fig. 1). However, levels of free 20:4 and docosahexaenoic acid (22:6) were reduced about 60%. One explanation of these results is that the basal levels of

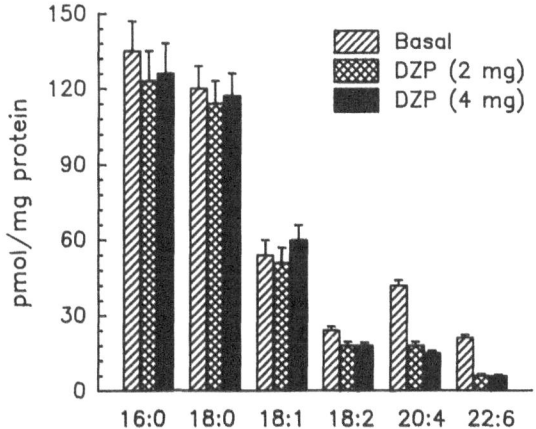

Figure 1. Diazepam (DZP) reduces the basal levels of 20:4 and 22:6 in the whole rat brain. Rats were treated with drug vehicle (basal) or 2 or 4 mg/kg diazepam (DZP), i.p., and sacrificed by head-focused microwave irradiation 1 hr after drug treatment. Values are means ± SE of 6 animals per group.

unsaturated FFA in brain can be reduced by enhancing the activity of GABA at $GABA_A$ receptors. Another possible explanation is that so-called basal levels of FFA in microwave-fixed rat brain are truly above normal because of the stress caused by a novel environment and by the 5 to 10 sec restraint in the microwave holder. Control animals were not acclimated to the microwave holder, so FFA levels in control rats may have been higher than true basal levels. Because diazepam is an anxiolytic drug, treated rats would not experience the stress, mild as it may be, and thus the FFA levels in treated rats would be reduced compared to the "stressed" controls.

Seizure-Induced Alterations in Lipid Metabolism in Specific Brain Regions

Most of the previous work on the effects of neuronal stimulation on brain lipids has been done in whole brain (Bazan et al., 1986). The following experiments examined the changes in FFA and DG in specific brain regions and found a remarkable heterogeneity in these responses.

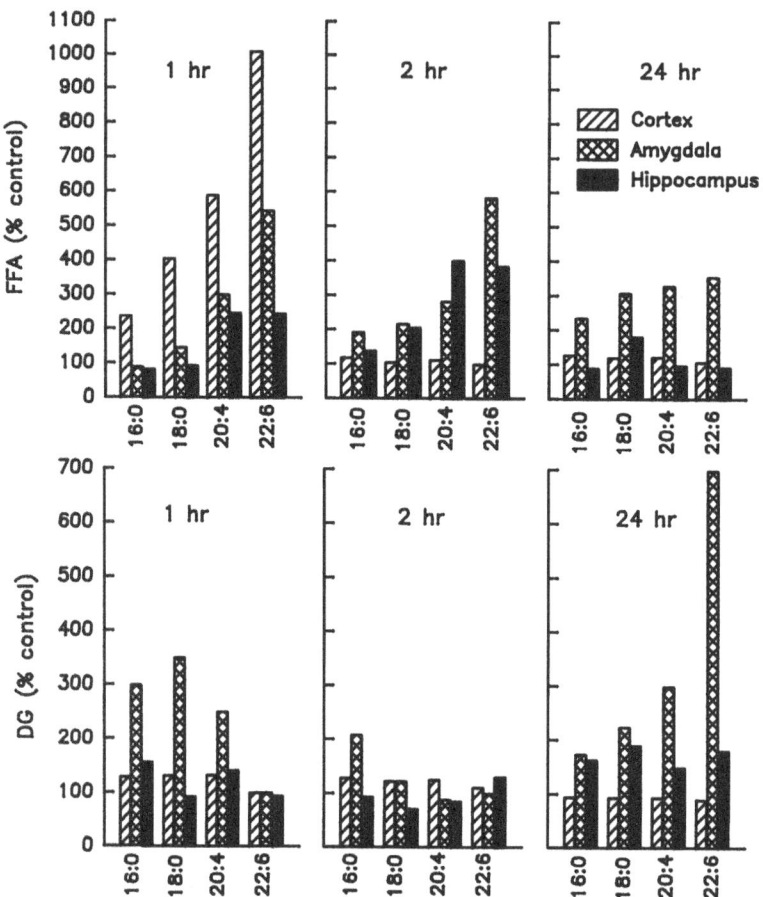

Figure 2. Alterations in FFA and DG in selected brain regions 1, 2, or 24 hr after kainic acid treatment *in vivo* (10 mg/kg, s.c.). Values are means of % of control (experimental/control), N = 6 to 8. Values greater than 120% of control were statistically significant (ANOVA, Tukey's protected t-test).

Effects of Kainic Acid

Kainic acid acts at the kainic acid subtype of glutamate receptors, induces limbic status epilepticus, and produces characteristic pathology in the brain (Sperk et al., 1983). Rats were treated with kainic acid (5 to 15 mg/kg, s.c.) or 0.9% NaCl. Behavioral responses were rated (no response to full limbic status epilepticus) over time, and rats were killed by head-focused microwave irradiation at 1, 2, or 24 hr after drug treatment. Brains were dissected to yield frontal cortex, hippocampus, and amygdala/pyriform cortex, areas rich in glutamate receptors (London and Coyle, 1979). FFA and DG were isolated from lipid extracts by TLC and quantified by capillary GLC. In all brain areas, an increase in FFA, particularly 20:4 and 22:6, was evident at 1 hr (Fig. 2, top panel), 20 min prior to any behavioral manifestations of seizure activity. FFA accumulation was most pronounced in the frontal cortex. By 2 hr, FFA levels in cortex had returned to normal values, but in the hippocampus and amygdala, FFA continued to increase. This time point was taken during the period of most intense seizure activity, which occurred

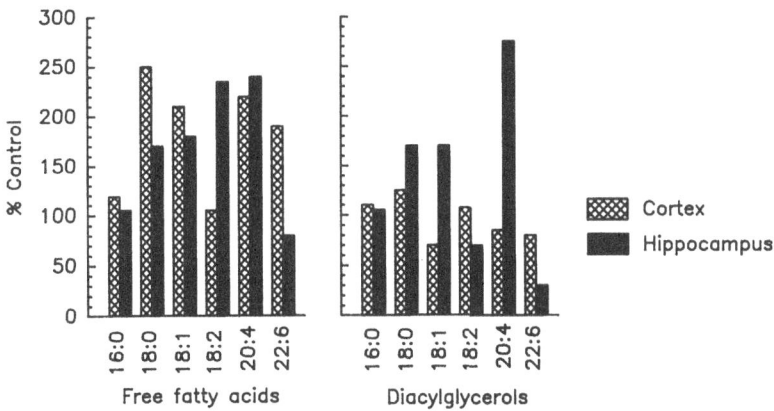

Figure 3. Effect of a single electroconvulsive shock on the accumulation of FFA and DG in the rat hippocampus and cortex. Values are % of control (experimental/control) and are means of 6 samples.

between 80 and 200 min post drug. Twenty-four hours later, when histological evidence of neuronal damage is apparent in the amygdala (data not shown), FFA levels had normalized in hippocampus, but remained elevated in amygdala.

DG levels were increased at 1 hr in amygdala, but there was a delayed increase in DG in the hippocampus, occurring at 24 hr (Fig. 2, bottom panel). DG did not change at any time point in cortex. DG in the amygdala remained elevated at 24 hr. The DG pool was not specifically enriched in 18:0/20:4; however, increases in 16:0 and 22:6 in the DG pool were observed. This suggests that the PI cycle was not the sole source of this lipid.

These data show a heterogeneity in the changes in lipid mediators in response to kainic acid. This suggests that at least some of the changes in lipids are not due directly to an action of kainic acid at its receptor, but rather to the release of other neurotransmitters induced by kainic acid. The neurotransmitters vary with the brain area, and so the lipid responses also vary.

Effects of Electroconvulsive Shock

Electroconvulsive shock causes a reversible increase in FFA and DG and a loss of phosphatidylinositol-4,5-bis-phosphate (PIP_2) in whole brain (Reddy and Bazan, 1987). These experiments investigated brain lipid metabolism in specific brain regions important in the generation and propagation of generalized seizures, the hippocampus and cerebral cortex (Birkle et al., 1988a), in response to a single tonic-clonic seizure induced by maximal electroconvulsive shock. Rats were subjected to a single electroconvulsive shock, then killed by microwave irradiation 30 sec later (during the clonic phase of the seizure). Hippocampus and cerebral cortex were dissected and the lipids extracted. FFA and DG were isolated by TLC and quantified by capillary GLC.

The data show that although FFA levels increase in both brain regions, DG levels increased only in the hippocampus (Fig. 3). This could indicate that seizure-induced stimulation of the PI cycle (the main pathway for the production of DG in this model) occurs mainly in the hippocampus, whereas activation of phospholipase A_2, leading to FFA release, occurs in both brain regions. Alternatively, DG lipase may be selectively

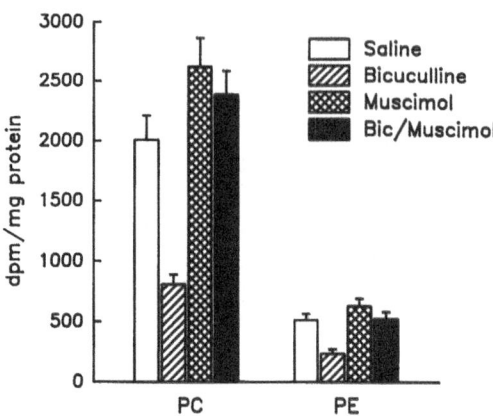

Figure 4. Loss of ^{14}C-22:6 from phosphatidylcholine (PC) and phosphatidylethanolamine (PE) in rat brain synaptosomes incubated 30 min with 1 μM bicuculline methiodide with or without 10 μM muscimol. Values are means ± SE of 6 samples.

activated in the cortex, so that DG are catabolized to produce FFA, and thus do not accumulate.

FFA RELEASE IN RAT BRAIN SYNAPTOSOMES

The following experiments were conducted in preparations of synaptosomes from the whole rat cerebrum. This model system consists largely of synaptic boutons and the biochemical events represent, for the most part, actions of drugs at presynaptic receptors. Synaptosomal preparations are highly purified in neuronal elements compared to preparations of the intact brain. Intact circuitry is absent from a synaptosomal preparation. However, depolarization of synaptosomes does cause the release of neurotransmitters (de Belleroche and Bradford, 1972), which must be considered in the interpretation of results in terms of direct vs. indirect effects of the stimuli. Synaptosomes are also a "damaged" preparation, and subject to postmortem alterations in lipids that occur during the isolation procedures. For this reason, high levels of endogenous FFA and DG are present (Birkle and Bazan, 1987; Birkle and Bazan, 1988). Therefore, to study FFA release mechanisms, radiotracer techniques must be used. Radiotracer studies do have the advantage of allowing measurement of the changes in phospholipid pools, which can be used to infer possible sources of released FFA.

Bicuculline-induced status epilepticus causes the release and accumulation of FFA in the brain *in vivo* (Rodriguez de Turco et al., 1983; Siesjo et al., 1982). Previous experiments have not targeted any particular neurotransmitter system as a putative link to FFA release. Bicuculline is a GABA$_A$ antagonist, so its effect *in vivo* to release FFA may be due to interaction at GABA$_A$ receptors. In the intact animal, however, the contribution of disinhibition of other neurotransmitter systems must be considered. These experiments were designed to investigate the direct effects of bicuculline on membrane lipid metabolism and to investigate the phospholipid sources of the released FFA.

Synaptosomes were isolated from whole cerebrum of naive rats (Birkle and Bazan, 1987). The membrane lipids were labeled with ^{14}C-22:6 by incubation at 37°C for 1 hr in a Tris buffer containing CoASH, ATP, Mg^{2+}, DTT, EGTA, and 1-^{14}C-22:6. The labeled membranes were washed in Tris buffer containing BSA to remove excess unincorporated ^{14}C-22:6, then re-incubated (pulse chased) in Tris buffer with 1 μM Ca^{2+}, 2 μM unlabeled 22:6, and varying concentrations of bicuculline methiodide (0.1, 1, 10 μM) for 5, 10, and 30 min. The incubations were stopped by addition of 2 vol chloroform:methanol (1:1), the lipids were extracted, and the radioactivity in the FFA pool and various glycerolipids was assessed chromatographically and by liquid scintillation counting.

Bicuculline methiodide caused a loss of ^{14}C-22:6 from PC and phosphatidylethanolamine (PE) (Fig. 4). There was a corresponding increase in the amount of free ^{14}C-22:6 (data not shown). The effect of bicuculline was both concentration and time dependent (data not shown). When synaptosomes were incubated with bicuculline (1 μM) after a 10 min pre-incubation in the presence of the GABA$_A$ agonist, muscimol (10 μM), the bicuculline-induced loss of radioactivity from PC and PE was inhibited (Fig. 4).

These experiments demonstrate that a GABA$_A$ antagonist can stimulate FFA release in a subcellular preparation devoid of intact neuronal circuitry. Synaptosomes contain substantial amounts of GABA, which exerts a tonic hyperpolarizing effect on the membrane potential. It has been shown previously that bicuculline causes a voltage-dependent influx of calcium into synaptosomes, presumably by antagonizing this tonic hyperpolarization by GABA (Straub et al., 1990; de Belleroche and Bradford, 1972). This depolarization-induced calcium influx may be responsible for activating FFA release mechanisms. When synaptosomes were exposed to the potent GABA$_A$ agonist, muscimol, the effects of bicuculline were blocked. This observation lends further support to the hypothesis that the effects of bicuculline are mediated via the GABA$_A$ receptor. Another possible (or additional) mechanism for bicuculline's effect is the depolarization-induced release of other neurotransmitters, which could then activate their receptors to induce FFA release. FFA release in these experiments is likely to be the result of the activation of phospholipase A$_2$, because the target lipids were PC and PE. Activation of the PI-specific phospholipase C would result in a loss of phosphoinositides, which was not observed (data not shown).

FATTY ACID RELEASE IN NEUROBLASTOMA CELLS IN CULTURE

Synaptosomes are somewhat limited in their usefulness for studies of FFA release because they are a damaged preparation that is leaky to ions. Furthermore, synaptosomes represent presynaptic membranes mainly, so postsynaptic effects cannot be observed. Neurons in cell cultures form synapses with each other, so both pre- and postsynaptic elements are present. Use of these cell cultures allows pharmacological manipulation of receptors with simultaneous measurement of phospholipid metabolism, without causing any damage to the cells and, therefore, with less artifactual activation of lipases. Moreover, tumor cell lines are easily maintained and quickly yield the large numbers of cells needed for biochemical studies.

Previous studies investigated the basal and norepinephrine-stimulated metabolism of 20:4 in Neuro-2A cells and found reasonable similarity in responses, as compared to the brain homogenates (Birkle and Ellis, 1983). Incubation of confluent, differentiated cultures of Neuro-2A with radiolabeled 20:4 resulted in labeling of various glycerolipid pools and the synthesis of cyclooxygenase and lipoxygenase products. Phospholipid labeling patterns were similar to the distribution of endogenous 20:4 in brain lipids. In a short term incubation (20 min), about 40% of the radiolabel was converted to prostaglandins and hydroxyeicosatetraenoic acids (HETE), a rate of conversion much greater than that usually observed in brain using exogenous substrates. Over a 24-hr period, about 75% of the fatty acid was incorporated in phospholipids; the remainder was in the form of free (10%) or oxygenated (15%) metabolites. When prelabeled cells were incubated in fresh medium, about 20% of the fatty acid was released from glycerolipids, mainly from PE. The loss of label from PE was stimulated by norepinephrine (5 μM), and the effect of norepinephrine was blocked by the ß-antagonist, propranolol (10 μM). These data, coupled with the report that Neuro-2A cells have binding sites for GABA and benzodiazepines that are similar in density and affinity to those found in the brain (Baraldi et al., 1979), led to the studies on the use of these cells as a model system for studying the interaction of GABA and phospholipase A_2.

Effects of Bicuculline and Calcium Ionophore (A23187) on
Fatty Acid Metabolism in Neuro-2A Cells

These experiments examined the release of FFA from Neuro-2A cells after stimulation with bicuculline or A23187 (Birkle and Wiley, 1991). The objectives were to characterize the system for use as a model to study the regulation of phospholipase A_2, and to further investigate the phospholipid sources of released FFA. Confluent monolayer cultures of Neuro-2A cells in 35-mm wells were prelabeled in growth medium by addition of 0.1 or 0.5 μCi 1-^{14}C-22:6 or 1-^{14}C-20:4. After 30 min, 60 min, or 24 hrs of labeling, cells were washed to remove serum-containing growth medium and unesterified fatty acid. Cells were then re-incubated for various times in fresh buffer with 10 μM A23187, 1 or 10 μM bicuculline methiodide, or drug vehicles (0.01% DMSO or normal saline). At the end of the incubation time, ice-cold methanol (2 vols) was added and the lipid extracts were prepared. Major glycerolipids and FFA were separated by TLC and quantified by liquid scintillation counting or capillary GLC.

Endogenous FFA were released to the medium during re-incubation of Neuro-2A cells for 10 or 30 min (data not shown). Treatment with either 10 μM A23187 or 1 μM bicuculline stimulated this release. The FFA pool was specifically enriched in 18:1 and 20:4 after the 10 min treatment with A23187 or bicuculline; at 30 min, drug treatment caused an enrichment of 20:4 only. This enrichment in unsaturated fatty acids in the free pool is an indication of the specific activation of phospholipase A_2.

FFA measured at 10 min represented about 2% of the total fatty acids esterified in phospholipids; therefore the source of released FFA could not be determined by measuring loss of particular glycerolipids. However, prelabeling cells with 1-^{14}C-22:6 or 1-^{14}C-20:4 provided an indication of the source of the released fatty acid. A time-dependent release of free 1-^{14}C-22:6 or 1-^{14}C-20:4 occurred during re-incubation, similar to the release of endogenous FFA (data not shown). Release was markedly stimulated by 1 μM bicuculline or 10 μM A23187. Loss of 1-^{14}C-22:6 from PC was observed, while 1-^{14}C-20:4 was lost from both PC and PE. Other glycerolipids were not significantly altered over time, with the exception of TG. Labeling of TG increased during the re-incubation period, possibly indicating that some of the FFA was rapidly re-esterified.

The results of this study suggest that FFA release induced by A23187 or bicuculline is mediated by phospholipase A_2. This mechanism is supported by the specific enrichment of unsaturated fatty acids (18:1, 18:2, and 20:4) in the free pool and labeling studies, which demonstrate a loss of 1-^{14}C-20:4 and 1-^{14}C-22:6 from PC and PE. In addition, no accumulation of DG or PA and no loss of radiolabel from PI were detected. These latter two observations argue against release of FFA via the phospholipase C/DG lipase pathway. The results of this study suggest that the Neuro-2A cell line, with its advantages of homogeneous cell type and easily controllable milieu, is a potentially useful system for determining the molecular mechanisms that control the normal and pathological regulation of phospholipase A_2 in neurons. However, this model is limited by the types of receptors that are present in this cell line and potentially by derangements in receptor-effector coupling related to the loss of control of cell division that characterizes immortal cell lines.

There are some valid concerns about the use of a clonal cell line for studies of receptor-mediated changes in lipase activity. While a clonal cell line has advantages in terms of availability of the cells and the studies described above suggest it may be a useful system, the question of what types of receptors are present and if the receptors change as a function of the degree of differentiation is a crucial consideration.

Primary cultures of hippocampal neurons have been well characterized electro-physiologically and pharmacologically, in terms of receptors for excitatory (Murphy and Miller, 1988; Furuya et al., 1989) and inhibitory amino acids (Zorumski and Yank, 1988). Recently some studies on the regulation of 20:4 release in hippocampal neurons have been reported. Release of 20:4 is stimulated by NMDA (Sanfeliu et al., 1990), muscarinic (M1) agonists (Kanterman et al., 1990), and agonists at $5HT_2$ receptors (Felder et al., 1990). In all cases the effects of the agonist were blocked by removing calcium from the incubation medium. The effects of other neurotransmitters have not been reported. The sources of released 20:4 have not been identified, but a dissociation from stimulation of the PI cycle has been well established (Kanterman et al., 1990; Felder et al., 1990; Sanfeliu et al., 1990).

Uptake of Radiolabeled Fatty Acids into Hippocampal Neurons

The purpose of these experiments was to determine the ability of hippocampal neuron cultures to incorporate radiolabeled fatty acid and to determine the pattern of labeling of glycerolipids. Hippocampal cultures in 35-mm, 6-well plates were prepared from rat embryos at E16 or E17 (Shahar et al., 1989; Conn, 1990). Cells were seeded at a density of 1 to 2 million cells per well. On the 10th day of culture, $1\text{-}^{14}C\text{-}20{:}4$ was added to the growth medium (0.2 to 0.5 μCi per well). Twenty-four hours later the cells were rinsed and the lipids were extracted to assess the uptake of the fatty acid into cellular glycerolipids. About 70% of the added fatty acid was incorporated into the cells. About 80% of the incorporated label was in the phospholipid fraction, 3% in DG, 11% in TG, and 6% in FFA. In the phospholipid fraction, the radiolabel was distributed as follows: phosphatidylserine (PS), 1.3%; PA, 1.8%; PI, 42.5%; PC, 49.7%; and PE, 4.7%. This pattern of distribution differs from Neuro-2A in that in the primary cells, there is more labeling of PI and DG and less labeling of PE, suggesting a more active PI cycle in the hippocampal cells.

Release of 20:4 in Response to Bicuculline

Cells prelabeled for 24 hrs were washed and re-incubated with 1 μM bicuculline. At 5, 10, 15, or 30 min, incubations were stopped by addition of methanol and lipid extracts prepared and analyzed. Bicuculline caused significant increases in radioactivity in the incubation medium, representing an increase in FFA released from the cells over time (Fig. 5). In untreated cells, FFA release increased rapidly over the first 10 min, then reached a plateau. In the bicuculline-treated cells, there was an enhanced release of FFA during the first 10 min, followed by a slower increase over the 10 to 30 min period. The release of FFA was accompanied by a decline in PC and a rapid increase in DG during the first 10 min. DG began to decrease during the 5 to 30 min period, perhaps indicating an activation of DG lipase. The level of label in PC increased slightly during the 10 to 30 min period, indicating a reacylation of this phospholipid. At no time were any alterations observed in PI (Fig. 5) or in any other glycerolipid (data not shown).

These results demonstrate several points. First, the response of hippocampal cells to bicuculline is similar to that of Neuro-2A in terms of accumulation of FFA and loss

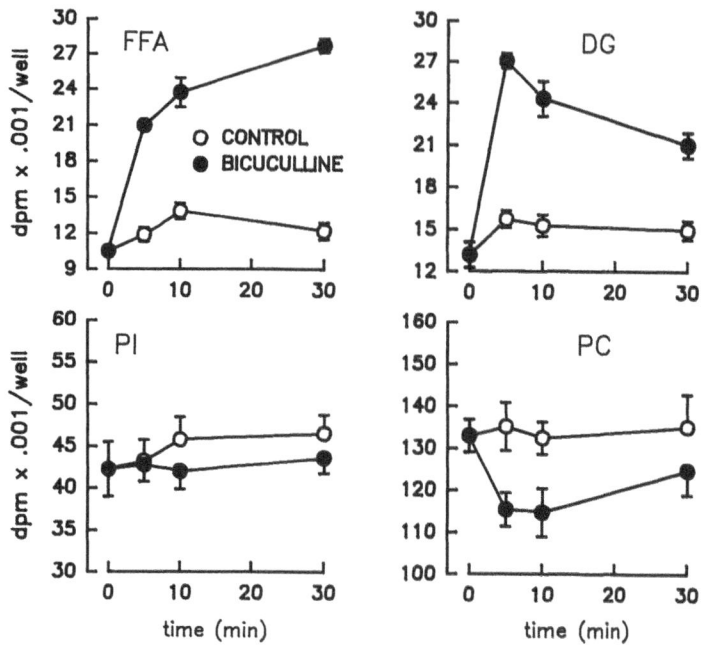

Figure 5. Activation of FFA release, DG accumulation, and loss of radiolabel from PC in hippocampal neurons in primary culture incubated with 1 µM bicuculline methiodide. Each well was seeded at a density of 1.5 x 10⁶ cells per well and data were normalized to a total radioactivity of 200,000 dpm/well. Values are means ± SE of 6 wells.

of PC. However, no accumulation of DG was observed in the Neuro-2A cells. Second, bicuculline's effects are most prominent during the early periods of incubation. In both Neuro-2A cells and hippocampal neurons, the effects of bicuculline appear to be on the release of FFA from PC via phospholipase A_2. In hippocampal cells phospholipase C may also be involved, since an accumulation of DG was observed. However, since no loss of PI was measured, phospholipase C may be acting on PC to produce DG.

Release of 20:4 in Response to Kainic Acid

In these experiments, cells were prelabeled for 24 hrs with [14]C-20:4, washed, then re-incubated in fresh buffer. At 10 min, 100 µM kainic acid was added, and the incubations were continued an additional 20 min. Aliquots of the incubation medium were taken every 2 min (250 µl out of 2 ml) and the medium replaced to maintain the volume and concentration of drug. Aliquots were placed in scintillation vials and counted to obtain cumulative counts present in the medium. At the end of the incubation, methanol was added to extract the lipids of the cells and the amounts of radioactivity present in the various lipid fractions were determined.

In untreated cells, an increase in FFA occurred over the first 10 min, and then reached a plateau (Fig. 6). This probably reflects a nonspecific stimulation of FFA release during the washing steps, as this rapid release was also observed in the experiments with bicuculline (see Fig. 5). When kainic acid was added, the release of FFA continued to increase over the remaining 20 min of incubation, instead of reaching a plateau (Fig. 6, top panel). In kainic acid-treated cells there was a substantial loss of label from PC, a minor loss of label from PS, and an increase in DG (Fig. 6, bottom

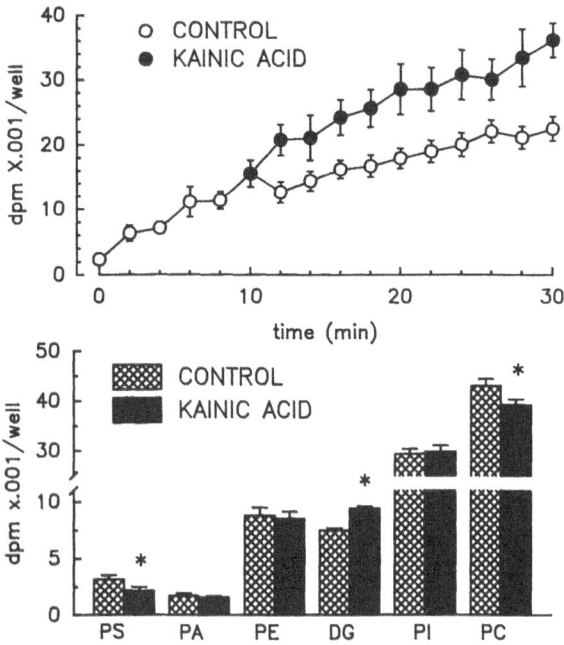

Figure 6. Effects of 100 μM kainic acid on FFA release and loss of label from phospholipids in hippocampal neurons in primary culture. Each well was seeded at a density of 2.5 x 10^6 cells per well. Values are means ± SE of 12 samples.

panel). There was no change in other phospholipids, including PI and PA.

These results, like those of the experiments with bicuculline, suggest that depolarization of hippocampal neurons by the glutamate agonist, kainic acid, causes an activation of phospholipase A_2 and a PC-specific phospholipase C.

SUMMARY AND CONCLUSIONS

Several model systems have been used to test the hypothesis that the release of FFA in the brain is regulated by depolarization of neurons. This FFA release is likely the result of the activation of phospholipase A_2. The increased neuronal activity that occurs due to synchronous depolarization during seizures causes activation of phospholipase A_2. Decreasing neuronal activity by administering the anxiolytic, diazepam, appears to decrease the activity of phospholipase A_2. The GABA antagonist, bicuculline, which causes depolarization by negating the hyperpolarizing tone imposed on neurons by GABA, causes FFA release in synaptosomes and in neurons in tissue culture. Likewise, the glutamate agonist, kainic acid, which depolarizes neurons by opening sodium channels, increases the activity of phospholipase A_2. PC-specific phospholipase C, another enzyme important in the generation of the second messenger, DG, is also activated by depolarization.

Several important questions remain to be answered. The site of FFA release, in terms of the pre- vs. postsynaptic membrane, is not clear, although the experiments with synaptosomes support the hypothesis that activation of phospholipase A_2 may be an important regulator of presynaptic events. This idea has also been suggested by studies on the phenomenon of long-term potentiation, where free 20:4 or its metabolites may be involved in presynaptic facilitation of neurotransmitter release (Freeman et al., 1990; Massicotte et al., 1990; Williams et al., 1989; also see Dorman, this volume). The activation of the PI cycle and subsequent stimulation of protein kinase C may be a postsynaptic event important in the integration of inputs at the dendrite and soma or a presynaptic event involved in the modulation of neurotransmitter release (Taniyama et al., 1990; El-Fakahany et al., 1990; also see Nishizuka, this volume). Therefore the stimulation of a PC-specific phospholipase C, which is capable of generating large amounts of DG over a prolonged period of time (Exton, 1990; Martinson et al., 1990; Diaz-Laviada et al., 1990), could occur at either site.

Another important question is the role of FFA and DG in affecting cell-cell signaling events, particularly with regard to ion fluxes. Modulation of an acetylcholine-linked K^+ channel in the heart by FFA and their oxygenation products has been reported (Kim and Clapham, 1989). The cardiac muscarinic receptor is linked to a hyperpolarizing K^+ channel via a G protein. Receptor-activated dissociation of the G protein releases ßγ subunits, which stimulate phospholipase A_2, resulting in the release of 20:4 and subsequent conversion to lipoxygenase products; these metabolites increase the activity of the K^+ channel (Kim et al., 1989). Lipoxygenase products also modulate responses of *Aplysia* sensory neurons to FRMF-amide (Piomelli et al., 1987; Volterra and Siegelbaum, 1988; also see Volterra, this volume) via changes in a K^+ channel. In the heart and in smooth muscle cells, there are also other K^+ channels that are activated by FFA directly (Kim and Duff, 1990). In hippocampus, 20:4 and its metabolites cause hyperpolarization and increased inhibitory postsynaptic potentials (IPSPs), which may be related to changes in K^+ conductance (Carlen et al., 1989). In the rat brain, $GABA_B$ and $5HT_{1a}$ receptors are linked via a G protein to a hyperpolarizing K^+ channel (Andrade et al., 1986; Innis et al., 1988), which may be modulated by phospholipase A_2 activation (Duman et al., 1986). These observations underscore the importance of determining the regulation of FFA release in the brain.

It is now clear that FFA and DG are important lipid-derived messengers in neurons, and that their metabolism is highly regulated. The techniques and model systems now available should provide the means for discerning regulatory factors, contributions to the normal function of neurons, and mechanisms of pathological alterations in diseases of the central nervous system.

ACKNOWLEDGMENTS

Supported by grants from Epilepsy Foundation of America, Pharmaceutical Manufacturers Association Foundation, NSF #BNS-8919943, and NIH-BMRS #2S07-RR05433-28. The technical assistance of B.J. Victor and K.S. Wiley is appreciated.

REFERENCES

Andrade R, Malenka RC, Nicoll RA (1986) A G protein couples serotonin and
    $GABA_B$ receptors to the same channels in hippocampus. Science 234:1261-1265.
Baraldi M, Guidotti A, Schwartz JP, Costa E (1979) GABA receptors in clonal cell
    lines: A model for study of benzodiazepine action at molecular level. Science
    205: 821-823.
Bazan NG (1970) Effects of ischemia and electroconvulsive shock on free fatty acid
    pool in the brain. Biochim Biophys Acta 218:1-10.
Bazan NG, Birkle DL, Tang W, Reddy TS (1986) The accumulation of free arachi-
    donic acid, diacylglycerols, prostaglandins and lipoxygenase reaction products in
    the brain during experimental epilepsy. In: Advances in Neurology, Vol 44

(Delgado-Escueta AV, Ward AA, Woodbury DM, Porter RJ, eds) pp 879-902. New York: Raven Press.

Birkle DL and Bazan NG (1984) Effect of K^+ depolarization on the synthesis of prostaglandins, hydroxyeicosatetraenoic acid (HETE) and other eicosanoids in the rat retina: Evidence for esterification of 12-HETE in lipids. Biochim Biophys Acta 795:564-573.

Birkle DL and Bazan NG (1987) Effect of bicuculline-induced status epilepticus on prostaglandins and hydroxyeicosatetraenoic acids in rat brain subcellular fractions. J Neurochem 48:1768-1778.

Birkle DL and Bazan NG (1988) Cerebral perfusion of metabolic inactivators: A new method for rapid fixation of labile lipid pools in brain. Neurochem Res 13: 849-852.

Birkle DL and Ellis EF (1983) Conversion of arachidonic acid to cyclooxygenase and lipoxygenase reaction products and incorporation into phospholipids in the mouse neuroblastoma clone, Neuro-2A. Neurochem Res 8:319-332.

Birkle DL and Wiley KS (1991) Bicuculline induces free fatty acid release from phospholipids in Neuro-2A cells in culture. Neurochem Res 16:1285-1293.

Birkle DL, Kurian P, Bazan NG (1988a) Seizure-induced alterations in lipid metabolism in hippocampus and cerebral cortex. Soc Neurosci Abstr 14:574.

Birkle DL, Kurian P, Braquet P, Bazan NG (1988b) The platelet activating factor antagonist, BN52021, decreases accumulation of free polyunsaturated fatty acids in mouse brain during ischemia and electroconvulsive shock. J Neurochem 51:1900-1905.

Carlen PL, Gurevich N, Wu PH, Su WG, Corey EJ, Pace-Asciak CR (1989) Actions of arachidonic acid and hepoxilin A3 on mammalian hippocampal CA1 neurons. Brain Res 497:171-176.

Channon JY and Leslie CC (1990) A calcium-dependent mechanism for associating a soluble arachidonoyl-hydrolyzing phospholipase A_2 with membrane in the macrophage cell line RAW 264.7. J Biol Chem 265, 5409-5413.

Conn PM (1990) Methods in neuroscience, Vol 2: Cell culture. San Diego: Academic Press Inc.

de Belleroche JS and Bradford HF (1972) Metabolism of beds of mammalian cortical synaptosomes: Response to depolarizing influences. J Neurochem 19:585-602.

Diaz-Laviada I, Larrodera P, Diaz-Meco MT (1990) Evidence for a role of phosphatidylcholine-hydrolysing phospholipase C in the regulation of protein kinase C by ras and src oncogenes. EMBO J 9:3907-3912.

Dorman RV (1991) $PGF_{2\alpha}$ synthesis in isolated cerebellar glomeruli: Effects of membrane depolarization, calcium availability and phospholipase activity. Prostaglandins Leukot Essent Fatty Acids 42:233-240.

Duman RS, Karbon EW, Harrington C, Enna SJ (1986) An examination of the involvement of phospholipases A_2 and C in the alpha-adrenergic and GABA receptor modulation of cAMP accumulation. J Neurochem 47:800-810.

Dumuis A, Pin JP, Oomagari K, Sebben M, Bockaert J (1990) Arachidonic acid released from striatal neurons by joint stimulation of ionotropic and metabotropic quisqualate receptors. Nature 347, 182-184.

Dumuis A, Sebben M, Haynes L, Pin J-P, Bockaert J (1988) NMDA receptors activate the arachidonic acid cascade system in striatal neurons. Nature 336:68-70.

El-Fakahany EE, Alger BE, Lai WS, Pitler TA, Worley PF, Baraban JM (1990) Neuronal muscarinic responses: Role of protein kinase C. FASEB J 2:2575-2583.

Exton JH (1990) Hormonal regulation of phosphatidylcholine breakdown. Adv Second Messenger Phosphoprotein Res 24:152-157.

Felder CC, Kanterman RY, Ma AL, Axelrod J (1990) Serotonin stimulates phospholipase A_2 and the release of arachidonic acid in hippocampal neurons by a type 2 serotonin receptor that is independent of inositolphospholipid hydrolysis. Proc Natl Acad Sci USA 87:2187-2191.

Flynn CJ, Birkle DL, Wecker L (1986) Diazepam prevents seizure-induced increases in free fatty acids and choline in rat cerebrum. Soc Neurosci Abstr 12:454-454.

Freeman EJ, Terrian DM, Dorman RV (1990) Presynaptic facilitation of glutamate release from isolated hippocampal mossy fiber nerve endings by arachidonic acid. Neurochem Res 15:743-750.

Furuya S, Ohmori H, Shigemoto T, Sugiyama H (1989) Intracellular calcium mobilization triggered by a glutamate receptor in rat cultured hippocampal cells. J Physiol 414:539-548.

Ho AK and Klein DC (1987) Activation of $alpha_1$-adrenoceptors, protein kinase C, or treatment with intracellular free Ca^{2+} elevating agents increases pineal phospholipase A_2 activity. J Biol Chem 262:11764-11770.

Innis RB, Nestler EJ, Aghajanian GK (1988) Evidence for G protein mediation of serotonin- and $GABA_B$-induced hyperpolarization of rat dorsal raphe neurons. Brain Res 459:27-36.

Kanterman RY, Ma AL, Briley EM, Axelrod J, Felder CC (1990) Muscarinic receptors mediate the release of arachidonic acid from spinal cord and hippocampal neurons in primary culture. Neurosci Lett 118:235-237.

Kim D and Clapham, DE (1989) Potassium channels in cardiac cells activated by arachidonic acid and phospholipids. Science 244:1174-1176.

Kim D and Duff RA (1990) Regulation of K^+ channels in cardiac myocytes by free fatty acids. Circ Res 67:1040-1046.

Kim D, Lewis DL, Graziadei L, Neer EJ, Bar-Sagi D, Clapham DE (1989) G-protein beta-gamma subunits activate the cardiac muscarinic K^+-channel via phospholipase A_2. Nature 337:557-560.

Lands W and Crawford C (1976) Enzymes of membrane phospholipid metabolism in animals. In: Enzymes of biological membranes (Martonosi A, eds) pp 3-85. New York: Plenum Press.

Lazarewicz JW, Wroblewski JT, Costa E (1990) N-methyl-D-aspartate-sensitive glutamate receptors induce calcium-mediated arachidonic acid release in primary cultures of cerebellar granule cells. J Neurochem 55:1875-1881.

London ED and Coyle JT (1979) Specific binding of [^{3}H]kainic acid to receptor sites in rat brain. Mol Pharmacol 15:492-505.

Martinson EA, Goldstein D, Heller Brown J (1990) Muscarinic receptor activation of phosphatidylcholine hydrolysis: Relationship to phosphoinositide hydrolysis and diacylglycerol metabolism. J Biol Chem 264:14748-14754.

Massicotte G, Oliver MW, Lynch G, Baudry M (1990) Effect of bromophenacyl bromide, a phospholipase A_2 inhibitor, on the induction and maintenance of LTP in hippocampal slices. Brain Res 537:49-53.

Morrison WR and Smith LM (1964) Preparation of fatty acid methyl esters and dimethylacetals from lipids with boron-trifluoride-methanol. J Lipid Res 5:600-608.

Moskowitz N, Schook W, Puszkin S (1984) Regulation of endogenous calcium-dependent synaptic membrane phospholipase A_2. Brain Res 290:273-280.

Murphy SN and Miller RJ (1988) A glutamate receptor regulates Ca^{2+} mobilization in hippocampal neurons. Proc Natl Acad Sci USA 85:8737-8741.

Piomelli D, Shapiro E, Feinmark SJ, Schwartz JH (1987) Metabolites of arachidonic acid in the nervous system of *Aplysia*: Possible mediators of synaptic modulation. J Neurosci 7:3675-3686.

Reddy TS and Bazan NG (1987) Arachidonic acid, stearic acid and diacylglycerol accumulation correlates with the loss of phosphatidylinositol 4,5-bisphosphate in cerebrum two seconds after electroconvulsive shock: Complete reversion of changes 5 minutes after stimulation. J Neurosci Res 18:449-455.

Rodriguez de Turco EB, Morelli de Liberti S, Bazan NG (1983) Stimulation of free fatty acid and diacylglycerol accumulation in cerebrum and cerebellum during bicuculline-induced status epilepticus: Effect of pre-treatment with alpha-methyl-p-tyrosine and p-chlorophenylalanine. J Neurochem 40:252-259.

Sanfeliu C, Hunt A, Patel AJ (1990) Exposure to N-methyl-D-aspartate increases release of arachidonic acid in primary cultures of rat hippocampal neurons and not in astrocytes. Brain Res 526:241-248.

Shahar A, de Vellis J, Vernadakis A, Haber B (1989) A dissection and tissue culture manual of the nervous system. New York: Alan R. Liss, Inc.

Siesjo BK, Ingvar M, Westerberg E (1982) The influence of bicuculline-induced seizures on free fatty acid concentrations in cerebral cortex, hippocampus and cerebellum. J Neurochem 39:796-802.

Sperk G, Lassman H, Baran H, Kish SJ, Seitelberger F, Hornykiewicz O (1983) Kainic acid induced seizures: Neurochemical and histopathological changes. Neuroscience 10:1301-1315.

Straub H, Speckmann EJ, Bingmann D, Walden J (1990) Paroxysmal depolarization shifts induced by bicuculline in CA3 neurons of hippocampal slices: Suppression by the organic calcium antagonist verapamil. Neurosci Lett 111:99-101.

Taniyama K, Saito N, Kose A, Matsuyama S, Nakayama S, Tanaka C (1990) Involvement of the gamma subtype of protein kinase C in GABA release from the cerebellum. Adv Second Messenger Phosphoprotein Res 24:399-404.

Volterra A and Siegelbaum SA (1988) Role of two different genuine nucleotide binding proteins in the antagonistic modulation of the S-type K^+ channel by cAMP and arachidonic acid metabolites in *Aplysia* sensory neurons. Proc Natl Acad Sci USA 85:7810-7814.

Williams JH, Errington ML, Lynch MA, Bliss TVP (1989) Arachidonic acid induces a long-term activity-dependent enhancement of synaptic transmission in the hippocampus. Nature 341:739742.

Zorumski CF and Yank J (1988) Non-competitive inhibition of GABA currents by phenothiazines in cultured chick spinal cord and rat hippocampal neurons. Neurosci Lett 92:86-91.

NMDA RECEPTOR-MEDIATED ARACHIDONIC ACID RELEASE IN NEURONS: ROLE IN SIGNAL TRANSDUCTION AND PATHOLOGICAL ASPECTS

Jerzy W. Lazarewicz,[1,2] E. Salinska,[2]
and J.T. Wroblewski[1]

[1]Fidia-Georgetown Institute for the Neurosciences
Georgetown University School of Medicine
Washington DC, USA

[2]Medical Research Centre
Polish Academy of Sciences
Warsaw, Poland

ABSTRACT

The N-methyl-D-aspartate (NMDA)-sensitive subtype of glutamate receptor, which gates Ca^{2+}-permeable ion channels, is known for its role in learning and memory formation, in the induction of long-term potentiation, and also in seizure activity and neurotoxicity. In primary cultures of cerebellar neurons, agonists of NMDA receptors induce a dose-dependent release of [^{3}H]arachidonic acid ([^{3}H]AA), which is potentiated by activation of the glycine-positive modulatory site and inhibited by NMDA receptor antagonists. NMDA receptor-induced [^{3}H]AA release is inhibited by quinacrine and partially depends on the presence of extracellular calcium. The [^{3}H]AA release is not sensitive, however, to pretreatment with pertussis or cholera toxin, which suggests a Ca^{2+}-dependent activation of phospholipase A_2 not employing G proteins. Pretreatment of cultures with the natural and semisynthetic sphingolipids GT1b and PKS 3, respectively, inhibits NMDA receptor-mediated [^{3}H]AA release. We also demonstrated glutamate-evoked [^{3}H]AA release from rat hippocampal slices, which is NMDA receptor mediated, calcium dependent and sensitive to quinacrine. Arachidonic acid and its metabolites have been shown to play a role as second messengers and to modulate neuronal activity. Moreover, they are thought to act as transsynaptic modulators in the mechanism of NMDA receptor-induced long-term potentiation in the hippocampus. Their role in ischemic brain pathology has also been postulated. Our experiments on cultured cerebellar granule cells, incubated in a Mg^{2+}-free medium deprived of glucose and oxygen, demonstrated a time-dependent stimulation of [^{3}H]AA release. This

Neurobiology of Essential Fatty Acids
Edited by N.G. Bazan *et al.*, Plenum Press, New York, 1992

release was inhibited by antagonists of NMDA receptors and by quinacrine. Stimulation of NMDA-sensitive glutamate receptors and the subsequent calcium-mediated activation of phospholipase A_2 may play a role in the *in vivo* release of arachidonic acid during brain ischemia. This hypothesis is supported by the observation that the enhanced level of thromboxane B_2 in the gerbil brain after 5 min of global ischemia is reduced by the systemic application of either the NMDA antagonist MK-801 or the ganglioside GM1.

INTRODUCTION

There is increasing interest in the mechanism of signal transduction in excitatory amino acid receptors. It is known that activation of receptors sensitive to N-methyl-D-aspartate (NMDA receptors), apart from their ionotropic responses (Mayer and Westbrook, 1988), leads to the induction of well-defined calcium-dependent metabolotropic responses such as translocation and stimulation of protein kinase C (PKC), nitric oxide (NO) release, cyclic GMP formation, and modification of gene expression (Novelli et al., 1987; Vaccarino et al., 1987; Garthwaite et al., 1988; Szekely et al., 1990). NMDA receptors constitute a functional complex with high-conductance cation channels permeable to Ca^{2+} (Wroblewski et al., 1987: Mayer and Westbrook, 1988). This association with calcium channels seems to determine their role in the induction of long-term potentiation, learning, memory formation, and in the mechanism of neurotoxicity evoked by excitatory amino acids (Choi, 1987; Lynch and Muller, 1990). Cerebellar granule cells in primary culture proved to be a suitable model for studies on the mechanism of signal transduction in glutamate receptors (Costa et al., 1988). Using this model we recently demonstrated stimulation of arachidonic acid release evoked by activation of NMDA receptors (Lazarewicz et al., 1988; 1990).

Activation of several different receptors as well as nonspecific cell stimulation by chemical or electrical depolarization has been shown to result in the release of AA from brain slices, synaptosomes, synaptoneurosomes, and from *in vitro* cultured neurons (Lazarewicz et al., 1983; Ho and Klein, 1987; Strosznajder and Strosznajder, 1989; Sartarelli et al., 1990). *In vivo*, a rapid increase in the brain content of free AA was noted in ischemia, hypoglycemia, and seizures (Bazan, 1989). Diversity of the effector enzymes and receptor-effector coupling mechanisms instrumental to this process have been suggested (Bazan, 1989). Arachidonic acid and products of AA metabolism have been implicated as putative second messengers in intracellular signaling, in the modulation of neurotransmission, and in the mechanism of neuronal damage (Chan et al., 1985; Piomelli et al., 1987; Axelrod et al., 1988; Williams et al., 1989; Shimizu and Wolfe, 1990). Thus, demonstration of the enhanced release of AA evoked by stimulation of NMDA receptors in cultured neurons raises the following questions. Which effector enzymes and coupling mechanisms are responsible? Does this mechanism also function in brain neurons *in situ*? What is the role of AA in signal transduction in NMDA receptors, and does NMDA receptor-mediated AA release participate in the ischemia-evoked increase in the brain content of free AA?

In this paper, we will investigate and discuss these questions. First, we will focus on the role of phospholipase A_2 (PLA_2) as a primary effector enzyme and the role of intracellular calcium as a regulator/coupling factor involved in NMDA receptor-mediated AA release in cerebellar granule cells. Next, the demonstration of excitatory amino acid-evoked release of AA using hippocampal slices of adult rats will be discussed. And finally, both *in vitro* experiments on cultured cerebellar granule cells and *in vivo* studies on a gerbil brain ischemic model will be used to evaluate a possible role for NMDA receptors in ischemic AA release in the brain.

A steady-state efflux of [³H]AA from prelabeled primary cultures of cerebellar granule cells incubated in Mg^{2+}-free medium was enhanced in a dose-dependent manner by agonists of excitatory amino acid receptors such as glutamate, aspartate, NMDA, kainate, and quisqualate. This effect was inhibited by 0.5 μM phencyclidine (PCP), a noncompetitive antagonist of NMDA receptors. Moreover, the release of AA evoked by the NMDA receptor agonist aspartate was potentiated by D-serine, an agonist of strychnine-insensitive glycine binding sites on the NMDA receptor complex and a positive modulator of these receptors, whereas magnesium ions and 2-amino-5-phosphonovaleric acid (APV), a competitive antagonist of NMDA receptors, inhibited this effect (Fig. 1). These data indicate that NMDA receptors are directly responsible for the glutamate-evoked arachidonic acid release, whereas the stimulation evoked by the non-NMDA agonists kainate and quisqualate was indirect and probably connected with depolarization and glutamate release from granule cells.

The release of AA from granule cells seems to be specifically dependent on stimulation of NMDA receptors. Agonists such as norepinephrine, GABA, or baclofen, which activate AA release in other cells (Burch et al., 1986; Duman et al., 1986; Enna and Karbon, 1987; Axelrod et al., 1988), failed to produce such an effect in granule cells. Carbachol, on the other hand, produced a relatively small stimulation of AA release that was sensitive to PCP inhibition (data not shown). The carbachol-mediated release may be explained by the indirect stimulation of NMDA receptors. Studies on striatal, cortical, and hypothalamic neuronal cultures (Dumuis et al., 1988; 1990; Sanfeliu et al., 1990; Tapia-Arancibia et al., 1990) confirmed activation of AA release by agonists of NMDA receptors in neurons but not in astrocytes (Sanfeliu et al., 1990).

PRIMARY EFFECTOR ENZYME INVOLVED IN NMDA RECEPTOR-INDUCED AA RELEASE

Two enzymatic mechanisms of phospholipid hydrolysis, which are known to lead to AA release, may be considered as the effector enzymes responsible for NMDA recep-

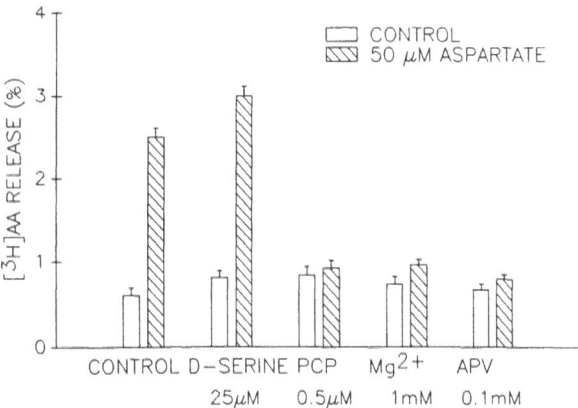

Figure 1. Effects of NMDA receptor modulators on aspartate-induced [³H]arachidonic acid release from cerebellar granule cells. [³H]Arachidonic acid release is expressed as percent of total radioactivity incorporated. The incubation was carried out for 10 min in Mg^{2+}-free Locke's solution with 0.2% BSA as described previously (Lazarewicz et al., 1988; 1990). Values are means ± SEM from four experiments. Results taken with permission from Lazarewicz et al., 1988.

tor-mediated AA release. Arachidonate is an immediate product of phospholipid and phosphatidic acid hydrolysis facilitated by PLA_2. On the other hand, diacylglycerol lipase may liberate AA from DG released from inositol phospholipids by phospholipase C (PLC). In granule cells, agonists of excitatory amino acid receptors were shown to stimulate PLC-mediated accumulation of the inositol phosphates (Nicoletti et al., 1986). Therefore, participation of PLC in NMDA receptor-mediated AA release can be postulated. PLA_2 and PLC display different sensitivities to several inhibitors, including quinacrine (mepacrine), which interferes with the enzyme-phospholipid interaction in membranes. Since PLA_2 is more sensitive to quinacrine inhibition than either PLC or DG lipase (Hofmann et al., 1982), this drug has been used as a tool to differentiate between the activities of these AA-releasing enzymes (Lapetina et al., 1981).

In granule cells, quinacrine, in a dose-dependent manner, potently inhibits the release of $[^3H]AA$ evoked by aspartic acid (Fig. 2), but does not influence the basal AA release (data not shown). However, quinacrine at a concentration of 25 μM (which exceeds its IC_{50} for inhibition of AA release) failed to inhibit glutamate- and quisqualate-induced inositol phospholipid hydrolysis (Fig. 2). Quinacrine at this concentration did inhibit the activity of PLA_2 but not that of PLC in membranes prepared from cultured cerebellar granule cells (Fig. 2). The selectivity of quinacrine-evoked inhibition of PLA_2 was thus additionally confirmed. A lack of substantial participation of PLC in NMDA receptor-mediated AA release in cerebellar granule cells may be also postulated based on our findings that activation of quisqualate-sensitive excitatory amino acid receptors, muscarinic receptors, or α_1-adrenergic receptors failed to induce AA release even though they are all known to enhance PLC-mediated phosphoinositide hydrolysis in cerebellar granule cells (Nicoletti et al., 1986; Xu and Chuang, 1987). Our results support the conclusion that PLA_2 is the primary effector enzyme responsible for NMDA receptor-evoked release of AA in cerebellar granule cells. Other studies on cultured cortical and striatal neurons also indicate the key role of PLA_2 in this phenomenon (Dumuis et al., 1988; Sanfeliu et al., 1990; Tapia-Arancibia et al., 1990).

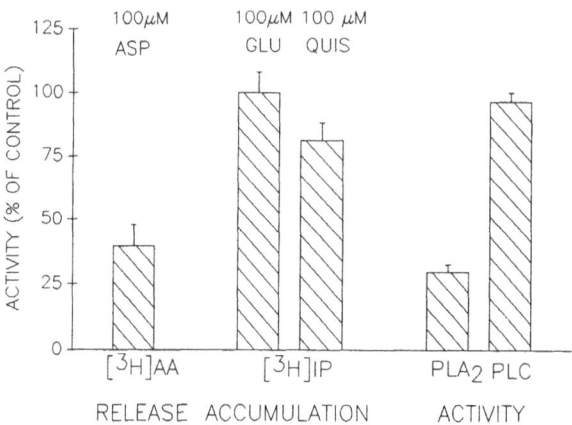

Figure 2. Effects of 25 μM quinacrine on aspartate-evoked $[^3H]$arachidonic acid release, glutamate- and quisqualate-induced inositol phospholipid hydrolysis, and activity of phospholipases A_2 and C in cerebellar granule cells. PCP, phencyclidine; APV, 2-amino-5-phosphonovaleric acid. For details of measurements see Lazarewicz et al., 1990. Results are expressed as percent of control activity, measured in the absence of quinacrine, and represent means from four to six experiments $\pm$ SEM. Results taken with permission from Lazarewicz et al., 1990.

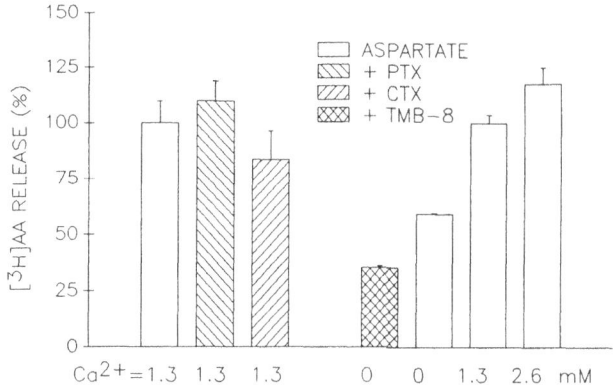

Figure 3. Sensitivity of aspartate-evoked arachidonic acid release in granule cells to pertussis (PTX) and cholera toxin (CTX) and to changes in extracellular calcium concentration. The cultures were treated with toxins (1 μg/ml) for 16 hr before the incubation. TMB-8 (100 μM) was present for 10 min before and during the incubation. For further details see Figure 1 and Lazarewicz et al., 1990. [^{3}H]Arachidonic acid release is expressed as percent of the total incorporated activity after subtracting the basal release in the absence of aspartate. Values are means from four experiments $\pm$ SEM. Results taken with permission from Lazarewicz et al., 1990.

PLA$_2$ COUPLING TO THE NMDA RECEPTOR

Phospholipases A$_2$ and C may be coupled directly to metabolotropic receptors by various GTP-binding proteins. A pertussis toxin-sensitive G protein was shown to be involved in the coupling of the α_1-adrenergic receptors to PLA$_2$ but not in their coupling to PLC in cultured rat thyroid cell line FRTL5 (Burch et al., 1986). Transducin, a G protein activating PLA$_2$ in the outer rod segment of the retina, is sensitive to both pertussis and cholera toxin (Xu and Wojcik, 1987). The coupling of PLA$_2$ to cholinergic muscarinic receptors in the membranous brain fraction was also demonstrated (Strosznajder and Strosznajder, 1989). It is known that in cultured cerebellar granule cells, pretreatment with pertussis toxin induces ADP-ribosylation of G protein (Jelsema, 1987) and inhibits the hydrolysis of inositol phospholipids evoked by stimulation of quisqualate-sensitive metabolotropic receptors (Nicoletti et al., 1988). On the other hand, stimulation of NMDA-sensitive receptors activates calcium flux into granule cells (Wroblewski et al., 1985). Therefore, Ca^{2+}, which binds to the catalytic center of phospholipase A$_2$ and facilitates enzyme-substrate interactions, may play a role as an intracellular signal transducer coupling NMDA receptors to phospholipase A$_2$.

The release of AA induced by aspartate was not changed significantly by the pretreatment of granule cells with either pertussis toxin or cholera toxin (Fig. 3), but was modified by changes in the extracellular calcium concentration (Fig. 3). The persistent stimulation of arachidonic acid release which was observed during incubation of granule cells with aspartate in a nominal calcium-free medium (Fig. 3) may also be attributed to calcium-stimulated events since it was sensitive to inhibition by TMB-8, which blocks the mobilization of intracellular calcium (Chiou and Malagodi, 1975). A massive quinacrine-sensitive and calcium-dependent release of AA from granule cells

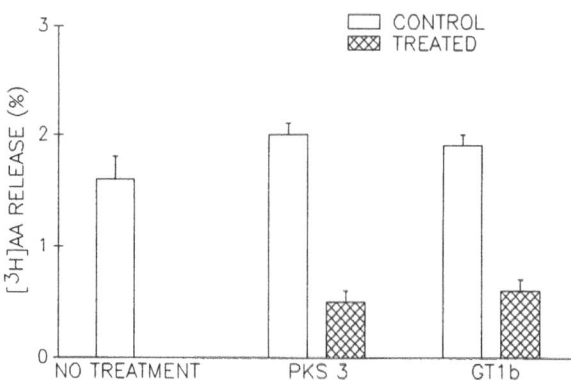

Figure 4. Protective effect of natural (GT1b) and semisynthetic (PKS 3) sphingolipids on aspartate-evoked [^{3}H]arachidonic acid release in granule cells. Primary cultures were pretreated with 30 μM PKS 3 or 60 μM GT1b for 2 hr. Measurements as in Figure 1. [^{3}H]Arachidonic acid release is expressed as percent of the total incorporated activity after subtracting the basal release in the absence of aspartate. Values are means from four experiments ± SEM.

was also observed during incubation in the presence of the Ca^{2+} ionophore ionomycin, which produces a strong stimulation of Ca^{2+} flux into neurons (data not shown).

Previous studies indicate that activation of PKC may result in the enhanced accumulation of arachidonate metabolites (Hartung and Toyka, 1987), presumably by PLA_2 activation. It is known that stimulation of NMDA receptors in cerebellar granule cells leads to Ca^{2+}-dependent translocation and stimulation of PKC (Vaccarino et al., 1987). Pretreatment of granule cells with gangliosides protects against glutamate-evoked PKC activation and neurotoxicity (Vaccarino et al., 1987; Manev et al., 1990). It was shown in previous studies on cultured cerebellar granule cells that prolonged or excessive stimulation of NMDA receptors results in a sustained increase of Ca^{2+} influx to neurons via channels that are not exclusively NMDA receptor coupled, and leads to a long-lasting increase in the intracellular Ca^{2+} concentration. This effect, which may be prevented by sphingolipids, seems to be dependent on the translocation and stimulation of PKC (Manev et al., 1990). Our unpublished results indicate that after excessive stimulation of NMDA-sensitive glutamate receptors, in parallel with sustained calcium uptake by neurons, an increased release of [^{3}H]AA is also observed (data not shown). Pretreatment of granule cells with two sphingolipids, GT1$_b$ and PKS 3, resulted in a potent inhibition of aspartate-evoked AA release (Fig. 4). A corresponding suppression of calcium influx to neurons was also noted (data not shown). This effect of gangliosides is most probably connected with their inhibition of glutamate-evoked stimulation of PKC in granule cells (Vaccarino et al., 1987), and illustrates a potent ability of sphingolipids to antagonize glutamate receptor abuse by interfering with the mechanisms of signal transduction utilized by NMDA receptors (Manev et al., 1990) in pathological conditions. However, further studies will be needed to differentiate between the ganglioside-evoked inhibition of PLA_2 activity resulting from suppression of Ca^{2+} influx through unselective ionic channels and blockage of the hypothetical PKC-evoked stimulation of PLA_2.

Our results did not demonstrate a contribution of G proteins in the coupling of NMDA receptors to PLA_2, but they do indicate that in cerebellar granule cells calcium ions may play a role as triggers that stimulate the activity of PLA_2 after NMDA receptor stimulation. Studies on other primary neuronal cultures also point to the role of calcium as a coupling factor in NMDA receptor-mediated activation of PLA_2 (Dumuis et al., 1988; Sanfeliu et al., 1990; Tapia-Arancibia et al., 1990). It remains to be investigated whether the elevated intracellular concentration of Ca^{2+} ions stimulates PLA_2 activity only by direct interaction with the enzyme or by additional calcium-dependent mechanisms of PLA_2 activation involving PKC as well.

NMDA RECEPTOR-DEPENDENT [^{3}H]AA RELEASE IN RAT HIPPOCAMPAL SLICES

Although AA release evoked by stimulation of NMDA receptors has not been directly demonstrated in experiments on material other than cultured neurons, it is highly improbable that this effect represents a feature unique to cultured neurons. To estimate the ability of NMDA receptors to stimulate AA release in brain neurons *in situ* we studied this effect using rat hippocampal slices, which are known to have a high density of NMDA receptors (Cotman et al., 1987).

After preincubation of hippocampal slices with [^{3}H]AA, which resulted in the incorporation of the fatty acid into brain phospholipids, the material was superfused with an ionic medium containing bovine serum albumin (BSA), as described previously (Lazarewicz et al., 1983). A spontaneous high release of [^{3}H]AA into the medium in the absence of stimulation was observed (Fig. 5A), in agreement with the results of other studies (Lazarewicz et al., 1983; Sartarelli et al., 1990). To demonstrate the effect of the activation of NMDA receptors, glutamate was applied in a magnesium-free medium enriched with glycine in order to minimize the antagonism of NMDA channels by magnesium ions and to potentiate the effect of the agonist. This procedure resulted in a small additional activation of [^{3}H]AA release (Fig. 5B). A glutamate-evoked [^{3}H]AA release was inhibited in the presence of Mg^{2+} and was blocked by 1 μM MK-801, a noncompetitive antagonist of NMDA receptors (Fig. 5C,D). Incubation with 50 μM quinacrine (Fig. 5E) and a calcium-free, EGTA-containing medium (Fig. 5F) also inhibited this effect. Thus, the basic features of glutamate-evoked [^{3}H]AA release in the hippocampal slices did not differ from those of NMDA receptor-evoked, Ca^{2+}-dependent, and PLA_2-mediated release in cultured cerebellar granule cells.

Our results, which correspond with previous data reporting a glutamate-induced release of some eicosanoids in brain slices *in vitro* (Wolfe and Pellerin, 1989), point to the functioning of an NMDA receptor-mediated mechanism of arachidonic acid release in the hippocampal neurons *in situ*.

ARACHIDONIC ACID AND SIGNAL TRANSDUCTION OF N-METHYL-D-ASPARTATE (NMDA) RECEPTORS

It was shown that arachidonic acid and its metabolites may function as second messengers in a variety of cells and also may interfere with other signaling mechanisms (Piomelli et al., 1987; Axelrod et al., 1988; Sartarelli et al., 1990). Free arachidonate has been shown to directly activate protein kinase C (Shearman et al., 1989). Arachidonate and some of its metabolic products also elevate the free intracellular calcium level (Ito et al., 1991). This effect may be attributed to the activation of calcium influx into the cell (Kandasamy and Hunt, 1990; Mochizuki-Oda et al., 1991) or to Ca^{2+} mobilization from intracellular stores (Ito et al., 1991). Arachidonic acid and some eicosanoids may interfere with cyclic nucleotide systems (Gilman and Nirenberg, 1971; Hamprecht and Schultz, 1973) and AA may also mediate signals for gene expression (Hannigan and Williams, 1991). Many of these interactions may apply to the NMDA receptor-mediated

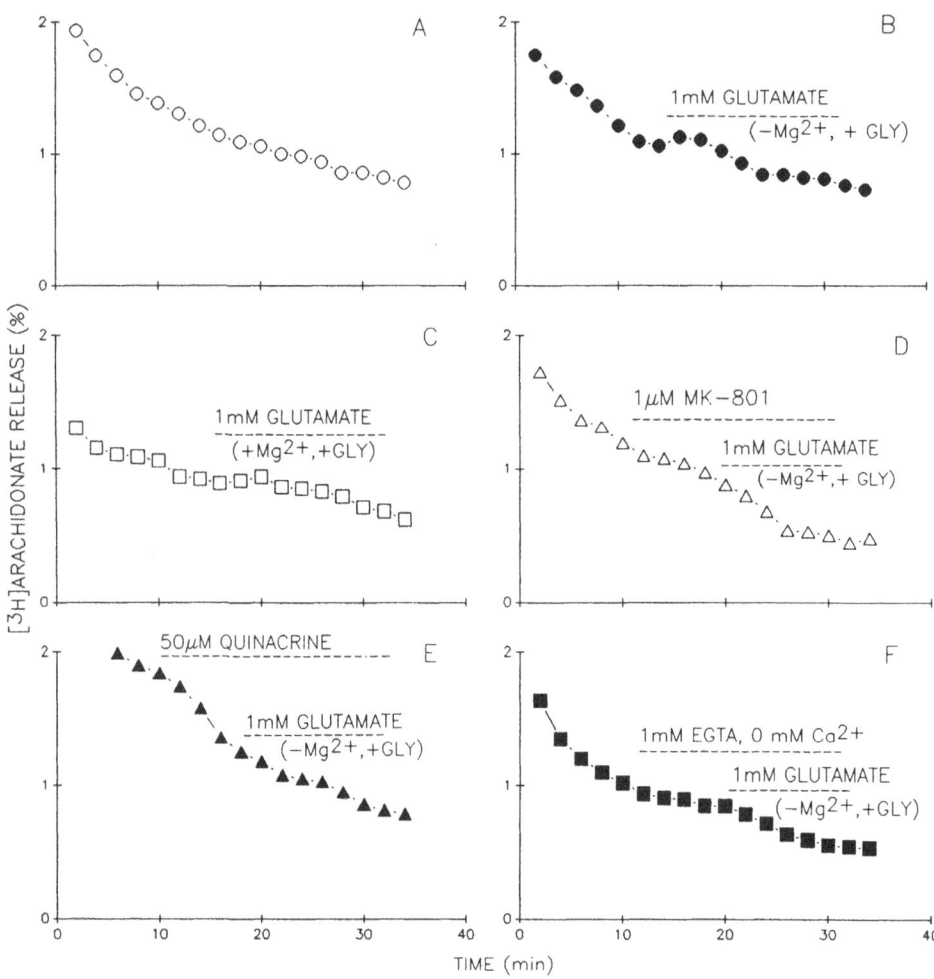

Figure 5. Glutamate-evoked [^{3}H]arachidonic acid release in rat hippocampal slices. A) Control efflux of [^{3}H]AA. B) Effect of glutamate applied in Mg^{2+}-free medium supplemented with 10 μM glycine (GLY). C) Inhibition by 1 mM Mg^{2+}. D) Inhibition by MK-801. E) Inhibition by quinacrine. F) Effect of glutamate in Ca^{2+} free medium. Hippocampal slices (0.3 x 0.3 mm), after preincubation for 40 min in Krebs bicarbonate medium containing 5 mM ATP and 0.1 mM CoA with 5 μCi [^{3}H]arachidonic acid, were superfused with 0.2% BSA-containing medium as described previously (Lazarewicz et al., 1983). The modified media were applied as indicated by the horizontal bars. The amount of [^{3}H]arachidonate released per fraction is expressed as a percentage of total radioactivity. Results represent means from four experiments, SEM < 5%.

signal transduction mechanisms. Experiments on cerebellar granule cells demonstrated NMDA receptor-mediated elevation of intracellular calcium (Manev et al., 1990), translocation and activation of PKC (Vaccarino et al., 1987; Manev et al., 1990), cGMP formation (Novelli et al., 1987), and enhanced c-fos gene expression (Szekely et al., 1990). Arachidonic acid and its metabolites may possibly participate in generation of these signals or may amplify them in NMDA receptors. Apart from the role of nitric oxide in the mediation of NMDA receptor-mediated cGMP formation in granule cells (Garthwaite et al., 1988), there are data that suggest a cooperative stimulatory role in this process by lipoxygenase products of AA metabolism (Wroblewski et al., 1991).

It is known that eicosanoids formed in stimulated cells may function as primary messengers interacting with specific receptors on other cells (Shimizu et al., 1982). Moreover, eicosanoids liberated in some cells may play a role as neuromodulators in these or other neurons. Such a mechanism of signaling within or between nerve cells was initially demonstrated in *Aplysia* sensory neurons (Piomelli et al., 1987) and is based on the stimulation of specific K^+ channels by arachidonate and lipoxygenase metabolites (Carlen et al., 1989; Premkumar et al., 1990). Recently AA and eicosanoids have been implicated in the modulation of neurotransmission in excitatory amino acid receptors associated with long-term potentiation (Williams and Bliss, 1988; Lynch and Muller, 1990). Arachidonic acid and its lipoxygenase metabolites were shown to stimulate the release of glutamate from hippocampal nerve endings. This possibly represents the facilitation of synaptic transmission in excitatory receptors via a presynaptic mechanism (Williams et al., 1989; Linden and Routtenberg, 1989; Lynch and Voss, 1990; Lynch and Voss, 1991). However, it is not clear what the role of lipoxygenase products of arachidonate metabolism may be in this process. 12-HETE was reported either to facilitate (Lynch and Voss, 1990) or to attenuate (Freeman et al., 1991) the release of glutamate from isolated nerve terminals. In addition the source of the AA and eicosanoids, which are possibly involved in the potentiation of glutamatergic transmission, is a matter of controversy. It is suggested that after activation of NMDA-sensitive glutamate receptors, which is a prerequisite for long-term potentiation, 12-HETE is formed at postsynaptic sites and may act as a transsynaptic retrograde messenger in the facilitation of glutamate release (Lynch and Voss, 1990). Others argue, however, that both the AA and the lipoxygenase products involved in the modulation of glutamatergic neurotransmission are presynaptic in origin (Linden and Routtenberg, 1989; Freeman et al., 1991). Despite these controversies, it may be that arachidonic acid acts as a signal transduction messenger used by NMDA-sensitive glutamate receptors.

POSSIBLE ROLE OF NMDA RECEPTORS IN ISCHEMIA-EVOKED ARACHIDONATE RELEASE

The release of arachidonic acid, reflecting degradation of brain phospholipids, is one of the earliest biochemical responses to ischemia of the mammalian brain (Bazan, 1970; 1989). This phenomenon is probably involved in the mechanisms of ischemic neuronal injury (Lazarewicz et al., 1972; Chan and Fishman, 1985; Hillared and Chan, 1988). It has been ascribed almost exclusively either to the cholinergic and/or adrenergic agonist-dependent activation of PLC and subsequent liberation of arachidonate from diacylglycerol (Ikeda et al., 1986; Yoshida et al., 1986; Abe et al., 1987; Strosznajder et al., 1987) or to the activation of PLA_2 coupled to these receptors by G proteins (Bazan, 1989; Strosznajder and Strosznajder et al., 1989). Similarly to ischemic brain injury, the pathological conditions of hypoglycemia and seizures are associated with both an elevation of free arachidonic acid and an excessive stimulation of excitatory amino acid receptors (Simon et al., 1984; Wieloch, 1985; Bazan, 1989; Siesjö and Bengtsson, 1989). These data and other circumstantial evidence (Westerberg and Wieloch, 1986; Abe et al., 1989) could suggest a possible involvement of excitatory amino acids in the release of arachidonic acid in the nervous system during hypoxia, ischemia and hypoglycemia. Although this supposition was supported by the demonstration of NMDA receptor-mediated AA release in cultured neurons (Dumuis et al., 1988; Lazarewicz et al., 1988; 1990; Sanfeliu et al., 1990; Tapia-Arancibia et al., 1990), evidence for the involvement of excitatory amino acid receptors in the mechanism of anoxic-ischemic arachidonic acid release is lacking.

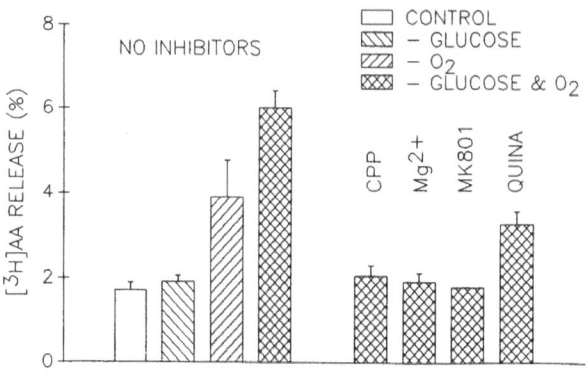

Figure 6. Release of [^{3}H]arachidonic acid from cerebellar granule cells incubated in glucose- and oxygen-deprived medium. Cells were incubated in Locke's-Tris, Mg^{2+}-free medium in the absence of glucose, oxygen or both. For the latter case the effect of 100 μM CPP, 1 mM Mg^{2+}, 1 μM MK-801 and 25 μM quinacrine (QUINA) was tested. Measurements as in Figure 1. [^{3}H]Arachidonic acid release is expressed as percent of total radioactivity incorporated. Results are means from eight experiments ± SEM.

Recently a model of ischemia-like conditions, produced by deprivation of both glucose and oxygen, was used in *in vitro* studies of cultured neurons and incubated brain slices (Goldberg et al., 1989; Pohorecki et al., 1990). In our experiments involving cultured cerebellar granule cells, a similar pathological model was used. In control experiments, glucose deprivation alone had no effect on [^{3}H]AA release, whereas anoxic incubation produced a noticeable stimulation of AA release as compared to the basal liberation of arachidonic acid (Fig. 6). Only the combination of anaerobic conditions and glucose-free medium produced a massive [^{3}H]AA release (Fig. 6). This effect lasted continuously for up to 30 min (data not shown). The [^{3}H]AA release observed during incubation of granule cells under ischemia-like conditions was similar to the arachidonate release evoked by agonists of NMDA receptors. The release was

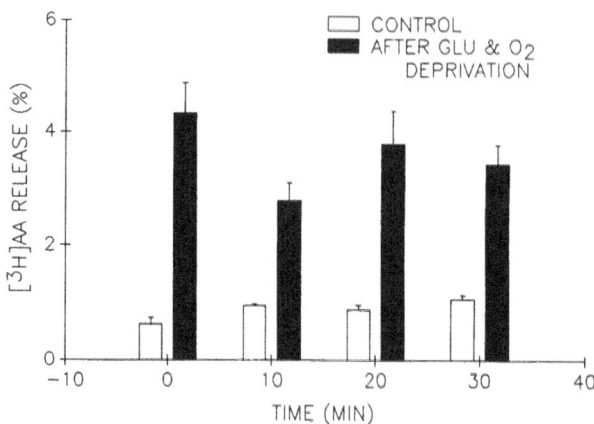

Figure 7. [^{3}H]Arachidonic acid release in cultured granule cells after exposure to glucose (GLU) and oxygen deprivation. Granule cells were incubated in Mg^{2+}-free medium during a 20-min exposure to the absence of glucose and oxygen and after the return to control conditions. [^{3}H]Arachidonate release was measured in three 10-min periods. Values are means from four experiments ± SEM.

completely blocked by both competitive and noncompetitive inhibitors of NMDA-sensitive receptors (CPP, MK-801, and Mg^{2+}) and was inhibited in the presence of quinacrine (Fig. 6). The enhanced release of $[^3H]AA$ was not limited only to the period of anoxic-hypoglycemic incubation but persisted for at least 30 min after the restoration of control conditions (Fig. 7). The release of $[^3H]AA$ was accompanied by sustained stimulation of calcium entry into granule cells (data not shown).

Although these data indicate that arachidonic acid release from cultured neurons incubated in glucose- and oxygen-deprived medium may be attributed to their stimulation by the endogenous agonists of NMDA-sensitive glutamate receptors, more direct evidence implicating involvement of excitatory amino acids in the mechanism of *in vivo* arachidonic acid release in ischemic brain was needed. We used a model of transient global cerebral ischemia in Mongolian gerbils subjected to bilateral common carotid occlusion in normothermic conditions. Employing this model we studied the effect of pretreatment with MK-801, a noncompetitive antagonist of NMDA-sensitive glutamate receptors, on the brain content of the AA metabolites thromboxane B_2 (TXB_2), prostaglandin D_2 (PGD_2) and 6-keto-prostaglandin $F_{1\alpha}$ (6-keto-$PGF_{1\alpha}$).

Pretreatment of gerbils with MK-801 had no effect on the basal level of TXB_2 (Fig. 8), PGD_2, and 6-keto-$PGF_{1\alpha}$ (data not shown) in the cerebral cortex of the control (naive) and sham-operated animals. Five min after 5 min of global cerebral ischemia, an increase in the brain content of TXB_2 (Fig. 8) and the other two eicosanoids tested was found. A similar early accumulation of various oxygenated AA metabolites in the brain, with a peak 5 min after global cerebral ischemia, had been observed previously (Kiwak et al., 1985; Dempsey et al., 1986; Minamisawa et al., 1988).

Application of MK-801 resulted in a statistically significant decrease in TXB_2 content in the brain after ischemia (Fig. 8), whereas a tendency toward suppressed accumulation of PGD_2 and 6-keto-$PGF_{1\alpha}$ was not significant (data not shown). This effect indicates that in the brain cortex during and after ischemia, a portion of arachidonic acid and its metabolites may be liberated via activation of NMDA receptor-gated ionic channels permeable to Ca^{2+}. These channels are sensitive to inhibition by MK-801 (Wong et al., 1986). A protective effect of MK-801 against neurotoxicity of NMDA and in ischemia in various animal models was described (Olney et al., 1987; Ozyurt et al., 1988; Ford et al., 1989; Schoepp et al., 1989). The mechanism of this protection was, however, disputed, since it may be partially attributed to a prolonged postischemic hypothermic effect of MK-801 (Buchan and Pulsinelli, 1990). However, this criticism does not apply to our short normothermic experiments. Although on the basis of presented data it is impossible to identify the enzymatic mechanism which may be directly responsible for AA release, our results may indicate that, similarly to the *in vitro* model (Dumuis et al., 1988; Lazarewicz et al., 1988; 1990; Sanfeliu et al., 1990; Tapia-Arancibia et al., 1990), the *in vivo* model also displays an excessive influx of calcium through NMDA channels (Salinska et al., 1991). The calcium influx may activate mainly phospholipase A_2. It is possible that in addition a nonspecific, Ca^{2+}-dependent stimulation of PI-specific PLC may take place in the initial phase of glutamate-evoked Ca^{2+} flux into neurons during ischemia (Abe et al., 1989). The MK-801-insensitive portion of the release may be attributed at least partially to the adrenergic or cholinergic stimulation of PLC (Ikeda et al., 1986; Yoshida et al., 1986; Abe et al., 1987; Strosznajder et al., 1987) and to the G protein-coupled activation of PLA_2 (Strosznajder and Strosznajder, 1989; Bazan, 1990).

Recently, ganglioside GM_1-lactone was reported to prevent accumulation of eicosanoids in the rat brain after ischemia (Petroni et al., 1989). This prompted us to study the effect of pretreatment of Mongolian gerbils with ganglioside GM1 on the postischemic accumulation of TXB_2 in the brain cortex (Fig. 8). In agreement with previous data (Petroni et al., 1989) and with our *in vitro* experiments (Fig. 4), a significant attenuation of the increase of TXB_2 in the brains of GM1-treated animals

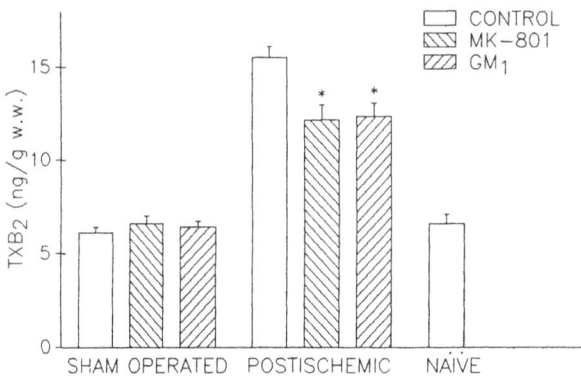

Figure 8. Effect of pretreatment with MK-801 and ganglioside GM1 on the accumulation of thromboxane B_2 in gerbil brain cortex after global cerebral ischemia. Mongolian gerbils were subjected to bilateral common carotid occlusion for 5 min, or to sham operation, which was followed by 5 min recirculation. The animals were injected i.p. with MK-801 (0.8 mg/kg body weight) 30 min before operation and with ganglioside GM1 (20 mg/kg body weight) for two days, twice a day, with the last injection 30 min before the operation. Thromboxane B_2 in the gerbil brain cortex was detected radioimmunologically after extraction, as described by Powell (1980) with modifications (Minamisawa et al., 1988). Values are means from six to eight experiments ± SEM. Asterisks mark results significantly different from untreated control. (p<0.05).

was shown (Fig. 8). These results indicate that the RADA therapeutic strategy (glutamate receptor abuse-dependent antagonism by gangliosides) proposed by Costa and colleagues (Manev et al., 1990) is effective in the postischemic brain on the level of receptor-mediated arachidonate release. Signal transduction mechanisms in the NMDA-sensitive glutamate receptors employing PKC were shown to be the primary targets of gangliosides in neurons. Thus, the inhibitory effect of ganglioside pretreatment on ischemia-evoked accumulation of eicosanoids in the postischemic brain offers both potential therapeutic implications and evidence for a contribution by NMDA receptors to the mechanism of AA release in the ischemic brain.

CONCLUSIONS

Our results, as well as data from the literature, point to the diversity of mechanisms leading to the receptor-mediated release of arachidonic acid. Various receptors may employ different primary effector enzymes (lipases) and receptor-effector coupling mechanisms. It is evident from our results using both *in vivo* and *in vitro* models that: i) The stimulation of excitatory amino acid receptors leads to a release of arachidonic acid by NMDA receptors. These receptors use phospholipase A_2 as a primary effector enzyme and Ca^{2+} as a coupling factor. We did not detect a direct coupling of NMDA receptors with phospholipase A_2 via cholera and pertussis toxin-sensitive G proteins. ii) The glutamate-evoked arachidonate release mediated by NMDA receptors may be demonstrated not only in cultured neurons but also in freshly prepared hippocampal slices. iii) This NMDA receptor-mediated mechanism of arachidonate release may participate also in ischemia-evoked accumulation of arachidonic acid and its metabolites in the brain.

Thus, arachidonic acid and products of its degradation may be involved in neurons in signal transduction of NMDA-sensitive glutamate receptors under physiological and pathological conditions.

REFERENCES

Abe K, Kogure K, Yamamoto H, Imazawa M, Miyamoto K (1987) Mechanism of arachidonic acid liberation during ischemia in gerbil cerebral cortex. J Neurochem 48:503.

Abe K, Yoshidomi M, Kogure K (1989) Arachidonic acid metabolism and ischemic neuronal damage. In: Arachidonic acid metabolism in the nervous system. Physiological and pathological significance (Barkai AI, Bazan NG, eds) p 259. New York: NY Acad Sci.

Axelrod J, Burch RM, Jelsema CL (1988) Receptor-mediated activation of phospholipase A_2 via GTP-binding proteins: Arachidonic acid and its metabolites as second messengers. Trends Neurosci 11:117.

Bazan NG (1970) Effects of ischemia and electroconvulsive shock on free fatty acid pool in the brain. Biochim Biophys Acta 218:1.

Bazan NG (1989) Arachidonic acid in the modulation of excitable membrane function and at the onset of brain damage. In: Arachidonic acid metabolism in the nervous system. Physiological and pathological significance (Barkai AI, Bazan NG, eds) p 1. New York: NY Acad Sci.

Buchan A and Pulsinelli WA (1990) Hypothermia but not the N-methyl-D-aspartate antagonist, MK-801, attenuates neuronal damage in gerbils subjected to transient global ischemia. J Neurosci 10:311.

Burch RM, Luini A, Axelrod J (1986) Phospholipase A_2 and phospholipase C are activated by distinct GTP-binding proteins in response to α_1-adrenergic stimulation in FRTL5 thyroid cells. Proc Natl Acad Sci USA 83:7201.

Carlen PL, Gurevich N, Wu PH, Su W-G, Corey EJ, Pace-Asciak CR (1989) Actions of arachidonic acid and hepoxilin A_3 on mammalian hippocampal CA_1 neurons. Brain Res 497:171.

Chan PH and Fishman RA (1985) Free fatty acids, oxygen free radicals and membrane alterations in brain ischemia and injury. In: Cerebrovascular diseases (Plum F and Pulsinelli W, eds), New York: Raven Press.

Chan PH, Fishman RA, Longar S, Chen S, Yu A (1985) Cellular and molecular effects of polyunsaturated fatty acids in brain ischemia and injury. Progr Brain Res 63:227.

Choi DW (1987) Ionic dependence of glutamate neurotoxicity in cortical cell culture. J Neurosci 7:369.

Chiou CY and Malagodi MH (1975) Studies on the mechanism of action of a new Ca^{++} antagonist 8-(N,N-diethylamino)-octyl 3,4,5-trimetoxybenzoate hydrochloride in smooth and skeletal muscles. Br J Pharmacol 53:279.

Costa E, Fadda E, Kozikowski AP, Nicoletti F, Wroblewski JT (1988) Classification and allosteric modulation of excitatory amino acid signal transduction in brain slices and primary cultures of cerebellar neurons. In: Neurobiology of amino acids, peptides and trophic factors. (Ferrendelli J, Collins R, Johnson E, eds) p 35. Boston: M. Nijhoff Publ.

Cotman CW, Monaghan DT, Otterson OP, Storm-Mathisen J (1987) Anatomical organization of excitatory amino acid receptors and their pathways. Trends Neurosci 10:273.

Dempsey RJ, Roy MW, Mayer K, Cowen DE, Tai HH (1986) Development of cyclooxygenase and lipoxygenase metabolites of arachidonic acid after transient cerebral ischemia. J Neurosurg 62:865.

Duman RS, Karbon EW, Harrington C, Enna SJ (1986) An examination of the involvement of phospholipases A_2 and C in the α-adrenergic and γ-aminobutyric acid receptor modulation of cyclic AMP accumulation in rat brain slices. J Neurochem 47:800.

Dumuis A, Pin J-P, Oomagari K, Sebben M, Bockaert J (1990) Arachidonic acid released from striatal neurons by joint stimulation of ionotropic and metabotropic quisqualate receptors. Nature 347:182.

Dumuis A, Sebben M, Haynes L, Pin J-P, Bockaert J (1988) NMDA receptors activate the arachidonic acid cascade system in striatal neurons. Nature 336:68.

Enna SJ and Karbon EW (1987) Receptor regulation: Evidence for a relationship between phospholipid metabolism and neurotransmitter receptor-mediated cAMP formation in brain. Trends Pharmacol Sci 8:21.

Ford LM, Sanberg PR, Norman AB, Fogelson MH (1989) MK-801 prevents hippocampal neurodegeneration in neonatal hypoxic-ischemic rats. Arch Neurol 46:1090.

Freeman EJ, Damron DS, Terrian DM, Dorman RV (1991) 12-Lipoxygenase products attenuate the glutamate release and Ca^{2+} accumulation evoked by depolarization of hippocampal mossy fiber nerve endings. J Neurochem 56:1079.

Garthwaite J, Charles SL, Chess-Williams R (1988) Endothelium-derived relaxing factor release on activation of NMDA receptors suggests role as intracellular messenger in the brain. Nature 336:385.

Gilman AG and Nirenberg M (1971) Regulation of adenosine 3',5'-cyclic monophosphate metabolism in cultured neuroblastoma cells. Nature 234:356.

Goldberg MP, Kurth MC, Giffard RG, Choi DW (1989) [45]Calcium accumulation and intracellular calcium during *in vitro* "ischemia." Soc Neurosci Abstr 15:803.

Hamprecht B and Schultz J (1973) Stimulation by prostaglandin E_1 of adenosine 3',5'-cyclic monophosphate formation in neuroblastoma cells in the presence of phosphodiesterase inhibitors. FEBS Lett 34:85.

Hannigan GE and Williams BRG (1991) Signal transduction by interferon-a through arachidonic acid metabolism. Science 251:204.

Hartung H-P and Toyka KV (1987) Phorbol diester TPA elicits prostaglandin E release from cultured rat astrocytes. Brain Res 417:347.

Hillared L and Chan PH (1988) Effects of arachidonic acid on respiratory activities in isolated brain mitochondria. J Neurosci Res 19:94.

Ho AK and Klein DC (1987) Activation of α_1-adrenoceptors, protein kinase C, or treatment with intracellular free Ca^{2+} elevating agents increases pineal phospholipase A_2 activity. J Biol Chem 262:11764.

Hofmann SL, Prescott SM, Majerus PW (1982) The effects of mepacrine and p-bromophenacyl bromide on arachidonic acid release in human platelets. Arch Biochem Biophys 215:237.

Ikeda M, Yoshida S, Busto R, Santiso M, Ginsberg MD (1986) Polyphosphoinositides as a probable source of brain free fatty acids at the onset of ischemia. J Neurochem 47:123.

Ito S, Mochizuki-Oda N, Hori L, Ozaki K, Miyakawa A, Negishi M (1991) Characterization of prostaglandin E_2-induced Ca^{2+} mobilization in single bovine adrenal chromaffin cells by digital image microscopy. J Neurochem 56: 531.

Jelsema CL (1987) Light activation of phospholipase A_2 in rod outer segments of bovine retina and its modulation by GTP-binding proteins. J Biol Chem 262:163.

Kandasamy SB and Hunt WA (1990) Arachidonic acid and prostaglandins enhance potassium-stimulated calcium influx into rat brain synaptosomes. Neuropharmacology 29:825.

Kiwak KJ, Moskowitz MA, Levine L (1985) Leukotriene production in gerbil brain after ischemic insult, subarachnoid hemorrhage, and convulsive injury. J Neurosurg 62:865.

Lapetina EG, Billah MM, Cuatracasas P (1981) The initial action of thrombin on platelets. Conversion of phosphatidylinositol to phosphatidic acid preceding the production of arachidonic acid. J Biol Chem 256:5037.

Lazarewicz JW, Leu V, Sun AY (1983) Arachidonic acid release from K^+-evoked depolarization of brain synaptosomes. Neurochem Int 5:471.

Lazarewicz J, Strosznajder J, Gromek A (1972) Effect of ischemia and exogenous fatty acids on the energy metabolism in brain mitochondria. Bull Acad Pol Sci [Biol] 20:599.

Lazarewicz JW, Wroblewski JT, Costa E (1990) N-methyl-D-aspartate-sensitive glutamate receptors induce calcium-mediated arachidonic acid release in primary cultures of cerebellar granule cells. J Neurochem 55:1875.

Lazarewicz JW, Wroblewski JT, Palmer ME, Costa E (1988) Activation of N-methyl-D-aspartate-sensitive glutamate receptors stimulates arachidonic acid release in primary cultures of cerebellar granule cells. Neuropharmacology 27:765.

Linden DJ and Routtenberg A (1989) The role of protein kinase C in long-term potentiation: A testable model. Brain Res Rev 14:279.

Lynch G and Muller D (1990) Steps between the induction and expression of long-term potentiation. In: Neurotoxicity of excitatory amino acids (Guidotti A, ed) p 125. New York: Raven Press.

Lynch MA and Voss KL (1990) Arachidonic acid increases inositol phospholipid metabolism and glutamate release in synaptosomes prepared from hippocampal tissue. J Neurochem 55:215.

Lynch MA and Voss KL (1991) Presynaptic changes in long-term potentiation: Elevated synaptosomal calcium concentration and basal phosphoinositide turnover in dentate gyrus. J Neurochem 56:113.

Manev H, Favaron M, Bertolino M, Brooker G, Guidotti A, Costa E (1990) Importance of sustained protein kinase C translocation and destabilization of Ca^{2+} homeostasis in glutamate-induced neuronal death. In: Neurotoxicity of excitatory amino acids (Guidotti A, ed) p 63. New York: Raven Press, Ltd.

Mayer LM and Westbrook GL (1988) The physiology of excitatory amino acids in the vertebrate central nervous system. Progr Neurobiol 28:197.

Minamisawa H, Terashi A, Katayama Y, Kanda Y, Shimizu J, Shiratori T, Inamura K, Kaseki H, Yoshino Y (1988) Brain eicosanoid levels in spontaneously hypertensive rats after ischemia with reperfusion: Leukotriene C_4 as a possible cause of cerebral edema. Stroke 19:372.

Mochizuki-Oda N, Mori K, Negishi M, Ito S (1991) Prostaglandin E_2 activates Ca^{2+} channels in bovine adrenal chromaffin cells. J Neurochem 56:541.

Nicoletti F, Wroblewski JT, Fadda E, Costa E (1988) Pertussis toxin inhibits signal transduction at a specific metabolotropic glutamate receptor in primary cultures of cerebellar granule cells. Neuropharmacology 27:551.

Nicoletti F, Wroblewski JT, Novelli A, Alho H, Guidotti A, Costa E (1986) The activation of inositol phospholipid metabolism as a signal transducing system for excitatory amino acids in primary cultures of cerebellar granule cells. J Neurochem 6:1905.

Novelli A, Nicoletti F, Wroblewski JT, Alho H, Costa E, Guidotti A (1987) Excitatory amino acid receptors coupled with guanylate cyclase in primary cultures of cerebellar granule cells. J Neurosci 7:40.

Olney J, Price M, Salles KS, Lebruyere J, Frierdich G (1987) MK-801 powerfully protects against N-methyl-aspartate neurotoxicity. Eur J Pharmacol 141:357.

Ozyurt E, Graham DI, Woodruff GN, McCulloch J (1988) Protective effect of the glutamate antagonist, MK-801 in focal cerebral ischemia in the cat. J Cereb Blood Flow Metabol 8:138.

Petroni A, Bertazzo A, Sarti S, Galli C (1989) Accumulation of arachidonic acid cyclo- and lipoxygenase products in rat brain during ischemia and reperfusion: Effects of treatment with GM1-lactone. J Neurochem 53:747.

Piomelli D, Volterra A, Dale N, Siegelbaum SA, Kandel ER, Schwartz JH, Belardetti F (1987) Lipoxygenase metabolites of arachidonic acid as second messengers for presynaptic inhibition of *Aplysia* sensory cells. Nature 328:38.

Pohorecki R, Becker GL, Reilly PJ, Landers DF (1990) Ischemic brain injury *in vitro*: Protective effects of NMDA receptor antagonists and calmidazolium. Brain Res 528:133.

Powell WS (1980) Rapid extraction of oxygenated metabolites of arachidonic acid from biological samples using octadecylsilyl silica. Prostaglandins 20:947.

Premkumar LS, Gage PW, Chung S-H (1990) Coupled potassium channels induced by arachidonic acid in cultured neurons. Proc R Soc Lond B 242:17.

Salinska E, Pluta R, Puka M, Lazarewicz JW (1991) Blockade of N-methyl-D-aspartate-sensitive excitatory amino acid receptors with 2-amino-5-phosphono-valerate reduces ischemia-evoked calcium redistribution in rabbit hippocampus. Exp Neurol 112:89.

Sanfeliu C, Hunt A, Patel AJ (1990) Exposure to N-methyl-D-aspartate increases release of arachidonic acid in primary cultures of rat hippocampal neurons and not in astrocytes. Brain Res 526:241.

Sartarelli MD, Yamada K, Coyle JT (1990) Phospholipase A_2 and ^{3}H-hemicholinum-3 binding sites in rat brain: a potential second-messenger role for fatty acids in the regulation of high-affinity choline uptake. J Neurosci 10:62.

Schoepp DD, Salhoff CR, Hillman CC, Ornstein PL (1989) CGS-19755 and MK-801 selectively prevent rat striatal cholinergic and gabaergic neuronal degeneration induced by N-methyl-D-aspartate and ibotenate *in vivo*. J Neural Transm 78:183.

Shearman MS, Naor Z, Sekiguchi K, Kishimoto A, Nishizuka Y (1989) Selective activation of the γ-subspecies of protein kinase C from bovine cerebellum by arachidonic acid and its lipoxygenase metabolites FEBS Lett 243:177.

Shimizu T and Wolfe LS (1990) Arachidonic acid cascade and signal transduction. J Neurochem 55:1.

Shimizu T, Yamashita A, Hayaishi O (1982) Specific binding of prostaglandin D_2 to rat brain synaptic membrane. Occurrence, properties and distribution. J Biol Chem 257:13570.

Siesjö BK and Bengtsson F (1989) Calcium fluxes, calcium antagonists, and calcium-related pathology in brain ischemia, hypoglycemia, and spreading depression: A unifying hypothesis J Cereb Blood Flow Metab 9:127.

Simon RP, Swan JH, Griffiths T, Meldrum BS (1984) Blockade of N-methyl-D-aspartate receptors may protect against ischemic damage in the brain. Science 226:850.

Strosznajder J and Strosznajder RP (1989) Guanine nucleotides and fluoride enhance carbachol-mediated arachidonic acid release from phosphatidylinositol. Evidence for involvement of GTP-binding protein in phospholipase A_2 activation. J Lipid Mediators 1:217.

Strosznajder J, Wikiel H, Sun GY (1987) Effects of cerebral ischemia on [^{3}H]inositol lipids and [^{3}H]inositol phosphates of gerbil brain and subcellular fractions. J Neurochem 48:943.

Szekely AM, Costa E, Grayson DR (1990) Transcriptional program coordination by N-methyl-D-aspartate-sensitive glutamate receptor stimulation in primary cultures of cerebellar neurons. Mol Pharmacol 38:624.

Tapia-Arancibia L, Rage F, Astier H (1990) Activation of N-methyl-D-aspartate receptors induces arachidonic acid release and somatostatin secretion in cortical and hypothalamic neurons. Neurochem Int 16 (suppl 1):70.

Vaccarino F, Guidotti A, Costa E (1987) Ganglioside inhibition of glutamate mediated protein kinase C translocation in primary cultures of cerebellar neurons. Proc Natl Acad Sci USA 84:8707.

Westerberg E and Wieloch T (1986) Lesions to the corticostriatal pathways ameliorate hypoglycemia-induced arachidonic acid release. J Neurochem 47:1507.

Wieloch T (1985) Hypoglycemia-induced neuronal damage prevented by an N-methyl-D-aspartate antagonist. Science 230:681.

Williams JH and Bliss TVP (1988) Induction but not maintenance of calcium-induced long-term potentiation in dentate gyrus and area CA1 of the hippocampal slice is blocked by nordihydroguaiaretic acid. Neurosci Lett 88:81.

Williams JH, Errington ML, Lynch MA, Bliss TVP (1989) Arachidonic acid induces a long-term activity-dependent enhancement of synaptic transmission in the hippocampus. Nature 341:739.

Wolfe LS and Pellerin L (1989) Arachidonic acid metabolites in the rat and human brain: New findings on the metabolism of prostaglandin D_2 and lipoxygenase products. In: Arachidonic acid metabolism in the nervous system. Physiological and pathological significance. (Barkai AI, Bazan NG, eds) p 74. New York: NY Acad Sci.

Wong EHG, Kemp JA, Priestley T, Knight AR, Woodruff GN, Iverson LL (1986) The anticonvulsant MK-801 is a potent N-methyl-D-aspartate antagonist. Proc Natl Acad Sci 83:7104.

Wroblewski JT, Lazarewicz JW, Wroblewska B, Costa E (1991) Role of arachidonic acid in signal transduction of N-methyl-D-aspartate-sensitive glutamate receptors. In: Excitatory amino acids. (Meldrum BS, Moroni S, Simon SP, Woods GH, eds) p 231. New York: Raven Press.

Wroblewski JT, Nicoletti F, Costa E (1985) Different coupling of excitatory amino acid receptors with Ca^{2+} channels in primary cultures of cerebellar granule cells. Neuropharmacology 24:919.

Wroblewski JT, Nicoletti F, Fadda E, Costa E (1987) Phencyclidine is a negative allosteric modulator of signal transduction at two subclasses of excitatory amino acid receptors. Proc Natl Acad Sci USA 84:5068.

Xu J and Chuang DM (1987) Serotonergic, adrenergic and histaminergic receptors coupled to phospholipase C in cultured cerebellar granule cells of rats. Biochem Pharmacol 36:2353.

Xu J and Wojcik WJ (1987) Gamma-aminobutyric acid ß receptor-mediated inhibition of adenylate cyclase in cultured cerebellar granule cells: Blockade by islet-activating protein. J Pharmacol Exp Ther 239:568.

Yoshida S, Ikeda M, Busto R, Santiso M, Martinez E, Ginsberg M (1986) Cerebral phosphoinositide, triacylglycerol, and energy metabolism in reversible ischemia: origin and fate of free fatty acids. J Neurochem 47:744.

NON-EICOSANOID FUNCTIONS OF ESSENTIAL FATTY ACIDS:
REGULATION OF ADENOSINE-RELATED FUNCTIONS
IN CULTURED NEUROBLASTOMA CELLS

Mary G. Murphy and Zenobia Byczko

Department of Physiology and Biophysics
Dalhousie University
Halifax, Nova Scotia
Canada B3H 4H7

ABSTRACT

Studies have demonstrated that augmenting the $\omega 6$ polyunsaturated-fatty-acid (PUFA) content of N1E-115 neuroblastoma cells by media supplementation with linoleic acid results in $\geq$ 2-fold increases in basal levels of intracellular cyclic AMP (cAMP). Data suggested some involvement of increased production of adenosine from endogenous metabolites; however, increases in adenosine were not related to increased activity of 5'-nucleotidase or decreased uptake of extracellular adenosine. PUFA-dependent elevations in basal cAMP were evident within 1 min of exposure to a phosphodiesterase inhibitor; this phenomenon did not appear to be due to PUFA-dependent changes in Ca^{2+} uptake or to increases in sensitivity of adenylate cyclase to Ca^{2+}. Forskolin-stimulated cAMP formation was 3-fold higher in PUFA-enriched cells than in control cells, which suggested a direct effect on the functioning of the catalytic unit. Linoleic acid supplementation resulted in a 2-fold increase in the maximum amounts of cAMP produced in response to the stable adenosine analogue, 5'-N'ethylcarboxy-amidoadenosine (NECA). The altered stimulatory response did not involve eicosanoid formation, but may have been related to an increase in the number of stimulatory adenosine receptors, as judged by binding of [³H]NECA. These studies indicate that membrane PUFA modulate adenosine-related functions in neuroblastoma cells, and suggest that a complex series of mechanisms is involved in this regulation.

INTRODUCTION

Polyunsaturated fatty acids (PUFA) derived from linoleic ($18:2\omega 6$) or linolenic ($18:3\omega 3$) acids are major constituents of neural membranes, and are distributed in brain in a highly specific manner with respect to phospholipid class, brain region, cell type, and even particular microdomains within cell membranes (Sastry, 1985). Select species of PUFA fulfill important roles as substrates for the bioactive eicosanoids, including prostaglandins, leukotrienes, and the epoxy-fatty acids. Far less well understood are the 'non-eicosanoid' functions of PUFA, including their roles as modulators of the activities of membrane-bound proteins, such as enzymes, transport molecules, and ion channels (Love et al., 1985; Spector and Yorek, 1985; Murphy, 1990). Activities such as adenylate cyclase, which plays a central role in cell signaling and function, could be regulated directly through specific/nonspecific effects of PUFA on membrane fluidity and/or the lipid microenvironment, or indirectly via the effects of PUFA on other related functions.

Studies in this laboratory are concerned largely with PUFA modulation of adenylate cyclase activity. We demonstrated previously that the ω6 PUFA content of N1E-115 neuroblastoma cells could be increased dramatically by supplementing the culture medium with linoleic acid (Murphy, 1986). This resulted in ≥ 2-fold increases in both basal (Murphy, 1986) and adenosine-stimulated (Murphy and Byczko, 1990) formation of intracellular cAMP. PUFA-dependent increases in 'basal' cAMP synthesis did not involve increased production of prostaglandins or lipid peroxides (Murphy, 1985). They did appear to involve endogenous adenosine (Murphy, 1990); however, to date, we have not confirmed PUFA-dependent increases in extracellular adenosine by direct measurement. Recently, studies have been carried out to determine whether the activities of other proteins that may be related to the availability and/or actions of adenosine are influenced by membrane PUFA composition. We now describe the results of these studies, and discuss their potential relationship to the neuromodulatory role of adenosine in the central nervous system.

MATERIALS AND METHODS

Materials

Linoleic acid was obtained from Supelco Canada, Inc. (Oakville, Canada). [2,8,5'-^{3}H]Adenosine (50-70 Ci/mmol), [^{3}H]adenosine 5'-monophosphate (5-15 Ci/mmol) and $^{45}Ca^{2+}$ (10-40 mCi/mg) were from Du Pont Canada Inc., Mississauga, Canada. The following chemicals were from Sigma Chemical Co. (St. Louis, MO): adenosine deaminase (ADA), Type VI (calf intestinal mucosa) (>750 U/mg protein), adenosine, and dipyridamole. Forskolin was purchased from Calbiochem Behring (LaJolla, CA), and 8-phenyltheophylline (8-PT) was from Research Biochemicals Inc. (Natick, MA). Ro 20-1724 (4-(3-butoxy-4-methoxybenzyl)imiazolidin-2-one) was a generous gift of Hoffmann-LaRoche Ltd. (Etobicoke, Canada).

Cell Culture and Fatty-Acid Supplementation

Murine neuroblastoma cells, clone N1E-115, were a gift from Dr. E. Richelson. The cells were seeded at a density of 5-7 x 10^4 cells/ml into 35 mm Primaria culture dishes (Becton Dickinson, Mississauga, Canada) in Dulbecco's modified Eagle's medium (DMEM) containing 10% (v/v) fetal bovine serum (Flow Laboratories, Mississauga, Canada). Cultures were maintained in a humidified atmosphere of 5% CO_2/95% air at 37°C; subculture was carried out every 5 days. Only cells between passages 20-35 were used for this study. Twenty-four hr following subculture, the cells were supplemented with linoleic acid (final concentration, 50 μM) complexed with fatty-acid-poor bovine serum albumin (BSA) as described previously (Murphy, 1986). Forty-eight hr later, the cultures were washed with serum-free, HEPES-buffered (25 mM, pH 7.6) DMEM and used for assays. Lipids were extracted and analyzed as described previously (Murphy, 1986).

Cyclic AMP Assays

Immediately prior to assay, the cells were washed and 1 ml of incubation medium (serum-free HEPES-buffered DMEM, with or without 0.7 mM Ro 20-1724) was added to each dish; assays were carried out at 30°C for a total of 40 min. In studies with adenosine deaminase (ADA), the cells were pretreated (1.5 U/ml) for 1-3 hr before the assay began, and ADA was also included in the incubation medium. Drugs (e.g., forskolin, 8-PT) or appropriate vehicle (e.g., ethanol) were added at the times indicated in Results. The reactions were terminated by aspirating the medium, and the cells were washed quickly with ice-cold HEPES-buffered DMEM. The dish contents were transferred to disposable tubes in small aliquots of ethanol, and membranes and supernatants were separated by centrifugation. The pellets were analyzed for protein according to Lowry et al. (1951), using BSA as the standard. Supernatants were dried under N_2, the residues were resuspended in 50 mM sodium acetate buffer (pH 4.5) containing 4 mM EDTA, and duplicate aliquots from each dish were analyzed for cAMP using a kit from Amersham (Oakville, Canada). Each assay was carried out in triplicate dishes.

$^{45}Ca^{2+}$ Uptake Assays

After the cells were washed, phosphate-buffered saline (PBS, 950 μl) was added to each dish and the cultures were preincubated for 20 min at 37°C. $^{45}Ca^{2+}$ (1 μCi/ml) was added in 50 μl aliquots; uptake was terminated by aspirating the dish contents and washing the cells with ice-cold PBS. The cells were digested in 0.2 N NaOH; after neutralization, aliquots were counted and protein was analyzed (Lowry et al., 1951).

Measurements of [^{3}H]Adenosine Uptake

The uptake of adenosine into the neuroblastoma cells *in situ* was determined essentially as described by Johnson and Geiger (1989). Briefly, the cells were washed with HEPES buffer (pH 7.4) that contained NaCl (110 mM), glucose (25 mM), sucrose (68.3 mM), KCl (5.3 mM), $CaCl_2$ (1.8 mM), $MgSO_4$ (1.0 mM), and HEPES (20 mM) and then were preincubated for 3 min (37°C) in 1 ml of this buffer that also contained BSA (3%, vol/vol). Uptake was initiated by addition of [^{3}H]adenosine (1.3-20 μM) and terminated after 20 sec by aspirating the medium and rapidly washing the cells with ice-cold HEPES buffer that contained the uptake inhibitor, dipyridamole (DPR; 10^{-4} M). Uptake assays were carried out in triplicate dishes.

5'-Nucleotidase Assays

The activity of 5'-nucleotidase (EC 3.1.3.5) was measured in the intact cells as the amount of [^{3}H]5'-AMP converted to [^{3}H]adenosine, essentially as described by Avruch and Wallach (1971). The cells were washed and preincubated for 10 min (30°C) in 1 ml of 0.9% NaCl that contained $MgCl_2$ (10 mM), and DPR (10^{-4}M). Reactions were initiated by addition of [^{3}H]5'-AMP (3-100 μM), and were continued for 20 min, at which time 0.5 ml aliquots of the medium were transferred to 1.5 ml tubes that contained 0.35 ml of ice-cold $ZnSO_4$ (5%, w/v). $Ba(OH)_2$ (0.3 M, 0.35 ml) was added to precipitate protein and unhydrolyzed AMP. The tube contents were centrifuged and duplicate aliquots (0.4 ml) of the supernatants were counted.

[^{3}H]NECA Binding Assays

Forty-eight hr after supplementation, the cells were washed and suspended in

HEPES-buffered DMEM that contained ADA (1.5 U/ml) and cyclopentyladenosine (50 nM) for 2-3 hr at 37°C. Binding assays were carried out using a modification of the assay of Bruns et al. (1986). The assays were set up in 1.5 ml microfuge tubes that contained the cell suspension, [³H]NECA (10-200 nM), and either 2-chloroadenosine (100 μM) or HEPES-buffered DMEM, in a final volume of 1 ml. The tubes were incubated on a shaking water bath at 30°C for 40 min, and free and bound [³H]NECA were separated by centrifugation (2 min, 12,000 × g). The supernatants were aspirated and the pellets were washed twice with ice-cold buffered DMEM. Cell protein was digested in 1 N NaOH, neutralized with HCl, and aliquots were taken for counting and protein analysis.

Statistical Analysis

The data are presented as means ± SEM. Student's t-tests for paired or unpaired data were used to determine levels of significance, and differences were judged to be significant when $p < 0.05$.

RESULTS

Addition of linoleic acid to cultures of N1E-115 neuroblastoma cells for 48 hr increased the membrane content of ω6 PUFA by approximately 3-fold; the most pronounced elevations were in the amounts of linoleic, eicosatrienoic and arachidonic acids (Murphy, 1986). These changes were associated with moderate (~20%) increases in cell protein, increased membrane fluidity (Murphy et al., 1987), and elevated production of prostaglandins (Murphy, 1986) and lipid peroxides (Murphy, 1985). Of particular interest was the fact that PUFA enrichment resulted in dose-dependent ≥ 2-fold increases in 'basal' intracellular levels of cAMP when the cells were incubated for 40 min in the presence of the phosphodiesterase inhibitor, Ro 20-1724 (Murphy, 1986), and altered the magnitude and kinetics of adenosine-stimulated cAMP formation (Murphy and Byczko, 1990). The experiments described below were carried out in an attempt to elucidate the mechanism(s) whereby membrane PUFA modulated adenylate cyclase activity.

Effects of Ro 20-1724 and Ca^{2+} on Short-Term, PUFA-Dependent Increases in 'Basal' cAMP Formation

The time course of cAMP accumulation in the N1E-115 cells was examined under basal assay conditions, to determine if the PUFA-dependent increases were rapid or occurred gradually. The data presented in Table 1 demonstrate that with as little as 1 min exposure to Ro 20-1724, there was a significant (1.5-fold) increase in the cAMP content of PUFA-enriched cells relative to the controls. The increase was not observed in the absence of the phosphodiesterase inhibitor. We then examined the effect of removing Ca^{2+}, to determine its requirement for cAMP formation in the intact cells. The data in Table 1 demonstrate that removal of extracellular Ca^{2+} resulted in approximate 70% reductions in cAMP levels in both control and linoleic acid-supplemented cells, and therefore, did not eliminate the difference due to PUFA enrichment. Since it was clear that adenylate cyclase activity in the neuroblastoma cells was sensitive to Ca^{2+}, we addressed the question of whether PUFA exerted these effects by increasing Ca^{2+} uptake from the medium. From time-course studies, it was evident that PUFA enrichment did not significantly alter net $^{45}Ca^{2+}$ influx for at least the first 2 min of incubation (Fig. 1). If anything, influx was reduced with longer incubation periods.

Table 1. Effects of Ro 20-1724 and extracellular Ca^{2+} on cAMP formation in control and linoleic-acid-supplemented neuroblastoma cells

Assay Conditions	cAMP Formed (pmol/mg protein)	
	Control (BSA) Cells	PUFA-enriched Cells
1 min, DMEM + Ro 20-1724	68.7 ± 2.2	107.6 ± 4.2
1 min, DMEM - Ro 20-1724	45.7 ± 4.0	49.8 ± 5.5
1 min, PBS + Ca^{2+} + Ro 20-1724	76.4 ± 0.8	110.7 ± 13.9
1 min, PBS - Ca^{2+} + Ro 20-1724	22.6 ± 4.7	33.9 ± 6.4

The cells were cultured in the presence of BSA or linoleic acid (50 μM)/BSA for 48 hr, then washed and incubated for 1 min in either HEPES-buffered DMEM or phosphate-buffered saline (PBS, ± Ca^{2+}). Where indicated, Ro 20-1724 (final concentration, 0.7 mM) was added at the beginning of the incubation. The assays were terminated by aspirating the medium and washing the cells with ice-cold buffer; intracellular cAMP levels were determined as described in Methods. Data represent the means ± SEM of values obtained from three separate experiments.

Effects of Adenosine Deaminase, 8-Phenyltheophylline and Forskolin on 'Basal' cAMP Formation

To determine whether the PUFA-dependent increases in 'basal' cAMP formation involved synthesis of adenosine from endogenous metabolites, the cells were exposed to ADA or 8-PT as described in Methods. As we reported earlier (Murphy and Byczko, 1990), both treatments significantly reduced the amounts of cAMP in the PUFA-enriched cells after 40 min incubation (Table 2). ADA- and 8-PT-dependent reductions were also observed after a 1 min incubation in the presence of Ro 20-1724 (data not shown).

We demonstrated previously that forskolin, the plant diterpene that directly activates adenylate cyclase (Seamon and Daly, 1986), stimulated cAMP formation in unsupplemented neuroblastoma cells, and that the stimulated activity was sensitive to treatment with ADA or 8-PT (Murphy and Byczko, 1989). When the ADA experiments were repeated with cells supplemented with linoleic acid, we found that ADA reduced cAMP in the controls by approximately 40% and in the PUFA-enriched cells by approximately 60% (Table 2). These findings suggested that in addition to a possible adenosine-mediated effect of forskolin in both sets of cultures, there was also a direct effect of PUFA enrichment on the activity of the catalytic unit of adenylate cyclase.

Effects of PUFA Enrichment on Functions Related to Adenosine Availability

A series of experiments were carried out to determine whether increasing the ω6 PUFA content of the neuroblastoma cells altered the functioning of several membrane proteins that determine, at least in part, the availability of extracellular adenosine.

Table 2. Effects of ADA or 8-PT on basal and forskolin-stimulated cAMP formation in control and PUFA-enriched neuroblastoma cells

Assay Conditions	cAMP Formed (pmol/mg protein)	
	Control (BSA) Cells	PUFA-enriched Cells
Control (no additions)	70.4 ± 5.2	183.9 ± 18.4[a]
+ ADA (1.5 U/ml)	66.8 ± 8.2	91.7 ± 4.9[b]
+ 8-PT (10^{-4} M)	62.9 ± 10.3	102.7 ± 8.8[b]
+ Forskolin (10^{-5} M)	802.2 ± 79.0	2656.2 ± 123.5[c]
+ Forskolin (10^{-5} M) + ADA (1.5 U/ml)	506.6 ± 21.1	1125.5 ± 58.2[d]

The cells were cultured for 48 hr in the presence of either BSA or linoleic acid (50 μM)/BSA; assays for cAMP were carried out as described in Methods. Where indicated, ADA was added 1-3 hr before the assay, and was included in the incubation mixture. 8-PT was added at the beginning of the 40 min assay. Data are the means ± SEM of values obtained in 3-10 experiments. [a]$p < 0.001$, relative to control (BSA) cells; [b]$p < 0.02$, relative to basal values for PUFA-enriched cells; [c]$p < 0.001$ relative to forskolin-stimulated values in control (BSA) cells; [d]$p < 0.001$, relative to values with forskolin-stimulated cAMP formation in PUFA-enriched cells.

Studies have shown that 5'-nucleotidase, the ectoenzyme that increases extracellular adenosine by catalyzing breakdown of 5'-AMP, is very sensitive to its lipid environment (Bernsohn and Spitz, 1974; Zuniga et al., 1989). However, when we measured the activity of 5'-nucleotidase activity in intact N1E-115 cells, we found that neither the maximum responses nor the values for K_a were altered in cells that were supplemented with linoleic acid (Fig. 2).

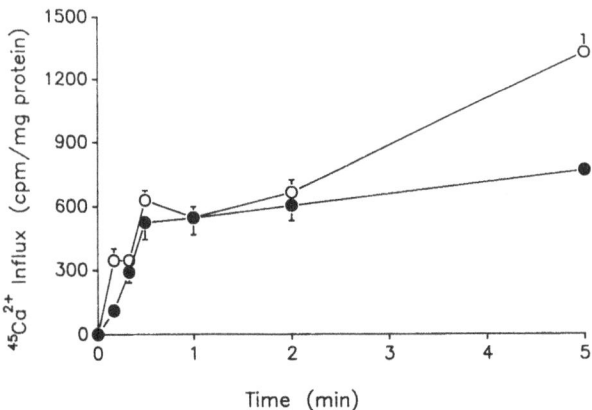

Figure 1. Effects of PUFA enrichment on uptake of $^{45}Ca^{2+}$ into N1E-115 neuroblastoma cells. Cells were supplemented with BSA (open circles) or linoleic acid (50 μM)/BSA (filled circles) for 48 hr. They were incubated in situ in PBS that contained Ca^{2+} (1.8 mM) for 20 min at 37°C prior to the addition of $^{45}Ca^{2+}$ (1 μCi/ml). Reactions were terminated at the times indicated. The data represent means ± SEM of values obtained in 3-8 separate experiments, with values for t=0 subtracted.

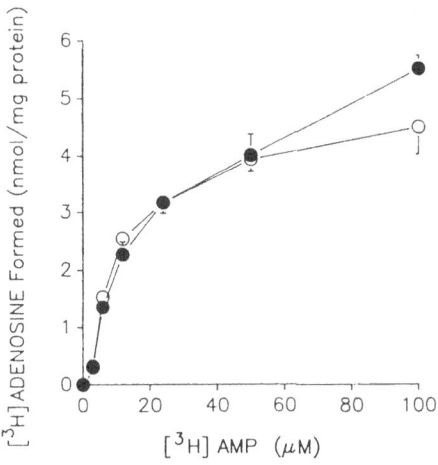

Figure 2. Effects of membrane PUFA content on 5'-nucleotidase activity. The cells were grown for 48 hr in the presence of either BSA (open circles) or linoleate (50 μM)/BSA (filled circles). The activity of 5'-nucleotidase was measured as described in Methods. Data are the means ± SEM of values obtained in three separate experiments.

[³H]Adenosine was taken up rapidly by the neuroblastoma cells; analysis of measurements of uptake of 1.25-20 μM adenosine in the presence of Na$^+$ indicated that nucleoside transport into the PUFA-enriched cells was generally higher than transport in the control cultures (Fig. 3); however, the differences were not statistically significant.

Effects of PUFA Enrichment on Adenosine Receptor-Mediated cAMP Formation

We demonstrated previously that addition of the stable adenosine analogue, NECA (10^{-7} to 10^{-4}M), to control or PUFA-enriched N1E-115 neuroblastoma cells produced

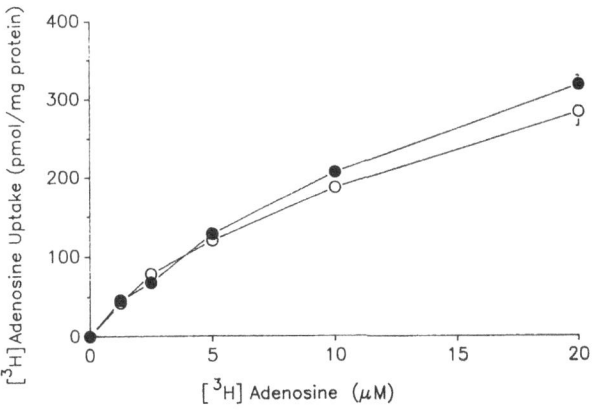

Figure 3. Effect of PUFA enrichment on [³H]adenosine uptake. The cells were cultured in the presence of either BSA (open circles) or linoleic acid (50 μM)/BSA (filled circles) for 48 hr. The cells were then washed and [³H]adenosine uptake during 20 sec was measured as described in Methods. Data represent the means ± SEM of values obtained in three separate experiments.

dose-dependent increases in intracellular cAMP formation (Murphy and Byczko, 1990). Even in cells treated with ADA to remove any effects of endogenous adenosine, values for maximum response were 2-fold higher in PUFA-enriched cells than in controls (1260 and 620 pmol/mg protein, respectively), and PUFA enrichment resulted in an almost 50% reduction in the amounts of NECA required to elicit half maximum response.

Several studies were then carried out to determine what mechanisms could be involved in the increased effectiveness of NECA at the adenosine receptor. We first determined whether metabolites of the ω6 PUFA were involved by carrying out the assays in the presence of agents known to inhibit eicosanoid formation or lipid peroxidation. The data in Figure 4 show that cAMP formation in the control cultures was not affected appreciably by any of the agents. The minor (15-20%) increases due to indomethacin and esculetin (a potent lipoxygenase inhibitor) in the PUFA-enriched cells were not statistically significant.

The binding of [^{3}H]NECA to the intact cells was analyzed to determine whether supplementation could be altering the number of adenosine receptors on the cells. At 60-70 nM [^{3}H]NECA, the mean (± SEM, N=4) amounts of ligand specifically bound were 383±24 and 507±41 fmol/mg protein for control and PUFA-enriched cells, respectively. For both, specific binding ranged from 40-50% of total binding. Analysis of the data from two separate saturation experiments indicated that the values for B_{max} (receptor density) and K_d (the concentration of ligand at which half maximal binding occurs) were 2-fold higher in the linoleic acid-supplemented cells than in controls (Table 3).

DISCUSSION

Adenylate cyclase has been shown to be particularly sensitive to its lipid/fatty-acid

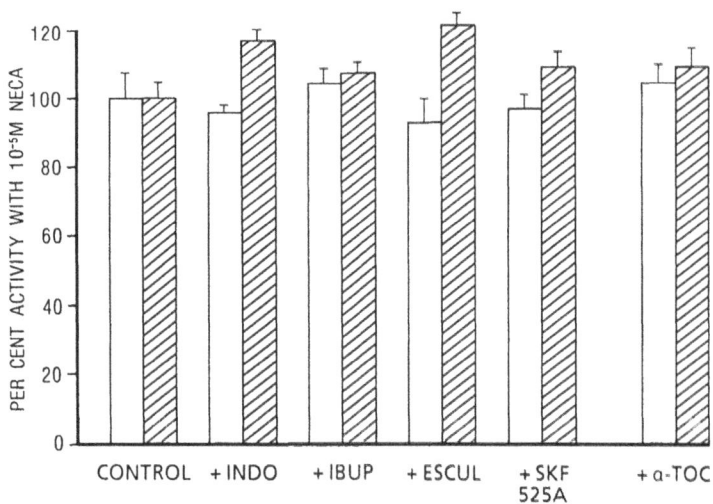

Figure 4. Effects of inhibitors of PUFA metabolism and lipid peroxidation on NECA-stimulated cAMP formation in control (open bars) and linoleate (50 μM)-supplemented (striped bars) neuroblastoma cells. The cells were cultured and assayed for cAMP as described in Methods. The inhibitors were added at the beginning of the assay as follows: +INDO, indomethacin (10^{-5}M); +IBUP, ibuprofen (10^{-5}M); +ESCUL, esculetin (10^{-5}M); +SKF 525A, SKF 525A (2x10^{-5}M); α-TOC, α-tocopherol (2x10^{-5}M). Values obtained in the presence of NECA (10^{-5}M) alone were 378.5 ± 21.5 and 679.1 ± 48.5 pmol/mg protein for control and PUFA-enriched cells, respectively (means ± SEM, N=3).

Table 3. Effect of PUFA enrichment on the binding of [³H]NECA to intact N1E-115 neuroblastoma cells

Culture Supplement	B_{max} (fmol/mg protein)		K_d (nM)	
	Expt. 1	Expt. 2	Expt. 1	Expt. 2
BSA (Controls)	1072	1226	111	166
Linoleic acid (50 μM)/BSA	2065	2330	220	270

After the cells were supplemented as indicated above and cultured for 48 hr, they were washed and assayed for binding of [³H]NECA as described in Methods. Values for receptor density (B_{max}) and K_d were obtained by Eadie-Hofstee analysis of data obtained in the two separate experiments.

environment in most tissues (Houslay and Gordon, 1983; Murphy, 1990); however, the molecular mechanism(s) of its modulation are far from well understood. Our previous findings—that ω6 PUFA-dependent increases in the 'basal' content of cAMP in N1E-115 neuroblastoma cells were reduced by treating the cells with ADA or 8-PT (Murphy and Byczko, 1990) and further increased in the presence of agents (DPR, 2'-DCF) that have the potential to regulate the availability of extracellular adenosine (Murphy and Byczko, in press)—suggested that augmenting cell PUFA content could result in increased production of endogenous adenosine. To date, we have not detected elevated adenosine levels in these cells; however, under our analytical conditions any adenosine formed could have been removed from the medium by uptake and metabolism or by degradation. Studies to resolve this problem are currently underway. As an alternative approach, we examined the effects of two membrane functions that could be altered by PUFA enrichment. An increase in 5'-nucleotidase activity, which catalyzes break-down of extracellular 5'-AMP, would result in increased adenosine which, even if only transient, could activate cell-surface A_2 receptors and elevate cAMP levels. However, the data in Figure 2 indicated this to be unlikely in our cells, which was consistent with the earlier report of Bernsohn and Spitz (1974) that the activity of this neural membrane enzyme is selectively sensitive to ω3 PUFA. We then examined the effects of linoleic acid supplementation on uptake of [³H]adenosine, since transport of several small molecules has been shown to be sensitive to PUFA environment (Spector and Yorek, 1985), and transient increases in adenylate cyclase could result from impaired uptake of the nucleoside into the cells. However, enrichment of the cells in ω6 PUFA either did not affect (e.g., at low concentrations) or slightly increased uptake over the brief period (20 sec) that we measured the process (Fig. 3). We did not examine uptake over longer periods, since the data would be very difficult to interpret due to extensive metabolism of adenosine within the cells (Johnson and Geiger, 1989).

One question that we addressed was whether PUFA-dependent elevations in cAMP involved alterations in the sensitivity of adenylate cyclase to Ca^{2+} and/or in the transport of Ca^{2+} into the cells. Adenylate cyclase activity in both types of cultures appeared to be largely dependent upon extracellular Ca^{2+}, since its removal yielded ~70% reductions in cellular cAMP content (Table 1). If, as has been suggested (Cooper et al., 1988; Eliot et al., 1989), most of the adenylate cyclase in neural cells is stimulated by Ca^{2+}, we would predict that elevations in cellular cAMP could be due to PUFA-mediated increases in intracellular Ca^{2+}. This could be achieved by increasing the uptake of Ca^{2+} from the medium, or by increasing the net release from intracellular stores. To date, we have examined only the former possibility, and have found that PUFA enrichment does not alter $^{45}Ca^{2+}$ uptake into the N1E-115 cells during brief incubations ($\leq$ 1 min) and even decreases it with longer periods of exposure (Fig. 1). Little is known regarding the effects of membrane PUFA on Ca^{2+} influx into neural cells. However, consistent with our findings, the results of studies with non-neural systems demonstrated that increasing membrane $\omega6$ PUFA results in inhibition of $^{45}Ca^{2+}$ influx (Mahfouz et al., 1989; Rustenbeck and Lenzen, 1989) and suggested that this may be related to changes in cell membrane potential (Mahfouz et al., 1989). Recent studies by Narahashi and coworkers (1987) indicated that a relationship exists between membrane potential, intracellular cAMP levels, and Ca^{2+} currents in the N1E-115 cell line; however, we have not examined the effect of PUFA enrichment on these relationships. An alternative explanation for PUFA-dependent elevations in 'basal' cAMP levels was raised by Boyajian et al. (1991), who recently reported that increases in cytosolic Ca^{2+} inhibit rather than stimulate cAMP formation in the hybrid neural cell line, NCB-20.

Previous studies from this (Murphy and Byczko, 1989) and other (Boyajian et al., 1991) laboratories have shown that forskolin stimulates cAMP formation in intact N1E-115 neuroblastoma cells. Linoleic acid supplementation of the cells augmented forskolin-dependent stimulation (Table 3), which suggested that adenylate cyclase was affected directly by the fatty-acid content of the membranes. However, we believe that this is an oversimplification of the situation. One reason is that acutely administered ethanol, which mimics the membrane 'fluidizing' effects of PUFA, has been reported to inhibit rather than stimulate forskolin-stimulated adenylate cyclase in this cell line (Stenstrom et al., 1985). A second is that treatment of both control and supplemented cells with ADA appreciably reduced forskolin-stimulated cAMP accumulation, which suggests involvement of endogenous adenosine and/or adenosine receptors. Mante and Minneman (1990) recently reported that forskolin-stimulated cAMP formation in slices of rat cerebral cortex was blocked by ADA and adenosine receptor antagonists; however, based on concentrations of the agents required to effect blockade, they concluded that only about 30% of the forskolin effect was mediated by endogenous adenosine. We do not yet know what (if any) proportion of our forskolin effects are due to endogenous adenosine, nor whether the proportion is higher in the PUFA-enriched than in the control cells.

In earlier studies, Anand-Srivastava and Johnson (1981) concluded that adenosine receptors were tightly coupled to adenylate cyclase, based on observations that treatment of striatal membranes with detergents had little effect on cAMP formation. Our findings that increasing the $\omega6$ PUFA content of the N1E-115 cells increased the maximum response to NECA and decreased the values for K_a (Murphy and Byczko, 1990) argue against this concept. Recently, Nanoff et al. (1991) provided data that might explain this discrepancy. They showed that in select systems (e.g., rabbit striatum), there are two possible forms of the adenosine receptor complex, and that only the minor species functions via a coupling-uncoupling (versus permanently coupled) mechanism. It is possible that it is the minor, more mobile form of the receptor that is present in the N1E-115 cells. We carried out studies to determine whether eicosanoids are involved in PUFA-mediated changes in the functioning of the neuroblastoma receptor; however, the data in Figure 4, in keeping with those of other studies (Dunwiddie et al., 1990), indicate that this is unlikely. The results of our binding studies (Table 3) suggest that PUFA enrichment of the neuroblastoma cells could have resulted in an increase in the number of adenosine receptors. While this may be the case, we feel that it is wise to be cautious in making this interpretation, based on recent reports that [^{3}H]NECA binds selectively and with high affinity to sites other than the classic A_2 receptor (Keen et al., 1989).

The high content and highly selective distribution of PUFA in central nervous system membranes and the fact that polyenes are preserved even under periods of prolonged dietary deprivation suggest that they play an essential role in maintaining normal brain function. We and others believe that they fulfill important modulatory functions aside from their roles as second messengers and as precursors of the eicosanoids. Based on the data described in this and previous reports, we chose to focus on the relationship(s) between PUFA and adenosine-related functions in neural membranes. Adenosine has many modulatory actions in the CNS, including inhibition of neurotransmitter release (presumably by decreasing Ca^{2+} uptake) and suppression of neuronal firing. At least some of these functions are related to the ability of adenosine to inhibit and stimulate cAMP formation via A_1 and A_2 adenosine receptors. Our studies have shown that membrane PUFA selectively alter A_2 receptor function, and suggest that this is mediated via a complex series of mechanisms that include not only direct effects on enzyme activities, but also effects on ion channel activities and synthesis/availability of key neural metabolites.

ACKNOWLEDGMENTS

This work was supported by the Medical Research Council of Canada.

REFERENCES

Anand-Srivastava MB and Johnson RA (1981) Role of phospholipids in coupling of adenosine and dopamine receptors to striatal adenylate cyclase. J Neurochem 36:1819-1828.

Avruch J and Wallach DFH (1971) Preparation and properties of plasma membrane and endoplasmic reticulum fragments from isolated rat fat cells. Biochim Biophys Acta 233:334-347.

Bernsohn J and Spitz FJ (1974) Linoleic- and linolenic acid dependency of some brain membrane-bound enzymes after lipid deprivation in rats. Biochem Biophys Res Commun 57:293-298.

Boyajian CL, Garritsen A, Cooper DMF (1991) Bradykinin stimulates Ca^{2+} mobilization in NCB-20 cells leading to direct inhibition of adenylylcyclase. A novel mechanism for inhibition of cAMP production. J Biol Chem 266:4995-5003.

Bruns RF, Lu GH, Pugsley TA (1986) Characterization of the A_2 adenosine receptor labeled by [3H]NECA in rat striatal membranes. Mol Pharmacol 29:331-346.

Cooper DMF, Ahlijanian MK, Perez-Reyes E (1988) Calmodulin plays a dominant role in determining neurotransmitter regulation of neuronal adenylate cyclase. J Cell Biochem 36:417-427.

Dunwiddie TV, Taylor M, Cass WA, Fitzpatrick FA, Zahniser NR (1990) Arachidonic acid metabolites do not mediate modulation of neurotransmitter release by adenosine in the rat hippocampus or striatum. Brain Res 527:76-80.

Eliot LS, Dudai Y, Kandel ER, Abrams TW (1989) Ca^{2+}/calmodulin sensitivity may be common to all forms of neural adenylate cyclase. Proc Natl Acad Sci USA 86:9564-9568.

Houslay MD and Gordon LM (1983) The activity of adenylate cyclase is regulated by the nature of its lipid environment. Curr Topics Membr Transport 18:179-231.

Johnson ME and Geiger JD (1989) Sodium-dependent uptake of nucleosides by dissociated brain cells from the rat. J Neurochem 52:75-81.

Keen M, Kelly E, Nobbs P, MacDermot J (1989) A selective binding site for 3H-NECA that is not an adenosine A_2 receptor, Biochem Pharmacol 38:3827-3833.

Love JA, Saum WR, McGee R Jr (1985) The effects of exposure to exogenous fatty acids and membrane fatty acid modification on the electrical properties of NG108-15 cells. Cell Mol Neurobiol 5:333-352.

Lowry OH, Rosebrough NJ, Farr AL, Randall RJ (1951) Protein measurements with the Folin phenol reagent. J Biol Chem 193:265-275.

Mahfouz MM, Smith TL, Kummerow FA (1989) Changes in phospholipid composition and calcium flux in LLC-PK cells cultured at low magnesium concentrations. Biochim Biophys Acta 1006:75-83.

Mante S and Minneman KP (1990) Is adenosine involved in inhibition of forskolin-stimulated cyclic AMP accumulation by caffeine in rat brain? Mol Pharmacol 38:652-659.

Murphy MG (1985) Membrane fatty acids, lipid peroxidation and adenylate cyclase activity in cultured neural cells. Biochem Biophys Res Commun 132:757-763.

Murphy MG (1986) Studies of the regulation of basal adenylate cyclase activity by membrane polyunsaturated fatty acids in cultured neuroblastoma. J Neurochem 47:245-253.

Murphy MG (1990) Dietary fatty acids and membrane protein function. J Nutr Biochem 1:68-79.

Murphy MG and Byczko Z (1989) Effects of adenosine analogues on basal, prostaglandin E_1- and forskolin-stimulated cyclic AMP formation in intact neuroblastoma cells. Biochem Pharmacol 38:3289-3295.

Murphy MG and Byczko A (1990) Effects of membrane polyunsaturated fatty acids on adenosine receptor function in intact N1E-115 neuroblastoma cells. Biochem Cell Biol 68:392-395.

Murphy MG and Byczko Z (in press) Further studies of the mechanism(s) of polyunsaturated-fatty-acid-mediated increases in intracellular cAMP formation in N1E-115 neuroblastoma cells. Neurochem Res.

Murphy MG, Moak CM, Rao BG (1987) Effects of membrane polyunsaturated fatty acids on opiate peptide inhibition of basal and prostaglandin E_1-stimulated cyclic AMP formation in intact N1E-115 neuroblastoma cells. Biochem Pharmacol 36:4079-4084.

Nanoff C, Jacobson KA, Stiles GL (1991) The A_2 adenosine receptor: Guanine-nucleotide modulation of agonist binding is enhanced by proteolysis. Mol Pharmacol 39:130-135.

Narahashi T, Tsunoo A, Yoshii M (1987) Characterization of two types of calcium channels in mouse neuroblastoma cells. J Physiol 383:231-249.

Rustenbeck I and Lenzen S (1989) Regulation of transmembrane ion transport by reaction products of phospholipase A_2. II. Effects of arachidonic acid and other fatty acids on mitochondrial Ca^{2+} transport. Biochim Biophys Acta 982:147-155.

Sastry PS (1985) Lipids of nervous tissue: Composition and metabolism. Prog Lipid Res 24:69-176.

Seamon KB and Daly JW (1986) Forskolin: Its biological and chemical properties. Adv Cyclic Nucleotide Protein Phosphorylation Res 20:1-150.

Spector AA and Yorek MA (1985) Membrane lipid composition and cellular function. J Lipid Res 26:1015-1035.

Stenstrom S, Seppala M, Pfenning M, Richelson E (1985) Inhibition by ethanol of forskolin-stimulated adenylate cyclase in a murine neuroblastoma clone (N1E-115). Biochem Pharmacol 34:3655-3659.

Zuniga ME, Lokesh BAR, Kinsella JE (1989) Disparate effects of dietary fatty acids on activity of 5'-nucleotidase of rat liver plasma membrane. J Nutr 119:152-160.

CONTRIBUTIONS TO ARACHIDONIC ACID RELEASE IN MOUSE CEREBRUM BY THE PHOSPHOINOSITIDE-PHOSPHOLIPASE C AND PHOSPHOLIPASE A$_2$ PATHWAYS

Grace Y. Sun

Biochemistry Department
University of Missouri School of Medicine
Columbia, MO 65212
USA

ABSTRACT

Recent studies have indicated two major mechanisms for the release of arachidonic acid (20:4) from membrane phospholipids: 1) activation of phospholipase A$_2$ and 2) stimulated hydrolysis of poly-phosphoinositides (PI) and diacylglycerols (DG) through phospholipase C and diacylglycerol lipase, respectively. In mammalian brain both mechanisms seem to be operable, although the relative contributions by these two pathways have not been carefully assessed. In this study three experimental protocols were used to examine 20:4 release in brain due to ischemia and agonist stimulation, as well as the metabolic relationship between this release and the increase in diacylglycerols, lysophospholipids, and inositol phosphates. The preferential release of arachidonic acid during the initial phase after decapitation was attributed mainly to the sequential hydrolysis of poly-PI to DG. During the second phase, the release of 20:4 along with other free fatty acids (FFA) correlated well with the increase in labeled lysophospholipids, suggesting the involvement of phospholipase A$_2$. Diacylglycerols in brain are enriched in 18:0 and 20:4. Decapitation induced a rapid increase in the level of DG, which remained elevated during the 30 min period under study. Between 5 sec and 5 min, the increase in FFA lagged behind that of DG. The parallel increases in 18:0 and 20:4 in the FFA pool further support the notion that, during the early phase, 20:4 could be derived from the sequential hydrolysis of poly-PI and DG. Decapitation also induced a sequential appearance of Ins(1,4,5)P$_3$, Ins(1,4)P$_2$, and Ins(4)P, which peaked at 30 sec, 1 min, and 2 min, respectively. The level of 20:4 in brain was also examined with respect to poly-PI turnover due to stimulation by cholinergic agonists. Administration of pilocarpine to lithium-treated mice resulted in increased accumulation of labeled inositol monophosphate (IP$_1$) compared to the amount in controls receiving lithium alone, as well as a less obvious increase in 20:4. Both pilocarpine-mediated increases (IP$_1$ and 20:4) could be blocked by atropine. These results point to the presence of an active mechanism for poly-PI turnover and for the recycling of 20:4 in brain.

INTRODUCTION

Despite the high levels of arachidonic acid (20:4) present in brain membrane phospholipids, only trace amounts of this fatty acid are found in the free form. This observation suggests the presence of stringent biochemical mechanisms for regulating the dynamic turnover of 20:4 in brain (Sun and MacQuarrie, 1989). In addition to serving as a precursor for the synthesis of prostanoids, which may have neuromodulatory functions (Shimizu and Wolfe, 1990), 20:4 is also known to influence a number of cellular activities including activating specific subtypes of protein kinase C (Shearman et al., 1991) and facilitating the translocation of diacylglycerol kinase from cytosol to membranes (Kelleher and Sun, 1985).

It has been demonstrated that cerebral ischemia and electroconvulsive shock induce the rapid release of FFA in brain (Bazan, 1970; Bazan et al., 1971). Nevertheless, the mechanism(s) contributing to the FFA release have not been clearly elucidated. Studies in which the release of FFA (including 20:4) in brain was examined using the global cerebral ischemia model provided convincing evidence for the involvement of stimulated hydrolysis of poly-PI (Sun and Huang, 1987) and DG in the initial phase of FFA release (Ikeda et al., 1986; Abe et al., 1987; Yoshida et al., 1986; Huang and Sun, 1986). Indeed, di- and monoacylglycerol lipases have been well studied in brain (Farooqui et al., 1985). Recent studies with cultured cells further suggest an increase in the activities of these enzymes following agonist stimulation of poly-PI breakdown (Farooqui et al., 1990). The rapid hydrolysis of poly-PI in brain after decapitation has been shown to occur in synaptic membranes but not in myelin (Sun et al., 1990b). Furthermore, this process has been shown to occur in brain tissue after focal ischemic insult (Lin et al., 1991).

There is also evidence for the role of phospholipase A_2 in FFA release in cerebral ischemia (Edgar et al., 1982; Sun and Foudin, 1984). However, the relative contributions of the phospholipase A_2 mechanism to the overall increase in FFA due to (i) ischemia and (ii) stimulation have not been investigated in detail. Several recent reviews describing the ischemia-induced changes in phosphoinositides and fatty acid release are available (Sun, 1989; 1990; in press).

In this study, three experimental protocols were used to examine the FFA in brain (especially 20:4) in relation to the release of DG, lysophospholipids, and inositol phosphates. Furthermore, an attempt was made to relate the increase in 20:4 in brain to the appearance of second messengers resulting from stimulated turnover of poly-PI (Berridge, 1987). The advantage of using the global cerebral ischemia model induced by decapitation is that the release of FFA and other second messenger compounds can be observed under cellular conditions in which ATP is depleted (Lowry et al., 1964; Novak et al., 1985) and thus the extensive recycling activity occurring in brain is eliminated. Results indicate that, in the intact brain, 20:4 is maintained in a dynamic equilibrium, and that extensive recycling activity is present to prevent the build-up of this fatty acid due to agonist stimulation.

METHODS

Experimental Protocols:

1) The first experimental protocol examined the release of 20:4, as well as the amounts of DG and labeled lysophospholipids, in brain at different times after decapitation. Balb/c mice were injected intracerebrally with either 5 μCi of [^{3}H-methyl]-choline chloride (specific radioactivity 60 mCi/mmol, NEN, Boston, MA) or 10 μCi of [^{3}H]myo-inositol (specific radioactivity 17 Ci/mmol, NEN, Boston, MA). The labeled

precursor was allowed to equilibrate in brain for 16 hr, after which animals were subjected to decapitation ischemia. At time points ranging from 30 sec (control) to 30 min after decapitation, brain tissue (mainly cerebrum) was dissected and lipids extracted for analysis of the levels of FFA and DG as well as radioactivity of the respective lysophospholipids and phospholipids. A more detailed description of this experimental protocol is found in Sun et al. (in press).

2) The second experimental protocol examined 20:4 release with respect to the increase in DG and inositol phosphates during the early phase after decapitation. C57BL/6J mice were subjected to decapitation ischemia. At various intervals from 5 sec (control) to 5 min after decapitation, the cerebrum was dissected and subjected to lipid extraction. The organic phase was used for analysis of FFA and DG; the aqueous phase was used for analysis of inositol mono- and bisphosphates using the ion chromatography procedure described by Sun et al. (1990a). Analysis of $Ins(1,4,5)P_3$ was carried out using the radioreceptor binding assay procedure described by Bredt et al. (1989).

3) The third experimental protocol was aimed at correlating 20:4 in brain with poly-PI turnover due to cholinergic stimulation. Adult C57BL/6J mice were injected intracerebrally with myo-$[2-^3H(N)]$inositol (specific activity, 14 Ci/mmol, American Radiolabeled Chemicals, St. Louis, MO), and the label was allowed to equilibrate in brain for 16-20 hr. At 4 hr prior to decapitation, animals were injected i.p. with LiCl (8 meq/kg body wt) and then divided into four groups. Each group received subcutaneous injection of one of the following: saline 30 min prior to dissection; atropine (100 mg/kg body wt) 1 hr prior to dissection; pilocarpine (30 mg/kg body wt) 30 min prior to brain dissection; or atropine 1 hr and then pilocarpine 30 min prior to brain dissection. At the appropriate time, animals were killed by decapitation and the heads were dropped into liquid nitrogen to rapidly cool the brain. The cerebrum was removed, weighed, and homogenized in 20 ml of chloroform-methanol (2:1, by vol). After the samples were partitioned with deionized water, the aqueous phase was taken for analysis of inositol phosphates and the organic phase for analysis of FFA and DG. A more detailed description of this experimental protocol has been published elsewhere (Sun et al., in press).

Lipid Extraction and Separation

Brain tissue homogenate was first extracted with 4 volumes of chloroform-methanol (2:1, by vol). The lower organic phase was removed and the upper aqueous phase was extracted again with 2 volumes of chloroform-methanol-12 N HCl (2:1:0.025, by vol). After removal of the second organic phase, samples were neutralized with 4 N NH_4OH prior to combining with the first organic extraction. The organic solvent was evaporated to dryness, and the lipid residues were redissolved in 3 ml of chloroform-methanol (2:1, by vol). Aliquots of the lipid extracts were taken for counting of radioactivity and for lipid analysis.

Analysis of FFA and DG

For analysis of FFA, 20 μg of heptadecanoic acid (17:0) was added to each sample as internal standard and for assessing the recovery of FFA. The lipid extract was applied to silica gel 60 HPTLC plates (Merck, Darmstadt, Germany) and developed with a solvent system containing petroleum ether-diethyl ether-acetic acid (60:40:1, by vol). After solvent development, the plates were sprayed with 2,7-dichlorofluorescein and the lipid bands were visualized under a UV lamp. The lipid bands corresponding to FFA, DG (together with cholesterol), and phospholipids (PL) were scraped into test tubes. FFA were derivatized by reacting the samples with 2 ml of BF_3-methanol reagent (Sigma Chem. Co., St. Louis, MO) at 60°C for 30 min. The acyl groups of DG and PL were converted to their methyl esters at room temperature with 2 ml of 0.5 M NaOH in methanol. After further purification by HPTLC, the fatty acid methyl esters were analyzed by a Hewlett-Packard 5890 gas chromatograph as described previously (Sun, 1988). Data are expressed as micrograms of fatty acids from FFA and DG per milligram of fatty acids from PL.

Analysis of Lysophospholipids

Lysophospholipids were separated from the PL by HPTLC developed in the same dimension twice with a solvent system containing chloroform-methanol-16 N NH_4OH (70:30:5, by vol). Visualization of the minor lipids was aided by adding lysophosphatidyl-choline (LPC) or lysopolyphosphoinositide (LPI) (Sigma Chem. Co., St. Louis, MO) to the sample prior to HPTLC separation. Lipids were visualized by exposing the plate to iodine vapors and the bands corresponding to phosphatidylcholine (PC) and LPC or PI and LPI were scraped from the plates and radioactivity measured using a Beckman LS 5801 liquid scintillation counter.

Separation of Inositol Phospholipids

For separation of the inositol phospholipids [PtdIns, PtdIns(4)P, PtdIns(4,5)P_2], a portion of the lipid extract was applied to oxalate-impregnated silica gel 60 HPTLC plates. Plates were developed successively (in the same direction), first with a solvent system containing chloroform-methanol-acetone-16 N NH_4OH (70:40:10:10, by vol), and then with a solvent system containing chloroform-methanol-16 N NH_4OH-H_2O (36:28: 2:6, by vol). Before applying samples to the plates, a standard mixture containing PtdIns(4)P and PtdIns(4,5)P_2 (Sigma) was routinely added to each sample to aid visualization. After solvent development, plates were exposed to iodine vapors and the respective lipid bands were scraped for radioactivity measurement.

Analysis of Inositol Phosphates by Dowex Ion Exchange Column and Ion Chromatography

For samples obtained from brains that were labeled with [^{3}H]inositol, aliquots of the aqueous phase were taken for analysis of the inositol phosphates by Dowex ion exchange column chromatography (AG1-X8, formate form, Bio-Rad Laboratories, Richmond, CA) according to the procedure described by Berridge et al. (1983). Aliquots of fractions were mixed with scintillation fluid and counted in a Beckman LS 5800 liquid scintillation spectrometer.

Analysis of Inositol Mono- and Bisphosphates by Ion Chromatography

For analysis of samples by ion chromatography, the upper aqueous phase was freeze-dried and redissolved in a known volume of deionized water. The aqueous suspension was then eluted through a Maxi-clean IC-H$^+$ cartridge (Alltech Associates, Deerfield, IL) to remove Cl$^-$ and reduce background conductivity. Samples (50 μl) were then injected into the ion chromatograph (Dionex BioLC, Sunnyvale, CA) and separation was carried out using a Dionex Ion Pac AS5A 5-μm column with guard column. Conditions for separation of the inositol mono- and bisphosphates using isocratic elution systems were similar to those described earlier (Sun et al., 1990a).

Determination of Ins(1,4,5)P_3 by the Radioreceptor Binding Assay

The procedure for mass determination of Ins(1,4,5)P_3 using the radioreceptor binding assay is similar to that described by Bredt et al., (1989) using rat cerebellar membranes as the source of IP$_3$ receptors. Binding assays were performed in 275 μl of buffer B (50 mM Tris-HCl, pH 8.4, 1 mM EDTA, 1 mM 2-mercaptoethanol) containing 1 nM [^{3}H]Ins(1,4,5)P$_3$ (17 Ci/mmol, NEN), 50 μl of brain extract, and 60-100 μg membrane protein. A standard curve was constructed using 1-100 nM Ins(1,4,5)P$_3$ for the binding assay. Non-specific binding was determined using 2 μM of Ins(1,4,5)P$_3$. The above binding mixtures were incubated at 4°C for 10 min and the binding was terminated by centrifugation at 12,000 × g for 5 min followed by aspiration of the supernatant. The pellets were suspended in 5 ml of scintillation cocktail and left overnight; radioactivities were determined by scintillation counting.

RESULTS AND DISCUSSION

Evidence for Contribution of 20:4 Release by the Poly-PI and Phospholipase A$_2$ Pathways

In the first experimental protocol, the levels of FFA and DG in mouse brain were analyzed between 30 sec (control) and 30 min after decapitation. Decapitation resulted in a parallel increase in FFA and DG during the first minute with rates corresponding to 3.1 and 3.2 μg FA/mg of FA in PL/min, respectively (Sun et al., in press). Thereafter, FFA continued to increase, but at a slower rate; no further increase in DG was observed.

Analysis of FFA content showed 18:0 and 20:4 to be the major fatty acid components. During the first minute after decapitation, there was a parallel increase in both fatty acids; the rates were roughly 1.0 μg FA/mg PL/min (Fig. 1). In agreement with the results observed previously (Tang and Sun, 1982; Yasuda et al., 1985), there were two distinct modes for the increase in 20:4—a rapid one during the first minute and a less rapid one after 5 minutes. The fatty acids of DG in brain are enriched in 18:0 and 20:4. As observed by Aveldaño and Bazan (1975), this DG species increased rapidly during the first min after decapitation (1.1-1.3 μg FA/mg PL/min), reached a plateau, and remained relatively stable for the remaining 30 min (Fig. 2). The high levels of 18:0 and 20:4 in DG strongly suggest that this pool is metabolically related to poly-PI. An advantage of using the decapitation model is the rapid depletion of ATP (Lowry et al., 1964; Novak et al., 1985), which prevents the replenishment of PtdIns(4)P and PtdIns(4,5)P$_2$. Under these conditions, the amount of PtdIns(4,5)P$_2$ degraded should be comparable to the amount of DG formed. We previously showed that more

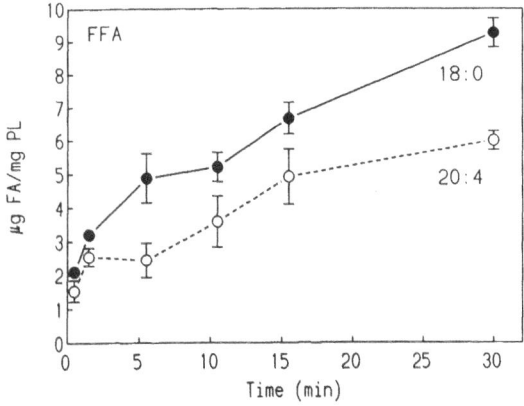

Figure 1. Levels of 18:0 and 20:4 in the FFA fraction of mouse cerebrum with respect to time after decapitation (30 sec to 30 min).

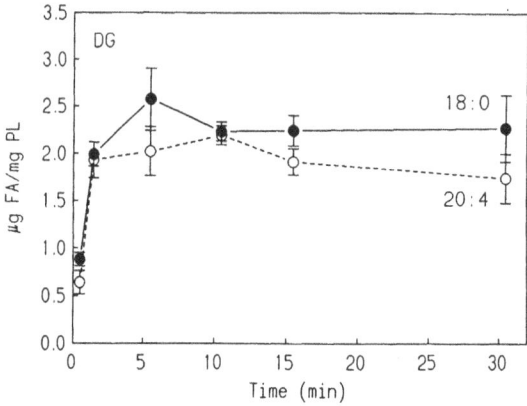

Figure 2. Levels of 18:0 and 20:4 in the DG fraction of mouse cerebrum with respect to time after decapitation (30 sec to 30 min).

than 60% of the poly-PI pool is depleted within 2 min after decapitation (Sun et al., 1988) and the amount of poly-PI degraded is enough to account for the DG and FFA released during the early time points (Huang and Sun, 1986).

After intracerebral injection of [^{3}H]choline or [^{3}H]inositol and equilibration of the label in the mouse brain for 16-20 hr, a substantial portion of the labeled precursor was incorporated into PI and PC. Under these conditions, it is possible to relate FFA release to the increase in lysophospholipids at various times after decapitation. As shown in Figure 3, decapitation ischemia resulted in a steady increase in the levels of labeled LPC and LPI, reaching a 2- to 3-fold increase by 30 min. The steady increase in the level of labeled lysophospholipids correlated well with the increase in FFA during the second phase of the ischemic insult. These results are in agreement with the results from previous studies suggesting the involvement of phospholipase A$_2$ (Edgar et al., 1982; Sun and Foudin, 1984). Furthermore, the results are in agreement with the notion

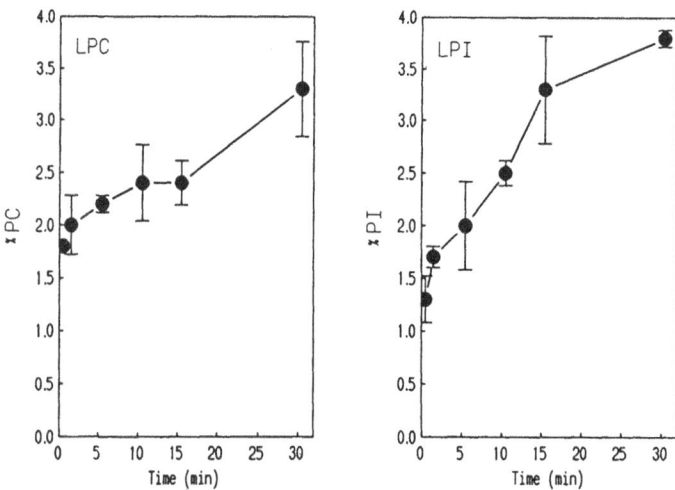

Figure 3. Increase in levels of labeled LPC and LPI in mouse cerebrum with respect to time after decapitation (30 sec to 30 min).

that both PC and PI can contribute to phospholipase A_2-mediated FFA release (Marion and Wolfe, 1979). Since PC constitutes approximately 40% of total phospholipids in brain (Sun, 1988), hydrolysis of a small pool of PC is probably enough to account for the observed increase in FFA. On the other hand, although PI is a minor phospholipid, constituting only 4% of the total phospholipids in brain (Sun, 1988), hydrolysis of this phospholipid would result in the preferential release of 18:0 and 20:4. The rapid depletion of ATP during decapitation further prevented the possibility of activation of the fatty acids for reacylation. However, because lysophospholipases are present in brain tissue to further degrade the lysophospholipids (Sun et al., 1987), the amount of lysophospholipids formed may not reflect the true amount of phospholipase A_2 activity.

Temporal Relationship Between 20:4 Released and Increase in DG and Inositol Phosphates

In the second experiment, the decapitation-induced release of 20:4 and DG in mouse brain was evaluated together with the appearance of inositol tri-, bis- and monophosphates. Since the level of $Ins(1,4,5)P_3$ can be more sensitively determined using the radioreceptor binding assay procedure, the HPLC (ion chromatography) procedure was confined to analysis of isomers of inositol mono- and bisphosphates, namely $Ins(1,4)P_2$, $Ins(1)P$, and $Ins(4)P$. Mass measurement of the $Ins(1,4,5)P_3$ in mouse brain indicated a rapid and transient increase in $Ins(1,4,5)P_3$, which reached a peak 30 sec after decapitation and declined precipitously to below the 5 sec level by 2 min (Fig. 4). After a slight delay (15 sec), the level of $Ins(1,4)P_2$ increased rapidly, reached a peak at 1 min, and then decreased to basal level by 5 min. $Ins(4)P$ showed a delay of 30 sec, reached a peak at 2 min, and declined to basal level by 5 min. As observed previously in rat brain (Lin et al., 1990), decapitation did not result in an appreciable increase in $Ins(1)P$ (Fig. 4). This observation supports the hypothesis that $Ins(1,4,5)P_3$ released from $PtdIns(4,5)P_2$ is either converted to $Ins(1,3,4,5)P_4$ by the 3-kinase or hydrolyzed to $Ins(1,4)P_2$ by the 5-phosphatase. Under these conditions, hydrolysis of $Ins(1,4,5)P_3$ through the 5-phosphatase route would result in the synthesis of $Ins(1,4)P_2$ and $Ins(4)P$, whereas operation of the 3-kinase route may result in synthesis of $Ins(1)P$ (Berridge and Irvine, 1989). In decapitation ischemia, however, operation of the 3-kinase route is suppressed due to ATP depletion, leaving the phosphatase route as the major mechanism for removal of $Ins(1,4,5)P_3$.

When DG and FFA in brain were analyzed at the same time intervals as the inositol phosphates, the increase in DG was found to be less rapid than that of $Ins(1,4,5)P_3$, reaching an apparent peak at 1 min (data not shown). Unlike the release of $Ins(1,4,5)P_3$ which was transient, DG levels remained elevated after the increase. During this period, there was a parallel increase in 18:0 and 20:4 after an apparent lag time of approximately 30 sec (Fig. 5). Both fatty acids reached a plateau after 2 min.

Effect of Pilocarpine and Atropine on 20:4, DG, and IP in Brain

With the mouse brain prelabeled with [^{3}H]inositol, lithium treatment (8 meq/kg body wt for 4 hr) resulted in a substantial increase in the level of labeled inositol monophosphates (IP_1), indicating effective inhibition of IP_1-phosphatase (Sherman et al., 1985). Consequently, the lithium-induced increase in labeled IP_1 can be used to assess the agonist-mediated poly-PI turnover in brain (Sun et al., 1992). As shown in Figure 6, injection of pilocarpine in the lithium-treated group resulted in a further increase in labeled IP_1 over that seen with lithium alone; this effect could be blocked by prior administration of atropine. Analysis of the IP_1 isomers in these brain samples

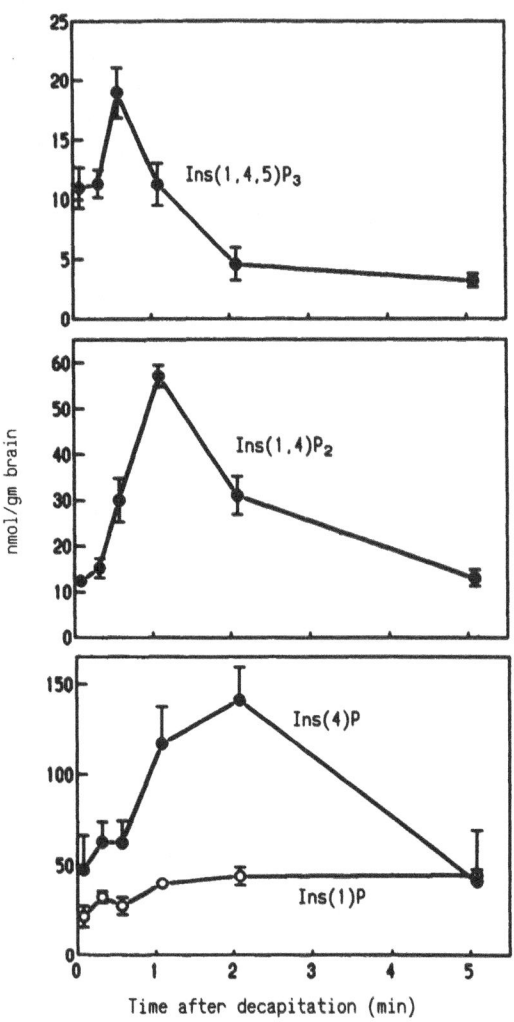

Figure 4. Levels of Ins(1,4,5)P_3, Ins(1,4)P_2, Ins(4)P, and Ins(1)P in mouse cerebrum with respect to time after decapitation (5 sec to 5 min).

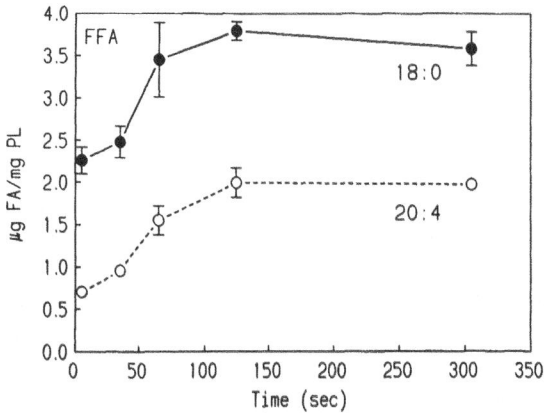

Figure 5. Levels of 18:0 and 20:4 in FFA fraction of mouse cerebrum with respect to time after decapitation (5 sec to 5 min).

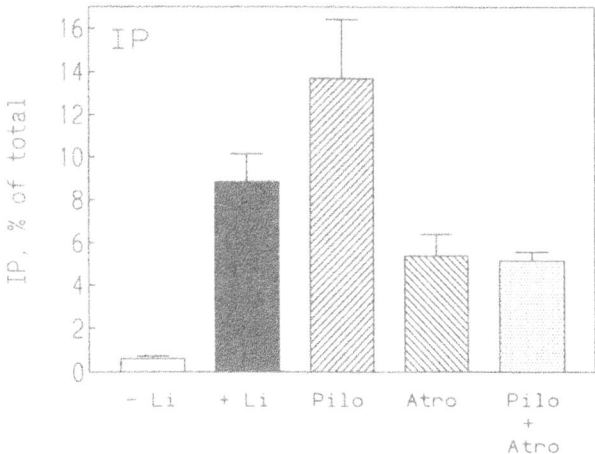

Figure 6. Levels of labeled inositol monophosphates (IP) in [^{3}H]inositol-labeled mouse cerebrum according to protocol #3 described in Methods: -Li, no lithium treatment; +Li, i.p. injection of lithium (8 meq/kg body wt) 4 hr prior to decapitation; Pilo, subcutaneous injection of pilocarpine (30 mg/kg body wt) in lithium-treated mice; Atro, subcutaneous injection of atropine (100 mg/kg body wt) in lithium-treated mice; Pilo + Atro, injection of atropine and then pilocarpine in lithium-treated mice.

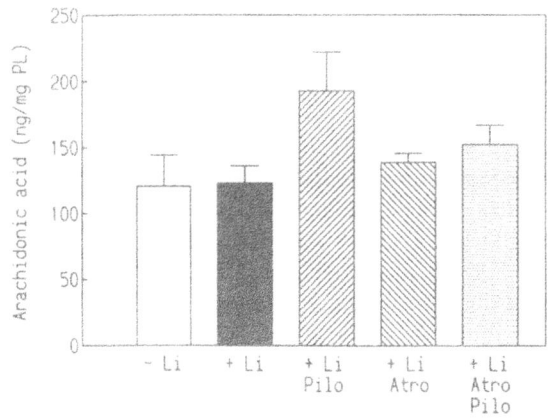

Figure 7. Levels of 20:4 in mouse cerebrum under the conditions described in Figure 6.

by ion chromatography indicated increases in both Ins(1)P and Ins(4)P after lithium administration, as well as a further increase in the levels of both isomers after administration of pilocarpine (data not shown).

When brain 20:4 was analyzed under the same treatment conditions involving injections of lithium, pilocarpine, and/or atropine, a small increase in the level of 20:4 was observed in the group treated with lithium and then pilocarpine as compared to the lithium control, and atropine was able to block this increase (Fig. 7). Failure to show a large increase in the level of 20:4 was probably due to the active recycling of this fatty acid in the intact brain, since there is sufficient supply of ATP to mediate the reacylation process (Sun and MacQuarrie, 1989).

CONCLUSIONS

1. Decapitation induced a biphasic increase in 20:4 in mouse cerebrum; the early phase was attributed to release resulting from the breakdown of poly-PI and the later phase to hydrolysis of phospholipids by phospholipase A_2.

2. The fatty acids of DG are enriched in 18:0 and 20:4. Decapitation resulted in a rapid increase in the level of DG in brain and this level remained elevated for at least 30 min.

3. Decapitation ischemia also induced transient increases in the levels of Ins(1,4,5)P_3, Ins(1,4)P_2, and Ins(4)P, peaking at 30 sec, 1 min, and 2 min, respectively. The time course of these events agreed with the notion that Ins(1,4,5)P_3 released from PtdIns(4,5)P_2 is degraded mainly through the 5-phosphatase pathway. During this time, the increase in 20:4 lagged behind that of DG.

4. An experimental protocol was developed to assess agonist-induced poly-PI turnover in brain. Pilocarpine induced an increase in poly-PI turnover and a small but significant increase in the level of 20:4. The small increase in 20:4 was attributed to the presence of an active mechanism for recycling FFA in the intact brain.

REFERENCES

Abe K, Kogure K, Yamamoto H, Imazawa M, Miyamoto K (1987) Mechanisms of arachidonic acid liberation during ischemia in gerbil cerebral cortex. J Neurochem 48:503-509.

Aveldaño MI and Bazan NG (1975) Rapid production of diacylglycerols enriched in arachidonate and stearate during early brain ischemia. J Neurochem 25:919-920.

Bazan NG (1970) Effects of ischemia and electroconvulsive shock on free fatty acid pool in the brain. Biochim Biophys Acta 218:1-10.

Bazan NG, De Bazan HEP, Kennedy WG, Joel CD (1971) Regional distribution and rate of production of free fatty acids in rat brain. J Neurochem 18:1387-1393.

Berridge MJ (1987) Inositol triphosphate and diacylglycerol: Two interacting second messengers. Annu Rev Biochem 56:156-193.

Berridge MJ and Irvine RF (1989) Inositol phosphates and cell signalling. Nature 341:197-205.

Berridge MJ, Dawson RMC, Downes CP, Heslop JP, Irvine RF (1983) Changes in the levels of inositol phosphates after agonist-dependent hydrolysis of membrane phosphoinositides. Biochem J 212:473-482.

Bredt DS, Mourey RJ, Snyder SH (1989) A simple, sensitive, and specific radioreceptor assay for inositol 1,4,5-triphosphate in biological tissues. Biochem Biophys Res Commun 159:976-982.

Edgar AD, Strosznajder J, Horrocks LA (1982) Activation of ethanolamine phospholipase A_2 in brain during ischemia. J Neurochem 39:1111-1116.

Farooqui AA, Anderson DK, Flynn C, Bradel E, Means ED, Horrocks LA (1990) Stimulation of mono- and diacylglycerol lipase activities by bradykinin in neural cultures. Biochem Biophys Res Commun 166: 1001-1009.

Farooqui AA, Pendley CE, Taylor WA, Horrocks LA (1985) In: Phospholipids in the nervous system, Vol II: Physiological role (Horrocks LA, Kanfer JN, Porcellati G, eds), pp 179-192. New York: Raven Press.

Huang S F-L and Sun GY (1986) Cerebral ischemia induced quantitative changes in rat brain membrane lipids involved in phosphoinositide metabolism. Neurochem Int 9:185-190.

Ikeda L, Yoshida S, Busto R, Santiso L, Ginsberg MD (1986) Polyphosphoinositides as a probable source of brain free fatty acids accumulated at the onset of ischemia. J Neurochem 47:123-132.

Kelleher JA and Sun GY, (1985) Enzymic hydrolysis of arachidonoyl-phospholipids by rat brain synaptosomes. Neurochem Int 7:825-831.

Lin TN, Liu TH, Su Y, Hsu CY, Sun GY (1991) Brain poly-phosphoinositide metabolism during focal ischemia in rat cortex. Stroke 22:495-498.

Lin TN, Sun GY, Premkumar N, MacQuarrie RA, Carter SR (1990) Decapitation-induced changes in inositol phosphates in rat brain. Biochem Biophys Res Commun 167:1294-1301.

Lowry OH, Passonneau JV, Hasselberger FY, Schulz DW (1964) Effect of ischemia on known substrates and cofactors of the glycolytic pathway in brain. J Biol Chem 239:18-30.

Marion J and Wolfe LS (1979) Origin of the arachidonic acid released post-mortem in rat forebrain. Biochim Biophys Acta 574:25-32.

Novak TS, Fried RL, Lust WD, Passonneau JV (1985) Changes in brain energy metabolism and protein synthesis following transient bilateral ischemia in the gerbil. J Neurochem 44:487-494.

Shearman MS, Shinomura T, Odd T, Nishizuka Y (1991) Protein kinase C subspecies in adult rat hippocampal synaptosomes. FEBS Lett 279:261-264.

Sherman WR, Munsell LY, Gish BG, Honchar MP (1985) The effect of systemic administration of lithium on phosphoinositide metabolism in rat brain, kidney and testis. J Neurochem 44:798-807.

Shimizu T and Wolfe LS (1990) Arachidonic acid cascade and signal transduction. J Neurochem 55:1-15.

Sun GY (1988) Preparation and analysis of acyl and alkenyl groups of glycerophospholipids from brain subcellular membranes. In: Neuromethods: Lipids and related compounds (Boulton AA, Baker GB, Horrocks LA, eds) vol 7, pp 63-82. Totowa NJ: Humana Press.

Sun GY (1989) Phospholipid metabolism in response to cerebral ischemia. In: Phospholipids in the nervous system: Biochemical and molecular pathology (Bazan NG, Horrocks LA, Toffano G, eds), Fidia Research Series, vol. 17, pp 133-149. Padova: Liviana Press.

Sun GY (1990) Mechanisms for ischemia-induced release of free fatty acids in brain. In: Cerebral ischemia and resuscitation (Schurr A, Regor BM, eds), pp 123-136. Boca Raton, FL: CRC Press.

Sun GY (in press) Cerebral ischemia and poly-phosphoinositide metabolism. In: Neurochemical correlates of cerebral ischemia (Bazan NG, Braquet P, Ginsberg ME, eds) New York: Pergamon Press.

Sun GY and Foudin LL (1984) On the status of lysolecithin in rat cerebral cortex during ischemia. J Neurochem 43:1081-1086.

Sun GY and Huang F-L (1987) Labeling of phosphoinositides in rat brain membranes: an assessment of changes due to post-decapitative ischemic treatment. Neurochem Int 10:361-369.

Sun GY and MacQuarrie RA (1989) Deacylation-reacylation of arachidonoyl groups in cerebral phospholipids. In: Arachidonic acid metabolism in the nervous system: Physiology and pathological significance (Barkah AI, Bazan NG, eds), Ann NY Acad Sci 559:282-295.

Sun GY, Huang F-L, Chandrasekhar R (1988) Turnover of inositol phosphates in brain during ischemia-induced breakdown of polyphosphoinositides. Neurochem Int 13:63-68.

Sun GY, Lin TN, Premkumar N, Carter SR, MacQuarrie RA (1990a) Separation and quantitation of isomers of inositolphosphates by ion chromatography. In: Methods in inositide research (Irvine R, ed), pp 135-144. London: Raven Press.

Sun GY, Lu FL, Lin SE, Ko MR (in press) Decapitation ischemia-induced release of free fatty acids in mouse brain: Relationship with diacylglycerols and lysophospholipids. Mol Chem Neuropathol.

Sun GY, Navidi M, Yoa FG, Lin TN, Orth OD, Stubbs EB, MacQuarrie RA (1992) Lithium effects on inositol phospholipids and inositol phosphates: Evaluation of an *in vivo* model for assessing polyphosphoinositide in brain. J Neurochem 58:290-297.

Sun GY, Tang W, Huang S F-L, MacQuarrie RA (1987) Lysophospholipase activity in rat brain subcellular fractions. Neurochem Res 12:451-458.

Sun GY, Yoa F-F, Lin T-N (1990b) Degradation of poly-phosphoinositides in brain subcellular membranes in response to decapitation insult. Neurochem Int 174:529-535.

Tang W and Sun GY (1982) Factors affecting the free fatty acids in rat brain cortex. Neurochem Int 4:269-273.

Yasuda H, Kishiro K, Izumi N, Nakamishi M (1985) Biphasic liberation of arachidonic and stearic acid during cerebral ischemia. J Neurochem 45:168-172.

Yoshida S, Ikeda M, Busto R, Santiso M, Martinez E, Ginsberg MD (1986) Cerebral phosphoinositide, triacylglycerol, and energy metabolism in reversible ischemia: origin and fate of free fatty acids. J Neurochem 47:744-757.

MODULATION OF ARACHIDONIC ACID METABOLISM IN CULTURED RAT

ASTROGLIAL CELLS BY LONG-CHAIN N-3 FATTY ACIDS

C. Galli, F. Marangoni, and A. Petroni

Institute of Pharmacological Sciences
University of Milan
Via Balzaretti, 9
20133, Milan, Italy

INTRODUCTION

Glial cells, in addition to providing mechanical and metabolic support to neurons, interact with neuronal function by binding and metabolizing several types of neuromediators. Glial cells are also involved in processes which are activated by brain injury, and participate in inflammatory events by producing and releasing molecules, such as the eicosanoids and other compounds derived from phospholipid hydrolysis, which are typical mediators of inflammation. Eicosanoids are generated in brain under various pathologic conditions, such as trauma, ischemia and convulsions, following the release of arachidonic acid (AA) from cell phospholipids.

Among the products that are generated from phospholipids through activation of phospholipase A2, platelet-activating factor (PAF) is a potent mediator of inflammation and is synthesized and released from various types of cells. PAF is also produced by nervous tissue, e.g., by chick retina after stimulation with neurotransmitters (Bussolino et al., 1986) and by rat cerebellar granule cells in culture (Yue et al., 1990). PAF also accumulates in brain during pharmacologically induced convulsions (Kumar et al., 1988), together with AA, diacylglycerol (DAG) and various eicosanoids.

Information on glial cell biochemistry and function has been largely obtained in the last two decades through *in vitro* studies on cell preparations. Primary cultures of astroglial cells are able to produce eicosanoids after stimulation but, despite the number of receptors that have been identified in these cells, the eicosanoid response is evoked by only a few types of stimuli. In fact the classic neurotransmitters that have been tested do not trigger the eicosanoid cascade in cultured glial cells, whereas potent stimulation is observed after incubation with the calcium ionophore A23187 and ATP (Gebicke-Haerter et al., 1988) or with substance P (Hartung et al., 1988).

Since it has been recently reported that PAF stimulates inositol phosphate production in cultured astrocytes (Murphy and Welk, 1990) and it is known that PAF stimulates AA metabolism in various cells, we proceeded to investigate the response of the eicosanoid system to PAF in astroglial cell cultures. The effect of PAF on AA metabolism in glial cells may be relevant in the activation of these cells after trauma or injuries resulting from circulatory events. In fact, as astroglial cells are closely associated with the cerebral vasculature and are highly responsive to trauma, they should be special targets of PAF action.

Studies on the activation of eicosanoids in various cells should take into account that the AA cascade is modulated by the relative levels of available substrate and by the levels of fatty acids which may interfere with AA metabolism. This is particularly relevant when cells grown *in vitro* are used, since they are strictly dependent upon the supply of polyunsaturated fatty acids in the incubation medium for the assembly of structural lipids.

The principal aim of our studies was to measure the formation of AA metabolites in cultured astroglial cells in comparison with the effects induced by the calcium ionophore A23187. An additional aspect of our investigation, which we believe to be of general relevance, was to evaluate the ability of glial cells to incorporate polyunsaturated fatty acids of the n-3 series, e.g., eicosapentaenoic acid (EPA) and docosahexaenoic acid (DHA), which we found to be selectively low in our preparations compared to in the brain, and to study the subsequent effects on the activation of the eicosanoid system.

This chapter presents results showing activation of the eicosanoid system in astroglial cells in culture by PAF and describes the modulation of the AA cascade in glial cells as a consequence of the accumulation of long-chain polyunsaturated fatty acids in cell lipids.

PAF AND EICOSANOID PRODUCTION IN ASTROGLIAL CELLS

The effects of PAF (1-0-octadecyl-2-acetyl-sn-glycero-3-phosphoryl choline) on the formation of eicosanoids in primary cultures of astroglial cells were compared with those induced by the calcium ionophore A23187. The cells, prepared from newborn rat cerebral hemispheres (> 95% astrocytes), were used at confluence (12-14 days). Thromboxane and prostacyclin produced after stimulation were measured as the stable metabolites TxB_2 and 6-keto-$PGF_{1\alpha}$ by enzyme immunoassay. The lipoxygenase (LO) products hydroxyeicosatetraenoic acids (HETE) were separated by HPLC and detected in UV. [^{3}H]-AA-labeled cells were also used and LO metabolites were separated by HPLC coupled with on-line radiodetection (Petroni et al., 1991).

In several experiments cells were incubated with PAF and A23187 at different concentrations and time periods of stimulation. At $1\mu M$ and after 15 min (Table 1) both agonists induced the accumulation of TxB_2 and 6-keto-$PGF_{1\alpha}$, A23187 being more potent than PAF (Petroni et al., 1991).

Table 1. Levels of TxB_2 and 6-keto-$PGF_{1\alpha}$ in astroglial cells 15 min after stimulation with PAF and A23187 (1 μM)

	TxB_2 (pg/mg P)	6-keto-$PGF_{1\alpha}$ (pg/mg P)
Non-stimulated	24±8	28±12
PAF	486±78	119±20
A23187	684±72	230±30*

Values are the average ± SEM of determinations carried out on six samples.
*Significantly greater (p<0.01) than values obtained with PAF stimulation.

The LO products (12- and 15-HETEs) were generated after A23187 stimulation of cells for periods of at least 30 min up to 2.5 hr, both when endogenous AA acted as precursor and also with the use of [³H]-AA-labeled cells. In contrast with A23187 stimulation, PAF stimulation did not result in detectable formation of HETEs from the endogenous AA pool, but when AA-labeled cells were used, radioactive peaks corresponding to 12-HETE and 15-HETE were produced. The discrepancy between the effects of A23187 and PAF on the formation of HETEs may be explained with the observation that PAF, in contrast to A23187, did not stimulate AA mobilization through the phospholipase A2 pathway, a condition which is essential for the release of adequate amounts of substrate for the LO activity (Petroni et al., 1991).

EFFECTS OF THE INCORPORATION OF 22:6n-3 ON EICOSANOID PRODUCTION IN GLIAL CELLS

Concentrations of long-chain polyunsaturated fatty acids (PUFA) in cells grown in culture are generally considerably lower than those measured in cells and tissues obtained *ex vivo*. This is particularly evident when cell lines are considered, but to some extent is observed also in primary cultures. The low levels of long-chain PUFAs in cultured cells are predictable on the basis of the relatively (and variably) low concentrations of the essential fatty acids linoleic and α-linolenic acid and/or of the long-chain products of the n-6 and n-3 series in the media used for growth. Conversion of the short-chain PUFAs, linoleic and α-linolenic, to the long chains through desaturation and elongation reactions may also be inadequate in cultured cells for optimal supply of these compounds to the cell membranes.

Some examples of the differences between cultured cells and the corresponding tissues concerning the relative levels of the long-chain PUFAs, AA of the n-6 series, and DHA of the n-3 series are shown in Figure 1. All cell preparations have considerably lower levels of AA and/or DHA than the corresponding tissues, but differences are observed among the various types of cell cultures. The hepatoma cells are depleted of both fatty acids, whereas the THP-1 cells are selectively low in 20:4 and the glial cells are low in 22:6 (DHA) but contain 20:4 in a concentration similar to that in brain tissue. DHA, together with AA, is the major long-chain PUFA in brain tissue

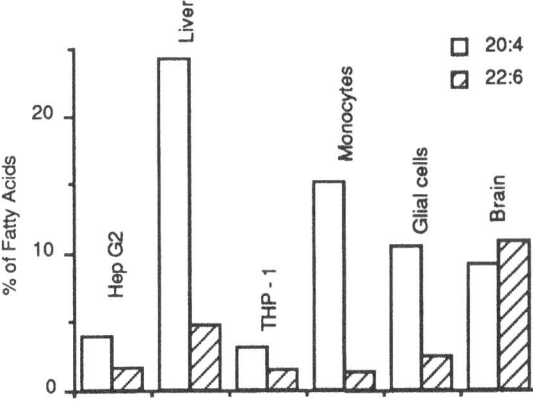

Figure 1. Levels of 20:4n-6 and 22:6n-3 in cultured cells and in tissues. Hep G2, hepatoma cell line; THP-1, monocytic cell line; glial cells, primary culture of astrocytes.

Table 2. Levels of n-3 fatty acids in glial cells incubated with EPA

	Control	EPA (µg/mg TL)	Δ (µg/mg TL)
20:5	0.86	1.32	0.46
22:5	1.18	2.60	1.42
22:6	7.80	9.40	1.60
Total			3.40

Conditions: EPA 10 µM dissolved as salt in the medium for 72 hr. TL, total lipids

and in specialized nervous structures, such as the retina. The low levels of DHA in our cell preparations prompted us to incubate the cultures in the presence of n-3 fatty acids, either 20:5n-3 (EPA), which is an intermediate in the conversion of α-linolenic to 22:6, 5 µM DHA, or 20 µM DHA. When cells were incubated with EPA (10 µM for 72 hr) the levels of the long-chain n-3 fatty acids measured by quantitative gas chromatography were appreciably elevated (Table 2). It appeared that products of EPA conversion accumulated more than EPA, with maximal increments for the elongation product 22:5n-3. DHA was also increased, suggesting that some desaturation of 22:5 to 22:6 occurred. Cells in culture have been generally reported to elongate more effectively than to desaturate polyunsaturated fatty acids (Garcia et al., 1990, Chapkin and Miller, 1990).

Incubation of cultured cells with DHA (5 or 20 µM) for periods up to 72 hr, either as an albumin complex or as free acid dissolved in fetal calf serum, enhanced the percentages of DHA in cell phospholipids from the initial low level of around 2.5%, up to 5-6%, still lower, however, than in brain tissue. From a quantitative point of view (Table 3), when cells were grown in the presence of 5 µM DHA the increase in DHA levels was much higher than when cells were supplemented with 20 µM DHA. Also, in both cases there was some increment of EPA, which was more pronounced with 20 µM DHA, suggesting retroconversion of 22:6 to 20:5.

Cells enriched with DHA produced significantly lower amounts of TxB_2 and 6-keto-$PGF_{1\alpha}$ than control cells after stimulation with the calcium ionophore A23187 (Table 4). In addition, when the same cells were labeled with [^{3}H]-AA before stimulation with the ionophore and the formation of radioactive LO products examined, it appeared that the pattern of HETEs was quite different in DHA-enriched cells, compared to control cells (Table 5). In fact, production of labeled 12-HETE was about 3-fold higher than

Table 3. Incorporation of DHA in glial cells

	Control (µg/mg TL)	DHA 5 µM	Δ	% change	Control (µg/mg TL)	DHA 20 µM	Δ	% change
20:5	1.32	1.58	0.26	+20	0.86	1.30	0.44	+51
22:6	4.45	12.30	7.85	+176	5.70	9.51	3.81	+66

Conditions : DHA 5 or 20 µM for 72 hr. TL, total lipids.

Table 4. Levels of TxB_2 and 6-keto-$PGF_{1\alpha}$ in preparations of control and DHA-enriched glial cells after stimulation with the calcium ionophore A23187

Eicosanoids	Control (pg/mg P)	+ DHA (pg/mg P)
TxB_2	918 ± 130	554 ± 31^a
6-keto-$PGF_{1\alpha}$	687 ± 80	475 ± 16^b

DHA enriched cells were obtained by growing the cells with 5 μM DHA for 72 hr. Then the medium was replaced and, after 1 hr, preparations were stimulated with 1 μM A23187 for 15 min. Values are the average $\pm$ SEM of determinations carried out on 6 samples. Values with a superscript are significantly different from controls at the following levels : a = $p<0.001$, b = $p< 0.01$ (Student's t test).

that of 15-HETE in control cells, whereas DHA-enriched cells produced more 15- than 12-HETE. No peak corresponding to the HETEs was observed in nonstimulated cells.

DISCUSSION AND CONCLUSIONS

Glial cells in culture represent a good model for studies of processes such as phospholipase activation and eicosanoid production. PAF, which may be released *in vivo* from cells in the vascular bed and/or from neurons and astrocytes, stimulates both polyphosphoinositide turnover (Murphy and Welk, 1990) and eicosanoid production. This appears as a reinforcing loop in cell activation, since initial release of PAF may contribute to reinforce the production of eicosanoids. Production of LO metabolites after PAF stimulation is less efficient since only the radioactive products generated from the small pool that was labeled during incubation with [^{3}H]-AA could be detected.

Table 5. Incorporation of radioactivity in 12- and 15-HETE in [^{3}H]-AA-labeled control and DHA-enriched cells after A23187 stimulation

	Control		DHA	
	(dpm/μg P)	% of cell radioactivity	(dpm/μg P)	% of cell radioactivity
12-HETE	105	7.54	29	3.47
15-HETE	35	2.55	134	15.75

Control and DHA-enriched cells (20 μM, 72 h) were labeled with [^{3}H]-AA for 6 hr, and stimulated with A23187 (1 μM, 2.5 hr). Values are the amounts of radioactivity in peaks separated by HPLC coupled with a radiodetector (Radiomatic Flow One Beta, Tampa, FL).

From our studies it appeared also that levels of long-chain n-3 PUFA are considerably lower in cultured glial cells than in brain tissue and that, when more physiologic levels of the 22:6 n-3 fatty acid, DHA, are restored, the stimulated production of eicosanoids is significantly modified from a quantitative and also from a qualitative point of view. This indicates that the AA cascade is modulated not only by the levels of the precursor in the cells, but also by long-chain n-3, such as DHA, which is present in high concentrations in the nervous system. In addition, our cell preparations appeared to be able to convert 20:5n-3 to 22:6n-3, and this may be an important function of glial cells for the supply of structurally and functionally relevant fatty acids to adjacent cells.

Generally speaking, since the fatty acid profiles of cells in culture differ from those in cells and tissue obtained *ex vivo*, the control of the fatty acid composition of these preparations becomes an important issue in studies of fatty acid metabolism and eicosanoid production in *in vitro* systems.

REFERENCES

Bussolino F, Gremo F, Tetta C, Pescarmona GP, Camussi G (1986) Production of platelet-activating factor by chick retina. J Biol Chem 261:16502-16508.

Chapkin RS and Miller CC (1990) Chain elongation of eicosapentaenoic acid in the macrophage. Biochim Biophys Acta 1042:265-267.

Garcia MC, Sprecher H, Rosenthal MD (1990) Chain elongation of polyunsaturated fatty acids by vascular endothelial cells: Studies with arachidonate analogues. Lipids 25:211-215.

Gebicke-Haerter PJ, Wurster S, Schobert A, Hertting G (1988) P2-purinoceptor induced prostaglandin synthesis in primary rat astrocyte cultures. Naunyn Schmiedebergs Arch Pharmacol 338:704-707.

Hartung HP, Heininger K, Schafer B, Toyka KV (1988) Substance P stimulates release of arachidonic acid cycloxygenation products from primary culture rat astrocytes. Ann NY Acad Sci 540:427-429.

Kumar R, Harvey SAK, Kester M, Hanahan D, Olson MS (1988) Production and effects of platelet-activating factor in rat brain. Biochim Biophys Acta 963:375-383.

Murphy S and Welk G (1990) Hydrolysis of polyphosphoinositides in astrocytes by platelet-activating factor. Eur J Pharmacol-Mol Pharmacol Section, 188:399-401.

Petroni A, Blasevich M, Visioli F, Zancocchia B, Caruso D, Galli C (1991) Arachidonic acid cyclo and lipoxygenase pathways are activated by platelet activating factor in a primary culture of astroglial cells. Brain Res (in press).

Yue TL, Lysko PG, Feuerstein G (1990) Production of platelet-activating factor from rat cerebellar granule cells in culture. J Neurochem 54:1809-1811.

MODULATION OF GLUTAMATE RELEASE FROM HIPPOCAMPAL MOSSY

FIBER NERVE ENDINGS BY ARACHIDONIC ACID AND EICOSANOIDS

R.V. Dorman, T.F.R. Hamm,
D.S. Damron, and E.J. Freeman

Department of Biological Sciences
Kent State University
Kent, OH 44242, USA

Arachidonic acid has been implicated in normal synaptic transmission processes, including those related to the development of hippocampal long-term synaptic potentiation. Hippocampal mossy fiber (MF) synaptosomes were used to investigate the role of arachidonate in the evoked accumulation of presynaptic Ca^{2+} and the release of endogenous glutamate, since these nerve terminals express long-term potentiation and selectively release glutamate as the excitatory transmitter. It was demonstrated that membrane depolarization evoked the accumulation of Ca^{2+}, the release of glutamate, and the production of unesterified arachidonic acid. These events may be functionally related, since exogenous arachidonate and phospholipase A_2 activation mimicked the effects of depolarization on Ca^{2+} availability and glutamate release, while secretion processes were attenuated in the presence of phospholipase A_2 inhibitors. In addition, pretreatment of the nerve terminals with arachidonate or melittin allowed for the facilitated release of glutamate in response to a subsequent depolarizing stimulus. Inhibition of cyclooxygenase or lipoxygenase activities also potentiated presynaptic responses to membrane depolarization. In contrast, 12-lipoxygenase products attenuated the depolarization-evoked accumulation of intraterminal free Ca^{2+} and glutamate release. It is suggested that arachidonic acid acts as a positive modulator of mossy fiber secretion processes, including those involved in the increased glutamate release required for the induction of long-term potentiation, while 12-lipoxygenase metabolites provide negative feedback signals designed to limit neurotransmitter secretion.

INTRODUCTION

Arachidonic acid (AA) has been implicated in stimulus-secretion coupling in a variety of cell types and there is a growing body of evidence that it modulates the evoked release of neurotransmitters. AA has been shown to stimulate the release of transmitters from a number of synaptosomal and brain slice preparations. In addition, the increased availability of unesterified AA has been correlated with the induction of long-term synaptic potentiation (LTP; Lynch et al., 1989), the onset of seizure activity (Birkle and Bazan, 1987) and the development of ischemic brain damage (Bazan, 1970;

Dorman et al., 1990). The common denominator for these diverse observations may be the ability of AA to evoke and enhance the release of L-glutamate, since glutamate has been implicated in the normal processes involved in LTP (Staubli et al., 1990) and the pathology associated with seizures (Neuman et al., 1989) and stroke (Buchan, 1990). Therefore, characterizing the involvement of AA in the modulation of glutamate, secretion should provide valuable insights into learning and memory processes, as well as the molecular mechanisms of glutamate-dependent neurotoxicity.

Depolarization of isolated nerve terminals stimulates neurotransmitter efflux (Bradford and Marinetti, 1983), phospholipid catabolism (Lazarewicz et al., 1983), and arachidonic acid accumulation (Asakura and Matsuda, 1984; Dorman et al., 1986). Unesterified AA and neurotransmitter secretion appear to be functionally related, since AA evokes the release of amino acid neurotransmitters (Rhoads et al., 1983), γ-aminobutyric acid (Asakura and Matsuda, 1984), and catecholamines (Frye and Holz, 1984) from nerve terminals. It also stimulates the release of glutamate from cortical (Chan et al., 1983) and hippocampal slices (Williams et al., 1989) and hippocampal mossy fiber synaptosomes (Freeman et al., 1990). These direct effects of AA on transmitter secretion argue in favor of a presynaptic site of action.

Unesterified AA may serve to facilitate the release of glutamate and ensure the induction of hippocampal LTP (Lynch, 1989). Consistent with this proposal, exogenous AA decreases the intensity of stimulus parameters required to induce LTP (Williams et al., 1989). However, the facilitating effects of AA may depend on its oxidation via the lipoxygenase pathway, since the lipoxygenase inhibitor nordihydroguaiaretic acid (NDGA) blocks the induction, but not the maintenance of LTP (Williams and Bliss, 1988). NDGA also attenuates the AA-dependent potentiation of excitatory transmission in the hippocampal CA1 subfield (Drapeau et al., 1990). Furthermore, 12-hydroxy-eicosatetraenoic acid (12-HETE) facilitates the depolarization-induced release of [^{3}H]glutamate from synaptosomes isolated from rat dentate gyrus (Lynch and Voss, 1990). However, the proposal that 12-lipoxygenase products potentiate presynaptic secretion events is not consistent with their established effects at the *Aplysia* sensory neuron or the effects we have observed at the rat hippocampal mossy fiber nerve terminal.

The presynaptic, inhibitory effect of FMRFamide at the *Aplysia* sensory neuron depends on the synthesis of 12-lipoxygenase products (Piomelli et al., 1987; 1988). This inhibition appears to be mediated by opening S-type K^+ channels, which reduces transmitter secretion by hyperpolarizing the presynaptic plasma membrane (Belardetti et al., 1989; Volterra and Siegelbaum, 1989). These effects are consistent with the prediction that K^+ channel conductances modulate neurotransmitter release from nerve terminals (Zoltay and Cooper, 1990). That the inhibitory influences of 12-lipoxygenase products may depend on opening presynaptic K^+ channels in mammalian nerve terminals is indicated by two observations. First, blocking K^+ channels with amino-pyridines, which depolarize synaptic plasma membranes, stimulates the release of glutamate from hippocampal mossy fiber synaptosomes (Freeman et al., 1990). Second, products of the 12-lipoxygenase pathway attenuate the accumulation of intraterminal free Ca^{2+} ($[Ca^{2+}]_i$) and the glutamate release evoked by depolarization of mossy fiber nerve endings (Freeman et al., 1991). These results are consistent with those described in *Aplysia*, but not those reported for rat dentate gyrus. Therefore, we attempted to further characterize the roles of arachidonic acid and its lipoxygenase products in the presynaptic modulation of neurotransmitter release at a central mammalian synapse.

Long-term potentiation is described as a lasting facilitation of synaptic efficacy and has been related to associative learning mechanisms. In particular, the activity-dependent induction of LTP in the hippocampus is an accepted correlate of learning and memory in mammals. As with other excitatory, monosynaptic pathways in the hippocampus, LTP can be induced at the MF-CA3 synapse with repetitive stimulation of the afferents (Barrionuevo et al., 1986) and requires the enhanced release of glutamate (Yamamoto, 1987). However, the induction of LTP at the MF-CA3 synapse does not involve glutamate-dependent activation of N-methyl-D-aspartate (NMDA) receptors (Harris and Cotman, 1986) or an increase in postsynaptic Ca^{2+} (Staubli et al., 1990; Zalutsky and Nicoll, 1990). Thus, the facilitation of the MF-CA3 synapse depends on presynaptic alterations that result in an enhanced release of glutamate upon depolarization of the MF nerve endings.

The presynaptic components of the MF-CA3 synapse can be isolated from hippocampal tissue and used for biochemical studies. The characterization of the hippocampal mossy fiber synaptosomal (HMFS) preparation has been described in detail (Terrian et al., 1988). The metabolic competence of these synaptosomes is established. They retain respiratory control, an active Na^+-K^+-ATPase system, and occluded lactate dehydrogenase (LDH) activity. They respond to membrane depolarization with the evoked accumulation of $[Ca^{2+}]_i$ and the release of endogenous glutamate and dynorphin peptides (Terrian et al., 1988; 1990a,b). In addition, glutamate is the excitatory transmitter that is selectively released from HMFS. The inability of membrane depolarization to induce the release of either γ-aminobutyric acid or glycine indicates that smaller, inhibitory nerve terminals are not present in the preparation (Terrian et al., 1990a). The exclusive localization of dynorphin peptides to the mossy fibers within the hippocampus (Hoffman and Zamir, 1984) and their release from HMFS reaffirm the enrichment of mossy fibers in this synaptosomal preparation. Therefore, HMFS can be used to examine the role of AA metabolism in the presynaptic modulation of glutamate release from a population of nerve endings known to express LTP.

METHODS

Hippocampal Mossy Fiber Synaptosomes

Isolated mossy fiber nerve endings are routinely prepared from 14 to 18 pairs of hippocampi dissected from male Sprague-Dawley rats. The procedures for isolation of the HMFS have been described (Terrian et al., 1988). The protein content of each preparation is determined according to Lowry et al. (1951). HMFS are resuspended in a calcium-free, oxygenated Krebs-bicarbonate medium to the appropriate concentration of synaptosomal protein, followed by preincubation for 5 min at 30°C. Calcium ions are omitted during the preincubation to allow the synaptosomal plasma membranes to repolarize and establish control of cytosolic calcium levels (Scott et al., 1980). The HMFS are then resuspended in Krebs-Ringer buffer containing 1.0 mM $CaCl_2$. Stirred incubations are used for investigations on lipid metabolism and Ca^{2+} availability. Studies on glutamate release are performed under superfusion conditions (Dorman et al., 1986; Freeman et al., 1990; 1991), in order to eliminate the confounding effects of high affinity reuptake systems.

Measurement of Endogenous Glutamate Release

HMFS are loaded onto glass fiber filters located at the bottom of superfusion chambers maintained at 30°C. The terminals are superfused with oxygenated Krebs-Ringer buffer containing the various treatments. Flow rates are maintained at 0.5 ml/min and fractions are collected from parallel chambers at 4 min intervals. The fractions are frozen at -70°C and concentrated by lyophilization. They are then resuspended in buffer and the glutamate contents are determined using an enzyme-coupled, microfluorometric assay developed by Graham and Aprison (1966). This method has proven to be specific, sensitive, rapid, and economical. Also, it provides estimates of L-glutamate release that are in close agreement with those obtained by HPLC analyses (Terrian et al., 1988; 1990a). The results are expressed as pmoles glutamate released/min/mg HMFS protein. Evoked release is determined by subtracting the amount of glutamate spontaneously released from HMFS superfused with control buffer in parallel chambers.

Quantification of Intraterminal Free Calcium

A fluorometric assay to quantify $[Ca^{2+}]_i$ in HMFS has been established in this laboratory (Damron and Dorman, 1989; Terrian et al., 1989; 1991; Freeman et al., 1991). HMFS (5 mg protein) are suspended in oxygenated (95% O_2, 5% CO_2) Krebs-bicarbonate buffer and incubated with 10 μM Fura-2/AM for 20 min at 37°C. Fura-2-loaded HMFS are washed twice with control buffer to remove excess Fura-2/AM and resuspended in control, Krebs-Ringer buffer containing 1 mM $CaCl_2$. The synaptosomes are then incubated for 15 min at 37°C to allow intraterminal esterases to deacylate the Fura-2/AM, since the free form of Fura-2 is the Ca^{2+} chelator with fluorescent properties. The Fura-2-loaded HMFS (250 μg protein) are transferred to a quartz cuvette in a spectrofluorometer. The synaptosomes are stirred and maintained at 30°C. Experimental treatments are added directly to the cuvette. $[Ca^{2+}]_i$ is estimated using the fluorescence ratio method, with excitations at 340 and 380 nm and emission at 505 nm (Grynkiewicz et al., 1985). Calibration of the dye is performed for each experiment using sodium dodecyl sulfate (0.1%; R_{max}) and EGTA (8 mM; R_{min}). The results are expressed in nM.

Radiolabeling of HMFS Lipids

HMFS are radiolabeled with [³H]arachidonate as previously described for cerebellar glomeruli (Dorman et al., 1986; 1988). The terminals (about 10 mg protein) are incubated at 37°C for 15 min in the presence of 0.32 M sucrose, 50 mM Tris, 1 mM $MgSO_4$, 3 mM dithiothreitol, 0.1 mM CoA, 2.5 mM ATP, and 4 μCi [³H]arachidonate, pH 7.4. The synaptosomes are washed three times with Krebs-Ringer buffer containing 1% fatty-acid-free bovine serum albumin, in order to remove unincorporated [³H]AA. The final washed pellets are then suspended in oxygenated Krebs-Ringer buffer prior to incubation at 37°C. Incubations are terminated and the total lipids extracted according to Folch et al. (1957). Nonpolar lipids are separated by thin-layer chromatography (Freeman and West, 1966) and counted. The distribution of label in each free fatty acid pool is determined and the results are expressed as dpm/mg HMFS protein.

RESULTS

The HMFS preparation provides a suitable model for examining the biochemical controls of glutamate release and relating them to the development of long-term potentiation. Two primary presynaptic events were monitored, in order to investigate the role of AA metabolism in the control of glutamate release. The events were i) the release of endogenous glutamate from superfused HMFS and ii) changes in the concentrations of intraterminal free Ca^{2+}. Manipulations of glutamate release and presynaptic Ca^{2+} accumulation were correlated with membrane polarity, phospholipase activities, the presence of unesterified AA, and the synthesis of eicosanoids.

Depolarization of HMFS with increased extracellular K^+ or veratridine stimulated the dose-dependent accumulation of $[Ca^{2+}]_i$ and the veratridine effect was blocked by

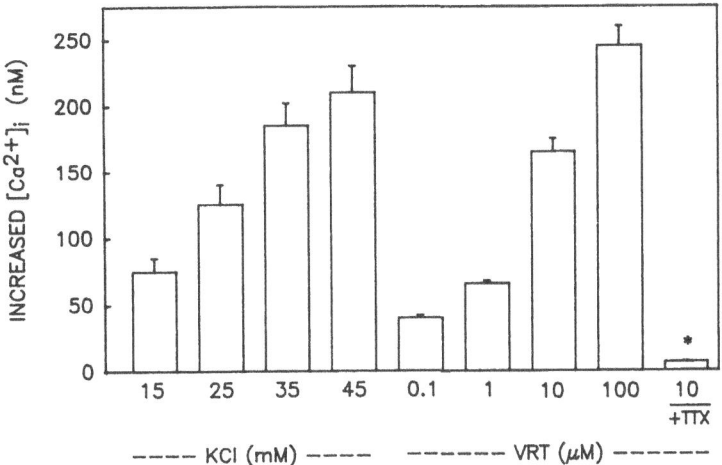

Figure 1. Effects of membrane depolarization on the accumulation of $[Ca^{2+}]_i$. HMFS were depolarized with 15-45 mM KCl or 0.1-100 μM veratridine (VRT). Ca^{2+} concentrations were determined as described. Tetrodotoxin (TTX; 1 μM) was applied in the presence of 10 μM VRT. Results are expressed as nM increases in $[Ca^{2+}]_i$ above resting levels ± SEM. * = different from 10 μM VRT alone; p < 0.05.

tetrodotoxin (TTX; Fig. 1). The concentration of intraterminal free Ca^{2+} increased 185 nM in the presence of 35 mM K^+, while 10 μM veratridine stimulated a 165 nM increase in $[Ca^{2+}]_i$, which was reduced to 6.6 nM in the presence of 1 μM TTX. The effects of TTX indicated a neuronal origin for the observed changes in Ca^{2+}. In addition, these changes appeared to be presynaptic alterations in $[Ca^{2+}]_i$, since postsynaptic receptor activation with glutamate or its agonists had no effect on Ca^{2+} concentrations (data not shown).

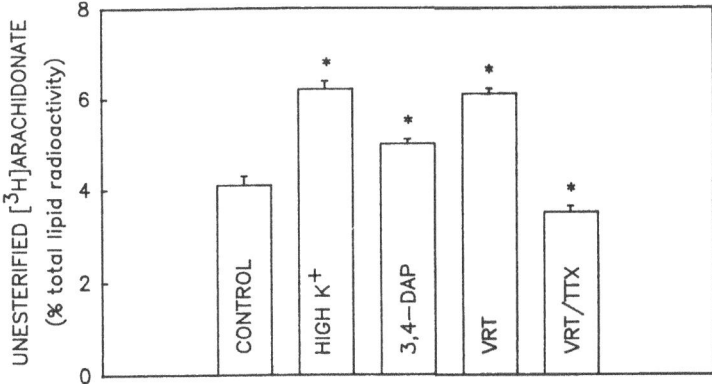

Figure 2. Effects of membrane depolarization on the accumulation of unesterified [³H]arachidonate. HMFS were radiolabeled as described prior to incubation for 10 min at 37°C. Control = Krebs-Ringer buffer; HIGH K^+ = + 45 mM KCl; 3,4-DAP = + 100 μM 3,4-diaminopyridine; VRT = + 10 μM veratridine; VRT/TTX = + 10 μM VRT plus 1 μM TTX. Results are expressed as % change in free [³H]arachidonate ± SEM. * = different from controls; p < 0.05.

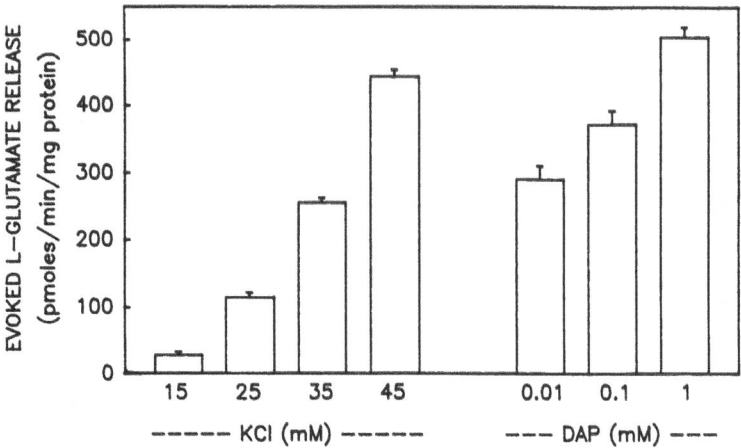

Figure 3. Depolarization-induced glutamate release from HMFS. Membranes were depolarized with 15-45 mM KCl or 0.01-1 mM 3,4-diaminopyridine (DAP). Results are expressed as pmoles glutamate released/min/mg HMFS protein ± SEM.

K^+-, veratridine-, and 3,4-diaminopyridine-dependent membrane depolarizations also stimulated the accumulation of unesterified AA; the veratridine effect was attenuated by TTX (Fig. 2). Raising the external K^+ concentration from 5 to 45 mM stimulated a 55% increase in the recovery of [^{3}H]AA in the free fatty acid pool of prelabeled HMFS. The K^+ channel blocker 3,4-diaminopyridine (0.1 mM) induced a 25% increase in the accumulation of [^{3}H]AA, while the Na^+ channel activator veratridine (10 μM) enhanced the recovery of unesterified AA by 53%. The presence of 1 μM TTX eliminated the veratridine-induced AA accumulation and even reduced the amount of recovered [^{3}H]AA below that observed in unstimulated controls.

K^+- and 3,4-diaminopyridine-dependent membrane depolarizations also stimulated the dose-dependent release of endogenous glutamate from the HMFS (Fig. 3; Freeman et al., 1990; 1991) and veratridine-evoked glutamate release was inhibited by TTX (Terrian et al., 1990a). The evoked release of glutamate was increased by 265 pmoles/min/mg protein in the presence of 35 mM KCl, while 0.01 mM 3,4-diaminopyridine stimulated the efflux of 286 pmoles/min/mg protein. In both cases, the release of endogenous glutamate returned to spontaneous levels upon removal of the stimuli. Therefore, the depolarization-evoked accumulation of unesterified AA correlated with presynaptic Ca^{2+} availability and the release of glutamate.

The phospholipase A_2 inhibitors, 4-bromophenacyl bromide (BPB), dibucaine, and luffariellolide, attenuated the K^+-evoked accumulation of intraterminal free Ca^{2+}. In contrast, phospholipase C inhibitors (compound 48/80; neomycin sulfate) and the diacylglycerol lipase inhibitor RHC80267 were ineffective (Fig. 4). The increased [Ca^{2+}]$_i$ induced by 45 mM KCl (215 nM) was inhibited 59% by 200 μM dibucaine, 53% by 200 μM BPB, and 49% by 200 μM luffariellolide. Compound 48/80, neomycin sulfate and RHC80267 had no effect on K^+-induced Ca^{2+} accumulation at any concentration tested, while compound 48/80 and RHC80267 also had no effect on the K^+-evoked release of glutamate (data not shown). Phospholipase A_2 inhibitors also reduced the K^+-evoked accumulation of unesterified [^{3}H]AA and the production of prostaglandins (data not shown). Unfortunately, these inhibitors interfered with the fluorometric glutamate assay, such that these experiments will be repeated with glutamate quantified by HPLC.

The effects of melittin on Ca^{2+} accumulation, AA availability, and glutamate release

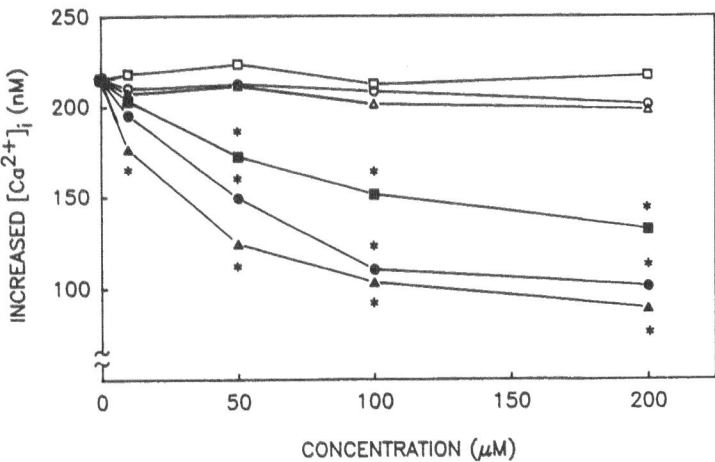

Figure 4. Effects of lipase inhibitors on K^+-evoked Ca^{2+} accumulation. HMFS were depolarized with 45 mM KCl in the presence of bromophenacyl bromide (filled circle), dibucaine (filled triangle), luffariellolide (filled square), neomycin sulfate (open circle), compound 48/80 (open triangle) or RHC80267 (open square) at the concentrations shown. $[Ca^{2+}]_i$ was determined as described. Results are expressed as nM ± SEM. * = different from untreated, stimulated controls; $p < 0.05$.

were determined, to further evaluate the role of phospholipase A_2 in secretion events. Low concentrations of this phospholipase A_2 activator (Conricode and Ochs, 1989) stimulated the accumulation of $[Ca^{2+}]_i$, as well as the release of glutamate (Fig. 5). The presence of 0.1 or 1 μM melittin-stimulated glutamate release of 135 or 280 pmoles/min/mg protein, respectively, while $[Ca^{2+}]_i$ was increased by 120 or 290 nM, respectively. In contrast, the presence of purified bee venom phospholipase A_2 at concentrations shown to contaminate 10 μM melittin had no significant effect on glutamate release or Ca^{2+} accumulation. The effects of melittin appeared not to be due to Ca^{2+} ionophore properties, since Ca^{2+} channel blockers inhibited 69% of the melittin-induced Ca^{2+} accumulation (Damron et al., 1990). Melittin effects were also not due to disruption of presynaptic plasma membranes, since the HMFS retained occluded LDH activity and

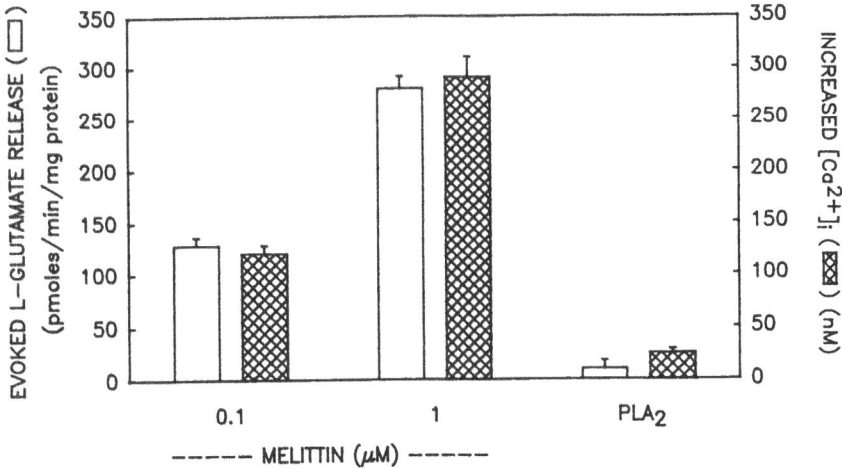

Figure 5. Effects of melittin and exogenous phospholipase A_2 (PLA_2) on glutamate release and Ca^{2+} accumulation. HMFS were treated with either melittin or purified bee venom phospholipase A_2 (0.6 units). Open bars, glutamate release; Hatched bars, $[Ca^{2+}]_i$.

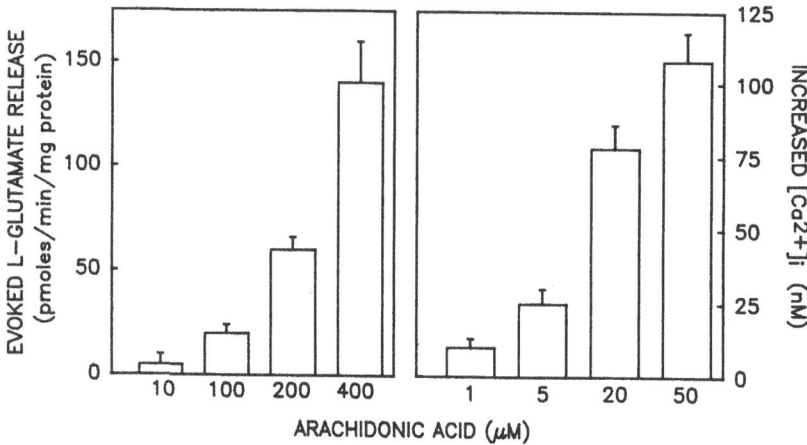

Figure 6. Arachidonic acid-stimulated glutamate release and Ca^{2+} accumulation. HMFS were treated with arachidonate at the concentrations shown. Effects on glutamate release are shown on the left; effects on $[Ca^{2+}]_i$ on the right.

glutamate release returned to spontaneous levels upon removal of the stimulus (Freeman et al., 1990). In addition, Melittin stimulated the accumulation of unesterified [^{3}H]AA in prelabeled HMFS and the mass of free AA as determined by gas-liquid chromatography (GLC; data not shown). It appears, therefore, that the effects of melittin were due to the activation of endogenous phospholipase A_2.

The effects of exogenous AA were consistent with those of melittin, since AA stimulated the dose-dependent release of glutamate (Freeman et al., 1990), as well as the accumulation of $[Ca^{2+}]_i$ (Fig. 6). The presence of 200 μM AA stimulated the efflux of 59 pmoles glutamate/min/mg protein, while $[Ca^{2+}]_i$ was increased 108 nM in the presence of 50 μM AA. The more potent effect of AA on Ca^{2+} accumulation was most likely due to differences in the availability of AA when HMFS are superfused versus stirred in a closed system. In fact, it was observed that about 3.5% of added AA gains access to HMFS membranes under superfusion conditions, such that the addition of 200 μM AA results in an actual concentration of 7 μM. Also, AA (200 μM) had no effect on occluded LDH activity and glutamate release returned to spontaneous levels when the AA was removed from the superfusion buffer (Freeman et al., 1990). The preservation of intact presynaptic membranes is also indicated by the ability of voltage-sensitive Ca^{2+} channel blockers to reduce the AA-evoked accumulation of $[Ca^{2+}]_i$ by 63% (Damron et al., 1990). Thus, exogenous AA and melittin did not perturb HMFS membranes and both phospholipase A_2 activation and exogenous AA mimicked the effects of depolarization on presynaptic release mechanisms, while phospholipase A_2 inhibition attenuated secretion processes.

The effects of other fatty acids on Ca^{2+} accumulation and glutamate release were determined to assess the specificity of the AA-induced responses (Fig. 7). AA was the most potent fatty acid at stimulating glutamate release from superfused HMFS (Freeman et al., 1990), as well as the accumulation of $[Ca^{2+}]_i$. Saturated fatty acids were ineffective, while linoleic acid and adrenic acid were less effective than AA. Although adrenate stimulated a substantial amount of glutamate release, GLC was used to show that HMFS did not accumulate significant amounts of linoleic, linolenic, or adrenic acids, while AA and palmitic acid accumulated in depolarized and melittin-treated synaptosomes (data not shown). It was suggested that other unsaturated fatty acids may partially mimic the effects of AA by interfering with the oxidation and reacylation of the AA (Freeman et al., 1990). Regardless, it appears that AA accumula-

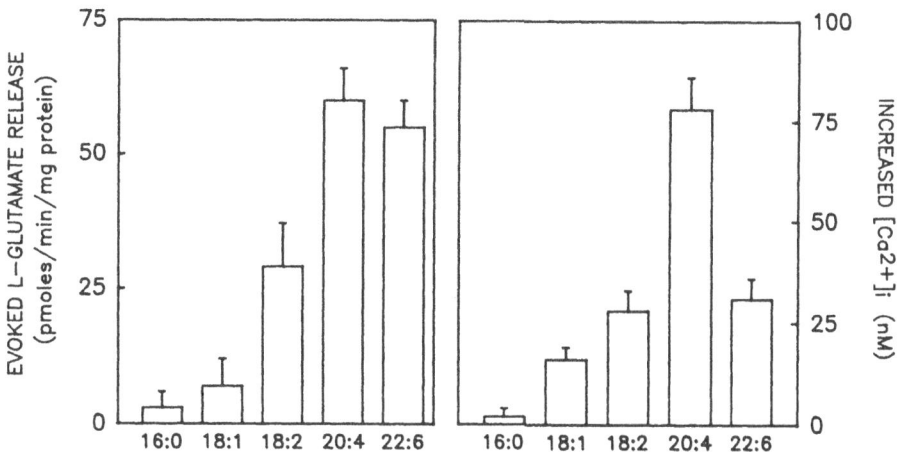

Figure 7. Effects of various fatty acids on glutamate release and $[Ca^{2+}]_i$. Glutamate release (left) was determined in the presence of 200 μM fatty acids, while Ca^{2+} accumulation (right) was assessed with 20 μM fatty acids: palmitate (16:0), oleate (18:1), linoleate (18:2), arachidonate (20:4), or adrenate (22:6).

tion plays an excitatory role in presynaptic secretion processes, in addition to acting as the precursor for a complex array of physiologically active derivatives.

A double-stimuli protocol was established, in order to investigate the involvement of AA in the facilitated release of glutamate that is associated with the development of LTP. Superfused HMFS were stimulated with AA followed by exposure to control buffer until glutamate release reached spontaneous levels. These terminals were then depolarized with 35 mM K^+. The K^+-induced glutamate release was increased following pretreatment with AA, when compared to HMFS that received only the K^+ stimulus (Freeman et al., 1990). The activation of phospholipase A_2 with melittin had similar effects, since a subsequent stimulation with K^+ also resulted in a significant enhancement of glutamate release. In fact, melittin was more potent at facilitating K^+-evoked transmitter release, because it activates endogenous phospholipase A_2, circumventing the limited accessibility of exogenous AA.

A triple-stimuli protocol was used to further compare the facilitated release of glutamate with that due to K^+-dependent depolarization. Superfused HMFS were exposed to 35 mM KCl three times, each time after glutamate release had returned to spontaneous levels. HMFS in parallel superfusion chambers were also exposed to K^+, followed by 1.0 μM melittin, and followed again by K^+. Glutamate release was diminished with the second and third exposures to KCl. Consistent with previous results, phospholipase A_2 activation allowed for a 37% increase in the K^+-evoked glutamate release (Fig. 8). This potentiation was the same as had been reported for the double-stimuli procedure (Freeman et al., 1990).

The effects of eicosanoids on Ca^{2+} accumulation and glutamate release, as well as the effects of cyclooxygenase and lipoxygenase inhibitors, were examined. Prostaglandins E_2 and $F_{2\alpha}$ had no direct effect on either Ca^{2+} concentrations or glutamate release, except for modest stimulations at high, non-physiologic concentrations ($\leq$ 100 μM). Also, these eicosanoids had no detectable effect on K^+-evoked responses, even at high concentrations (data not shown). In contrast, inhibition of cyclooxygenase activities with ibuprofen (10 μM) potentiated the K^+-induced release of glutamate by 35%, while the AA-evoked release of glutamate was increased 33% (Freeman et al., 1990). Ibuprofen (10 μM) had a similar effect on $[Ca^{2+}]_i$, since it increased the K^+-evoked accumulation of Ca^{2+} by 38%. Thus, prostaglandins had no direct effect on mossy fiber secretion

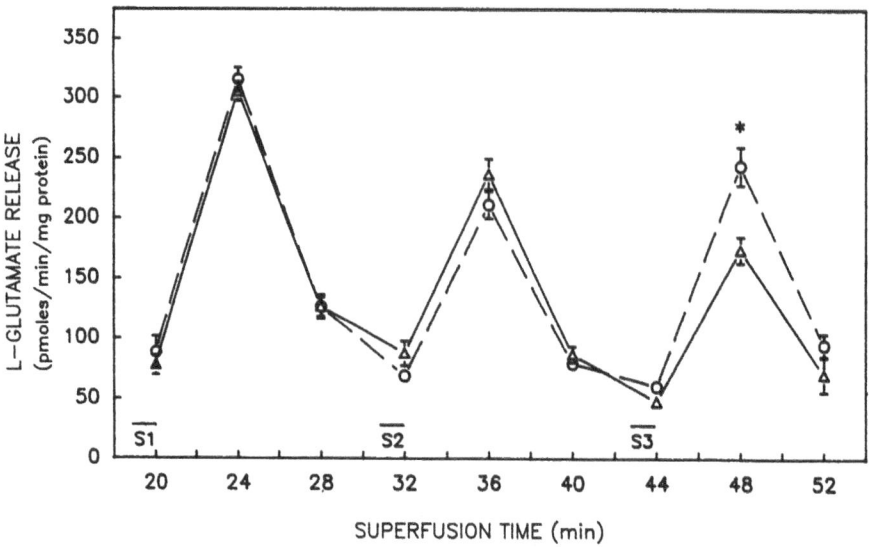

Figure 8. Facilitation of K^+-evoked glutamate release by melittin. Controls (open triangles, solid line) were depolarized with 35 mM KCl at the times indicated (S1,S2,S3), while experimentals (open circles, dashed line) received 1 μM melittin as S2. Results are shown as pmoles released/ min/mg protein $\pm$ SEM. * = different from stimulus controls; $p < 0.05$.

mechanisms and the potentiated release due to cyclooxygenase inhibition may depend on blocking a major pathway for the removal of unesterified AA.

Similarly, blocking AA removal by inhibiting lipoxygenase activities with low concentrations of NDGA (0.1-10 μM) potentiated the K^+-evoked release of glutamate and accumulation of Ca^{2+} (Fig. 9). The presence of 0.1, 1, or 10 μM NDGA facilitated the K^+-evoked glutamate release by 44, 66, or 83%, respectively, while Ca^{2+} accumulation

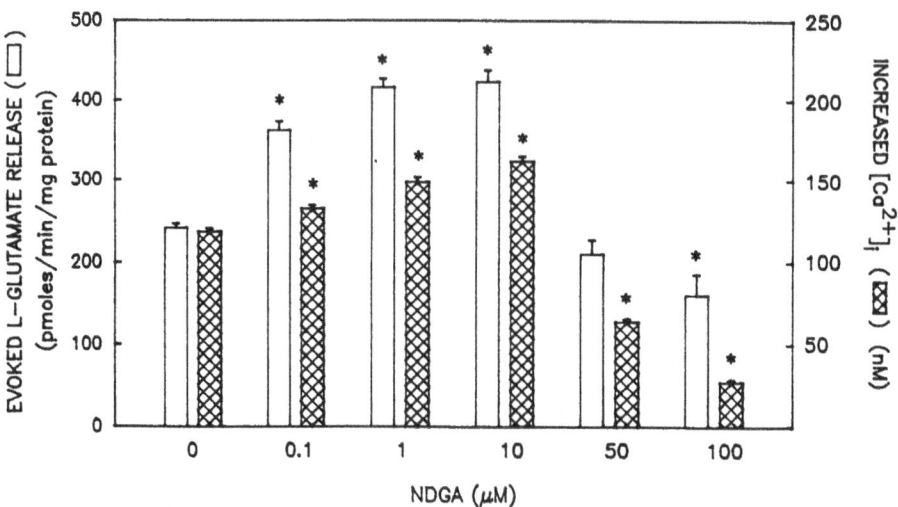

Figure 9. Effects of NDGA on K^+-evoked Ca^{2+} accumulation and glutamate release. HMFS were depolarized with KCl in the presence of NDGA at the concentrations shown. Effects on glutamate release are shown with open bars; $[Ca^{2+}]_i$ with hatched bars. * = different from untreated, stimulus controls; $p < 0.05$.

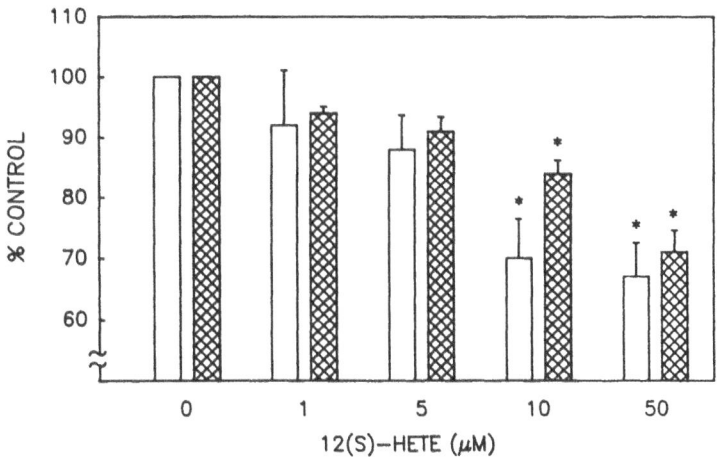

Figure 10. 12(S)-HETE inhibition of K^+-induced glutamate release and Ca^{2+} accumulation. HMFS were depolarized with KCl in the presence of 12(S)-HETE at the concentrations shown. Effects on glutamate release are shown with open bars; $[Ca^{2+}]_i$ with hatched bars. * = different from stimulus controls; $p < 0.05$.

was potentiated by 12, 25, or 36%, respectively. These concentrations of NDGA were shown to facilitate the AA-induced release of glutamate and to inhibit 12-HETE synthesis, but not the accumulation of unesterified [^{3}H]AA (data not shown). Paradoxically, high concentrations of NDGA (50-100 μM) attenuated the K^+-evoked responses, as well as the accumulation of AA. The K^+-induced release of glutamate was inhibited by 14 or 38% in the presence of 50 or 100 μM NDGA, respectively, while the evoked accumulation of $[Ca^{2+}]_i$ was attenuated by 44 or 76%, respectively (Fig. 9).

Observations that low concentrations of NDGA potentiated K^+-evoked Ca^{2+} accumulation and glutamate release were consistent with the ability of lipoxygenase products to attenuate K^+-induced responses in HMFS (Freeman et al., 1991). Although 5(S)- and 15(S)-HETE had no effect, 12(S)-HETE, 12(R)-HETE and 12(S)-HPETE reduced the K^+-evoked accumulation of Ca^{2+} and glutamate release (Freeman et al., 1991). Dose-response curves for the attenuating effects of 12(S)-HETE are shown in Figure 10. The presence of 50 μM 12(S)-HETE reduced the K^+-evoked release of glutamate by 33%, while the evoked accumulation of Ca^{2+} was attenuated by 28%.

DISCUSSION

A hippocampal mossy fiber synaptosomal preparation was used to relate arachidonic acid metabolism to evoked changes in presynaptic Ca^{2+} concentrations and the release of endogenous glutamate. This preparation was used because the mossy fiber nerve terminals express glutamate-dependent long-term synaptic potentiation. It was observed that membrane depolarization stimulated the accumulation of unesterified AA, as well as the release of glutamate and an increase in $[Ca^{2+}]_i$. These effects were mimicked by phospholipase A_2 activation or the presence of exogenous AA and they were attenuated by phospholipase A_2 inhibitors. In addition, pretreatment of HMFS with melittin or AA potentiated the K^+-induced release of glutamate. Inhibition of cyclooxygenase or lipoxygenase also facilitated the evoked release of glutamate and the accumulation of $[Ca^{2+}]_i$, while products of the 12-lipoxygenase pathway attenuated the evoked responses.

Veratridine- and K^+-dependent membrane depolarizations stimulated increases in $[Ca^{2+}]_i$ and the veratridine effect was blocked by TTX. In contrast, receptor activation with glutamate or its agonists had no effect on free Ca^{2+} concentrations. These results can be used to suggest that the observed changes in $[Ca^{2+}]_i$ are of neuronal origin and reflect presynaptic responses. Membrane depolarization also stimulated the release of glutamate. In addition, depolarization of HMFS by clamping membrane potentials with increased extracellular K^+, opening voltage-gated Na^+ channels with veratridine, or blocking K^+ channels with 3,4-diaminopyridine stimulated the accumulation of unesterified AA. Therefore, the availability of free AA paralleled the depolarization-dependent stimulation of presynaptic secretion mechanisms.

Phospholipase A_2 inhibition attenuated the K^+-evoked accumulation of intraterminal free Ca^{2+}. BPB, dibucaine, and luffariellolide all caused dose-dependent reductions in the increased $[Ca^{2+}]_i$ induced by 45 mM KCl. In addition, these inhibitors attenuated the K^+-dependent accumulation of unesterified AA. In contrast, putative inhibitors of phospholipase C and DGase had no effect on the evoked accumulation of Ca^{2+} or the release of endogenous glutamate. It appeared, therefore, that depolarization-stimulated secretion mechanisms were correlated with phospholipase A_2 activity.

An involvement of phospholipase A_2 activity in secretion processes was also indicated by the ability of melittin to stimulate the accumulation of Ca^{2+}, the release of glutamate, and the production of unesterified AA. These effects appeared to be related to the activation of endogenous phospholipase A_2, since exogenous phospholipase A_2 did not alter glutamate efflux or Ca^{2+} accumulation. That the effects of phospholipase A_2 activation on secretion processes might be due to the production of free AA was supported by the ability of exogenous AA to mimic the effects of membrane depolarization and melittin on glutamate efflux and Ca^{2+} accumulation. Addition of AA to the superfusion buffer stimulated the dose-dependent release of glutamate. AA also induced the dose-dependent accumulation of $[Ca^{2+}]_i$. Therefore, both phospholipase A_2 activation and added AA were correlated with the stimulation of secretion processes.

The effects of AA on glutamate efflux and Ca^{2+} accumulation appeared to be relatively specific for that fatty acid, since other exogenous fatty acids were either ineffective or less effective at stimulating secretion processes. The exception was adrenic acid, which was almost as potent as AA at inducing glutamate release. However, the concentration of adrenate in high K^+- or melittin-treated synaptosomes was about 25% of the AA concentration. Therefore, AA would be the more effective agent at modulating secretion mechanisms. In addition, depolarization and melittin treatment had no effect on the accumulation of free linoleate or linolenate, which were detected as trace amounts using GLC. It is possible that exogenous adrenate and other unsaturated fatty acids have some effect on glutamate release and Ca^{2+} accumulation by interfering with the reacylation and/or oxidation of unesterified arachidonate.

The suggestion that interference with the removal of unesterified AA can affect presynaptic mechanisms was supported by the observation that the cyclooxygenase inhibitor ibuprofen potentiated the depolarization-induced release of glutamate and accumulation of $[Ca^{2+}]_i$. It is possible that the ibuprofen effects were due to inhibition of prostanoid synthesis. However, exogenous prostaglandins had no effect on the K^+-evoked release of glutamate or Ca^{2+} accumulation and produced slight direct effects only at high concentrations. It would be expected that exogenous prostaglandins would inhibit presynaptic events, if the effects of ibuprofen were due to the inhibition of prostaglandin synthesis. Therefore, we propose that the potentiating effects of cyclooxygenase inhibition are due to blocking the removal of free arachidonate and not the attenuation of prostaglandin production.

Inhibition of lipoxygenase activities with low concentration of NDGA also potentiated the K^+-evoked release of glutamate and the accumulation of $[Ca^{2+}]_i$, as well as the glutamate release induced by exogenous AA. However, the effects of

NDGA were not limited to its ability to block the removal of free AA, since lipoxygenase products affected secretion processes. 12(S)-HETE, 12(R)-HETE, and 12(S)-HPETE attenuated the K^+-evoked release of glutamate and accumulation of $[Ca^{2+}]_i$ (Freeman et al., 1991). These effects appeared to be specific for 12-lipoxygenase products, since 5- and 15-HETE did not alter presynaptic responses to membrane depolarization. Our results were consistent with those obtained at *Aplysia* sensory neurons (Piomelli et al., 1987; 1988) and may reflect a 12-HETE-dependent opening of K^+ channels (Belardetti et al., 1989; Volterra and Siegelbaum, 1989), which would hyperpolarize presynaptic membranes and reduce neurotransmitter release and Ca^{2+} accumulation. Consistent with this suggestion, we found that 3,4-diaminopyridine, which blocks K^+ channels and depolarizes nerve terminals, stimulated the efflux of glutamate from HMFS, as well as the accumulation of unesterified AA.

The effects of lipoxygenase inhibition on presynaptic secretion processes were not consistent with some previous results obtained in other regions of the rat hippocampus. For example, it was reported that NDGA blocks the induction of hippocampal LTP (Williams and Bliss, 1988) and this effect may be due to presynaptic inhibition of glutamate release (Lynch et al., 1989). In addition, it was reported that NDGA attenuates the arachidonate-dependent facilitation of excitatory transmission in the hippocampal CA1 subfield (Drapeau et al., 1990). The discrepancies between these reports and our investigations may be related to the use of excessive concentrations of NDGA, since we observed that NDGA had a biphasic effect on the evoked release of glutamate from HMFS, as well as the accumulation of $[Ca^{2+}]_i$. Low concentrations of NDGA (0.1-10 μM) potentiated the evoked responses while higher concentrations (50-100 μM) were inhibitory. These inhibitory effects were consistent with the suggestion that NDGA blocks phospholipase A_2 activity (Billah et al., 1985; Robison et al., 1990), which would diminish the depolarization-induced accumulation of unesterified AA. Unfortunately, the reports that NDGA blocks glutamate release and the induction of LTP were based on the use of high concentrations of NDGA (50-200 μM). However, our observations that low concentrations of NDGA potentiated the evoked release of glutamate and Ca^{2+} accumulation appeared to be related to the inhibition of lipoxygenase activities, since 12-lipoxygenase products attenuated presynaptic responses. Opposing effects of NDGA and 12-HETEs would be expected, if NDGA is reducing the production of the inhibitory 12-HETE signal.

It appeared, therefore, that 12-lipoxygenase products provided inhibitory signals at the mossy fiber nerve terminal to limit secretion events, while the parent compound arachidonic acid acted as an excitatory messenger. It is also possible that the presynaptic accumulation of unesterified AA is involved in the enhanced release of glutamate that is associated with the development of LTP, since pretreatment of HMFS with either AA or melittin allowed for a facilitation of glutamate release upon subsequent depolarization of the nerve terminals (Freeman et al., 1990). These effects were consistent with the ability of exogenous AA to reduce the intensity of stimulus parameters required to induce hippocampal LTP (Williams et al., 1989). Further work is required to fully characterize the involvement of AA in LTP events. Such investigations should help clarify not only the biochemical controls of learning and memory, but also the relationships between AA and glutamate-dependent neurotoxicity.

ACKNOWLEDGMENTS

The technical assistance of Nancy Edgehouse is greatly appreciated, as well as the patience shown by Chris, Casey, and Dan during the preparation of this manuscript. We would also like to acknowledge the continued support and encouragement of Dr. David M. Terrian. This work was supported by AFOSR grant 89-0245.

REFERENCES

Asakura T and Matsuda M (1984) Efflux of gamma-aminobutyric acid from and appearance of free arachidonic acid inside synaptosomes. Biochim Biophys Acta 773: 301-307.

Barrionuevo G, Kelso SR, Johnston D, Brown TH (1986) Conductance mechanism responsible for long-term potentiation in monosynaptic and isolated excitatory synaptic inputs to hippocampus. J Neurophysiol 55:540-550.

Bazan NG (1970) Effects of ischemia and electroconvulsive shock on free fatty acid pool in the brain. Biochim Biophys Acta 218:1-10.

Belardetti F, Campbell WB, Falck JR, Demontis G, Rosolowsky M (1989) Products of heme-catalyzed transformation of the arachidonate derived 12-HPETE open S-type K+ channels in *Aplysia*. Neuron 3:497-505.

Billah MM, Bryant RW, Siegel MI (1985) Lipoxygenase products of arachidonic acid modulate biosynthesis of platelet-activating factor (1-*O*-alkyl-2-acetyl-*sn*-3-phosphocholine) by human neutrophils via phospholipase A_2. J Biol Chem 260: 6899-6906.

Birkle DL and Bazan NG (1987) Effect of bicuculline-induced status epilepticus on prostaglandins and hydroxyeicosatetraenoic acids in rat brain subcellular fractions. J Neurochem 48:1768-1778.

Bradford PG and Marinetti GV (1983) Stimulation of phospholipase A2 and secretion of catecholamines from brain synaptosomes by potassium and A23187. J Neurochem 41:1684-1693.

Buchan AM (1990) Do NMDA antagonists protect against cerebral ischemia: Are clinical trials warranted? Cerebrovasc Brain Metab Rev 2:1-26.

Chan PH, Kerlan R, Fishman RA (1983) Reductions of gamma-aminobutyric acid and glutamate uptake and (Na^+-K^+)-ATPase activity in brain slices and synaptosomes by arachidonic acid. J Neurochem 40:309-316.

Conricode KM and Ochs RS (1989) Mechanism for the inhibition and stimulatory actions of proteins on the activity of phospholipase A_2. Biochim Biophys Acta 1003:36-43.

Damron DS and Dorman RV (1989) Calcium mobilization in hippocampal mossy fiber terminals. Trans Am Soc Neurochem 20:134.

Damron DS, Freeman EJ, Terrian DM, Dorman RV (1990) Arachidonic acid-induced calcium mobilization in hippocampal mossy fiber synaptosomes. Neurosci Abst 16:166.

Dorman RV, Damron DS, Hamm TFR (1990) Description and manipulation of ischemia-induced alterations of cerebral arachidonic acid metabolism. In: Lipid mediators in ischemic brain damage and experimental epilepsy. (Bazan NG, ed) pp 36-66. Basel: Karger.

Dorman RV, Schwartz MA, Terrian DM (1986) Prostaglandin involvement in the evoked release of D-aspartate from cerebellar mossy fiber terminals. Brain Res Bull 17:243-248.

Dorman RV, Schwartz MA, Terrian DM (1988) Depolarization-induced [^{3}H]arachidonic acid accumulation: Effects of external Ca^{2+} and phospholipase inhibitors. Brain Res Bull 21:445-450.

Drapeau C, Pellerin L, Wolfe LS, Avoli M (1990) Long-term changes of synaptic transmission induced by arachidonic acid in the CA1 subfield of the rat hippocampus. Neurosci Lett 115:286-292.

Folch J, Lees M, Sloane-Stanley GH (1957) A simple method for the isolation and purification of total lipids from animal tissues. J Biol Chem 226:497-509.

Freeman CP and West D (1966) Complete separation of lipid classes on a single thin-layer plate. J Lipid Res 7:324-327.

Freeman EJ, Damron DS, Terrian DM, Dorman RV (1991) 12-Lipoxygenase products attenuate the glutamate release and Ca^{2+} accumulation evoked by depolarization of hippocampal mossy fiber nerve endings. J Neurochem 56:1079-1082.

Freeman EJ, Terrian DM, Dorman RV (1990) Presynaptic facilitation of glutamate release from isolated hippocampal mossy fiber endings by arachidonic acid. Neurochem Res 15:743-750.

Frye RA and Holz RW (1984) The relationship between arachidonic acid release and catecholamine secretion from cultured bovine adrenal chromaffin cells. J Neurochem 43:146-150.

Graham LT and Aprison MH (1966) Fluorometric determination of aspartate, glutamate and gamma-aminobutyrate in nerve tissue using enzymic methods. Anal Biochem 15:487-497.

Grynkiewicz G, Poenie M, Tsien Y (1985) A new generation of calcium indicators with greatly improved fluorescence properties. J Biol Chem 260:3440-3450.

Harris EW and Cotman CW (1986) Long-term potentiation of guinea pig mossy fiber responses is not blocked by N-methyl-D-aspartate antagonists. Neurosci Lett 70:132-137.

Hoffman AL and Zamir N (1984) Localization and quantitation of dynorphin B in the rat hippocampus. Brain Res 324:353-357.

Lazarewicz JW, Leu V, Sun GY, Sun AY (1983) Arachidonic acid release from K^+-evoked depolarization of brain synaptosomes. Neurochem Int 5:471-478.

Lowry OH, Rosebrough NJ, Farr AL, Randall RJ (1951) Protein measurements with Folin phenol reagent. J Biol Chem 193:265-275.

Lynch MA (1989) Mechanisms underlying induction and maintenance of long-term potentiation in the hippocampus. Bioessays 10:85-90.

Lynch MA and Voss KL (1990) Arachidonic acid increases inositol phospholipid metabolism and glutamate release in synaptosomes prepared from hippocampal tissue. J Neurochem 55:215-221.

Lynch MA, Errington ML, Bliss TVP (1989) Nordihydroguaiaretic acid blocks the synaptic component of long-term potentiation and the associated increases in release of glutamate and arachidonate: An *in vivo* study in the dentate gyrus of the rat. Neuroscience 30:693-701.

Neuman RS, Cherubini E, Ben-Ari Y (1989) Endogenous and network bursts induced by N-methyl-D-aspartate and magnesium-free medium in the CA3 region of the hippocampal slice. Neuroscience 28:393-399.

Piomelli D, Feinmark SJ, Shapiro E, Schwartz JH (1988) Formation and biological activity of 12-ketoeicosatetraenoic acid in the nervous system of *Aplysia*. J Biol Chem 263:16591-16596.

Piomelli D, Volterra ND, Siegelbaum SA, Kandel ER, Schwartz JH, Belardetti F (1987) Lipoxygenase metabolites of arachidonic acid as second messengers for presynaptic inhibition of *Aplysia* sensory cells. Nature 328:38-43.

Rhoads DE, Osburn LD, Peterson NA, Raghupathy E (1983) Release of neurotransmitter amino acids from synaptosomes: Enhancement of calcium-independent efflux by oleic and arachidonic acids. J Neurochem 41:531-537.

Robison TW, Sevanian A, Forman HJ (1990) Inhibition of arachidonic acid release by nordihydroguaiaretic acid and its antioxidant action in rat alveolar macrophages and Chinese hamster lung fibroblasts. Toxicol Applied Pharmacol 105:113-122.

Scott ID, Akerman KEO, Nicolls DG (1980) Calcium-ion transport by intact synaptosomes. Biochem J 192:873-880.

Staubli U, Larson J, Lynch G (1990) Mossy fiber potentiation and long-term potentiation involve different expression mechanisms. Synapse 5:333-335.

Terrian DM, Damron DS, Dorman RV, Gannon RL (1989) Effects of calcium antagonists on the evoked release of dynorphin A(1-8) and the availability of intraterminal calcium in rat hippocampal mossy fiber synaptosomes. Neurosci Lett 106:322-327.

Terrian DM, Dorman RV, Gannon RL (1990b) Characterization of the presynaptic calcium channels involved in glutamate exocytosis from rat hippocampal mossy fiber synaptosomes. Neurosci Lett 119:211-214.

Terrian DM, Gannon RL, Rea MA (1990a) Glutamate is the endogenous amino acid selectively released by rat hippocampal mossy fiber synaptosomes concomitantly with prodynorphin-derived peptides. Neurochem Res 15:1-5.

Terrian DM, Johnston D, Claiborne BJ, Ansah-Yiadom R, Strittmatter WJ, Rea MA (1988) Glutamate and dynorphin release from a subcellular fraction enriched in hippocampal mossy fiber synaptosomes. Brain Res Bull 21:343-351.

Volterra A and Siegelbaum SA (1989) Antagonistic modulation of S-K^+ channel activity by cyclic AMP and arachidonic acid metabolites. Ann NY Acad Sci 559:219-236.

Williams JH and Bliss TVP (1988) Induction but not maintenance of calcium-induced long-term potentiation in dentate gyrus and area CA1 of the hippocampal slice is blocked by nordihydroguaiaretic acid. Neurosci Lett 88:81-85.

Williams JH, Errington ML, Lynch MA, Bliss TVP (1989) Arachidonic acid induces a long-term activity-dependent enhancement of synaptic transmission in the hippocampus. Nature 341:739-742.

Yamamoto C (1987) Modulation of synaptic transmission in the hippocampus—A quantal analysis study. Neurosci Abst 22:S581.

Zalutsky RA and Nicoll RA (1990) Comparison of two forms of long-term potentiation in single hippocampal neurons. Science 248:1619-1624.

Zoltay G and Cooper JR (1990) Ionic basis of inhibitory presynaptic modulation in rat cortical synaptosomes. J Neurochem 55:1008-1012.

METABOTROPIC GLUTAMATE RECEPTORS AND NEURONAL TOXICITY

G. Aleppo,[1] A. Pisani,[1] A. Copani,[1]
V. Bruno,[1] E. Aronica,[1] V. D'Agata,[1]
P.L. Canonico,[2] and F. Nicoletti[1]

[1]Institute of Pharmacology
University of Catania
School of Medicine
Catania, Italy

[2]Chair of Pharmacology
University of Pavia
School of Dentistry
Pavia, Italy

INTRODUCTION

Specific glutamate receptors coupled to polyphosphoinositide (PPI) hydrolysis have been described in brain slices, cultured neurons, and astrocytes, and in amphibian oocytes injected with rat brain mRNA (Sladeczek et al., 1985; Nicoletti et al., 1986a,b; Sugiyama et al., 1987). In most of the systems, metabotropic receptors are activated by 1S,3R-aminocyclopentandicarboxylic acid (ACPD), quisqualate, ibotenate, and L-glutamate, but not by α-amino-3-hydroxy-5-methylisoxazolepropionate (AMPA), kainate, and N-methyl-D-aspartate (NMDA) (Nicoletti et al., 1986a; Schoepp and Johnson, 1988; 1989; Palmer et al., 1989). Trans-ACPD has been described as the most selective agonist of metabotropic receptors (Palmer et al., 1989), although it is less potent than quisqualate in stimulating inositolphosphate formation. In brain slices, stimulation of PPI hydrolysis by metabotropic receptor agonists is extremely high at the earlier stages of postnatal development (within the first 2 weeks after birth) and progressively declines during maturation (Nicoletti et al., 1986a). In adult tissue, the activation of metabotropic receptors is amplified in response to deafferentation (Nicoletti et al., 1987), as well as after induction of long-term potentiation (Aronica et al., 1991) or electrical kindling (Iadarola et al., 1986; Akiyama et al., 1987). Hence, it is likely that metabotropic receptors contribute to the synaptic events involved in the regulation of neuronal plasticity. However, based on the toxic effects of quisqualate in hippocampal slices (Garthwaite and Garthwaite, 1989) and cultured cortical neurons (Patel et al., 1990), a role for metabotropic receptors in the mechanism of neuronal degeneration has been suggested. We have addressed this problem in primary cultures of cerebellar neurons.

Table 1. Potency of excitatory amino acid receptor agonists in activating PPI hydrolysis in cultured cerebellar granule cells

Agonist	EC_{50} (μM)
L-glutamate	10
Quisqualate	20
Trans-ACPD	120
Ibotenate	150
NMDA	200
AMPA	inactive

TOXICITY OF METABOTROPIC RECEPTORS IN CULTURED CEREBELLAR NEURONS

Primary cultures of cerebellar neurons were prepared from 8-day-old rats, as described previously (Nicoletti et al., 1986b). These cultures contain mostly granule cells (> 90 %) with few GABAergic neurons, glial cells, and endothelial cells (Nicoletti et al., 1986b). Cultured cerebellar granule cells release glutamate as a neurotransmitter and express a variety of excitatory amino acid receptor subtypes coupled to membrane ion channels or PPI hydrolysis. In cultures at 7-9 days *in vitro* (DIV), metabotropic receptor agonists stimulate PPI hydrolysis with the following rank order of potency: glutamate > quisqualate > trans-ACPD > ibotenate (Table 1).

Stimulation of [^{3}H]inositolmonophosphate (InsP) formation by trans-ACPD peaks after 4 days in culture (Aronica et al., unpublished data), whereas the expression of ionotropic receptors (NMDA, AMPA, and kainate receptors) is maximal at later stages of maturation. Activation of metabotropic receptors by trans-ACPD may support neuronal survival in cultures at 4-5 DIV (unpublished observation). This action is reminiscent of that induced by NMDA receptor agonists, which support cell survival and exert trophic activity during the early stages of development of cerebellar granule cells (Balàzs and Jorgensen, 1987; Balàzs et al., 1988). However, as in the case of NMDA receptor agonists (Novelli et al., 1988), trans-ACPD promotes neuronal degeneration when mature granule cells are incubated under conditions that lead to energy depletion. Thus, when cultures at 9 DIV were incubated in the absence of glucose, addition of trans-ACPD induced neuronal degeneration, as reflected by a rapid increase in the number of cells labeled with propidium iodide, which specifically stains dead cells (Table 2). The action of trans-ACPD was mimicked by NMDA, which gates large-conductance ion channels permeable to Ca^{2+}, but not by AMPA, which activates low-conductance ion channels permeable to Na^+ (reviewed in Collingridge and Lester, 1989). Interestingly, trans-ACPD and NMDA acted synergistically in inducing neuronal degeneration in glucose-free buffer (Table 2).

The potential toxicity of a combined activation of NMDA and metabotropic receptors is supported by results obtained by incubating the cultures with ouabain (10 μM), an inhibitor of Na^+/K^+ ATPase. Addition of trans-ACPD (100 or 500 μM) to cultures treated with ouabain did not produce any appreciable sign of neuronal degeneration within 40 min of incubation. However, when combined with concentrations of NMDA (100 μM) that were *per se* devoid of activity, trans-ACPD dramatically increased the number of cells stained with propidium iodide (Fig. 1).

Two mechanisms may account for the permissive action of ouabain on neuronal toxicity: (i) the induction of membrane depolarization, which releases the Mg^{2+} blockade of NMDA receptors; and (ii) a sustained increase in intracellular calcium due

Table 2. Toxicity of excitatory amino acid receptor agonists in cultured cerebellar neurons incubated in the absence of glucose

	Percent of cells stained with propidium iodide
Control	5 ± 3
Glucose-free (GF)	20 ± 7
GF + NMDA, 100 μM	48 ± 11
GF + trans-ACPD, 100 μM	50 ± 7
GF + AMPA, 100 μM	10 ± 8
GF + NMDA + trans-ACPD	95 ± 3
GF + AMPA + trans-ACPD	85 ± 7

Values are means ± SEM of four individual determinations. Cultures at 8 DIV were pre-incubated in Locke's solution (for ionic composition, see Favaron et al., 1988) with (Control) or without glucose (GF) for 10 min. Thereafter, agonists were added and incubations were continued for 40 min. Cell viability was determined as follows: monolayers were washed with phosphate buffered saline (PBS, pH 7.4) and stained for 3 min at room temperature with a mixture of fluorescein diacetate (15 μg/ml), which yields yellow-green fluorescence in viable cells, and propidium iodide (4.6 μg/ml), which colors dead cells red.

to a secondary decrease in the activity of the Na^+/Ca^{2+} exchange pump. The latter mechanism may become particularly relevant when activation of membrane ion channels is accompanied by mobilization of intracellular Ca^{2+} (see below).

In another set of experiments, we studied the formation of free radicals in cultured neurons incubated in the absence of extracellular Mg^{2+}, a condition that allows the activation of NMDA receptors. A single pulse with toxic concentrations of glutamate (100 μM) induced a substantial increase in the formation of free radicals within 30 min of incubation. NMDA and trans-ACPD (both at 100 μM) displayed low activity when added alone, but the combination was as effective as glutamate in increasing free radical formation (Fig. 2, left). A synergistic activity of metabotropic and NMDA receptors was also observed in cultured cortical neurons incubated in the presence of NMDA or trans-ACPD (Fig. 2, right).

Activation of NMDA receptors leads to a massive but transient influx of extracellular Ca^{2+} through large conductance ion channels (reviewed in Collingridge and Lester, 1989). In contrast, activation of metabotropic receptors generates two independent classes of second messenger molecules: (i) inositol-1,4,5-trisphosphate ($InsP_3$), which mobilizes intracellular Ca^{2+}, thus producing long-lasting and oscillatory increases in cytosolic free Ca^{2+}; and (ii) diacylglycerol, which contributes to the activation of protein kinase C in the presence of Ca^{2+} and phosphatidylserine (reviewed in Berridge, 1987). Activation of NMDA and metabotropic receptors may converge into a sustained increase in the activity of a variety of Ca^{2+}-dependent enzymes, including protein kinase C and phospholipase A_2. Both enzymes have been implicated in the pathophysiology of excitotoxin-induced neuronal damage (Favaron et al., 1988; reviewed in Choi, 1988). In cultured cerebellar granule cells, the toxic action of glutamate is attenuated by dantrolene, which inhibits the release of Ca^{2+} from intracellular stores (Frandsen and Schousboe, 1991).

We speculate that activation of metabotropic receptors contributes to excitotoxin-induced neuronal degeneration in a variety of pathological conditions, including cerebral ischemia. In collaboration with Seren et al. (Fidia Research Laboratories, Abano Terme,

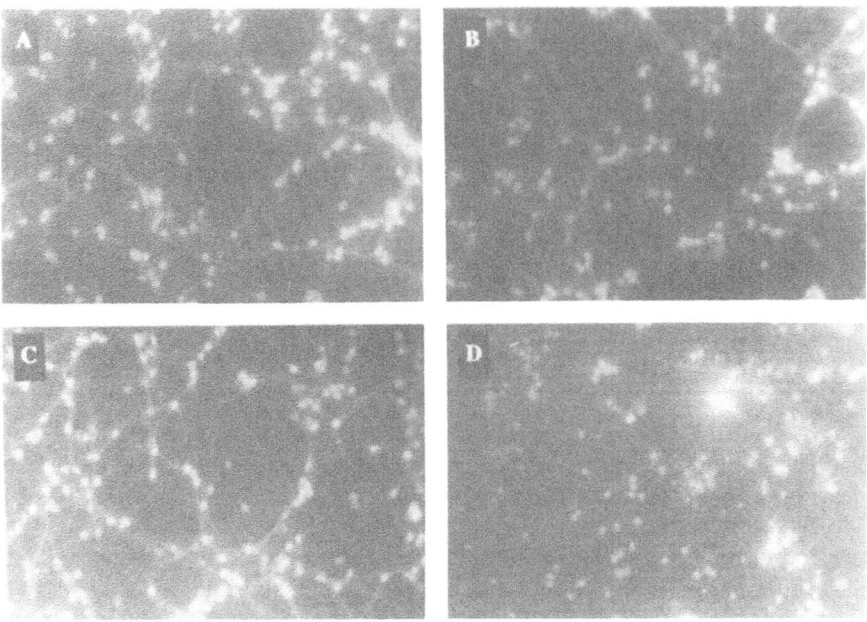

Figure 1. Double staining with fluorescein diacetate and propidium iodide in cultured cerebellar neurons incubated in the presence of ouabain and exposed to NMDA and/or trans-ACPD. Cultures were pre-incubated with Locke's solution containing 10 μM ouabain for 10 min. Thereafter, NMDA (100 μM) and trans-ACPD (100 μM) were added alone or in combination and incubations were continued for 40 min. Cells were stained with fluorescein diacetate and propidium iodide, as described in Table 2. A = ouabain: B = ouabain + transACPD; C = ouabain + NMDA; D = ouabain + trans-ACPD + NMDA.

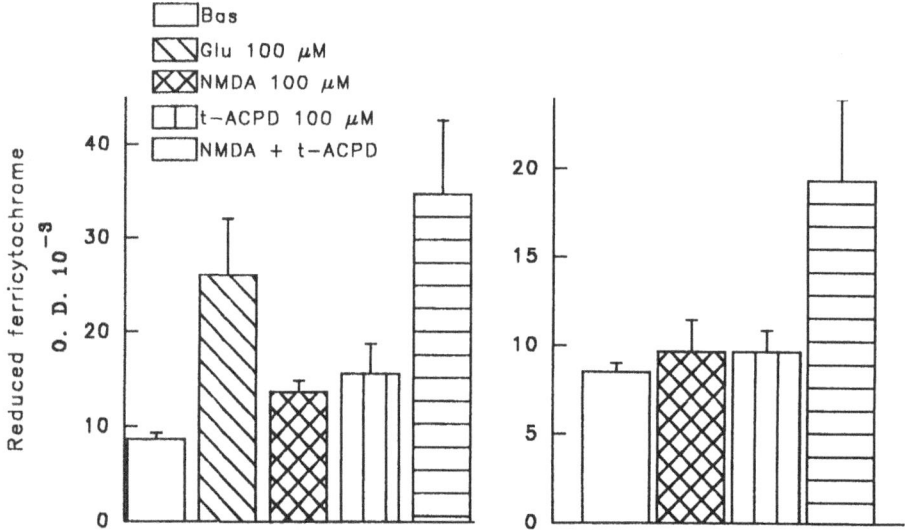

Figure 2. Formation of free radicals in cultured cerebellar granule cells (left) and cortical neurons (right) exposed to excitatory amino acid receptor agonists. Primary cultures of cortical neurons were prepared as described by Alho et al. (1988). Cultures of cerebellar or cortical neurons were pre-incubated with ferricytochrome c for 14 hours, then washed with Locke's solution and exposed for 30 min to glutamate, NMDA, and/or trans-ACPD (all at 100 μM). All incubations were performed in the absence of Mg^{2+}. Free-radical formation was estimated by spectrophotometric measurement of the reduced ferricyto-chrome c (adsorption at 550 nm) released into the incubation buffer.

Italy), we have studied the functional expression of metabotropic receptors in rats subjected to global transient ischemia, as described by Pulsinelli and Brierley (1979). Neuronal degeneration induced by the four-vessel occlusion model is relatively resistant to the protective action of NMDA receptor antagonists (Choi, 1988; Siesjo and Bengtsson, 1989;). Global transient ischemia leads to a dramatic increase in the sensitivity of metabotropic receptors, as reflected by an enhanced formation of [^{3}H]InsP in response to ibotenic acid. This effect develops within 24 hours after postischemic recirculation and is confined to the brain regions that are vulnerable to the ischemic damage, such as the hippocampus and cerebral cortex (Seren et al., 1990). This delayed increase in the sensitivity of metabotropic receptors suggests that the latter may contribute to the "maturation phenomenon," i.e., the delayed degeneration of vulnerable neurons after ischemia.

METABOTROPIC RECEPTORS AS A TARGET FOR THE ACTION OF ENVIRONMENTAL EXCITOTOXINS

The correlation between neurolathyrism (a form of spastic paraparesis resulting from degeneration of pyramidal neurons in the motor cortex) and consumptive ingestion of the chick pea *Lathyrus sativus* has raised the possibility that exposure to environmental excitotoxins may cause chronic degenerative diseases of the CNS. Accordingly, the seeds of *Lathyrus sativus* contain high concentrations of ß-N-oxaly-lamino-L-alanine (BOAA) (Spencer et al., 1986), which behaves as a potent agonist of AMPA-sensitive ionotropic glutamate receptors (Bridges et al., 1988). A role for an environmental excitotoxin has also been implicated in the etiology of the amyotrophic lateral sclerosis/parkinsonism-dementia (ALS/PD) complex among the Chamorro population of the western Pacific islands of Guam and Rota (Spencer et al., 1987). The high incidence of ALS/PD has been related to the chronic ingestion of ß-N-methyl-amino-L-alanine (BMAA), a non-protein amino acid present in the seed of the false sago palm *Cycas circinalis*. A neurological syndrome resembling the ALS/PD complex has been induced in macaques fed with high concentrations of BMAA (Spencer et al., 1987). Although BMAA lacks the Ω-carboxyl group typical of excitatory amino acids, interaction between the ß-amino group of the toxin and bicarbonate ions results in the formation of a carbamate adduct, which has the molecular configuration appropriate to interact with excitatory amino acid receptors (Weiss and Choi, 1988; Weiss et al., 1989). In electrophysiological studies, BMAA activates large conductance ion channels (Weiss et al., 1989), suggesting that the toxin interacts with NMDA receptors. To support this hypothesis, we have measured cGMP formation in cultured cerebellar neurons as a biochemical index of NMDA receptor activation. In cultures incubated in Mg^{2+}-free buffer containing 25 mM bicarbonate, BMAA increased the formation of cGMP and its action was antagonized by MK-801, which selectively inhibits the NMDA-gated ion channel (Table 3). However, BMAA was active at very high concentrations and stimulated cGMP formation to a much lesser extent than glutamate, the endogenous agonist of NMDA receptors (Table 3).

Hence, it seems unlikely that the toxic action of BMAA results exclusively from the activation of NMDA receptors. Accordingly, BMAA was much more potent in displacing specifically bound [^{3}H]glutamate than in displacing [^{3}H]CPP, a selective ligand of NMDA receptors (Copani et al., 1991). Under conditions that favor the labeling of metabotropic receptors, BMAA mimics quisqualate and trans-ACPD in displacing specifically bound [^{3}H]glutamate (Cha et al., 1990). This evidence raises the possibility that BMAA interacts with metabotropic receptors in the CNS. Accordingly, the toxin was more potent and efficacious than glutamate in stimulating [^{3}H]InsP formation in hippocampal slices from 8-day-old rats (Table 4). As with other metabo-

Table 3. Stimulation of cGMP formation by BMAA or glutamate (GLU) in cultured cerebellar neurons incubated in modified Locke's solution containing 25 mM bicarbonate

	cGMP (pmol/mg protein)
Basal	1.1 ± 0.1
BMAA, 100 μM	0.9 ± 0.2
BMAA, 1 mM	2.6 ± 0.4*
GLU, 100 μM	11 ± 0.8*
MK-801, 1 μM	1.3 ± 0.1
BMAA, 1 mM + MK-801, 1 μM	1.5 ± 0.2
GLU, 100 μM + MK-801, 1 μM	1.0 ± 0.04

Values are means ± SEM of 6 determinations. *$P < 0.01$ (one way ANOVA + Dunnett's t test), compared with basal values. Experiments were performed and cGMP levels were estimated as described by Novelli et al. (1987).

tropic receptor antagonists, the action of BMAA was antagonized by L-2-amino-3-phos-phonopropionate (AP3) and was markedly reduced in slices from adult rats (Copani et al., 1990). Interestingly, BOAA (such as AMPA or NMDA) did not stimulate PPI hydrolysis. As expected, BMAA was virtually inactive on PPI hydrolysis in slices incubated in the absence of bicarbonate (Copani et al., 1990), suggesting that the formation of a carbamate derivative is necessary for the interaction of the toxin with metabotropic receptors.

Based on the low concentration of BMAA in processed cycad flours from Guam (Duncan et al., 1990) and on the long latency between exposure to the toxin and onset of ALS/PD (Garruto et al., 1985; Rodgers-Johnson et al., 1986), it has been argued that BMAA is an unlikely cause of ALS/PD in Guam (Duncan et al., 1990). However, even small concentrations of BMAA can concentrate at the active site, as the toxin is a poor

Table 4. Stimulation of PPI hydrolysis by excitatory amino acid receptor agonists in hippocampal slices from 8-day-old rats

	[^{3}H]InsP formation (dpm/mg protein) x 10^{-3}
Basal	2.0 ± 0.3
GLU, 200 μM	6.2 ± 0.1
GLU, 500 μM	22 ± 0.9
GLU, 1 mM	37 ± 2.4
BMAA, 200 μM	11 ± 0.2
BMAA, 500 μM	33 ± 1.8
BMAA, 1 mM	46 ± 2.1
ibotenate, 500 μM	62 ± 3.2
trans-ACPD, 200 μM	70 ± 4.1
BOAA, 1 mM	2.1 ± 0.1

GLU = L-glutamate. Values are means ± SEM of 6-9 determinations. Stimulation of PPI hydrolysis in hippocampal slices was estimated as described by Nicoletti et al. (1986b).

substrate for the high-affinity transport system for excitatory amino acids (Copani et al., 1990). A combined activation of NMDA and metabotropic receptors by BMAA may result in a massive and sustained increase in the activity of Ca^{2+}-dependent enzymes, which may cause degeneration of more receptive neurons and sensitize resistant neurons to age-related damage. This may explain the occurrence of ALS/PD in some Chamorros two or three decades after they leave Guam, although this hypothesis may be untenable for such a massive and dramatic pathology (see Duncan et al., 1990).

CONCLUSIONS

Activation of metabotropic glutamate receptors generates a cascade of reactions (oscillatory increase in intracellular Ca^{2+} and activation of protein kinase C) that regulates neuronal maturation and plasticity, supports cell survival and, under appropriate conditions, may even protect neurons against the toxicity of NMDA receptor agonists (Koh et al., in press). However, these intracellular mechanisms are potentially toxic and may be transformed by a disease or by a condition of energy depletion into an instrument of cell destruction. In particular, the combined activation of metabotropic and ionotropic receptors may exert synergistic effects, thus reaching the threshold to promote cell toxicity.

The interaction between BMAA and metabotropic receptors suggests that the latter may be a specific target for endogenous or environmental excitotoxins and encourages the search for metabotropic receptor antagonists as possible neuroprotective agents against acute and chronic degenerative diseases of the CNS.

REFERENCES

Akiyama K, Norihito Y, Mitsumoto S (1987) Increase in ibotenate-stimulated phosphatidylinositol hydrolysis in slices of the amygdala/pyriform cortex and hippocampus of rat by amygdala kindling. Exp Neurol 98:499.

Alho H, Ferrarese C, Vicini S, Vaccarino F (1988) Subsets of Gabaergic neurons in dissociated cell cultures of neonatal rat cerebral cortex show co-localization with specific modulator peptides. Dev Brain Res 39:193.

Aronica E, Grey U, Wagner M, Schroeder H, Krug M, Ruthrich H, Catania MV, Nicoletti F, Reymann KG (1991) Enhanced sensitivity of "metabotropic" glutamate receptors after induction of long-term potentiation in rat hippocampus. J Neurochem 57:376.

Balàzs R and Jorgensen OS (1987) Trophic function of excitatory transmitter amino acids. Neuroscience 22:S41.

Balàzs R, Jorgensen OS, Hack N (1988) N-methyl-D-aspartate promotes the survival of cerebellar granule cells in culture. Neuroscience 27:437.

Berridge MJ (1987) Inositol trisphosphate and diacylglycerol: two interacting second messengers. Annu Rev Biochem 56:159.

Bridges RJ, Kadri MM, Monaghan DT, Nunn PB, Watkins JC, Cotman CW (1988) Inhibition of [3H]α-amino-3-hydroxy-5-methyl-4-isoxazolepropionic acid binding by the excitotoxin ß-N-oxalyl-L-αß-diaminopropionic acid. Eur J Pharmacol 145:357.

Cha JH-J, Makowiec RL, Penney JB, Young AB (1990) AP$_3$ and LBHAA displace [^{3}H]glutamate binding to the metabotropic receptor. Proc. 20th Annual Meeting of the American Society for Neuroscience. October 28-November 2, St. Louis, MO, Abstract #231.19.

Choi DW (1988) Glutamate neurotoxicity and diseases of the nervous system. Neuron 1:623.

Collingridge GL and Lester RA (1989) Excitatory amino acid receptors in the vertebrate central nervous system. Pharmacol Rev 40:145.

Copani A, Canonico PL, Catania MV, Aronica E, Bruno V, Ratti E, van Amsterdam FTM, Gaviraghi G, Nicoletti F (1991) Interaction between ß-N-methylamino-L-alanine and excitatory amino acid receptors in brain slices and neuronal cultures. Brain Res 558:79.

Copani A, Canonico PL, Nicoletti F (1990) Beta-N-methylamino-L-alanine (L-BMAA) is a potent agonist of 'metabolotropic' glutamate receptors. Eur J Pharmacol 181:327

Duncan MW, Steele JC, Kopin IJ, Marker SP (1990) 2-Amino-3-(methylamino)-propanoic acid (BMAA) in cycad flour: An unlikely cause of amyotrophic lateral sclerosis and parkinsonism-dementia of Guam. Neurology 40:767.

Favaron M, Money H, Alho H, Bertolino M, Ferret B, Guidotti A, Costa E (1988) Gangliosides prevent glutamate and kainate neurotoxicity in primary neuronal cultures of neonatal rat cerebellum and cortex. Proc Natl Acad Sci USA 85:7351.

Frandsen A and Schousboe A (1991) Dantrolene prevents glutamate cytotoxicity and Ca^{2+} release from intracellular stores in cultured cerebral cortical neurons. J Neurochem 56:1075.

Garruto RM, Yanagihara R, Gajdusek DC (1985) Disappearance of high-incidence amyotrophic lateral sclerosis and parkinsonism dementia on Guam. Neurology 35:193.

Garthwaite G and Garthwaite J (1989) Quisqualate neurotoxicity: a delayed, CNQX-sensitive process triggered by a CNQX-insensitive mechanism in young rat hippocampal slices. Neurosci Lett 99:113.

Iadarola MJ, Nicoletti F, Naranjo JR, Putnam F, Costa E (1986) Kindling enhances the stimulation of inositol phospholipid hydrolysis elicited by ibotenic acid in rat hippocampal slices. Brain Res 374:174.

Koh J, Palmer E, Cotman CW (in press) Activation of the metabotropic glutamate receptors attenuates N-methyl-D-aspartate in cortical neurons. Proc Natl Acad Sci USA.

Nicoletti F, Iadarola MJ, Wroblewski JT, Costa E (1986a) Excitatory amino acid recognition sites coupled with inositol phospholipid metabolism: Developmental changes and interaction with α_1-adrenoceptors. Proc Natl Acad Sci USA 83:1931.

Nicoletti F, Wroblewski JT, Alho H, Eva C, Fadda E, Costa E (1987) Lesions of putative glutamatergic pathways potentiate the increase in inositol phospholipid hydrolysis elicited by excitatory amino acids. Brain Res 436:103.

Nicoletti F, Wroblewski JT, Novelli A, Alho H, Guidotti A, Costa E (1986b) The activation of inositol phospholipid metabolism as a signal transducing system for excitatory amino acids in primary cultures of cerebellar granule cells. J Neurosci 6:1905.

Novelli A, Nicoletti F, Wroblewski JT, Alho H, Costa E, Guidotti A (1987) Excitatory amino acids receptors coupled with guanylate cyclase in primary cultures of cerebellar granule cells. J Neurosci 7:40.

Novelli A, Reilly JA, Lysko PG, Henneberry RC (1988) Glutamate becomes neurotoxic via the N-methyl-D-aspartate receptor when intracellular energy levels are reduced. Brain Res 451:205.

Palmer E, Monaghan DT, Cotman CW (1989) Trans-ACPD, a selective agonist of phosphoinositide-coupled excitatory amino acid receptors. Eur J Pharmacol 166:585.

Patel J, Zinland WC, Klika AB, Mangano TJ, Keith RA, Salama AI (1990) 6,7-Dinitroquinoxaline-2,3-dione blocks the cytotoxicity of N-methyl-D-aspartate and kainate, but not quisqualate, in cortical cultures. J Neurochem 55:114.

Pulsinelli WA and Brierley JB (1979) A new model of bilateral hemispheric ischemia in unanesthetized rat. Stroke 10:267.

Rodgers-Johnson P, Garruto RM, Yanagihara R, Chen KM, Gajdusek DC, Gibbs CJ Jr, (1986) Amyotrophic lateral sclerosis and parkinsonism-dementia on Guam: A 30 year evaluation of clinical and neuropathologic trends. Neurology 36:7.

Schoepp DD and Johnson BJ (1988) Excitatory amino acid agonist-antagonist interactions at 2-amino-4-phosphonobutyric acid-sensitive quisqualate receptors coupled to phosphoinositide hydrolysis in slices of rat hippocampus. J Neurochem 50:1605.

Schoepp DD and Johnson BJ (1989) Inhibition of excitatory amino acid-stimulated phosphoinositide hydrolysis in the neonatal rat hippocampus by 2-aminophosphonopro- pionate. J Neurochem 53:1865.

Seren MS, Aldinio C, Zanoni R, Leon A, Nicoletti F (1990) Stimulation of inositol phospholipid hydrolysis by excitatory amino acids is enhanced in brain slices from vulnerable regions after transient global ischaemia. J Neurochem 53:1700.

Siesjo BK and Bengtsson F (1989) Calcium fluxes, calcium antagonists, and calcium-related pathology in brain ischemia, hypoglycemia, and spreading depression: A unifying hypothesis. J Cereb Blood Flow Metab 9:127.

Sladeczek F, Pin J-P, Recasens M, Bockaert J, Weiss S (1985) Glutamate stimulates inositol phosphate formation in striatal neurons. Nature 317:717.

Spencer P, Boy DN, Ludolph A, Hugon J, Dwivedi MP, Schaumberg HH (1986) Lathyrism: evidence for role of the neuroexcitatory amino acid BOAA. Lancet ii:1066.

Spencer PS, Nunn PB, Hugon J, Ludolph AC, Boss SM, Boy DN, Robertson RC (1987) Guam amyotrophic lateral sclerosis: dementia linked to a plant excitant neurotoxin. Science 237:517.

Sugiyama H, Ito I, Hirono C (1987) A new type of glutamate receptor linked to inositol phosphate metabolism. Nature 325:531.

Weiss JH and Choi DW (1988) ß-N-methylamino-L-alanine neurotoxicity: requirement for bicarbonate as a co-factor. Science 241:973.

Weiss JH, Christine CM, Choi DW (1989) Bicarbonate dependence of glutamate receptor activation by ß-N-methylamino-L-alanine: Channel recording and study with related compounds. Neuron 3:321.

A ROLE FOR THE ARACHIDONIC ACID CASCADE IN FAST SYNAPTIC MODULATION: ION CHANNELS AND TRANSMITTER UPTAKE SYSTEMS AS TARGET PROTEINS

A. Volterra, D. Trotti, P. Cassutti,
C. Tromba, R. Galimberti,
P. Lecchi, and G. Racagni

Center of Neuropharmacology and
Institute of Pharmacological Sciences,
University of Milan
Italy

ABSTRACT

Recent evidence indicates that arachidonic acid (AA) and its metabolites play a fast messenger role in synaptic modulation in the CNS. 12-Lipoxygenase derivatives are released by *Aplysia* sensory neurons in response to inhibitory transmitters and directly target a class of K^+ channels, increasing the probability of their opening. In this way, hyperpolarization is achieved and action potentials are shortened, leading to synaptic depression. Other types of K^+ channels in vertebrate excitable cells have been found to be sensitive to arachidonic acid, lipoxygenase products, and polyunsaturated fatty acids (PUFA). In the mammalian CNS, arachidonic acid is released upon stimulation of N-methyl-D-aspartate (NMDA)-type glutamate receptors. We found that arachidonic acid inhibits the rate of glutamate uptake in both neuronal synaptic terminals and astrocytes. Neither biotransformation nor membrane incorporation are required for arachidonic acid to exert this effect. The phenomenon, which is rapid and evident at low μM concentrations of AA, may involve a direct interaction with the glutamate transporter or its lipidic microenvironment on the outer side of the cell membrane. Polyunsaturated fatty acids mimic arachidonate with a rank of potency parallel to the degree of unsaturation. Since the effect of glutamate on the synapses is terminated by diffusion and uptake, a slowing of the termination process may potentiate glutamate synaptic efficacy. However, excessive extracellular accumulation of glutamate may lead to neurotoxicity.

K^+ CHANNEL MODULATION BY ARACHIDONIC ACID CASCADE AND POLYUNSATURATED FATTY ACIDS

12-Lipoxygenase Metabolites Open S-type K^+ Channels in Aplysia Sensory Neurons

The neuropeptide Phe-Met-Arg-Phe-NH_2 (FMRFa) has an inhibitory action on the sensory neurons of the marine mollusk *Aplysia californica*. FMRFa-induced inhibition is, at least in part, achieved through the activation of a class of background K^+ channels termed the S-type K^+ channels (S channels). Functional analysis of single S channels with the patch-clamp technique has shown that FMRFa increases the probability of channel opening and acts through a diffusible second messenger (Belardetti et al., 1987). Lipoxygenase derivatives of arachidonic acid are produced by sensory neurons upon FMRFa stimulation and are able to mimic, at the cellular level, the inhibitory action of the peptide and, at the molecular level, the increased opening of the S channels. Moreover, FMRFa action is inhibited by both phospholipase and lipoxygenase inhibitors. These findings led to the conclusion that lipoxygenase metabolism of arachidonic acid acts as a second messenger of FMRFa action (Piomelli et al., 1987). 12-Lipoxygenase metabolites are significantly more effective than other lipoxygenase derivatives in modulating sensory cell resting and action potentials and S channel opening. Particularly potent is 12-HPETE, which opens S channels in both intact cells and excised patches that lack the intracellular medium. This observation suggests that 12-HPETE acts directly on the S channel molecule. 12-HPETE is effective at nanomolar concentrations when applied from the outer side of the membrane, whereas it acts in the micromolar range from the inner side (Buttner et al., 1989). Interestingly, if 12-HPETE is applied together with the heme-containing compound hematin, it becomes much more effective in opening S channels from the inner side of the membrane (Belardetti et al., 1989). Nonenzymatic metabolism of 12-HPETE is observed in this experiment, with formation of downstream products, possibly mimicking the cytochrome P_{450} pathway, leading to hepoxilins (Pace-Asciak, 1984). Hepoxilin A_3 can be formed from 12-HPETE in the nervous system of *Aplysia* (Piomelli et al., 1989). On the other hand, neither arachidonic acid nor the reduced product 12-HETE is effective in opening the S channels in excised patches, suggesting that 12-lipoxygenase metabolism at some intracellular location is required for the modulatory action. Therefore, in the case of S channel modulation, arachidonic acid acts as a precursor and 12-HETE as a by-product. 12-HPETE and/or a downstream derivative is the putative bioactive messenger.

Other K^+ Channels Are Opened by Lipoxygenase Derivatives, Arachidonic Acid, and PUFA in Vertebrate Cells

Since the discovery that K^+ channels represent a molecular target for the activity of arachidonic acid derivatives, much information has accumulated about the modulation of ion channels (including Na^+ channels, Ca^{2+} channels, Cl^- channels and gap-junctions) by these agents in various excitable and non-excitable cells of vertebrates and invertebrates (see Ordway et al., 1991, for a review). In many of the studies arachidonic acid was utilized as the test compound. The fatty acid was applied to the cells in the external medium, and the authors postulated subsequent membrane penetration and metabolic conversion to bioactive products. However, a distinction among the observed effects of arachidonic acid should be made. In some cases arachidonic acid metabolism through lipoxygenase pathways is required for ion channel modulation. In addition to *Aplysia* S channels, at least two other types of K^+ channels are opened by lipoxygenase products: K_{Ach} channels in heart myocytes (Kurachi et al., 1989; Kim et al., 1989) and K_m channels in hippocampal neurons (Schweitzer et al., 1990). On the other hand, some of the actions of arachidonic acid are not blocked by cascade inhibitors and are mimicked by other fatty acids, in particular polyunsaturated fatty acids. In excised patches, the effects of fatty acids persist, in contrast to what is observed in *Aplysia* S channels, indicating that arachidonic acid and PUFA themselves may directly modulate channel function (Kim and Clapham, 1989; Ordway et al., 1989). Finally, some of the actions of arachidonic acid have been blocked by free radical scavenger enzymes, such as superoxide dismutase (SOD), and may be achieved indirectly following protein kinase C activation (Keyser et al., 1990).

A point of convergence among the effects reported for lipoxygenase products, arachidonic acid, and PUFA is the general trend for these agents to modulate K^+ channels in terms of increased opening. In physiological situations, K^+ channels act for the most part by extruding K^+ ions from the cells and increasing internal electro-negativity. They participate in the setting of the resting potential of the cells and in the shaping of the action potential of excitable cells. In particular, they control the repolarization and the after-potential hyperpolarization phases. By increasing the contribution of K^+ channels to the resting or action potential, lipoxygenase derivatives and fatty acids appear to provide a general modulatory function that opposes electrical excitation. This idea seems to be supported by evidence for an inhibitory action of arachidonic acid and PUFA on Na^+ and Ca^{2+} channels, which, in turn, contribute to cell depolarization and increased excitation (Ordway et al., 1991).

GLUTAMATE UPTAKE MODULATION BY ARACHIDONIC ACID IN NEURONS AND ASTROCYTES

Arachidonic Acid Is Released upon Glutamate Receptor Stimulation in Neurons but Not in Glial Cells

Studies from different laboratories indicate that in mammalian CNS, the arachidonic acid cascade is coupled to excitatory amino acid receptors, in particular to the NMDA subtype. Following pre-incorporation of [³H]arachidonic acid in membrane phospholipids and application of glutamate or NMDA, radiolabeled material is released into the extracellular medium in both striatal neurons (Dumuis et al., 1988) and cerebellar granule cells (Lazarewicz et al., 1990) in primary culture. Most of the released material has been identified as arachidonic acid itself. In cortical slices without pre-labeling, NMDA receptor stimulation leads to 12-HETE formation (Wolfe et al., 1990). Arachidonic acid release is induced by Ca^{2+} dependent activation of PLA_2, probably due to Ca^{2+} entry via the NMDA receptor-channel (Lazarewicz et al., 1990).

In parallel neuronal-enriched and astrocyte-enriched primary cultures from rat cerebral cortex, we found that administration of the Ca^{2+} ionophore, ionomycin (1 μM, 5 min), was followed by an indiscriminate influx of Ca^{2+}, which resulted in a massive release of [³H]arachidonic acid-derived material. However, glutamate (0.1-1 mM, 15 min) and NMDA (100 μM, 15 min) selectively induced [³H]arachidonic acid release only from the neuronal cultures. These data indicate that both neuronal and glial cells possess the enzymatic machinery to trigger arachidonate release, but only neurons respond to glutamate with an extracellular liberation of arachidonic acid. In particular, NMDA receptors at glutamatergic synapses are probably located only in the post-synaptic dendrites. Therefore, the possibility exists that arachidonic acid is liberated by postsynaptic neurons in the synaptic and perisynaptic space to act as an intercellular communicator.

Arachidonic Acid Reduces the Rate of Glutamate Uptake in Both Neurons and Astrocytes

The activity of glutamate at the synapses is terminated by diffusion and uptake via high-affinity transport systems located in neuronal terminals and perisynaptic astrocytes. We have used both synaptosomes and cultured astrocytes from rat cerebral cortex to investigate the modulatory role of arachidonic acid on glutamate uptake. As shown in Figure 1, a substantial reduction in the V_{max} of basal glutamate uptake, as well as a

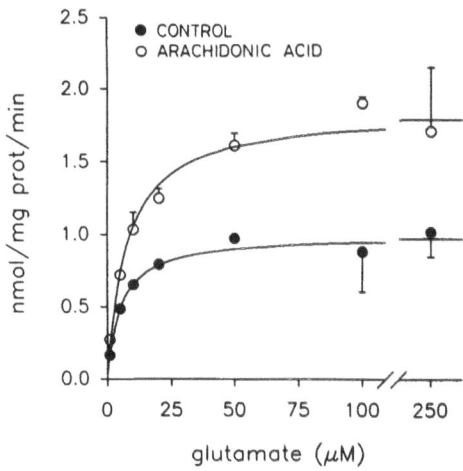

Figure 1. Arachidonic acid-induced inhibition of high-affinity glutamate uptake in cortical synaptosomes. Assay carried out for 1 min at 25°C in absence or presence of 100 μM AA. Estimated parameters: K_m (μM): 7.47 ± 1.88 (CONTROL) vs 5.11 ± 1.34 (AA) (n.s.); V_{max} (nmol/mg prot/min): 1.84 ± 0.21 (CONTROL) vs 0.99 ± 0.11 (AA) ($P<0.05$, F-test).

trend toward reduction in the K_m value, was seen in the presence of arachidonic acid. Arachidonic acid acts rapidly: significant inhibition is seen 30 s after its application and maximal inhibition is reached within 5-10 min. The threshold effective concentration is 1 μM (5-6% inhibition) whereas maximal inhibition (40-50%) is seen at 100 μM. This profile is consistent with a fast messenger role for arachidonic acid. Synaptosomal and astrocytic glutamate transport systems are affected in identical fashion by the fatty acid. Under arachidonate modulation, the uptake function proceeds at reduced speed and the effect, being noncompetitive in nature, cannot be overcome by the increasing concentration of glutamate accumulating in the extracellular space.

Arachidonic Acid Does Not Require Biotransformation or Membrane Incorporation to Act on Glutamate Uptake

Based on the observation that arachidonic acid inhibits glutamate uptake, a series of questions arises about the mechanisms of such action. First of all: is arachidonic acid the actual effector? Both synaptosomes and astrocytes have been reported to be capable of transforming arachidonic acid via the cyclooxygenase and the lipoxygenase pathways (Birkle and Bazan, 1987; Murphy et al., 1988). We have explored this point by incubating either synaptosomes or cultured astrocytes with 100 μM cold arachidonic acid plus 1 μCi [^{3}H]arachidonic acid under the same conditions (time, temperature, buffers, mechanical operations) utilized for the glutamate uptake assay. Biological preparations were then sonicated and both intracellular and extracellular media extracted and processed by reversed-phase high performance liquid chromatography (RP-HPLC) coupled to UV and radiochemical detectors (see Powell, 1985, for method, with slight modifications). As shown in Figure 2, only one major peak was detected; the retention time matched that of arachidonic acid. Automatic integration showed no other peak with an area ≥ 1% of the peak area of arachidonic acid peak area. Moreover,

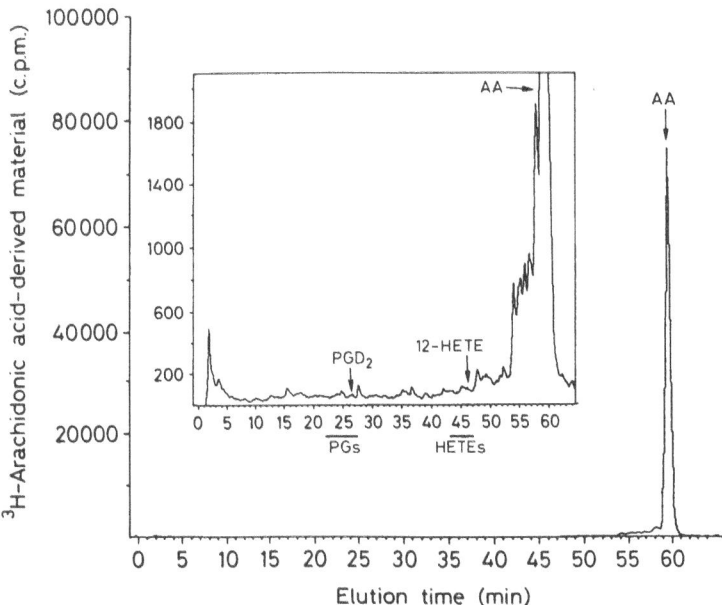

Figure 2. Reversed-phase HPLC chromatogram of [^{3}H]arachidonic acid-derived material following 10 min incubation with intact astrocytes. Insert: same chromatogram at 50-fold higher magnification. Arrows: retention times of standard [^{3}H]PGD$_2$, 12-HETE, AA. Solid lines below insert: elution periods for PGs (PGD$_2$, PGE$_2$, PGF$_{2\alpha}$, 6-keto-PGF$_{1\alpha}$, TXB$_2$) or 5,11,12,15-HETEs. Shoulder at left of AA peak due to nonmetabolic oxidation.

almost no changes were found in the total radioactivity recovered in the time-frames of elution of major cyclooxygenase products, i.e., prostaglandins (PGs) and thromboxane (TX), and lipoxygenase products, i.e., HETEs, in comparison with blank samples incubated in the absence of biological tissue (see Table 1). Therefore, we concluded that arachidonic acid is not significantly transformed into its metabolic products in synaptosomes and astrocytes during the glutamate uptake assay.

Table 1. Time-frame distribution by RP-HPLC of [^{3}H]arachidonic acid-derived material incubated in absence or presence of cortical synaptosomes or astrocytes

Elution Period (min)	Blank	Synaptosomes	Astrocytes
	(% of [^{3}H]AA-derived radioactivity ± SD)		
0 - 15	1.72 ± 0.26	1.64 ± 0.13	2.04 ± 1.10
15 - 30*	1.79 ± 0.59	2.06 ± 0.35	1.61 ± 0.84
30 - 43	2.46 ± 0.37	2.51 ± 0.55	2.95 ± 0.37
43 - 48**	2.11 ± 0.51	1.78 ± 0.18	3.29 ± 0.61
48 - 62***	91.23 ± 1.22	91.11 ± 2.38	89.56 ± 2.41
62 - 65	0.61 ± 0.16	0.68 ± 0.48	0.64 ± 0.08

*Elution period for major cyclooxygenase derivatives (PGs, TX); **elution period for lipoxygenase HETEs derivatives; ***elution period for arachidonic acid. Data are the average ± SD of two experiments in duplicate.

Recently, arachidonic acid was shown to modulate ion channels via free radical formation; the effect was prevented by the addition of the scavenger enzyme, superoxide dismutase (Keyser and Alger, 1990). We investigated this possibility with regard to the glutamate uptake assay and found the opposite result. As shown in Figure 3, arachidonic acid-induced inhibition of glutamate uptake in synaptosomes is not counteracted in the presence of superoxide dismutase. However, when the action of arachidonic acid was tested in the presence of the free fatty acid binding-protein albumin (BSA), inhibition was completely prevented. Moreover, if BSA was added when uptake was already inhibited by arachidonate, it could completely reverse the fatty acid effect. Since BSA chelates only free fatty acids and does not cross cell membranes, we conclude that arachidonic acid requires neither incorporation into the target membrane nor intracellular metabolic transformation to exert its inhibitory effect: it probably acts in its free form at some site on the outer side of the membrane.

PUFA but Not 12-Lipoxygenase Metabolites Mimic the Arachidonic Acid Effect

Arachidonic acid is the major product of glutamate-induced PLA_2 activation (Dumuis et al., 1988). However, 12-lipoxygenase metabolism and 12-HETE formation have also been reported in response to glutamate and NMDA (Wolfe et al., 1990). Therefore, we tested the effect of 12-lipoxygenase metabolites, namely 12-HPETE and 12-HETE, on glutamate uptake and found that they do not exert a significant effect. 12-HPETE at 1-2 μM has some inhibitory efficacy in astrocytes but not in synaptosomes. However, this effect is not always reproducible and does not show clear dose dependency. 12-HETE at 1-10 μM is devoid of any effect on glutamate uptake in both astrocytes and synaptosomes.

Theoretically, excitatory amino acids, by inducing PLA_2 activation, may release other unsaturated fatty acids present at position 2 in membrane phospholipids, such as oleic or docosahexaenoic acid. We have tested the effects of several long-chain saturated and unsaturated fatty acids on glutamate uptake. Figure 4 shows their efficacy in inhibiting glutamate transport in comparison to that of arachidonic acid. It is clear that both saturated (stearic acid) and trans-unsaturated (elaidic acid) molecules are completely ineffective, whereas all *cis*-unsaturated fatty acids, starting with oleic acid, reduce glutamate uptake. A pattern of potency increasing with the degree of unsaturation is similarly observed in synaptosomes and astrocytes. Linolenic acid (18:3), eicosapentaenoic acid (20:5), and docosahexaenoic acid (22:6) are the only unsaturated fatty acids that inhibit glutamate uptake to an extent comparable to that of arachidonic acid. This observation indicates that the effect has a certain specificity and that some structural and/or chemico-physical properties are required for its achievement.

Possible Molecular Sites and Mechanisms of the Action of Arachidonic Acid

Several different mechanisms of action can be hypothesized to explain the effect of arachidonic acid and PUFA on glutamate transport at the membrane level. Past work has indicated a parallel reduction in amino acid uptake and Na^+-K^+-ATPase activity (Chan et al., 1983). Therefore, the possibility exists that unsaturated fatty acids exert a general depressor effect on Na^+-dependent high-affinity uptake systems secondary to altered Na^+ co-transport. We have investigated this possibility by comparing the effect of the Na^+-K^+-ATPase inhibitor, ouabain, with that of arachidonic acid on glutamate uptake. Ouabain was tested at a concentration reported to completely inhibit Na^+/K^+ATPase. As shown in Figure 5, ouabain produced a reduction in glutamate

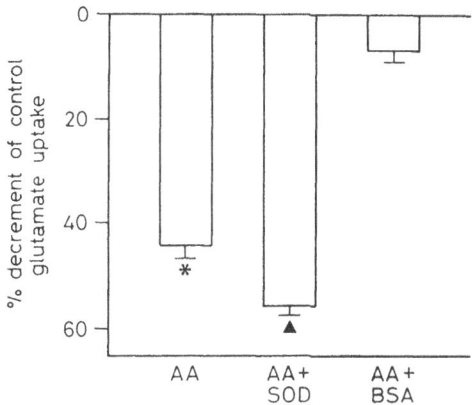

Figure 3. Arachidonic acid-induced inhibition of glutamate uptake in synaptosomes in absence or presence of superoxide dismutase (SOD, 90 U/ml) or bovine serum albumin (BSA, 0.1% w/v). Uptake assay: 10 μM glutamate, 5 min, 25°C. Data are the mean ± SD of two experiments in triplicate. Statistical analysis: AA vs BASAL and AA+SOD vs SOD (P<0.001); AA+BSA vs BSA (n.s., F-test).

uptake. However, at least three different observations militate against the hypothesis that the effect of arachidonic acid is secondary to Na^+-K^+-ATPase blockade: i) arachidonic acid is significantly more potent than ouabain; ii) the time-course of arachidonic acid inhibition is more rapid than that of ouabain; and iii) the two effects are not mutually exclusive and are partially additive, indicating that they involve different mechanisms.

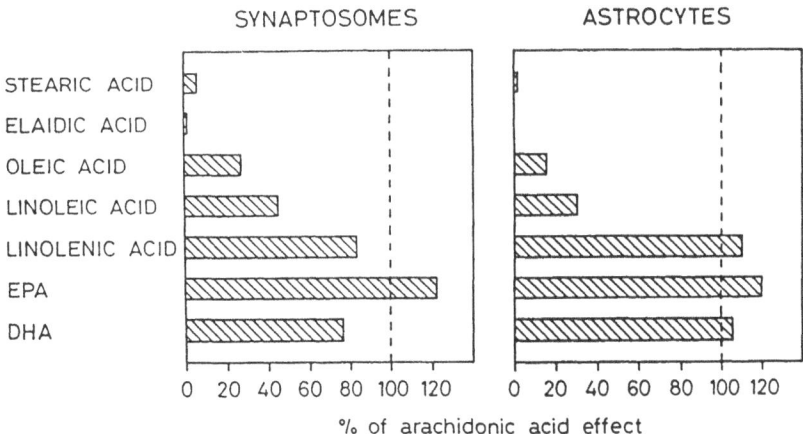

Figure 4. Glutamate uptake inhibition in cortical synaptosomes and astrocytes by different fatty acids: comparison with the arachidonic acid effect. All fatty acids, 100 μM. Synaptosome uptake assay: glutamate 10 μM, 1 min, 25°C; AA inhibition: 36.9%. Astrocyte uptake assay: glutamate 40 μM, 5 min, 25°C; AA inhibition: 51.8%. EPA = eicosapentaenoic acid; DHA = docosahexaenoic acid.

As noted above, arachidonic acid and PUFA need not be incorporated into membrane phospholipids to act on glutamate transport. Therefore, general changes in membrane fluidity following phospholipid remodeling are not required for their fast action. As reported for ion channels, PUFA could act "directly" on their membrane target proteins. Several proteins have been shown to possess fatty acid-binding domains. Serum albumin and lipoproteins are two established examples. Particularly attractive is the case of protein kinase C, which exists in several isoforms, some of which seem to have a recognition site for single unsaturated fatty acids together with the diacylglycerol binding site (Shearman et al., 1989). Since cloning and sequencing of the glutamate transporter is in progress, it will be of interest to see if the protein displays sequence homology with protein kinase C at those sites.

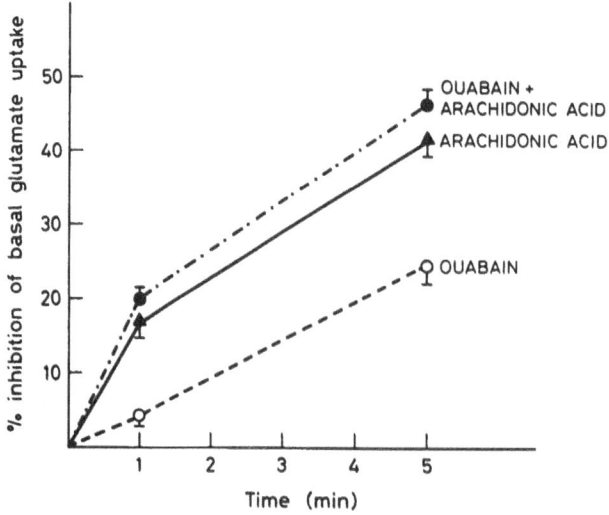

Figure 5. Time-course of glutamate uptake inhibition by ouabain (200 μM), arachidonic acid (100 μM), and ouabain + arachidonic acid in cortical synaptosomes. Uptake assay: 10 μM glutamate, 1 or 5 min, 25°C. Data are the mean ± SD of two experiments in triplicate. AA vs ouabain, $P<0.01$ at 1 and 5 min; ouabain+AA vs AA, $P<0.05$ at 5 min (F-test).

Another possible mode of action for arachidonic acid and related unsaturated fatty acids is to alter integral membrane protein function by interfering with the specific lipidic microenvironment. Cell membranes possess lipid domains with different fluidity. Free fatty acids can readily intercalate in these domains and induce changes in the packing of the lipid molecules. According to their structural and chemico-physical properties, different fatty acids target different domains. In particular, cis-unsaturated fatty acids, which have a rigid folded structure, target more fluid domains, whereas linear trans-unsaturated and unsaturated fatty acids prefer gel-like domains (Klausner et al., 1980). On the other hand, membrane-spanning proteins such as ion channels, pumps, transmitter receptors, and carriers have specific hydrophobic motifs in the trans-membrane regions allowing them to interact with the surrounding lipid moieties. It is therefore possible that, according to the amino acid organization of the protein in the trans-membrane region, a specific and appropriate lipidic microenvironment is selected. Free fatty acids, by intercalation in this microenvironment, may cause a structural reorganization of the lipid-protein interaction that can, in turn, lead to modifications in the functional properties of the protein.

Central glutamatergic synapses may undergo short- and long-term plasticity phenomena. Of particular interest and widely studied as a cellular model of memory storage is long-term potentiation (LTP; see Kennedy, 1989, for a review). LTP is induced by short high-frequency stimulation of the presynaptic neuron which results in enhanced excitatory responsiveness of the postsynaptic cell to normal stimulation lasting for hours or days. Both pre- and postsynaptic adaptive phenomena that could explain this change in synaptic efficiency have been reported, including enhanced glutamate release and increased responsiveness of postsynaptic glutamate receptors. At least in the case of LTP in area CA1 and the dentate gyrus of the hippocampus, it is known that the molecular event triggering LTP is Ca^{2+} entry into the postsynaptic neuron through the NMDA receptor-channel. How can this event be related to adaptive changes in the releasing system located in the presynapse? The existence of a retrograde trans-synaptic messenger has been postulated, which could be released by the postsynaptic site to bring adaptive information to the presynaptic pole. Early efforts to identify this messenger were concentrated on small peptides. More recently, following the discovery that amphiphilic 12-lipoxygenase derivatives of arachidonic acid are involved in synaptic modulation, the arachidonic acid cascade has become an attractive candidate. Indeed, a series of findings confirms the possible involvement of the arachidonic acid system in LTP: i) NMDA receptors release arachidonic acid and 12-lipoxygenase metabolites; ii) arachidonic acid levels are enhanced following LTP generation *in vivo*; iii) arachidonic acid cascade inhibitors prevent LTP induction; iv) exogenous arachidonate induces slow and persistent enhancement of synaptic efficacy; and v) arachidonic acid *in vitro* enhances glutamate release. However, there is conflicting evidence showing both increased and decreased glutamate release following the administration of 12-lipoxygenase products.

Our data suggest that another very simple mechanism for enhancing glutamate synaptic availability would be uptake inhibition. Thus, arachidonic acid released extracellularly from neurons carrying NMDA receptors can act directly on the outer side of the presynaptic terminal to reduce the rate of glutamate uptake. Following uptake inhibition, more glutamate would accumulate in the synaptic space. Still, it could be removed by diffusion. However, at the same time, arachidonic acid may also reduce perisynaptic astrocytic uptake, thus leading to a significant impairment of total glutamate clearance. Therefore, an increased amount of glutamate would remain in the synaptic area and be available to activate glutamate receptors. Function studies show that glutamate-induced postsynaptic depolarization is enhanced in hippocampal neurons in the presence of glutamate uptake inhibitors such as dihydrokainate and 3-hydroxy-DL-aspartate (Sawada et al., 1985). Moreover, the NMDA-dependent excitatory postsynaptic current elicited by afferent fiber stimulation in CA1 pyramidal cells is increased in the presence of dihydrokainate (Hestrin et al., 1990).

As mentioned above, free arachidonic acid levels are slightly increased following LTP generation (Lynch et al., 1989). In contrast, large amounts of arachidonic acid and PUFA are liberated in pathological conditions such as seizures and ischemia (Bazan, 1970). At the same time, a massive build-up of extracellular glutamate is also seen (Benveniste et al., 1984). This latter phenomenon leads to glutamate receptor overstimulation accompanied by the influx of excessive amounts of Ca^{2+} into neuronal cells which, supposedly, is responsible for the neurotoxic damage accompanying these pathologies. Our data suggest that the extracellular accumulation of arachidonic acid, PUFA, and glutamate may be interconnected, in that glutamate can release arachidonic acid which, in turn, both stimulates the release and inhibits the uptake of glutamate. In this way, a vicious cycle could be set up, ending in neuronal cell death.

In conclusion, it is possible that arachidonic acid and PUFA, by targeting a single mechanism (i.e., glutamate transport) and exerting the effect to different degrees, may either contribute to enhancement of synaptic efficacy or induce cell damage by excitotoxicity.

CONCLUSIONS

The information presented here leads to some general conclusions. The arachidonic acid cascade and polyunsaturated fatty acids may act as fast messengers in the CNS. Upon specific neurotransmitter receptor activation, neuronal cells release these compounds not only into their own internal medium, but also into the extracellular milieu which, *in vivo*, is the synaptic space and the perisynaptic area. Membrane-spanning proteins seem to possess sites—either in their own amino acid structure or in their lipid microenvironment—that directly bind these amphiphilic molecules. Our studies show examples of an ion channel and a neurotransmitter transporter protein which may be "directly" modulated in this way. These findings suggest the intriguing possibility that, in the CNS, these small amphiphilic molecules also serve to send information rapidly from a cell to its neighbors and, thereby, to integrate groups of cells to act in a concerted fashion. For example, amino acid neurotransmitters such as glutamate and GABA are taken up by both neuronal terminals and astrocytes and it would seem that coordinated action by these cells is necessary in order to control the extracellular levels of the two transmitters.

Modulation of ion channels and neurotransmitter transporters by arachidonic acid and related molecules has important functional implications, such as a role for these compounds in the regulation of both electrical and chemical neurotransmission. It is now an important goal to better understand this role and, in particular, to discover differences and similarities of action among different fatty acid molecules, as well as arachidonic acid and its many bioactive products.

REFERENCES

Bazan NG Jr (1970) Effects of ischemia and electroconvulsive shock on free fatty acid pool in the brain. Biochim Biophys Acta 218:1.

Belardetti F, Campbell WB, Falck JR, DeMontis GC, Rosolowsky M (1989) Products of heme-catalyzed transformation of the arachidonate derivative 12-HPETE open S-type K^+ channels in *Aplysia*. Neuron 3:497.

Belardetti F, Kandel ER, Siegelbaum SA (1987) Neuronal inhibition by the peptide FMRFamide involves opening of S K^+ channels. Nature 325:153.

Benveniste H, Drejer J, Schousboe A, Diemer NH (1984) Elevation of the extracellular concentrations of glutamate and aspartate in rat hippocampus during transient cerebral ischemia monitored by intracerebral microdialysis. J Neurochem 43:1369.

Birkle DL, Bazan NG (1987) Effect of bicuculline-induced status epilepticus on prostaglandins and hydroxyeicosatetraenoic acids in rat brain subcellular fractions. J Neurochem 48:1768.

Buttner N, Siegelbaum SA, Volterra A (1989) Direct modulation of *Aplysia* S-K^+ channels by a 12-lipoxygenase metabolite of arachidonic acid. Nature 342:553.

Chan PH, Kerlan R, Fishman RA (1983) Reductions of γ-aminobutyric acid and glutamate uptake and $(Na^+ + K^+)$-ATPase activity in brain slices and synaptosomes by arachidonic acid. J Neurochem 40:309.

Dumuis A, Sebben M, Haynes L, Pin JP, Bockaert J (1988) NMDA receptors activate the arachidonic acid cascade system in striatal neurons. Nature 336:68.

Hestrin S, Sah P, Nicoll RA (1990) Mechanisms generating the time course of dual component excitatory synaptic currents recorded in hippocampal slices. Neuron 5:247.

Kennedy MB (1989) Regulation of synaptic transmission in the central nervous system: Long-term potentiation. Cell 59:777.

Keyser DO and Alger BE (1990) Arachidonic acid modulates hippocampal calcium current via protein kinase C and oxygen radicals. Neuron 5:545.

Kim D and Clapham DE (1989) Potassium channels in cardiac cells activated by arachidonic acid and phospholipids. Science 244:1174.

Kim D, Lewis DL, Graziadei L, Neer EJ, Bar-Sagi D, Clapham D (1989) G-protein βγ subunits activate the cardiac muscarinic K^+ channel via phospholipase A_2. Nature 337:557.

Klausner RD, Kleinfeld AM, Hoover RL, Karnowsky MJ (1980) Lipid domains in membranes. J Biol Chem 255:1286.

Kurachi Y, Ito H, Sugimoto T, Shimizu T, Miki I, Ui M (1989) Arachidonic acid metabolites as intracellular modulators of the G protein-gated cardiac K^+ channel. Nature 337:555.

Lazarewicz JW, Wroblewski JT, Costa E (1990) N-methyl-D-aspartate-sensitive glutamate receptors induce calcium-mediated arachidonic acid release in primary cultures of cerebellar granule cells. J Neurochem 55:1975.

Lynch MA, Errington ML, Bliss TVP (1989) Nordihydroguaiaretic acid blocks the synaptic component of long-term potentiation and the associated increases in release of glutamate and arachidonate: An *in vivo* study in the dentate gyrus of the rat, Neuroscience. 30:693.

Murphy S, Pearce B, Jeremy J, Dandona P (1988) Astrocytes as eicosanoid-producing cells. Glia 241.

Ordway RW, Singer JJ, Walsh JV Jr (1991) Direct regulation of ion channels by fatty acids. Trends Neurosci 14:96.

Ordway RW, Walsh JV Jr, Singer JJ (1989) Arachidonic acid and other fatty acids directly activate potassium channels in smooth muscle cells. Science 244:1176.

Pace-Asciak CR (1984) Arachidonic acid epoxides: Demonstration through [18]oxygen studies of an intramolecular transformation of the terminal hydroxyl group of 12(S)-hydroperoxy-5,8,10,14-tetraenoic acid to form hydroxy epoxides. J Biol Chem 259:8332.

Piomelli D, Shapiro E, Zipkin R, Schwartz JH, Feinmark SJ (1989) Formation and action of 8-hydroxy-11,12-epoxy-5,9,14-eicosatrienoic acid in *Aplysia*: A possible second messenger in neurons. Proc Natl Acad Sci USA 86:1721.

Piomelli D, Volterra A, Dale N, Siegelbaum SA, Kandel ER, Schwartz JH, Belardetti F (1987) Lipoxygenase metabolites of arachidonic acid as second messengers for presynaptic inhibition of *Aplysia* sensory cells. Nature 328:38.

Powell WS (1985) Reversed-phase high-pressure liquid chromatography of arachidonic acid metabolites formed by cyclooxygenase and lipoxygenases. Anal Biochem 148:59.

Sawada S, Higashima M, Yamamoto C (1985) Inhibitors of high-affinity uptake augment depolarizations of hippocampal neurons induced by glutamate, kainate and related compounds. Exp Brain Res 60:323.

Schweitzer P, Madamba S, Siggins GR (1990) Arachidonic acid metabolites as mediators of somatostatin-induced increase of neuronal M-current. Nature 346:464.

Shearman MS, Naor Z, Sekiguchi K, Kishimoto A, Nishizuka Y (1989) Selective activation of the γ-subspecies of protein kinase C from bovine cerebellum by arachidonic acid and its lipoxygenase metabolites. FEBS Lett 243:177.

Wolfe LS, Pellerin L, Drapeau C, Rostworowski K (1990) Formation of 12-lipoxy-genase metabolites in rat cerebral cortical slices: Stimulation by calcium ionophore, glutamate and N-methyl-D-aspartate. J Neural Transm [Suppl] 29:29.

APLYSIA CALIFORNICA CONTAINS A NOVEL 12-LIPOXYGENASE WHICH GENERATES BIOLOGICALLY ACTIVE PRODUCTS FROM ARACHIDONIC ACID

Steven J. Feinmark,[1] Douglas J. Steel,[2]
Anoopkumar Thekkuveettil,[3] Mayumi Abe,[3]
Xiang-Duan Li,[4] and James H. Schwartz[3]

[1]Department of Pharmacology
[2]Department of Pathology
[3]Center for Neurobiology and Behavior
Howard Hughes Medical Institute
Columbia University
New York, NY 10037

[4]Institute of Clinical Pharmacology
Shanghai Medical University
Shanghai 200032 CHINA

ABSTRACT

Physiologic stimulation of identified neurons in ganglia of the marine mollusk, *Aplysia californica*, leads to the generation of arachidonic acid metabolites. Using various preparations of *Aplysia* nervous tissue, we have identified 12-lipoxygenase products including the inactive 12-hydroxyeicosatetraenoic acid (12-HETE) and the biologically active 12-ketoeicosatetraenoic acid (12-KETE) and 8-hydroxy-11(12)-epoxyeicosatrienoic acid (8-HEpETE). Each of these metabolites can be derived from the intermediate 12-hydroperoxyeicosatetraenoic acid (12-HPETE), which can itself activate several identified neurons in *Aplysia*. In spite of conflicting results in studies of mammalian brain 12-lipoxygenase, *Aplysia* nervous tissue clearly contains an enzymatic activity which generates stereochemically pure 12(S)-HETE. This activity is destroyed by boiling and is sensitive to nonspecific lipoxygenase inhibitors but not to agents specific for other lipoxygenases or the cyclooxygenase enzyme. The *Aplysia* 12-lipoxygenase is highly enriched in neural tissue and is almost completely absent in the neural sheath, which is composed primarily of connective tissue and muscle. Preliminary purification has shown that, in contrast to the previously characterized 12-lipoxygenases, the *Aplysia* enzyme is associated with membrane fractions and is not found in the cytosol. Further studies are in progress to determine the kinetic properties and to define the cellular and subcellular distribution of this novel lipoxygenase.

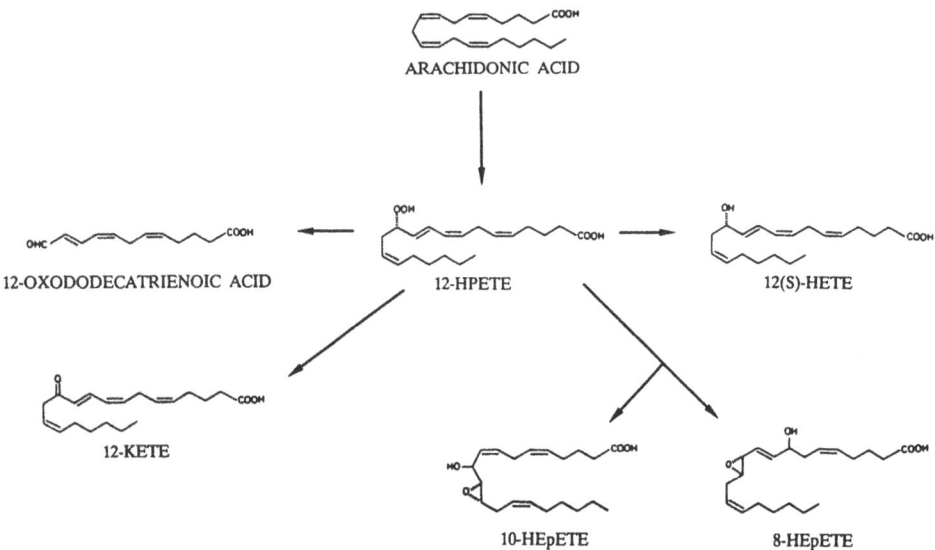

ARACHIDONIC ACID

12-OXODODECATRIENOIC ACID

12-HPETE

12(S)-HETE

12-KETE

10-HEpETE

8-HEpETE

Figure 1. 12-Lipoxygenase metabolites. Modified from Feinmark et al., 1989.

INTRODUCTION

12-Lipoxygenase catalyzes the insertion of molecular oxygen into a number of polyunsaturated fatty acids, producing biologically active hydroperoxides. One prominent substrate for this enzyme is arachidonic acid (eicosa-5Z,8Z,11Z,14Z-tetra-enoic acid) which is converted to 12-hydroperoxyeicosa-5Z,8Z,10E,14Z-tetraenoic acid (12-HPETE). Biologically generated hydroperoxides such as 12-HPETE are fairly transient because of the high levels of peroxidase in many tissues which reduce 12-HPETE to the corresponding hydroxy acid, 12-HETE. 12-HPETE, but not 12-HETE, has been shown to mimic the application of neurotransmitters in several identified *Aplysia* neurons (Piomelli et al., 1987a; 1989a). The hydroperoxide appears to cause an opening of potassium channels which leads to hyperpolarization in both cultured sensory cells and the motor neuron, L14. In spite of its ephemeral nature, 12-HPETE can also undergo a variety of rearrangements which yield biologically active ketones such as 12-ketoeicosa-5Z,8Z,10E,14Z-tetraenoic acid (12-KETE) and epoxy alcohols such as 8-hydroxy-11(12)epoxy-5Z,9E,14Z-tetrienoic acid (8-HEpETE). This pathway is outlined in Figure 1 (reviewed in Feinmark et al., 1989; Piomelli et al., 1989b; Shimizu and Wolfe, 1990). All of these metabolites have been isolated from neural preparations. In addition, a smaller subset has been found to possess various biological activities. For example, 12-KETE induces a depolarization in L14 while 8-HEpETE induces a slow hyperpolarization in this cell. 8-HEpETE-methyl ester has also been found to induce hyperpolarization of hippocampal CA1 neurons (Carlen et al., 1989). Some data suggest that the conversion of 12-HPETE to these metabolites is necessary to produce the observed biological effects (Belardetti et al., 1989) although direct effects have also been detected (Buttner et al., 1989).

At least three distinct forms of the 12-lipoxygenase have now been purified and cloned. The first was isolated from human platelets in 1974 by Hamberg and Samuelsson and later cloned (Funk et al., 1990). Yamamoto and his co-workers isolated a 12-lipoxygenase from porcine leukocytes, which was distinct from the platelet enzyme (Yokoyama et al., 1986; Takahashi et al., 1988). A third type of 12-lipoxygenase, immunologically related to the porcine leukocyte enzyme, has been purified (Hansbrough et al., 1990) and cloned (DeMarzo et al., 1991) from bovine tracheal epithelium. The tracheal enzyme is 87% homologous with the porcine leukocyte enzyme and appears to be more closely related to the human reticulocyte 15-lipoxygenase than to the human platelet 12-lipoxygenase (DeMarzo et al., 1991). These enzymes have several common features including their localization to the cytosol, the apparent lack of dependence on divalent cations for full activity, and the requirement for low levels of hydroperoxide to activate the enzyme or, in the absence of this priming, a characteristic lag phase prior to reaching maximum velocity (Yamamoto, 1989). In spite of these commonalities and the 60-90% homologies at the amino acid level, there are important differences in substrate specificities and inactivation kinetics. One goal of the studies described here is to determine whether the *Aplysia* 12-lipoxygenase is similar to one of the previously characterized enzymes or if it constitutes an entirely new (perhaps a neuronal-type?) form of this enzyme.

Recently, data supporting a role for the 12-lipoxygenase as a source of inter- and intracellular signaling molecules in both vertebrate and invertebrate nervous tissue have begun to accumulate (see Shimizu and Wolfe, 1990, for a review). Although a lipoxygenase product 12-HETE was identified in mammalian brain homogenates in 1978 (Sautebin et al., 1978), the possibility that formed elements of the blood were the true enzymatic source of this lipid continues to be debated today. Wolfe and his collaborators have recently shown that rat cortical slices generate stereochemically pure 12(S)-HETE in response to exposure to specific neurotransmitters (Pellerin et al., 1990). Others, however, have found that 12(S)-HETE production is associated with platelet contamination of the brain slice (Kim et al., 1991) and demonstrated that platelet-free brain homogenates generate only racemic mixtures of the hydroxy acids. This result suggests that the observed products arise by autoxidation and implies that the neural tissue does not even contain a functional 12-lipoxygenase. We have addressed this question directly in *Aplysia* where the neural tissue is not contaminated by blood elements. In these studies, intact *Aplysia* ganglia release 12-HETE after exposure to the neurotransmitter histamine or after intracellular stimulation of an identified neuron (Piomelli et al., 1987b). Cultured sensory neurons respond to the peptide transmitter FMRFamide (Piomelli et al., 1987a) in a similar way. To substantiate further the hypothesis that there is a neural 12-lipoxygenase, we have determined the tissue distribution of the invertebrate 12-lipoxygenase activity and the stereochemistry of its product.

METHODS

Tissue Preparations

Aplysia californica (70-100 gm) were anesthetized with magnesium chloride, the ganglia were removed, and in some cases the neural sheath was dissected as previously reported (Piomelli et al., 1987b). The tissue was placed in cold artificial sea water containing protease inhibitors (DTT, PMSF, benzamidine, aprotinin, and leupeptin) and homogenized in a Brinkmann Polytron (speed 10, 1 s). In some experiments, the tissue was placed in 1.1 M sucrose buffer and disrupted manually in a glass-glass homogenizer as described elsewhere (Chin et al., 1989). Generally the homogenates were spun at 80 × g for 5 min, the resulting supernatant was spun at 10,000 × g for 20 min, and the next supernatant was spun at 150,000 × g for 40 min. The resulting pellets were resuspended in artificial seawater for protein determination and assessment of enzyme activity. Abdominal ganglia were prepared for electrophysiological recording as described in detail elsewhere (Piomelli et al., 1989a).

Enzyme Assay and Chiral Analysis

Crude homogenate or the centrifugal fractions were incubated with [^{3}H]arachidonic acid (2 μCi; 76 Ci/mmol) in the presence of the activating hydroperoxide 13-HPODE (5 μM). The incubations were allowed to proceed for 3 min at 15° C and were stopped by the addition of ice cold acetone (2 vol). This mixture was allowed to stay at -20° C for at least 30 min before the precipitated proteins were removed by centrifugation. Trimethylphosphite (3.4 mM) was added to reduce any remaining hydroperoxides to the corresponding hydroxy compounds and the lipids were extracted either with ethyl acetate or diethyl ether. The extracts were dried, reconstituted in HPLC mobile phase (acetonitrile/water, 50:50, v/v; pH 4.5), and fractionated on a Waters Nova-Pak C$_{18}$ column eluted at 0.7 ml/min. This permitted complete separation of 12-HETE from other known hydroxy acids. Fractions were collected and the radioactivity was measured by liquid scintillation counting. Recovery was corrected by normalization to the 13-HPODE peak.

Experiments designed to determine the chirality of the *Aplysia* lipoxygenase product were carried out without added 13-HPODE. After the incubation, racemic 12-HETE (1 μg) was added to serve as carrier and internal standard. After 12-HETE was HPLC-purified from the lipid extract as described above, it was converted to the methyl ester by treatment with diazomethane. 12-HETE-ME was resolved into its component stereoisomers by chiral phase HPLC on two DNBPG columns in series (each column was 4.6 × 250 mm) eluted with hexane/isopropanol (100:0.5, v/v) at a flow rate of 0.5 ml/min. Since each sample contained standard racemic 12-HETE to serve as a retention marker, UV absorbing material was monitored at 235 nm. Fractions were collected for liquid scintillation counting.

Electrophysiological Recording

The details of these experiments have been described elsewhere (Piomelli et al., 1988; 1989a). In brief, abdominal ganglia were desheathed and pinned in a superfusion chamber flushed with artificial seawater. The cells were impaled for intracellular recording with standard glass microelectrodes filled with potassium citrate (1-20 megaohm). Test samples were applied by pressure ejection from glass micropipettes directly above the impaled cell and into the superfusion stream.

RESULTS AND DISCUSSION

Tissue and Subcellular Localization of Aplysia 12-Lipoxygenase

Aplysia ganglia are composed of neural tissue surrounded by a muscular and connective tissue sheath. Homogenates of intact ganglia were incubated with [^{3}H]arachidonic acid and the products were recovered by extractive isolation. Analysis of the lipids generated by the homogenates showed two major products (Fig. 2A). The initial peak (labeled peak I) eluted at the expected retention time of 12-HETE as previously reported (Piomelli et al., 1987b). The second peak (labeled peak II), which predominated in these incubations, has not yet been identified. Ganglia were manually dissected into two portions. The first contained neural tissue, while the second was

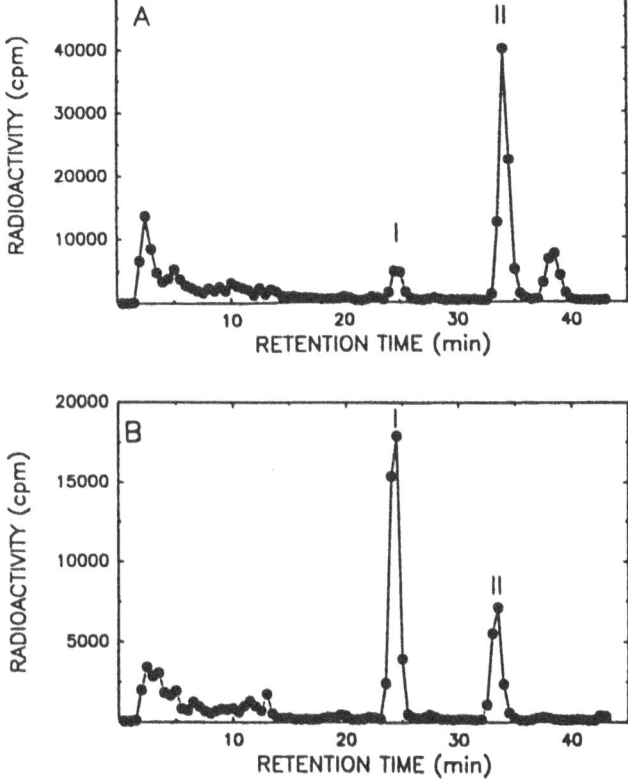

Figure 2. Metabolism of [³H]arachidonic acid by *Aplysia* nervous tissue. Ganglia were either homogenized intact (A) or after dissection of the neural sheath (B). Homogenates were incubated with [³H]arachidonic acid and the products were isolated and fractionated by reverse phase HPLC as described in the Methods. Peak I eluted at the expected retention time of 12-HETE. Peak II does not correspond to known standards and has not yet been characterized further.

composed mainly of the neural sheath. Homogenates of these preparations were made and incubated with [³H]arachidonic acid in a manner similar to the whole tissue homogenates. 12-Lipoxygenase activity was found in both tissues but was primarily concentrated in the neural tissue (Fig. 2B). The large unidentified arachidonic acid metabolite was a minor product in this fraction but continued to be the major product of the isolated neural sheath (data not shown). Because the unidentified metabolite appears to be formed outside of the isolated neurons, we have not yet attempted to characterize its structure. Based solely on HPLC retention, however, this peak does not correspond to any known arachidonate-derived hydroxy acid.

In a preliminary attempt to characterize and purify the neural 12-lipoxygenase from *Aplysia*, we have undertaken various differential centrifugation protocols modeled on the studies of Chin et al. (1989) and Piomelli et al. (1987b). It is clear from these efforts that the lipoxygenase activity is lost fairly quickly. This instability may relate to catalytic inactivation which appears to be common in the leukocyte-type enzymes (Takahashi et al., 1988; Hansbrough et al., 1990). A representative purification protocol is presented in Table 1.

In spite of the steady loss of activity described in this table, one important characteristic of the invertebrate neural enzyme is clear. In contrast to the three mammalian, non-neuronal enzymes, all of which are found in the soluble fraction, the

Table 1. Partial Purification of *Aplasia* 12-Lipoxygenase by Differential Centrifugation

Subcellular Fraction	Volume (ml)	Total Units (pmol/min × 10^{-3})	Protein (μg/ml)	Specific Activity (units/μg)	Yield (%)
Supernatant					
80 × g	1.3	5.4	850	4.9	100
10K × g	1.1	1.5	600	2.2	28
150K × g	0.65	0.1	178	0.8	2
Pellet					
10K × g	1.0	1.1	180	6.1	20
150K × g	1.0	0.4	80	5.0	7

12-Lipoxygenase activity was determined in various subcellular fractions. Incubations contained [^{3}H]arachidonic acid (2 μCi; 76 Ci/mmol) as substrate. The dilution of the label by endogenous lipids in the enzyme preparation was ignored. One unit was arbitrarily defined as that amount of enzyme which generated one pmol of 12-HETE/min under standard conditions described in the Methods.

Aplysia 12-lipoxygenase is concentrated in the particulate fractions. Both the 10,000 × g and 150,000 × g pellets had equal or higher specific activity than the starting homogenate. Similar fractions were prepared and analyzed by electron microscopy. The high speed pellet was composed primarily of small membranous organelles which were also abundant in the 10,000 × g pellet. In addition to the small vesicles, the low speed pellet contained larger structures including lysosomes and synaptosomes (Chin et al., 1989). Although it will be necessary to solubilize and stabilize this enzyme activity, these findings clearly distinguish the invertebrate enzyme from the known mammalian lipoxygenases.

Stereochemical Analysis of the Aplysia 12-Lipoxygenase Product

Consistent with what is known about most mammalian lipoxygenases, physiological experiments have tested the activity of 12(S)-HPETE (Belardetti et al., 1989; Buttner et al., 1989; Piomelli et al., 1989a). However, it is not safe to assume that an *Aplysia* enzyme actually generates the same product. This became particularly clear when Brash and his co-workers described an invertebrate enzyme with the opposite stereochemistry (Hawkins and Brash, 1987). Furthermore, it is not always possible to detect differences in biological activity between 12(S)- and 12(R,S)-HPETE (Belardetti et al., 1989; Feinmark et al., 1990; Marc Klein, personal communication). Recently, Kim et al. (1991) have shown that rat brain homogenates readily catalyze the production of racemic hydroxy acids, suggesting that the auto-oxidative processes could be responsible for much of the apparent lipoxygenase activity observed in earlier work. Since it is clear that polyunsaturated fatty acids are readily oxidized by nonenzymatic processes to yield racemic hydroperoxides, the demonstration of stereochemical purity is generally accepted as an indication of a specific enzyme activity. In addition to substantiating the presence of a 12-lipoxygenase in *Aplysia*, stereochemical information is also critical to interpreting the physiologic results adequately.

Aplysia ganglia were homogenized and incubated with [^{14}C]arachidonic acid and the lipoxygenase products were isolated and purified by reverse phase HPLC. In other experiments, intact abdominal ganglia were prelabeled with [U-^{14}C]arachidonate and the 12-lipoxygenase was activated by the application of histamine. In both cases the purified 12-HETE was converted to the methyl ester and analyzed by chiral phase HPLC. Both neural homogenates and histamine stimulated intact ganglia to generate a stereochemically pure product which was identified as 12(S)-HETE (Fig. 3).

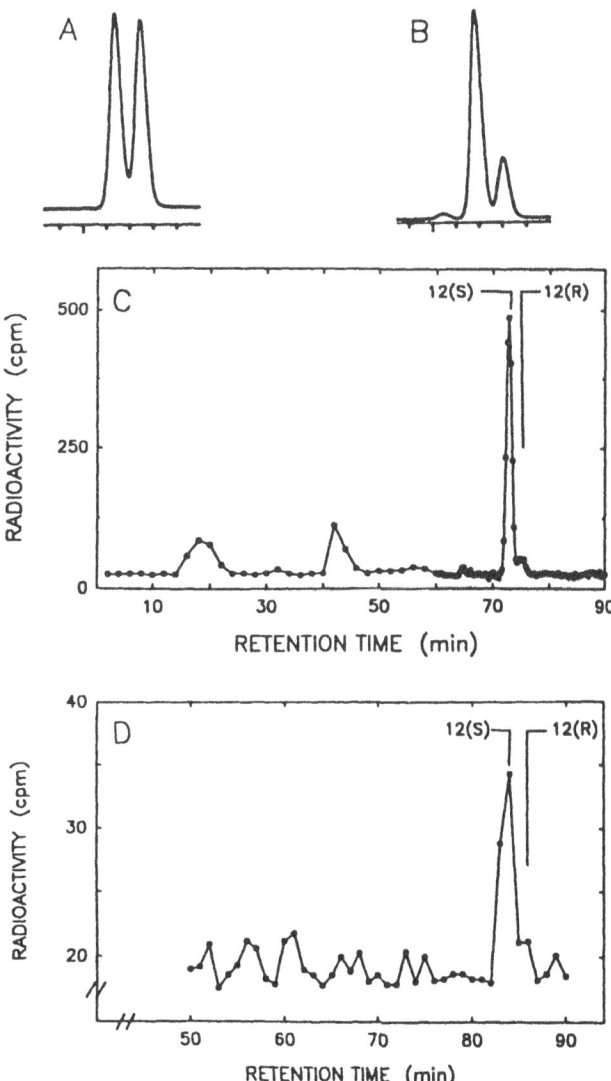

Figure 3. Chiral analysis of *Aplysia* 12-lipoxygenase products. The two stereoisomers of 12-HETE methyl ester (ME) were resolved by a chiral phase HPLC. Either standard or reverse phase purified lipids were resolved on DNBPG columns as described in the Methods. A. Partial chromatogram of racemic 12-HETE-ME. B. Partial chromatogram of racemic 12-HETE-ME with added 12(S)-HETE-ME. C. Radioactivity profile of [^{14}C]12-HETE-ME derived from *Aplysia* neural homogenate incubated with exogenous arachidonate. D. Partial radioactivity profile of [^{14}C]12-HETE-ME derived from a histamine-stimulated abdominal ganglion.

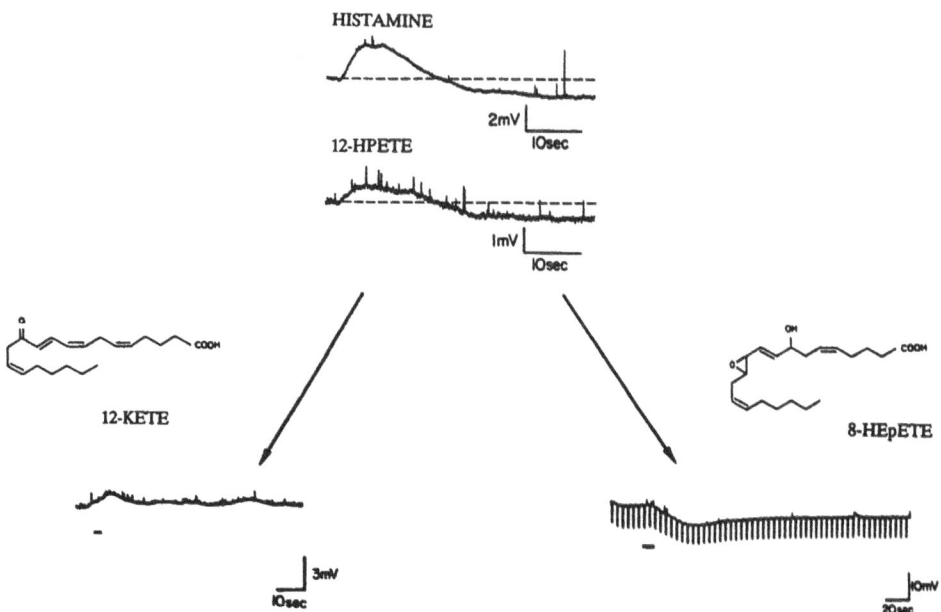

Figure 4. Biological responses of an identified *Aplysia* neuron L14 to histamine and several 12-lipoxygenase metabolites. The membrane potential of L14 neurons was recorded intracellularly as detailed elsewhere (Piomelli, Shapiro et al., 1989). Stimuli were applied to the cells by pressure ejection into the superfusion stream as indicated in the figure. Histamine (100 pmol) was applied over 1 s to a cell with a resting potential of -70 mV. 12(S)-HPETE (100 pmol over 3 s) was applied to the same cell. 12-KETE (1 nmol over 5 s) and 8-HEpETE (24 nmol over 10 s; resting potential -40 mV) were each applied to different cells. These traces are adapted from Piomelli et al., 1988; 1989a and Feinmark et al., 1989.

Biological Activity of 12-Lipoxygenase Products in Aplysia

The first intermediate that can be isolated from the 12-lipoxygenase pathway is 12(S)-HPETE, which is rapidly converted to either 12(S)-HETE or one of the products described in Figure 1. We have tested these metabolites for biological activity on the identified neuron L14 in *Aplysia* abdominal ganglion. L14 responds to histamine with a dual action response (depolarization followed by a slow hyperpolarization; Fig. 4). This response is mimicked by the application of 12(S)-HPETE (Fig. 4) while application of 12(S)-HETE has no effect (Piomelli et al., 1989a). This result suggested that 12-HPETE may serve as an intermediate in the synthesis of other biologically active lipids. This hydroperoxy acid is known to yield several epoxy alcohols (Pace-Asciak et al., 1983; German and Kinsella, 1986), as well as a ketone and an aldehyde (Glasgow et al., 1986; Fruteau de Laclos et al., 1987). The production of these metabolites in *Aplysia* was confirmed and they were also tested for activity on L14 (Piomelli et al., 1988; 1989a). It was found that the two phases of the L14 dual action response could be separately induced by application of these 12-HPETE metabolites. 12-KETE caused a depolarization of L14 while 8-HEpETE induced the hyperpolarization (Fig. 4).

CONCLUSIONS

Aplysia nervous tissue contains a 12-lipoxygenase which generates 12(S)-HPETE. This tissue also has the ability to convert 12-HPETE into 12-KETE and 8-HEpETE, among other products. These three arachidonic acid metabolites are generated by intact abdominal ganglia after the application of neurotransmitter or after physiological stimulation of identified neurons and appear to be responsible for the well-characterized dual action response of L14 neurons. 12(S)-HPETE causes both phases of this response and thus wholly mimics the effect of the neurotransmitter histamine on this cell, i.e., 12-KETE induces the depolarization and 8-HEpETE induces the hyperpolarization.

The enzyme that initiates this cascade of neuromodulators appears to be localized in the neural tissue and not to arise from the surrounding muscle or connective tissue. Interestingly, the neural sheath also actively metabolizes arachidonate but generates a distinct and as yet unidentified product. The *Aplysia* 12-lipoxygenase seems to differ from other known lipoxygenases in that it localizes with membrane fractions during differential centrifugation. This observation, coupled with the concentration of the enzyme activity in the neuronal tissue of the ganglion, raises the possibility that this is a novel, neuron-specific enzyme which could participate in both inter- and intraneuronal signaling.

In the future, it will be important to purify and characterize this 12-lipoxygenase and determine how it is related to similar enzymes in other species. This will also permit complete cellular and subcellular localization of this enzyme in *Aplysia* neurons and aid in defining the precise physiologic role of the pathway.

ACKNOWLEDGMENTS

We thank Alice Elste for performing the electron micrographic studies of the subcellular fractions.

REFERENCES

Belardetti F, Campbell WB, Falck JR, Demontis G, Rosolowsky M (1989) Products of heme-catalyzed transformation of the arachidonate derivative, 12-HPETE, open S-type K^+ channels in *Aplysia*. Neuron 3:497.

Buttner N, Siegelbaum SA, Volterra A (1989) Direct modulation of *Aplysia* S-K^+ channels by a 12-lipoxygenase metabolite of arachidonic acid. Nature 342:553.

Carlen PL, Gurevich N, Wu PH, Su W-G, Corey EJ, Pace-Asciak CR (1989) Actions of arachidonic acid and hepoxilin A_3 on mammalian hippocampal CA1 neurons. Brain Res 497:171.

Chin GJ, Shapiro E, Vogel SS, Schwartz JH (1989) *Aplysia* synaptosomes. I. Preparation and biochemical and morphological characterization of subcellular membrane fractions. J Neurosci 9:38.

DeMarzo N, Sloane DL, Dicharry S, Highland E, Sigal E (1991) Molecular cloning and expression of a new isoform of 12-lipoxygenase: Hypothesis for positional specificity. XIth Washington International Spring Symposium, p 76 (Abstract).

Feinmark SJ, Abe M, Shapiro E, Brezina V, Schwartz JH (1990) Identification of neuromodulators produced by the 12-lipoxygenase in *Aplysia*. In: Advances in prostaglandin, thromboxane and leukotriene research, Vol 21 (Samuelsson B, Ramwell PW, Paoletti R, Folco G, Granström E, eds) p 715, New York: Raven Press.

Feinmark SJ, Piomelli D, Shapiro E, Schwartz JH (1989) Biologically active metabolites of the 12-lipoxygenase pathway are formed by *Aplysia* nervous tissue. Ann NY Acad Sci 559:121.

Fruteau de Laclos B, Maclouf J, Poubelle P, Borgeat P (1987) Conversion of arachidonic acid into 12-oxo derivatives in human platelets. A pathway possibly involving the heme-catalyzed transformation of 12-hydroperoxyeicosatetraenoic acid. Prostaglandins 33:315.

Funk CD, Furci L, FitzGerald GA (1990) Molecular cloning, primary structure, and expression of the human platelet/erythroleukemia cell 12-lipoxygenase. Proc Natl Acad Sci USA 87:5638.

German JB and Kinsella JE (1986) Production of trihydroxy derivatives of arachidonic and docosahexaenoic acid by lipoxygenase activity in trout gill tissue. Biochim Biophys Acta 877:290.

Glasgow WC, Harris TM, Brash AR (1986) A short-chain aldehyde is a major lipoxygenase product in arachidonic acid-stimulated porcine leukocytes. J Biol Chem 261:200.

Hamberg M and Samuelsson B (1974) Prostaglandin endoperoxides: Novel transformations of arachidonic acid in human platelets. Proc Natl Acad Sci USA 71:3400.

Hansbrough JR, Takahashi Y, Ueda N, Yamamoto S, Holtzman MJ (1990) Identification of a novel arachidonate 12-lipoxygenase in bovine tracheal epithelial cells distinct from leukocyte and platelet forms of the enzyme. J Biol Chem 265:1771.

Hawkins DJ and Brash AR (1987) Eggs of the sea urchin, *Strongylocentrotus purpuratus*, contain a prominent (11R) and (12R) lipoxygenase activity. J Biol Chem 262:7629.

Kim H-Y, Sawazaki S, Salem Jr N (1991) Lipoxygenation in rat brain? Biochem Biophys Res Commun 174:729.

Pace-Asciak CR, Granström E, Samuelsson B (1983) Arachidonic acid epoxides. Isolation and structure of two hydroxy epoxide intermediates in the formation of 8,11,12- and 10,11,12-trihydroxy eicosatrienoic acids. J Biol Chem 258:6835.

Pellerin L, Drapeau C, Avoli M, Wolfe LS (1990) Formation of 12-lipoxygenase metabolites in brain: Possible modulators of synaptic transmission. Proceedings VII International Prostaglandin Conference, p 355 (Abstract).

Piomelli D, Feinmark SJ, Shapiro E, Schwartz JH (1988) Formation and biological activity of 12-ketoeicosatetraenoic acid in the nervous system of *Aplysia*. J Biol Chem 263:16591.

Piomelli D, Feinmark SJ, Shapiro E, Schwartz JH (1989b) 12-Keto-eicosatetraenoic acid: A biologically active eicosanoid in the nervous system of *Aplysia*. Ann NY Acad Sci 559:208.

Piomelli D, Shapiro E, Feinmark SJ, Schwartz JH (1987b) Metabolites of arachidonic acid in the nervous system of *Aplysia*: Possible mediators of synaptic modulation. J Neurosci 7:3675.

Piomelli D, Shapiro E, Zipkin R, Schwartz JH, Feinmark SJ (1989a) Formation and action of 8-hydroxy-11,12-epoxy-5,9,14-icosatrienoic acid in *Aplysia*: A possible second messenger in neurons. Proc Natl Acad Sci USA 86:1721.

Piomelli D, Volterra A, Dale N, Siegelbaum SA, Kandel ER, Schwartz JH, Belardetti F (1987a) Lipoxygenase metabolites of arachidonic acid as second messengers for presynaptic inhibition of *Aplysia* sensory cells. Nature 328:38.

Sautebin L, Spangnuolo C, Galli C, Galli G (1978) A mass fragmentographic procedure for the simultaneous determination of HETE and $PGF_{2\alpha}$ in the central nervous system. Prostaglandins 16:985.

Shimizu T and Wolfe LS (1990) Arachidonic acid cascade and signal transduction. J Neurochem 55:1.

Takahashi Y, Ueda N, Yamamoto S (1988) Two immunologically and catalytically distinct arachidonate 12-lipoxygenases of bovine platelets and leukocytes. Arch Biochem Biophys 266:613.

Yamamoto S (1989) Mammalian lipoxygenases: Molecular and catalytic properties. Prostaglandins Leukot Essent Fatty Acids 35:219.

Yokoyama C, Shinjo F, Yoshimoto T, Yamamoto S, Oates JA, Brash AR (1986) Arachidonate 12-lipoxygenase purified from porcine leukocytes by immunoaffinity chromatography and its reactivity with hydroperoxyeicosatetraenoic acids. J Biol Chem 261:16714.

ESSENTIAL FATTY ACID DEFICIENCY IN CULTURED SK-N-SH HUMAN

NEUROBLASTOMA CELLS

E.B. Stubbs, Jr., R.O. Carlson,
C. Lee, S.K. Fisher, A.K. Hajra, and
B.W. Agranoff

Mental Health Research Institute
Departments of Psychiatry,
Biological Chemistry and Pharmacology
Neuroscience Laboratory Building
University of Michigan
Ann Arbor, MI 48104-1687
USA

SUMMARY

SK-N-SH neuroblastoma cells grown under standard culture conditions contain significant amounts of Mead acid ($20:3\omega9$) in phospholipids, indicating essential fatty acid (EFA) deficiency. The amount of esterified $20:3\omega9$ was augmented by growth in a chemically defined EFA-free medium, whereas its presence could be virtually eliminated by supplementation of the culture medium with either arachidonic ($20:4\omega6$; AA), eicosapentaenoic ($20:5\omega3$; EPA), or linolenic ($18:3\omega3$) acids. Substitution of Mead acid for $\omega6$ fatty acids, particularly evident in phosphatidylinositol (PI), indicates a compensatory replacement of $\omega9$ for $\omega6$ fatty acids during EFA deficiency. Studies evaluating [^{3}H]scopolamine binding to the M_3 muscarinic acetylcholine receptors (mAChRs) present in these neurotumor cells as well as effects of carbachol on phosphoinositide turnover and intracellular Ca^{2+} mobilization, indicate that the biosubstitution of $20:4\omega6$ with $20:3\omega9$ does not detectably impair these measures of signal transduction. Stimulation of mAChRs with carbachol increased the cellular mass of diacylglycerol (DAG) approximately 60%. On the basis of distinctive fatty acid "signatures" of each of the phospholipid classes, it is concluded that the DAG initially released following muscarinic stimulation is derived from phosphoinositide breakdown. After several minutes, however, a significant amount of DAG comes from phosphatidylcholine (PC) as well. In contrast to DAG, the composition of phosphatidate (PA) following receptor stimulation closely resembles that of the phosphoinositides, even at the later time points examined. These results support a selective phosphorylation of DAG arising from the stimulated breakdown of phosphoinositides, favoring the conservation of the 1-stearoyl, 2-arachidonoyl (or $20:3\omega9$) moiety.

Neurobiology of Essential Fatty Acids
Edited by N.G. Bazan *et al.*, Plenum Press, New York, 1992

INTRODUCTION

The ω6 family of EFAs, including linoleic acid (18:2ω6) and elongated derivatives such as AA, are characteristically enriched in the central nervous system (CNS) and serve as critically important precursors for many cerebral bioactive compounds, including prostaglandins, hydroxyeicosatetraenoic acid, leukotrienes, and lipoxins (Wolfe, 1982; Samuelsson et al., 1987; Needleman et al., 1979). The ω3 series of EFAs is also well represented in brain lipids. Omission of EFAs from the mammalian diet reduces the content of both ω6 and ω3 polyunsaturated fatty acids (PUFAs) in peripheral tissue lipids resulting in a variety of pathological conditions characterized by impaired growth, infertility, dermatological lesions and kidney necrosis (Holman, 1968). The phospholipid fatty acid composition of the adult brain is uniquely resistant to dietary manipulation (Chaudiere et al., 1987; Neuringer et al., 1988), reflecting structural stringencies and/or essential physiological functions of cerebral eicosanoids (White and Hagen, 1982). An abnormality frequently associated with EFA deficiency is a reduction in the content of ω6 fatty acids and the accumulation of ω9 fatty acid metabolites (18:1ω9 and 20:3ω9) in cellular lipids (Holman, 1968). Early studies (Dhopeshwarkar and Mead, 1961) showed that ω9 and ω6 fatty acids compete for the same desaturase and chain elongation enzyme systems, suggesting that the *de novo* synthesis of Mead acid during EFA deficiency results from a lack of ω6 substrate competition.

In contradistinction to the CNS, the fatty acid composition of phospholipids recovered from neurotumor cells maintained in culture often contain small amounts of PUFAs, characteristic of EFA deficiency (Robert et al., 1977; McGee, 1981; Murphy, 1984; Spector and Yorek, 1985). Supplementation of the growth medium with essential fatty acids restores the phospholipid fatty acid profile of cultured cells to a more physiologic degree of unsaturation (McGee, 1981; Hyman and Spector, 1982; Park et al., 1990). Modification of phospholipid PUFA content has been shown to alter membrane fluidity (Stubbs and Smith, 1984) and affect membrane functions, such as carrier-mediated transport of excitatory and inhibitory amino acids (Balcar et al., 1980; Hyman and Spector, 1982; Yorek et al., 1984), catecholamine uptake (Brenneman and Rutledge, 1979), and insulin and opioid-receptor binding (Ginsberg et al., 1982; Ho and Cox, 1982; Remmers et al., 1990). Effector enzyme systems may also be sensitive to changes in membrane fatty acid composition. Murphy (1986) reported a significant increase in basal adenylate cyclase activity independent of prostanoid synthesis in EFA-enriched N1E-115 neurotumor cells. Ligand-receptor mediated exocytosis (Williams and McGee, 1982) and phagocytosis (Schroit and Gallily, 1979) are also affected by the degree of membrane unsaturation. These studies suggest that the degree of phospholipid polyunsaturation may be an important modulator of ligand-receptor-effector coupling in neural membranes. The present study evaluates mAChR-stimulated phosphoinositide-associated transduction events in cultured human neurotumor cells grown in EFA-deficient or AA-enriched medium. The results of a preliminary study have been published (Stubbs et al., 1990).

EXPERIMENTAL PROCEDURES

Human SK-N-SH neuroblastoma cells were grown at 37°C in a humidified atmosphere (10% CO_2) in Dulbecco's modified Eagle medium (DMEM) supplemented with 10% fetal bovine serum (FBS) or chemically defined (Bottenstein and Sato, 1979) DMEM as specified. Subcultures were maintained for 24-48 hr prior to fatty acid supplementation. Fatty acids (50-100 μM) were added to the culture medium as a complex with fatty acid-poor bovine serum albumin (BSA) at a molar ratio of 4:1 (McGee, 1981). Control cultures received uncomplexed BSA. Cultures were typically supplemented for 48 hr, harvested, washed, and immediately assayed. Cellular phospholipids were extracted in the presence of 50 μg/ml of butylated hydroxytoluene with chloroform/methanol (C/M; 2:1) followed by reextraction of the upper phase and interphase with acidified C/M (4:1). The lower phases were neutralized with NH_4OH, combined, dried under N_2, and the lipids separated by 2D TLC (Sun et al., 1988). Fatty methyl esters (FAMEs) were prepared by base methanolysis of the individual phospholipids, separated by GLC, and quantified using heptadecanoate as an internal standard. Intracellular calcium was monitored with Fura-2/acetoxy methylester (Fisher et al., 1989). [^{3}H]Scopolamine equilibrium binding studies were performed by incubating intact cells (0.3 mg protein) for 120 min at 37°C in the presence or absence of atropine (25 μM). Resting and PGE_1-stimulated levels of cAMP in intact neuroblastoma cells were determined from $HClO_4$ extracts by ^{125}I radioimmunoassay. Extraction, HPLC separation, and quantitative determination of DAG and PA molecular species were determined as previously described (Lee and Hajra, 1991).

RESULTS AND DISCUSSION

Phospholipid Composition and EFA Deficiency

A variety of cells cultured under standard conditions contain Mead acid as a component of cellular membrane phospholipids, signaling an EFA deficiency. The presence of bovine sera in the culture medium is unlikely to alleviate the EFA deficiency (Robert et al., 1977), since the bovine rumen effectively hydrogenates $\omega3$ and $\omega6$ fatty acids ingested from plant sources. As discussed below, the cellular EFA deficiency may or may not lead to detectable physiological consequences, depending on the cell type used and the physiological parameter measured.

The phospholipid profile of human SK-N-SH neuroblastoma cells cultured to confluence in a standard growth medium (DMEM) supplemented with 10% FBS is shown in Table 1.

Phosphatidylcholine (PC) is the predominant phospholipid present in these neuro-tumor cells, representing approximately 63% (by weight) of the five phospholipid groups analyzed. This apparent predominance of PC relative to the other phospholipids differs from brain tissue and can be accounted for by the specific lack in these cells of phosphatidalethanolamine (PE plasmalogens; Table 1). Brain is particularly enriched in PE plasmalogens, which represent approximately 14-15% of the major phospholipids,

Table 1. Phospholipid composition of EFA-deficient or arachidonic acid-enriched human SK-N-SH neuroblastoma cells

Phospholipid	EFA-Deficient	AA-Enriched
	(μg/mg protein)	
Phosphatidylcholine	32.8 ± 0.5	35.1 ± 2.4
Phosphatidylethanolamine	8.2 ± 0.2	10.6 ± 1.2
Phosphatidylserine	4.6 ± 0.2	4.8 ± 0.6
Phosphatidylinositol	3.9 ± 0.6	4.4 ± 0.7
Phosphatidalethanolamine (Plasmalogen)	3.0 ± 0.1	3.2 ± 0.2

Freshly passaged cells were cultured in EFA-deficient growth medium supplemented with AA(93 μM)/BSA or BSA (control) for 48 h at 37°C, 10% CO_2. Tabulated data are the means ± SEM from three cell preparations.

Table 2. Phospholipid fatty acid composition from EFA-deficient or AA-enriched cells

Acyl Groups	EFA-Deficient	AA-Enriched
Diacyl-GPC, % wt		
14:0	3.1 ± 0.3	4.8 ± 0.1**
16:0	38.5 ± 1.0	45.1 ± 1.0**
16:1	12.3 ± 0.8	6.7 ± 0.3**
18:0	5.2 ± 0.3	7.1 ± 0.2**
18:1	34.9 ± 2.1	22.2 ± 1.0**
18:2	< 0.1	< 0.1
18:3	< 0.1	< 0.1
20:3	1.3 ± 0.1	0.6 ± 0.1**
20:4	1.3 ± 0.1	9.2 ± 0.4**
22:4	0.3 ± 0.1	3.0 ± 0.1**
22:5	0.1	0.1 ± 0.0
22:6	0.6 ± 0.2	0.4 ± 0.1
Others	2.4	0.8
Mono/Sat	1.01	0.51
Poly/Sat	0.13	0.25
Diacyl-GPS, % wt		
14:0	< 0.1	< 0.1
16:0	4.0 ± 0.9	3.5 ± 0.6
16:1	3.9 ± 0.3	1.7 ± 0.1**
18:0	40.6 ± 1.0	45.2 ± 0.3**
18:1	36.3 ± 1.4	23.0 ± 1.3**
18:2	< 0.1	< 0.1
18:3	< 0.1	< 0.1
20:3	1.9 ± 0.1	0.9 ± 0.1**
20:4	2.0 ± 0.2	5.8 ± 0.6**
22:4	1.7 ± 0.6	12.8 ± 1.3**
22:5	0.8 ± 0.1	0.9 ± 0.1
22:6	2.0 ± 0.3	1.3 ± 0.1
Others	6.8	4.9
Mono/Sat	0.90	0.51
Poly/Sat	0.34	0.55
Diacyl-GPI, % wt		
14:0	< 0.1	< 0.1
16:0	5.6 ± 1.6	4.0 ± 0.8
16:1	2.0 ± 1.0	0.8 ± 0.2
18:0	37.1 ± 1.5	45.6 ± 2.4*
18:1	16.3 ± 0.6	7.5 ± 0.4**
18:2	< 0.1	< 0.1
18:3	< 0.1	< 0.1
20:3	20.6 ± 2.5	2.1 ± 0.1**
20:4	16.7 ± 1.8	38.0 ± 3.0**
22:4	0.2 ± 0.1	1.7 ± 0.0**
22:5	< 0.1	< 0.1
22:6	0.1 ± 0.1	< 0.1
Others	1.4	0.3
Mono/Sat	0.43	0.17
Poly/Sat	0.91	0.85
Diacyl-GPE, % wt		
14:0	0.2 ± 0.2	0.1 ± 0.1
16:0	9.2 ± 1.2	8.5 ± 0.4
16:1	3.6 ± 1.1	1.5 ± 0.5
18:0	27.7 ± 1.1	35.0 ± 0.2**
18:1	33.8 ± 1.5	18.4 ± 2.0**
18:2	< 0.1	< 0.1
18:3	< 0.1	< 0.1
20:3	5.8 ± 0.4	2.6 ± 0.1**
20:4	9.6 ± 0.8	19.0 ± 1.9**
22:4	1.5 ± 0.0	9.7 ± 0.6**
22:5	0.6 ± 0.1	0.5 ± 0.1
22:6	2.9 ± 0.3	1.6 ± 0.1*
Others	5.1	3.1
Mono/Sat	1.01	0.46
Poly/Sat	0.69	0.84
Alkenylacyl-GPE, % wt		
14:0	< 0.1	< 0.1
16:0	4.3	6.2± 0.4
16:1	3.6	2.3 ± 0.5
18:0	2.8 ± 0.8	3.3
18:1	14.7 ± 1.1	9.9 ± 1.3*
18:2	< 0.1	< 0.1
18:3	< 0.1	< 0.1
20:3	7.6 ± 1.3	3.4 ± 0.3*
20:4	20.4 ± 1.4	30.0 ± 1.1**
22:4	7.4 ± 1.3	24.4 ± 5.1*
22:5	3.5 ± 0.6	2.2 ± 0.3
22:6	14.1 ± 1.5	4.8 ± 0.4**
Others	22.1	13.5
Mono/Sat	2.58	1.28
Poly/Sat	10.58	8.24

Data shown are the means ± SEM (n=3). Phospholipid fatty acids were quantitated as described under Experimental Procedures. Significance determined by Student's t test with *, $p < 0.05$, **, $p < 0.01$.

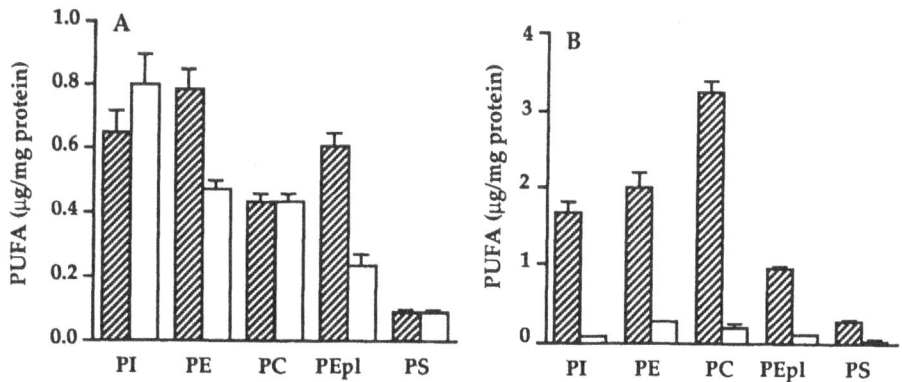

Figure 1. Distribution of Mead acid and AA in phospholipids extracted from SK-N-SH neuroblastoma cells cultured in EFA-deficient (A) or AA (93 μM)-enriched (B) growth medium. In each group, left bar represents AA and right bar represents Mead acid. Data shown are the means ± SEM from three separate cell preparations. PEpl= PE plasmalogen (plasmenyl PE).

whereas only 6% of the phospholipids extracted from SK-N-SH cells are PE plasmalogens. The percentages of phosphatidylinositol (PI), phosphatidylserine (PS), and PE are in agreement with reported values from rat brain (Sun et al., 1988). Although there were no significant differences between culture conditions tested, a trend toward an increase in the amount of each phospholipid class was observed (Table 1).

A detailed analysis of the acyl composition of the individual phospholipids extracted and purified from SK-N-SH cells cultured in EFA-deficient or AA-enriched growth medium is shown in Table 2.

Arachidonic acid supplementation dramatically increased the percent mass of esterified ω6 fatty acids (20:4 and 22:4) while reducing the content of ω9, ω7, and ω3 fatty acids in all five phospholipid classes. The content of esterified saturated fatty acids was also affected by AA supplementation. The percentages of myristate (14:0) and palmitate (16:0) were significantly increased in PC, whereas that of stearate was increased in PS, PI, and PE following AA supplementation, suggesting a concerted regulation of both saturated and unsaturated fatty acid esterification to phospholipids in SK-N-SH cells.

The esterification of Mead acid to phospholipids was highly selective. In SK-N-SH cells cultured in EFA-deficient medium, PI was notably enriched in 20:3ω9 (0.8 μg/mg protein), whereas PE (diacyl and alkenylacyl), PC, and PS contained substantially less 20:3ω9 (Fig. 1A). A substantial remodeling of the 20:3ω9 and 20:4ω6 profiles was seen following supplementation of the culture medium with AA (Fig. 1B).

The percent distribution of 20:3ω9 among the individual phospholipid PUFAs was similarly selective (Table 2). The PUFA contents of PI and PE were predominantly 20:3ω9 and 20:4ω6, whereas that of PE plasmalogens was predominantly 20:4ω6 and 22:6ω3. The deficit of esterified PUFAs in PC and PS was compensated for by an increase in the percentage of oleic acid (18:1ω9).

Substitution of ω6 fatty acids for 20:3ω9 during AA supplementation is evident from the prominent reduction in the percent mass of 20:3ω9 coincident with a significant increase in 20:4ω6 and 22:4ω6 fatty acids, particularly noteworthy in PI (Table 2; Fig. 2). In no case was Mead acid completely replaced. Chronic (1 week) maintenance of cells in a fatty acid-free chemically defined medium greatly reduced the percent content

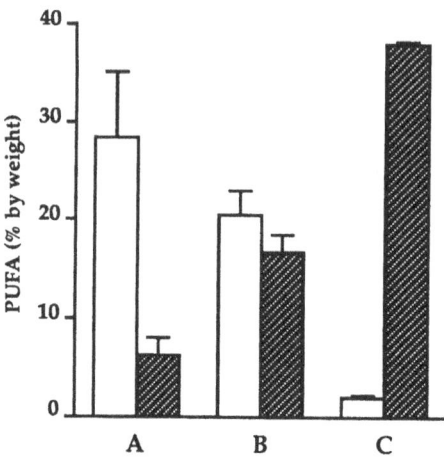

Figure 2. Compensatory substitution of Mead acid for AA in phosphatidylinositol. SK-N-SH neuroblastoma cells were cultured in chemically defined (A), EFA-deficient (B), or AA (93 μM)-enriched growth medium (C) for 3-7 days, and the acyl composition of PI determined as described under Experimental Procedures. In each group, left bar represents Mead acid and right bar represents AA. Data shown are the means ± SEM from three to five separate cell preparations.

of 20:4ω6 while exclusively increasing the content of 20:3ω9 in PI (Fig. 2), suggesting that under conditions of rigorous EFA deprivation, SK-N-SH cells may completely replace 20:4ω6 with 20:3ω9.

Supplementation of the growth medium with EPA (20:5ω3) or with linolenate (18:3ω3) also substantially reduced the content of Mead acid in PI by replacement with the elongated and desaturated metabolites of the supplement used (Fig. 3). The metabolism of ω3 fatty acids to 22:6ω3 strongly suggests the presence of an active Δ4

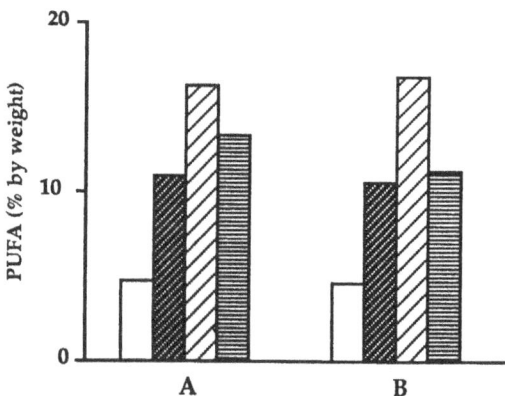

Figure 3. Replacement of Mead acid by ω3 fatty acids in phosphatidylinositol. SK-N-SH cells were cultured in chemically defined growth medium supplemented with 50 μM of either EPA (A) or linolenic acid (B) for 7 days. The acyl composition of PI was determined as described under Experimental Procedures. In each group of bars, from left to right: Mead acid; AA; EPA; docosahexaenoic acid. Data shown are the means of two cell preparations.

desaturase in SK-N-SH neuroblastoma cells. The desaturase activity has been reported in one other tumor cell line, the Y79 retinoblastoma cell (Spector and Yorek, 1985), and in the case of primary cell cultures, in astrocytes but not in neurons (Moore et al., 1991).

Cholinergic and Prostanoid Receptor-Stimulated Events Appear Refractory to Membrane Fatty Acid Remodeling

The significance of altered membrane fatty acid composition on carrier-mediated transport of small molecules, on insulin-, and on opioid-receptor pharmacology and certain effector enzyme systems has been investigated (for review see Spector and Yorek, 1985) and correlated, in part, to the perturbation of membrane physical properties (Stubbs and Smith, 1984). Recent evidence suggests that signal transduction across membranes may also be sensitive to changes in phospholipid acyl composition. Murphy et al. (1987) reported an approximately 2-fold enhancement of PGE_1-mediated cAMP formation in ω6-PUFA-enriched N1E-115 neuroblastoma cells, which was accompanied by a small, but statistically significant decrease in the EC_{50} for PGE_1. This suggests a PUFA-dependent enhancement in the efficacy of PGE_1 receptor-G protein-adenylate cyclase coupling. In contrast to N1E-115 cells, however, neither resting nor PGE_1-stimulated formation of cAMP was affected by AA enrichment of SK-N-SH cell membranes [cAMP (pmols/mg protein, n = 3) IBMX-treated EFA-deficient: basal = 7.0 ± 1.6; PGE_1 stimulated = 81.2 ± 24.8; IBMX-treated AA-enriched: basal = 6.8 ± 0.9, PGE_1-stimulated = 100.1 ± 18.1]. The compensatory elevation of esterified monounsaturated fatty acids observed in the phospholipids extracted from EFA-deficient SK-N-SH cells serves to maintain a relatively constant membrane fatty acid unsaturated/saturated ratio (see Table 2), perhaps necessary for preserving membrane function. A similar refractory response to changes in membrane fatty acid composition was observed with mAChR function. AA supplementation of EFA-deficient cells had no effect on M_3 mAChR density [(fmol/mg protein, n = 4) EFA-deficient: 169 ± 23; AA-enriched: 164 ± 23] or on the affinity for [^{3}H]scopolamine [(n = 4) EFA-deficient: 0.32 ± 0.04 nM; AA-enriched: 0.30 ± 0.04 nM]. Moreover, intracellular calcium transients seen following carbachol stimulation were quantitatively and qualitatively similar between EFA-deficient (20:3ω9-enriched) and AA-enriched (20:3ω9-poor) SK-N-SH cells (Fig. 4), suggesting that changes in membrane fatty acid composi-

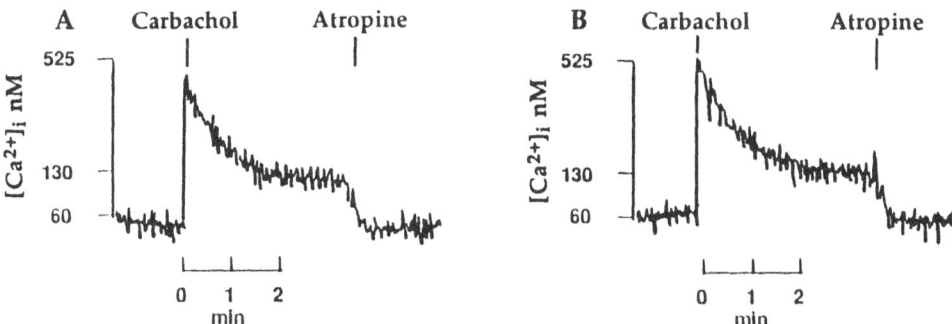

Figure 4. Intracellular calcium transients in Fura-2-loaded SK-N-SH neuroblastoma cells. Cells cultured in EFA-deficient (A) or AA (93 μM)-enriched (B) growth medium were incubated with Fura-2/AM for 15 min at 37°C as described under Experimental Procedures. Shown are two representative tracings of calibrated fluorescence signals. EFA-deficient (n=3): resting $[Ca^{2+}]_i$ = 57 ± 6 nM, peak = 578 ± 19 nM, plateau = 130 ± 19 nM; AA-enriched (n = 3): resting $[Ca^{2+}]_i$ = 70 ± 5 nM, peak = 608 ± 51 nM, plateau = 140 ± 10 nM.

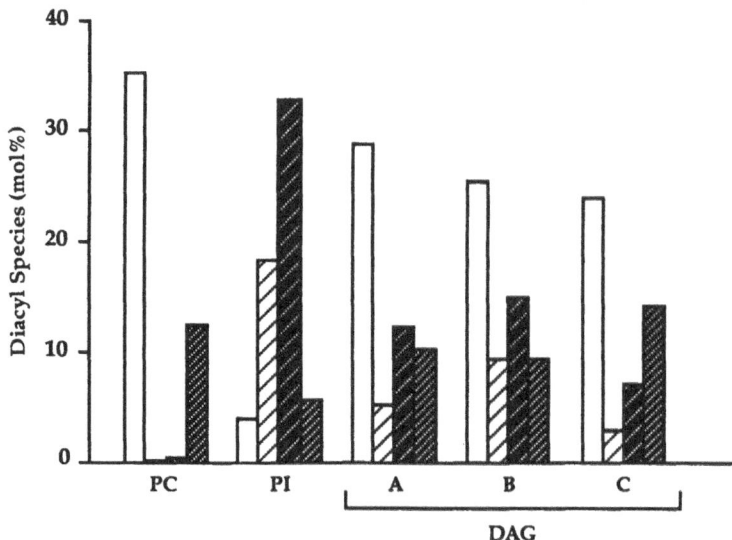

Figure 5. Major molecular species of PC, PI, and DAG from EFA-deficient SK-N-SH neuroblastoma cells. A, 0 s; B, 10 s; C, 300 s following carbachol (10 mM) stimulation. In each group, from left to right, bars represent 16:0-18:1ω9; 18:0-20:4ω6; 18:0-20:3ω9; 18:1ω9-18:1ω9. Diacyl molecular species were determined by reverse-phase HPLC of the benzoylated derivatives as described under Experimental Procedures. Data shown are the means of two or three experiments.

tion made by these neurotumor cells, including the substitution of Mead acid for AA, preserve critical membrane functions such as transmembrane signaling and calcium homeostasis.

Metabolism of DAG and PA Molecular Species Following Cholinergic Stimulation

The molecular species of PC, PI, and DAG in resting EFA-deficient SK-N-SH neuroblastoma cells are shown in Figure 5. Consistent with the fatty acyl composition presented in Table 2, the major diacyl fatty acid molecular species of PCs are the pairs 16:0-18:1ω9 and 18:1ω9-18:1ω9, whereas those of PIs are characteristically 18:0-20:4ω6 and 18:0-20:3ω9. The diacyl fatty acid composition of DAG in resting cells represents a composite of both PI and PC. Activation of the SK-N-SH M_3 mAChRs with carbachol (10 mM) elicits a 50-60% increase in the total mass of DAGs (resting, 1.4-1.5 nmols/mg protein). The diacyl fatty acid profile of the DAGs produced following carbachol stimulation appears biphasic. After 10 s of stimulation, an increase in both 18:0-20:3ω9 and 18:0-20:4ω6 DAG diacyl species is observed (Fig. 5), reflective of polyphosphoinositide breakdown. By 5 min, however, the mol % of these two diacyl species is reduced, whereas the stimulated increase in DAG content is sustained, suggesting a lipid source of the DAG other than the phosphoinositides. The diacyl composition of DAG at 5 min of carbachol stimulation is enriched in 16:0-18:1ω9 and 18:1ω9-18:1ω9, similar to that of PC.

Stimulated increases in PA are also observed in SK-N-SH cells responding to carbachol but differ significantly from those of DAG. The diacyl fatty acid composition of PAs in resting EFA-deficient cells is abundant in 16:0-18:1ω9, but also contains detectable amounts of 18:0-20:4ω6 and 18:1-20:3ω9 (Fig. 6A), representing a composite

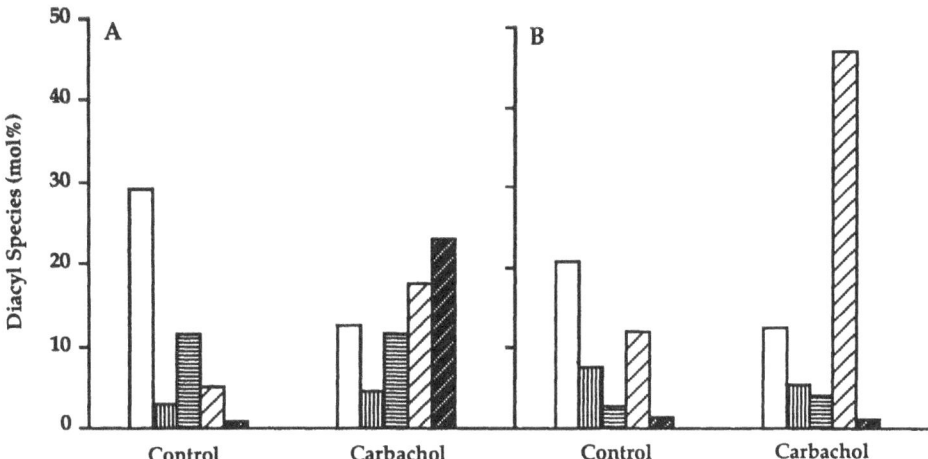

Figure 6. Diacyl fatty acid molecular species of phosphatidic acid (PA) in resting or carbachol (10 mM)-stimulated SK-N-SH cells cultured in EFA-deficient (A) or AA (93 μM)-enriched (B) growth medium. In each group, from left to right, bars represent 16:0-18:1ω9; 18:1ω9-20:4ω6; 18:1ω9-20:3ω9; 18:0-20:4ω6; 18:0-20:3ω9. Diacyl molecular species of PA were determined as described in Figure 5. Data shown are the means of triplicate determinations from a single preparation of cells.

of PI and PC. A similar profile of PAs is observed in AA-enriched cells with the exception of 18:0-20:3ω9 which is substantially reduced, and is replaced by 18:0-20:4ω6. Stimulation of EFA-deficient or AA-enriched SK-N-SH cells with carbachol elicits specific increases in 18:0-20:4ω6 and 18:0-20:3ω9 PAs (but not 18:1ω9-20:3ω9 or 18:1ω9-20:4ω6) consistent with the diacyl fatty acid profile of PI. This profile is sustained in PAs even after 5 min of stimulation (Fig. 6), a time when the diacyl fatty acid profile of DAG matches that of PC (Fig. 5). These data support a selective phosphorylation of DAGs arising from the stimulated breakdown of phosphoinositides, but not of PC. Studies in hepatocytes by Augert et al. (1989) indicate that DAG arising from PC breakdown may not be phosphorylated to PA. MacDonald et al. (1988) have reported selectivity of a membrane-bound DAG-kinase from 3T3 cells for AA-containing DAG, while cytosolic DAG-kinase from the same cells is not selective. This selectivity could explain both the findings in hepatocytes and the present findings in SK-N-SH cells. Our results further suggest that a membrane-bound DAG-kinase in neuroblastoma cells accepts Mead acid-containing DAG as a substrate.

CONCLUSION

Human SK-N-SH neuroblastoma cells cultured in a standard growth medium supplemented with FBS exhibit a phospholipid fatty acid composition characteristic of EFA deficiency. The reversible exchange of Mead acid for AA in phospholipids examined, particularly PI, suggests a compensatory mechanism by which these cells maintain a relatively constant degree of membrane lipid unsaturation during EFA deficiency. Biosubstitution of Mead acid for AA does not impair PGE$_1$- or carbachol-stimulated signal transduction events in SK-N-SH cells. Cholinergic stimulation of EFA-deficient cells elicits a biphasic accumulation of DAGs which initially are derived from phosphoinositide breakdown, and at longer incubation times, from the breakdown of PC. PAs produced following cholinergic stimulation are preferentially derived from phosphoinositide breakdown, even when the predominant molecular species of DAGs generated are from the breakdown of PC, a result suggesting that the phosphorylation of DAGs to PAs is highly selective in these neurotumor cells. Thus stearoyl-Mead DAG is recognized and conserved in the phosphoinositide cycle during EFA deficiency in complete analogy to stearoyl-AA DAG conservation in AA-enriched cells. While the bioconservation of Mead acid in SK-N-SH cells indicates its satisfactory substitution for AA in the phosphoinositide cycle, it remains possible that it is not able to serve as a replacement for EFAs in the case of brain function.

ACKNOWLEDGMENTS

The authors thank Dr. T. Ford-Holevinski for providing graphics.

REFERENCES

Augert G, Bocckino SB, Blackmore PF, Exton JH (1989) Hormonal stimulation of diacylglycerol formation in hepatocytes: Evidence for phosphatidylcholine breakdown. J Biol Chem 264:21689.

Balcar VJ, Borg J, Robert J, Mandel P (1980) Uptake of 1-glutamate and taurine in neuroblastoma cells with altered fatty acid composition of membrane phospholipids. J Neurochem 34:1678.

Bottenstein J and Sato G (1979) Growth of a rat neuroblastoma cell line in serum-free supplemented media. Proc Natl Acad Sci USA 76:514.

Brenneman DE and Rutledge CO (1979) Alteration of catecholamine uptake in cerebral cortex from rats fed a saturated fat diet. Brain Res 179:295.

Chaudiere J, Clement M, Driss F, Bourre JM (1987) Unaltered brain membranes after prolonged intake of highly oxidizable long-chain fatty acids of the (n-3) series. Neurosci Lett 82:233.

Dhopeshwarkar GA and Mead JF (1961) Role of oleic acid in the metabolism of essential fatty acids. J Am Oil Chem Soc 38:297.

Fisher SK, Domask LM, Roland RM (1989) Muscarinic receptor regulation of cytoplasmic Ca^{2+} concentrations in human SK-N-SH neuroblastoma cells: Ca^{2+} requirements for phospholipase C activation. Mol Pharmacol 35:195.

Ginsberg BH, Jabour J, Spector AA (1982) Effect of alterations in membrane lipid saturation on the properties of the insulin receptor of Ehrlich ascites cells. Biochim Biophys Acta 690:157.

Ho WKK and Cox BM (1982) Reduction of opioid binding in neuroblastoma X glioma cells grown in medium containing unsaturated fatty acids. Biochim Biophys Acta 668:211.

Holman RT (1968) Essential fatty acid deficiency. In: Progress in the chemistry of fats and other lipids. Vol 9 (Holman RT, ed) p 275.

Hyman BT and Spector AA (1982) Choline uptake in cultured human Y79 retinoblastoma cells: Effects of polyunsaturated fatty acid compositional modifications. J Neurochem 38:650.

Lee C and Hajra AK (1991) Molecular species of diacylglycerols and phosphoglycerides and the postmortem changes in the molecular species of diacylglycerols in rat brains. J Neurochem 56:370.

MacDonald ML, Mack KF, Williams BW, King WC, Glomset JA (1988) A membrane-bound diacylglycerol kinase that selectively phosphorylates arachidonoyl-diacylglycerol. J Biol Chem 263:1584.

McGee R Jr (1981) Membrane fatty acid modification of the neuroblastoma X glioma hybrid, NG108-15. Biochim Biophys Acta 663:314.

Moore SA, Yoder E, Murphy S, Dutton GR, Spector A (1991) Astrocytes, not neurons, produce docosahexaenoic acid (22:6ω-3) and arachidonic acid (20:4ω-6). J Neurochem 56: 518.

Murphy MG (1984) Increasing membrane polyunsaturated fatty acid content augments cyclic AMP formation and prostaglandin production in N1E-115 neuroblastoma. Prog Neuropsychopharmacol Biol Psychiatry 8:529.

Murphy MG (1986) Studies of the regulation of basal adenylate cyclase activity by membrane polyunsaturated fatty acids in cultured neuroblastoma. J Neurochem 47:245.

Murphy MG, Moak CM, Rao BG (1987) Effects of membrane polyunsaturated fatty acids on opiate peptide inhibition of basal and prostaglandin E_1-stimulated cyclic AMP formation in intact N1E-115 neuroblastoma cells. Biochem Pharmacol 36:4079.

Needleman P, Raz A, Minkes MS, Ferrendelli JA, Sprecher H (1979) Triene prostaglandins: Prostacyclin and thromboxane biosynthesis and unique biological properties. Proc Natl Acad Sci USA 76:944.

Neuringer M, Anderson GJ, Connor WE (1988) The essentiality of n-3 fatty acids for the development and function of the retina and brain. Annu Rev Nutr 8:517.

Park CC, Hennessey T, Ahmed Z (1990) Manipulation of plasma membrane fatty acid composition of fetal rat brain cells grown in a serum-free defined medium. J Neurochem 55:1537.

Remmers AE, Nordby GL, Medzihradsky F (1990) Modulation of opioid receptor binding by cis and trans fatty acids. J Neurochem 55:1993.

Robert J, Rebel G, Mandel P (1977) Essential fatty acid metabolism in cultured astroblasts. Biochimie 59:417.

Samuelsson B, Dahlen S-EL, Lindgren JA, Rouzer CA, Serhan CN (1987) Leukotrienes and lipoxins: Structures, biosynthesis, and biological effects. Science 237:1171.

Schroit AJ and Gallily R (1979) Macrophage fatty acid composition and phagocytosis: Effect of unsaturation on cellular phagocytic activity. Immunology 36:199.

Spector AA and Yorek MA (1985) Membrane lipid composition and cellular function. J Lipid Res 26:1015.

Stubbs CD and Smith AD (1984) The modification of mammalian membrane polyunsaturated fatty acid composition in relation to membrane fluidity and function. Biochim Biophys Acta 779:89.

Stubbs EB Jr, Lee C, Fisher SK, Hajra AK, Agranoff BW (1990) Modulation of phospholipid fatty acid composition in human SK-N-SH neuroblastoma cells. Trans Am Soc Neurochem 21:141.

Sun GY, Huang H-M, Kelleher JA, Stubbs EB Jr, Sun AY (1988) Marker enzymes, phospholipids and acyl group composition of a somal plasma membrane fraction isolated from rat cerebral cortex: A comparison with microsomes and synaptic plasma membranes. Neurochem Int 12:69.

White RP and Hagen AA (1982) Cerebrovascular actions of prostaglandins. Pharmacol Ther 18:313.

Williams TP and McGee R Jr (1982) The effects of membrane fatty acid modification of cloned pheochromocytoma cells on depolarization-dependent exocytosis. J Biol Chem 257:3491.

Wolfe LS (1982) Eicosanoids: Prostaglandins, thromboxanes, leukotrienes, and other derivatives of carbon-20 unsaturated fatty acids. J Neurochem 38:1.

Yorek MA, Hyman BT, Spector AA (1983) Glycine uptake by cultured human Y79 retinoblastoma cells: Effect of changes in phospholipid fatty acid unsaturation. J Neurochem 40:70.

Yorek MA, Strom DK, Spector AA (1984) Effect of membrane polyunsaturation on carrier-mediated transport in cultured retinoblastoma cells: Alterations in taurine uptake. J Neurochem 42:254.

PHOSPHOLIPID METABOLISM AND SECOND MESSENGER SYSTEM AFTER

BRAIN ISCHEMIA

Koji Abe, Tsutomu Araki, Jun-ichi Kawagoe,
Masashi Aoki, and Kyuya Kogure

Department of Neurology
Tohoku University School of Medicine
Sendai, Japan

ABSTRACT

To evaluate possible involvement of phospholipid metabolism and related second messenger systems in the selective neuronal damage after ischemia, we measured changes of polyphosphoinositides (PPIs) and free fatty acids (FFAs) in a model of 5-min or 10-min ischemia and reperfusion in gerbils. The binding activity of ^{3}H-phorbol 12,13-dibutyrate (PDBu) for protein kinase C (PKC) and ^{3}H-inositol 1,4,5-triphosphate (IP$_3$) for IP$_3$ receptors was demonstrated autoradiographically. Induction of 70 KDa heat shock protein (HSP70) mRNA and amyloid precursor protein (APP) mRNA was also examined using Northern blot analysis.

In the parietal cortex (an area resistant to transient ischemia), PPIs decreased during ischemia and recovered rapidly after reperfusion. However, recovery did not occur in the hippocampal CA1 area (an area more vulnerable to transient ischemia). In the cortex, arachidonic acid (AA) increased during ischemia and returned to baseline by 7 days after reperfusion; in the CA1 area, the AA level remained elevated even after 7 days of reperfusion.

PDBu binding decreased in CA1 cells after 2 days of reperfusion. IP$_3$ binding began to decrease at 5 hr of reperfusion, which is far earlier than either the onset of decreased PDBu binding or the observation of neuronal damage by light microscopy.

The induction of HSP70 mRNA occurred, but the induction of APP mRNA did not. Regional differences in the induction of HSP70 mRNA were found; CA1 cells produced less HSP70 mRNA than cortical cells 8 hr after transient ischemia.

These results suggest that CA1 cell membranes may not recover after transient ischemic attack, and that the membranes of the endoplasmic reticulum, which have IP$_3$ receptors, may undergo alterations earlier than cytoplasmic membranes. The variable induction of HSP70 mRNA may be related to regional differences in vulnerability in cortical and hippocampal CA1 cells after transient ischemia. Involvement of excitatory neurotransmission in the induction of HSP70 has been suggested. The combined data may support a role for inositol phospholipid metabolism, changes in related second messenger systems, and induction of HSP70 in the excitotoxic mechanism of hippocampal CA1 neuronal damage, death, and repair.

Neurobiology of Essential Fatty Acids
Edited by N.G. Bazan *et al.*, Plenum Press, New York, 1992

183

INTRODUCTION

Transient cerebral ischemia results in a selective pattern of neuronal degeneration within the central nervous system in both humans and experimental animals (Kirino, 1982; Rothman and Olney, 1986). The hippocampal CA1 neurons are especially susceptible to ischemia; neurons are lost after as little as 3 min of ischemia. Previous studies suggest that excitatory neurotransmission plays an important role in the development of ischemic neuronal damage (Rothman and Olney, 1986; Wieloch, 1985). However, the intracellular mechanisms triggering the degenerative process are not fully understood.

Transient ischemia liberates arachidonic acid (AA) and platelet-activating factor (PAF) from membrane phospholipids. Polyphosphoinositides (PPIs) are generally recognized as the main source of the increase in free AA during ischemia, especially in the early stages (Abe and Kogure, 1986; Abe et al., 1987). This liberation of AA from PPIs is suspected to be the result of the transneuronal breakdown of PPIs triggered mainly by the release of glutamate in the synaptic cleft during ischemia (Benveniste et al., 1984). A coupling of inositol phospholipid metabolism with excitatory amino acid receptors has recently been reported in the rat hippocampus (Nicoletti et al., 1986).

Alterations in cell signaling systems and related gene expression may play an important role in the selective neuronal damage after transient ischemia. FFAs may directly regulate ion channels of excitable membranes, and may indirectly regulate N-methyl-D-aspartate (NMDA) receptor function (Aizenman et al., 1990; Dumuis et al., 1988; Ordway et al., 1991). Therefore, we investigated changes in phospholipid and FFA metabolism, second messenger systems, and HSP70 gene expression in relation to possible changes in APP gene expression.

MATERIALS AND METHODS

Mongolian gerbils *(Meriones unguiculatus)*, aged 8-10 weeks and weighing 60-80 g, were lightly anesthetized by inhalation of a nitrous oxide/oxygen/halothane (70%:30%; 1%) mixture or ether. As the animals began to awaken from the anesthesia, both common carotid arteries were occluded for 5 or 10 min using surgical clips. The clips were then removed, and the animals were allowed to recover until the time of sampling *(in situ* freezing for lipid analysis, and decapitation for receptor autoradiography and Northern analysis). Sham-operated animals were sacrificed after exposing the carotid arteries without clamping the vessels.

The methods for the analysis of lipids and receptor autoradiography (ARG) have been described elsewhere (Abe and Kogure, 1986; Abe et al., 1987; Worley et al., 1987). Northern blot analysis was performed according to our previous reports (Abe et al., 1991a,b). Briefly, for analysis of lipids, parietal cortices and hippocampal CA1 sectors were dissected from brains that had undergone 30 min, or 1, 2, or 7 days of reperfusion after 5-min ischemia. The amounts of FFAs and PPIs were measured according to previously reported methods. For receptor ARG, the animals were decapitated 1 or 5 hr, 1, 2, or 7 days, or 1 month after 10-min ischemia. Autoradiographic localization of ^{3}H-phorbol 12,13-dibutyrate (PDBu) and ^{3}H-inositol 1,4,5-triphosphate (IP$_3$) binding sites in the brain was performed. Samples of parietal cortex or CA1 subfields from 10 animals were combined, and the amounts of HSP70 mRNA and APP mRNA were determined by Northern blot (Abe et al., 1991b).

RESULTS

As shown in Figure 1, PPI levels in the cortex rapidly recovered during reperfusion. In the CA1 area, however, a partial recovery of triphosphoinositide (TPI) occurred in the first 30 min of reperfusion; no further recovery in PPI levels was observed thereafter.

Figures 2 and 3 show changes in FFAs and AA, respectively. The amount of total FFAs in the cortex increased during ischemia, decreased to basal level during the first 30 min of reperfusion, then slowly increased again over the next 2 days and finally recovered to control levels by 7 days of reperfusion. In contrast, in the CA1 areas, the amount of total FFAs continued to increase during the first day of reperfusion, declined precipitously on the second day, and finally increased again by day 7. The pattern of

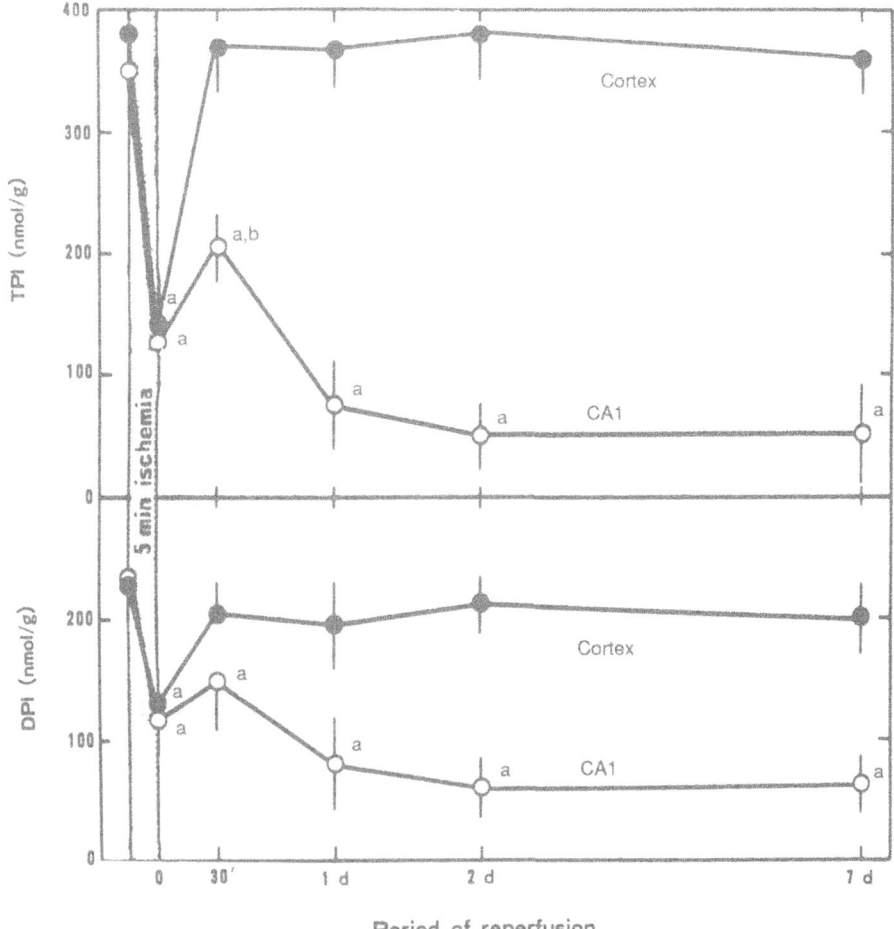

Figure 1. Changes in the amount of tri- and diphosphoinositides (TPI and DPI) during and after 5 min of forebrain ischemia in the parietal cortex (Cortex) or CA1 sub-field (CA1). Open and closed circles represent mean ± SD expressed as nmol/g wet weight. n=6. a: p=0.01, compared to sham control; b: p=0.05, compared to value immediately after 5 min of ischemia.

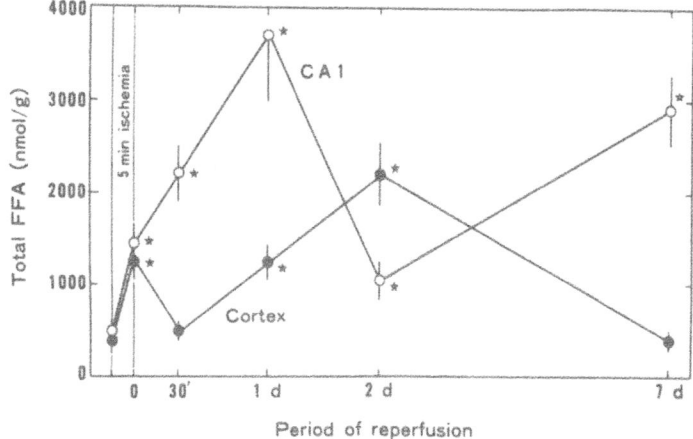

Figure 2. Changes in the amount of total free fatty acids (FFA) during and after 5 min of forebrain ischemia in the parietal cortex (Cortex) or CA1 sub-fields (CA1). Open and closed circles represent mean ± SD expressed as nmol/g wet weight. n=6. *p=0.01, compared to sham control.

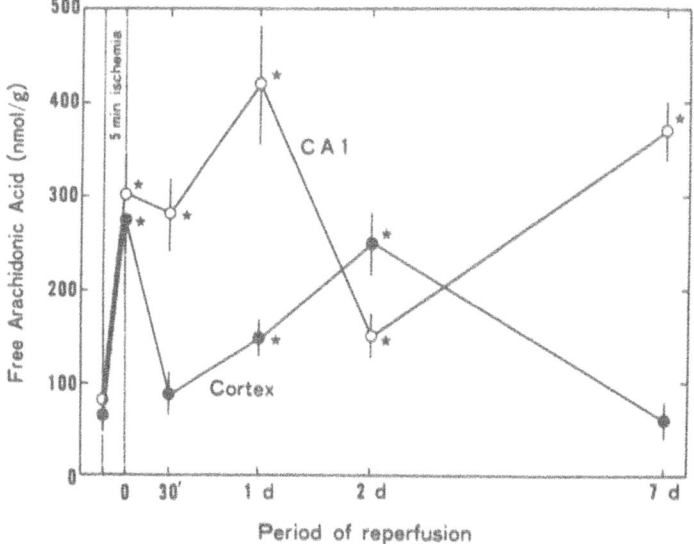

Figure 3. Changes in the amount of free arachidonic acid during and after 5 min of forebrain ischemia. Symbols and bars are the same as in Figure 2.

changes in AA levels in the cortex and the CA1 areas (Fig. 2) was similar to the pattern of changes in the levels of total FFAs (Fig. 3).

Post-ischemic alterations in ^{3}H-PDBu and ^{3}H-IP$_3$ binding sites are summarized in Tables 1 and 2, and representative autoradiograms are shown in Figures 4 and 5, respectively. ^{3}H-PDBu binding in selectively vulnerable areas showed no significant change 1-24 hr after ischemia, except for a transient decline in a few regions, followed by significant decreases 48 hr and 7 days after ischemia. One month after ischemia, however, ^{3}H-PDBu binding was reduced significantly only in the striatum and the hippocampal CA1 sector, where severe neuronal damage occurs. The dentate molecular layer, which is resistant to ischemia, showed a significant elevation in ^{3}H-PDBu binding sites as late as 1 month after ischemia. In contrast, ^{3}H-IP$_3$ binding showed significant reduction in the selectively vulnerable regions 1-24 hr after ischemia, followed by marked decreases for up to 1 month. In the dentate molecular layer, ^{3}H-IP$_3$ binding also showed significant reduction during recirculation except for a slight recovery at 48 hr and 7 days of reperfusion. One month after ischemia, ^{3}H-IP$_3$ binding in all regions was still significantly reduced.

Analysis of Northern blots revealed hybridization of the ^{32}P-labeled HSP70 probe to two sizes of mRNA corresponding to 2.8 Kb and 2.4 Kb (Fig. 6). In agreement with previous reports (Wu et al., 1985; Lowe and Moran, 1986; Marini et al., 1990), the 2.8 Kb mRNA is thought to represent HSP70 mRNA (a major inducible species in the HSP70 family) and the 2.4 Kb mRNA may correspond to heat shock protein-related (heat shock "cognate" protein 70, HSC70) mRNA. The HSC70 gene is constitutively expressed in normal development and differentiation in cells. In addition to the HSP70 mRNA, this HSP70 probe (pH 2.3) also hybridizes HSC70 mRNA under hybridizing and washing conditions (Abe et al., 1991b). Under control conditions, CA1 cells had more 2.4 Kb mRNA (HSC70 mRNA) and less APP mRNA, compared with cortical cells, while the 2.8 Kb mRNA (HSP70 mRNA) could not be clearly detected. With reperfusion, both HSP70 mRNA and HSC70 mRNA were induced, but no significant induction of APP mRNA was observed. The levels of both HSP70 and HSC70 mRNA reached maximum after 8 hr of reperfusion and then declined. The amount of increase of HSP70 mRNA was relatively smaller in the CA1 cells than in the parietal cortical cells at the maximum (8 hr).

DISCUSSION

The "phosphatidylinositol (PI) response" has been understood to be stimulated mainly by glutamatergic, cholinergic, or noradrenergic stimulation. This response induces PI breakdown and produces both diacylglycerol (DG) and inositol-triphosphate (IP$_3$). IP$_3$ can increase cytosolic calcium concentration by releasing calcium from the intracellular calcium store, and a part of IP$_3$ is converted by PI$_3$ kinase to PI$_4$, which can potentiate calcium influx from the extracellular space (Irvine et al., 1986), also resulting in an increase in cytosolic calcium concentration. DG can activate protein kinase C (PKC) if both phosphatidylserine and calcium are present (Abe and Kogure, 1986; Irvine et al., 1986). Thus, the "PI response" involving the production of DG and IP$_3$ from triphosphoinositides may result in the activation of PKC.

A breakdown of PPIs and an increase in the amount of DG are also seen in the model of brain ischemia (Abe and Kogure, 1986), and AA may be released from the increased DG by DG lipase in the early stage of ischemia (Abe and Kogure, 1986; Abe et al., 1987). Thus AA release is closely related to the breakdown of PPIs in the model of brain ischemia.

Table 1. Time course of [³H]-PDBu binding in the gerbil brain after transient cerebral ischemia

	Sham-operated	Recirculation time					
		1 h	5 h	24 h	48 h	7 days	1 month
Frontal cortex	1013 ± 63	780 ± 51*	958 ± 40	870 ± 34	804 ± 79	963 ± 77	826 ± 45
Striatum (lateral)	910 ± 52	745 ± 38	815 ± 40	768 ± 41	600 ± 26**	758 ± 98	595 ± 31**
(medial)	1003 ± 45	932 ± 39	961 ± 53	1018 ± 49	982 ± 90	895 ± 132	865 ± 34*
Hippocampus							
CA1 stratum oriens	1264 ± 76	1329 ± 77	1313 ± 39	1381 ± 42	1276 ± 83	637 ± 124**	620 ± 77**
stratum radiatum	1241 ± 78	1184 ± 85	1182 ± 60	1220 ± 40	1191 ± 84	816 ± 99**	496 ± 39**
stratum lacunosum-moleculare	1070 ± 62	872 ± 48*	1117 ± 49	901 ± 61	965 ± 83	625 ± 76**	580 ± 61**
CA3 average	1257 ± 69	1191 ± 78	1300 ± 64	1280 ± 89	1161 ± 64	804 ± 132*	1104 ± 106
Dentate gyrus (molecular layer)	1056 ± 38	961 ± 37	1137 ± 62	1020 ± 79	1163 ± 44**	1520 ± 113**	1339 ± 67**
Thalamus	536 ± 28	485 ± 27	592 ± 26	478 ± 48	427 ± 24*	602 ± 90	627 ± 62

Optical densities were converted to fmol/mg tissue using the [³H] micro-scale. Values are expressed as mean ± SEM. *p< 0.05, **p<0.01 vs. sham-operated group (Mann-Whitney U-test). n=8-14 hemispheres. Striatum (lateral): the dorsolateral part of the striatum; Striatum (medial): the ventromedial part of the striatum.

Table 2. Time course of $[^3H]$-IP_3 binding in the gerbil brain after transient cerebral ischemia

	Sham-operated	Recirculation time					
		1 h	5 h	24 h	48 h	7 day	1 month
Frontal cortex	69 ± 4.8	35 ± 4.8**	44 ± 4.3**	43 ± 4.3**	52 ± 4.1*	63 ± 4.9	44 ± 6.4*
Striatum (lateral)	121 ± 7.1	67 ± 4.0**	74 ± 2.3**	63 ± 3.5**	53 ± 4.8**	54 ± 3.3**	44 ± 7.2**
(medial)	143 ± 7.2	72 ± 4.4**	85 ± 3.4**	82 ± 4.4**	70 ± 7.5**	69 ± 11.1**	80 ± 12.2**
Hippocampus							
CA1 average	177 ± 18.4	59 ± 4.3**	63 ± 6.5**	58 ± 4.4**	72 ± 5.8**	45 ± 3.1**	36 ± 3.8**
CA3 average	79 ± 12.0	30 ± 3.4**	40 ± 4.2**	37 ± 4.5*	53 ± 4.7	45 ± 2.5*	41 ± 6.7*
Dentate gyrus (molecular layer)	87 ± 14.6	38 ± 4.9*	43 ± 5.4*	38 ± 3.6*	55 ± 2.4	46 ± 4.0	40 ± 7.6*
Thalamus	70 ± 9.9	31 ± 3.8**	35 ± 3.6**	36 ± 2.4*	53 ± 3.3	52 ± 2.7	37 ± 3.2*

Optical densities were converted to fmol/mg tissue using the $[^3H]$ micro-scale. Values are expressed as mean ± SEM. *P<0.05, **p<0.01 vs. sham-operated group (Mann-Whitney U-test). n = 8-14 hemispheres.

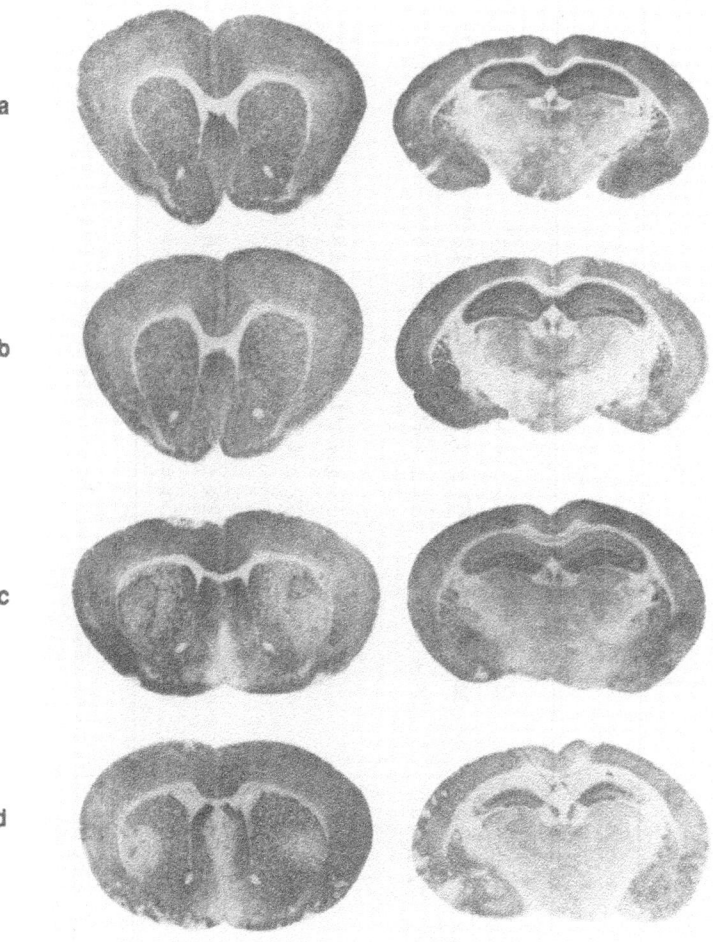

Figure 4. Representative [³H]-PDBu autoradiograms from sham-operated gerbil brain (a) and ischemic gerbil brain (b-d). In sham-operated animals (a), [³H]-PDBu binding sites were the most highly concentrated in the hippocampus, followed by the neocortex and the striatum. In the thalamus, the concentration of binding sites was very low. At 5 hr of recirculation (b), marked alterations in [³H]-PDBu binding were not observed in selectively vulnerable areas. At 7 days (c) and 1 month (d) of recirculation, marked reduction of [³H]-PDBu binding was seen in the hippocampal CA1 sector, the hippocampal CA3 sector, and the striatum. However, a significant increase in [³H]-PDBu binding was noted in the dentate molecular layer, which is resistant to ischemia.

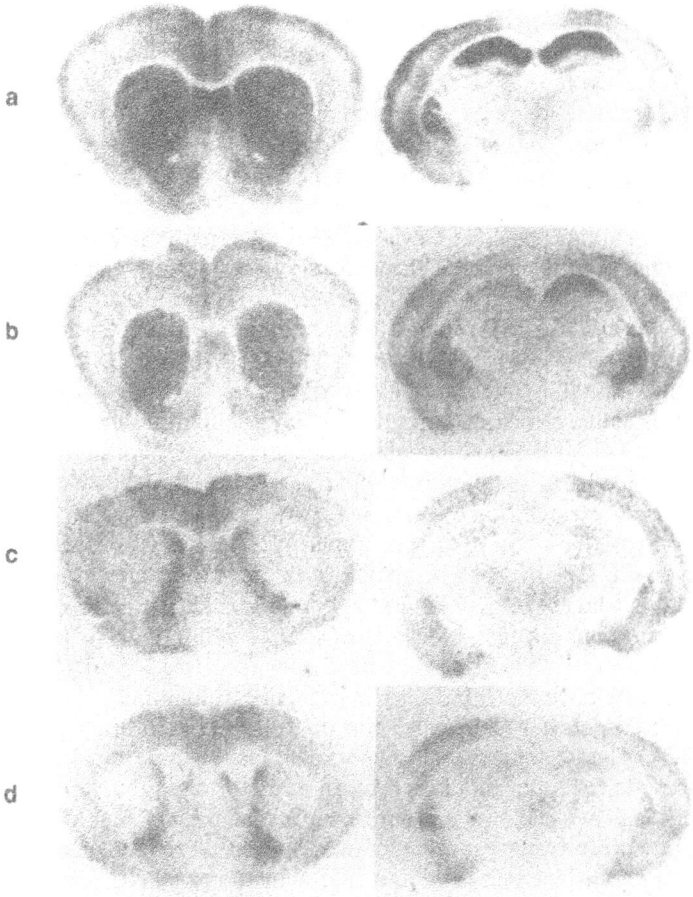

Figure 5. Representative [^{3}H]-IP$_3$ autoradiograms from sham-operated gerbil brain (a) and ischemic gerbil brain (b-d). In sham-operated animals (a), [^{3}H]-IP$_3$ binding sites were highly concentrated in the striatum and the hippocampal CA1 sector. At 5 hr of recirculation (b), a significant reduction in [^{3}H]-IP$_3$ binding was seen in the striatum and the hippocampal CA1 sector. Note also the decrease in grain density in the other regions. At 7 days (c) and 1 month (d) of recirculation, binding sites were markedly reduced in the selectively vulnerable areas. The grain densities in the dentate molecular layer were also markedly reduced.

In our studies, the changes in the amounts of total FFAs and AA were similar in the cortex and CA1 areas during ischemia (Figs. 2 and 3). During reperfusion, however, the changes were remarkably different. During reperfusion in the cortex, FFAs and AA decreased transiently, then slowly increased until the end of the second day, and finally returned to control levels (Figs. 2 and 3). In CA1, FFAs and AA continued to increase over the first day of reperfusion (Figs. 2 and 3), which is compatible with the continuous electrophysiological firing of CA1 neurons; the amounts declined sharply on the second day, by which time no electroactivity was found in the CA1 area. The continuous firing of CA1 cells has been understood to be the result of glutamatergic excitation (Benveniste et al., 1984; Westerberg and Wieloch, 1986; Wieloch, 1985). Therefore, our results indicate a close relationship between AA release and excitatory neurotransmission. The amounts increased again by the end of 7 days, which suggests active gliosis. In the activation of phospholipases, PI-specific phospholipase C is far more sensitive to the calcium concentration than phospholipase A_2. Therefore, a specific activation of phospholipase C may occur if the calcium concentration remains low. However, a massive influx of calcium due to glutamatergic excitation may induce non-specific activation of phospholipases during these periods (from 30 min to 1 day of reperfusion in the CA1 area), when the level of PPIs has already decreased and can no longer provide sufficient fatty acids to be liberated in the CA1 area (Fig. 2).

Excitability of hippocampal CA1 pyramidal cells is regulated mainly by potassium currents. Hippocampal pyramidal cells in rats have a voltage-dependent chloride current, which is active at resting potential and inhibited either by membrane depolarization or by activation of PKC by phorbol esters. Thus, blockade of this chloride current by activation of PKC potentiates the transmission of dendritic excitatory events by increasing dendritic membrane resistance. Furthermore, PKC may play a major role in the long-term potentiation of hippocampal pyramidal cells (Rothman and Olney, 1986; Worley et al., 1987). These results indicate that PKC may play a key role in the regulation of neuronal excitability of CA1 cells under normal conditions and also under pathological conditions. In fact, blockade of PKC ameliorated the selective neuronal damage.

Previous reports have indicated that an excessive influx of calcium during ischemia may play a role in delayed neuronal death of CA1 cells (Rothman and Olney, 1986; Wieloch, 1985). The validity of this hypothesis can be ascertained by using a few selective glutamate receptor antagonists or an inhibitor of glutamate-operated calcium channels. Although we did not measure changes in the amount of DG, which can activate PKC, our results suggest that glutamate-induced breakdown of PPIs may occur during and after transient ischemia, and that activation of PKC may potentiate CA1 excitability by inhibiting chloride channels; the result would be a glutamate-induced increase in excitotoxicity leading to delayed neuronal damage of CA1 pyramidal neurons.

The receptor ARG study indicates that post-ischemic alteration of the binding sites of two second messengers (PKC and IP_3) was produced by different processes in the striatum and the hippocampus. Post-ischemic alteration of ^{3}H-PDBu binding was particularly distinct from ^{3}H-IP_3 binding during the early stage of recirculation, at a time when no morphological damage was yet obvious. ^{3}H-PDBu binding showed no significant changes in the striatum and the hippocampal areas, whereas ^{3}H-IP_3 binding showed significant reduction in these regions. Thereafter, the binding activity of both ligands in the hippocampal CA1 sector showed a significant reduction at 7 days after ischemia, when severe neuronal damage was observed. However, the striatum, where severe neuronal damage was also seen, exhibited no significant change in ^{3}H-PDBu

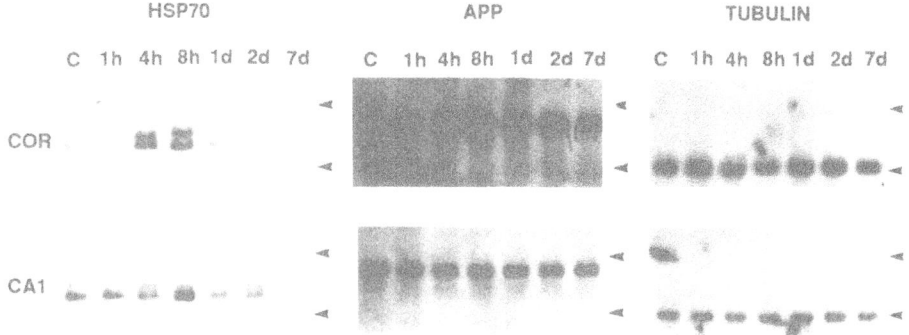

Figure 6. Northern blot analyses of HSP70, APP, and tubulin mRNA in parietal cortex (COR) and hippocampal CA1 (CA1) with 1-8 hr and 1-7 days of reperfusion after 10 min of ischemia. Each lane contains 15 μg of total RNA. C represents the sham-operated control. Arrowheads indicate ribosomal RNA.

binding sites. For this reason, a massive increase in glial cells should be considered, since the glial cells possess ^{3}H-PDBu binding sites. Interestingly, ^{3}H-PDBu binding showed a significant elevation in the dentate molecular layer. Although the reason for this is presently unclear, it might be explained by the possibility of sprouting from residual terminals following limited lesioning. In contrast, ^{3}H-IP$_3$ binding at 7 days after ischemia markedly deceased in the striatum, the hippocampal CA1 sector, and the hippocampal CA3 sector. A recent study suggests that the lasting decline in ^{3}H-IP$_3$ binding could be due to a downregulation of the intracellular receptors following excessive stimulation with IP$_3$ or calcium during and after ischemia.

Inductions of immediate-early genes such as Fos and Jun were reduced by a PAF antagonist (BN 50730) in rodent hippocampus and also in cultured cells (Bazan, in press). Treatment with PAF (which is derived from membrane phospholipid during ischemia) induces cell cycle regulatory gene expression such as calcyclin production. Thus, relationships between membrane phospholipid metabolism and immediate-early genes have recently been suggested (Bazan, in press). Our previous report shows that a zinc-finger gene (an immediate-early gene that uses the zinc-finger motif to bind DNA, in contrast to Fos/Jun, which use the leucine zipper motif) is induced in relation to HSP70 induction. Therefore, we looked for possible changes in HSP70 gene expression in this model.

The relatively high level of HSC70 mRNA expression found normally in the hippocampal CA1 region suggests an active turnover of clathrin in this area. Clathrin coats pinocytotic vesicles in cells (Abe et al., 1987). HSC70 has a role in disassembly of the clathrin cage and is involved in intracellular transportation. The induction of HSC70 mRNA found in this experiment may relate to the induction of an immunoreactive clathrin in this area (Yoshimi et al., 1990), suggesting a role of clathrin in the selective neuronal damage of CA1 cells. Although the promoter area of the APP gene has the sequence of heat shock consensus element (Salbaum et al., 1988), our results indicate that further investigations are required regarding the role of heat shock response in the induction of APP gene expression.

CA1 cells are under the most stress after reperfusion (Vass et al., 1988), a time when cells should have high levels of HSP70 due to post-transcriptional mechanisms (Rothman and Olney, 1986). However, the relatively low mRNA levels (at around 8 hr after reperfusion) observed in our experiments suggest a low level of induction in these cells. An impairment of protein ubiquitination has been immunohistochemically shown in rat hippocampal CA1 neurons after an ischemic insult (Magnussen and Wieloch, 1989). Taken together with our results, this is of interest because CA1 cells do not produce enough HSP70 mRNA and ubiquitin after ischemia although both are essential to the stress response.

A sustained induction of HSP70 mRNA has been reported in a gerbil hippocampal CA1 sector after 5 min of transient forebrain ischemia using a synthetic oligonucleotide as a probe in an *in situ* hybridization study (Nowak, 1991). The data seem to differ from our results. In general, Northern blot analysis shows the relative amount of a message in the total mRNA pool. Therefore, signal intensity depends on the ratio of the hybridizing mRNA to the total mRNA. On the other hand, *in situ* hybridization detects the absolute amount of a message, which should not be affected by the size of the total mRNA pool. Therefore, if the size of the total mRNA pool changes significantly after ischemia, the results of Northern and *in situ* analysis could be different. In fact, most proteins have translational difficulty in CA1 cells after ischemia, which could potentiate general mRNA syntheses. However, further study will be necessary to resolve this problem. A careful examination of our results shows that the induction of HSP70 mRNA continued for up to 2 days, by which time the amount of the message in cortical cells had already decreased below the level in CA1 cells. This result may partly correspond to the result obtained by *in situ* hybridization (Nowak, 1991).

CONCLUSION

Our results collectively support the hypothesis that inositol phospholipid metabolism, changes in the related second messenger systems, and HSP70 induction play a role in the excitotoxic mechanism of hippocampal CA1 neuronal damage.

REFERENCES

Abe K and Kogure K (1986) Accurate evaluation of 1,2-diacylglycerol in gerbil forebrain using HPLC and *in situ* freezing technique. J Neurochem 47:577-582.

Abe K, Kawagoe J, Sato S, Kogure K (1991a) Induction of zinc-finger gene after transient focal ischemia in rat cerebral cortex. Neurosci Lett 123:248-250.

Abe K, Kogure K, Yamamoto H, Imazawa M, Miyamoto K (1987) Mechanisms of arachidonic acid liberation during ischemia in gerbil cerebral cortex. J Neurochem 48:503-509.

Abe K, Tanzi RE, Kogure K (1991b) Induction of HSP70 mRNA after transient ischemia in gerbil brain. Neurosci Lett 125:166-168.

Aizenman E, Hartnett KA, Reynolds IJ (1990) Oxygen free radicals regulated NMDA receptor function via a redox modulatory site. Neuron 5:841-846.

Bazan NG (in press) Significance of phospholipase A_2 activation and second messengers in brain damage. Prog Brain Res.

Benveniste H, Drejer J, Schousboe A, Diemer NH (1984) Elevation of the extracellular concentrations of glutamate and aspartate in rat hippocampus during transient cerebral ischemia monitored by intracerebral microdialysis. J Neurochem 43:1369-1374.

Dumuis A, Sebben M, Haynes L, Pin J-P, Bockert J (1988) NMDA receptors activate the arachidonic acid cascade system in striatal neurons. Nature 336:68-70.

Irvine RF, Letcher AJ, Heslop JP, Berridge MJ (1986) The inositol tris/tetrakisphosphate pathway demonstration of Ins(1,4,5)P$_3$ 3-kinase activity in animal tissues. Nature 320:631-634.

Kirino T (1982) Delayed neuronal death in the gerbil hippocampus following ischemia. Brain Res 239:57-69.

Lowe DG and Moran LA (1986) Molecular cloning and analysis of DNA complementary to three mouse Mr=68000 heat shock protein mRNA. J Biol Chem 261:2102-2112.

Magnussen KG and Wieloch TW (1989) Impairment of protein ubiquitination may cause delayed neuronal death. Neurosci Lett 96:264-270.

Marini AM, Kozuka M, Lipsky RH, Nowak TS Jr. (1990) 70-kilodalton heat shock protein induction in cerebellar astrocytes and cerebellar granule cells *in vitro*: Comparison with immunocytochemical location after hyperthermia *in vivo*. J Neurochem 54:1509-1516.

Nicoletti F, Meek, JL, Iadarola Mj, Chuang DM, Roth BL, Costa E (1986) Coupling of inositol phospholipid metabolism with excitatory amino acid recognition sites in rat hippocampus. J Neurochem 46:40-46.

Nowak TS Jr (1991) Localization of 70 KDa stress protein mRNA induction in gerbil brain after ischemia. J Cereb Blood Flow Metab 11:432-439.

Ordway RW, Singer JJ, Walsh JV Jr (1991) Direct regulation of ion channels by fatty acids. Trends Neurosci 14:96-100.

Rothman SM and Olney JW (1986) Glutamate and pathophysiology of hypoxic-ischemic brain damage. Ann Neurol 19:105-111.

Salbaum JM, Weidermann A, Lemaire HG, Masters CL, Beyreuter K (1988) The promoter of Alzheimer's disease amyloid A4 precursor gene. EMBO J 7:2807-2819.

Vass K, Welch WJ, Nowak TS Jr (1988) Localization of 70kDa stress protein induction in gerbil brain after ischemia. Acta Neuropathol 77:128-135.

Westerberg E and Wieloch T (1986) Lesions to the corticostriatal pathways ameliorate hypoglycemia-induced arachidonic acid release. J Neurochem 47:1507-1511.

Wieloch T (1985) Neurochemical correlates to selective neuronal vulnerability. Prog Brain Res 63:69-85.

Worley PF, Baraban JM, Snyder SH (1987) Beyond receptors: Multiple second-messenger systems in brain. Ann Neurol 217-229.

Wu B, Hunt C, Morimoto R (1985) Structure and expression of the human gene encoding major heat shock protein HSP70. Mol Cell Biol 5:330-341.

Yoshimi K, Kudo T, Iwata N, Nishimura K (1990) Change of clathrin precedes delayed neuronal death of CA1 cells after ischemia. Jpn J Neurochem 29:456-457.

IMPACT OF DIETARY FATTY ACID BALANCE ON MEMBRANE

STRUCTURE AND FUNCTION OF NEURAL TISSUES

M. T. Clandinin,[1] M. Suh,[1] and K. Hargreaves[2]

[1]Nutrition and Metabolism Research Group
Department of Foods & Nutrition and Department of Medicine
533 Newton Research Building
University of Alberta
Edmonton, Alberta, Canada T6G 2C2

[2]Neurological Research Unit
Department of Surgery
146 Stuart Street, LaSalle Building
Queen's University
Kingston, Ontario, Canada K7L 3N6

ABSTRACT

Neural tissue has generally been viewed as resistant to structural changes induced by exogenous factors. Research has shown that the brain responds to changes in diet by altering neurotransmitter synthesis, and by shifting neuroendocrine controls over a variety of physiological events. Animal model research also indicates that fatty acid constituents and synthesis of brain structural lipid in membranes undergoing turnover can be altered by changing the composition of dietary fat. In growing animals, the balance between dietary ω6 and ω3 fatty acids influences brain phospholipid fatty acid composition, phosphatidylethanolamine methyltransferase activity, and rate of phosphatidylcholine biosynthesis via the CDP-choline pathway. It is concluded that biosynthetic control mechanisms regulating synthesis of brain structural lipid, in particular phosphatidylcholine, respond to exogenous factors and represent a normal physiological response by the brain. This response may provide a mechanism for therapeutic treatment of disorders involving degeneration of brain structural lipid.

INTRODUCTION

The Membrane

Increasing interest in membrane models has advanced understanding of membrane structure and its relationship to membrane function (Singer and Nicolson, 1972; Stubbs and Smith, 1984). It has become a generally accepted notion that biological membranes are not of constant composition but are changing, dynamic, and responsive structures in terms of membrane constituents (Clandinin et al., 1985; McMurchie, 1988; Spector and Yorek, 1985). This diversity results in differences between cell types and between similar membrane types of different cells. The overwhelming complexity characteristic of biological membranes, complicated by our still incomplete window of knowledge, makes it difficult to describe definitively how specific changes in membrane structure or within individual structural constituents alter membrane dependent functions. As most sites of metabolic regulation are to some degree membrane dependent, the perplexing nature of interpreting the functional-metabolic implications of a change in membrane composition is apparent. Thus, it is not yet possible to define membrane composition that is most desirable or optimal in terms of membrane functions.

The composition of lipids in individual cell membranes is somewhat characteristic of the membrane type. While membrane polar lipid composition and membrane cholesterol content are no doubt carefully regulated by the cell, the content of these constituents also varies with cell cycle, age, and in response to a variety of stimuli or changes in environment and physiological state, for example, with diet and in disease states. Clearly, as membrane phospholipid and cholesterol content may regulate activity of individual membrane proteins (Clandinin et al., 1985; Yeagle, 1989), this type of transition in membrane composition will have functional consequences.

Most biological membranes contain a variety of polar lipids (phosphatidylcholine, phosphatidylethanolamine, phosphatidylserine, phosphatidylinositol, sphingomyelin, and cardiolipin in mitochondria), some of which may be in alkyl-acyl or glycosylated forms. A wide variety of individual phospholipids are present in membrane; some phospholipid species predominate, but others are also present. Up to 40 different fatty acids may be incorporated into the *sn*-1 or *sn*-2 position of the phospholipid molecule. This vast array of diversity in the structural constituents requires complex analysis that normally results in assessment of only a few, often simple, structural characteristics when change in membrane structure is compared with change in membrane function. Measurement of properties and components of the bulk membrane lipid isolated will need to become much more specific to reveal any kind of integrated perspective on how the heterogeneous mixture of phospholipid species interacts with membrane protein to alter biological functions.

Understanding of subcellular sites for synthesis of membrane lipids and targeting of these constituents to membrane sites could easily form the basis for another article. Current research indicates that while phospholipid synthesis occurs for membrane assembly, phospholipid is also synthesized and modified extensively *in situ* in some cases to provide a role in second messenger pathways in response to specific stimuli, e.g., regulation of protein kinase by diacylglycerol (Kikkawa et al., 1989) and sphingomyelin (Hannun and Bell, 1989), and perhaps to provide molecular species for special purposes (Hargreaves and Clandinin, 1989). Thus research will need to differentiate between assessment of new membrane synthesis and the dynamic processes involved in routine maintenance of a wide variety of subcellular membrane lipid pools undergoing change in relation to regulation of subcellular events. While the former may appear to be quantitatively most important, the latter may be more significant in terms of understanding relationships to function.

Several forms of noncovalent and covalent forces are involved in the interaction of membrane protein with membrane lipid (Capaldi, 1977; Tanford, 1978; Marinetti and Cattieu, 1982). The specificity of these interactions in terms of a functional protein requiring a specific lipid has been the subject of many reviews (Clandinin et al., 1985; Yeagle, 1989) and has led to the concept of an annulus of lipid providing a specific microenvironment around hydrophobic regions of membrane proteins. Lipid polar head group specificity for membrane-bound enzymes supports this concept of specific lipid affinities and sequestration of lipid. This micropolymorphism of membrane structure enables lipid to independently influence membrane proteins, permitting precise control of membrane function and enabling the protein to retain integrity of specific functions while moving through a heterogeneous lipid environment. Tight or close association of a proportion of membrane lipids with functional proteins also provides for sequences of interacting enzymes or multienzyme assays. This concept explains why some lipid-dependent functions may not respond to change in bulk lipid properties such as fluidity, but it does not provide the overall mechanism for membrane order and the fundamental answer to why cell membranes and organelle membranes have characteristic shapes, domains with typical phospholipid composition, site-specific placement (e.g., sidedness), and distinctive structural morphology. The answer to these questions will require determination of subcellular traffic in phospholipids and specific cytoskeletal peptides and functional proteins to reveal how membrane elements are placed in specific regions in the cell and the manner by which the cell determines the overall heterogeneity and polarity of this traffic.

DIET AS A DETERMINANT OF SUBCELLULAR STRUCTURE AND FUNCTION

In animals capable of adaptive hypothermia (hibernators), reduction in body temperature to facilitate survival is associated with change in membrane composition and associated functions, reducing these functions to a new set point (Aloia and Raison, 1989). By analogy, for homeotherms normally having a constant high body temperature, extrinsic or dietary influence on membrane lipid fatty acyl tail and polar head group composition has only relatively recently been recognized as a consequential physiological mechanism for alteration of membrane structural lipid and thus membrane-dependent functions. This relationship with diet occurs as a consequence of the essential nature of linoleic acid (18:2ω6) and linolenic acid (18:3ω3) in the diet and the fact that through *de novo* membrane phospholipid synthesis and acyl group turnover in membrane phospholipids, new fatty acids of dietary origin can be incorporated into membrane lipids (Clandinin et al., 1985). Within the range of adequate nutritional status, changes in the balance of fatty acids forming the dietary fat consumed result in changes in membrane structural lipid constituents *in vivo* and in the activity of a wide variety of membrane functions.

Mechanisms by which dietary fat has a potential impact on biological processes are many and varied. From the physiological perspective of the homeotherm, initial responses to change in dietary fat intake occur at the level of the enterocyte. Adaptation to diet by the intestinal tract may buffer or modify composition of fatty acids delivered to other tissues and cell types for structural lipid synthesis or metabolism. Response to diet by other tissues may occur at several levels (Clandinin et al., 1991). The first, most obvious function to be identified involved synthesis of structural lipids of altered fatty acid composition, which results in alteration in membrane functions by specific integral catalytic proteins that control functions which may regulate biosynthesis of membrane components. Another level of biological function potentially altered by a change in dietary fat composition involves a change in hormone binding or responsiveness that alters either the magnitude of metabolism or synthesis of the cell's metabolic products, which may have an impact on the activity of other cell types (Clandinin et al., 1991). Finally as cellular activity is compartmented, it is logical that a change in dietary fat intake induces a change in membrane altering expression of nuclear function, either by changing receptor-mediated stimulation of gene products or by changing the transport of gene products out of the nucleus (Clandinin et al., 1991).

IMPACT OF DIETARY FAT ON MEMBRANE STRUCTURE
AND FUNCTION IN BRAIN

The brain is generally viewed as the organ most resistant to structural change by endogenous or exogenous factors. However, in recent years, research has shown the brain to be more responsive to exogenous manipulation than was previously believed (Cohen and Wurtman, 1976; Jope and Jenden, 1979; Lee, 1985; Hannun and Bell, 1989). Our research has also shown that nutritionally adequate diets differing in dietary fat composition influence the content and fatty acid composition of polar lipids in rat brain synaptosomal and microsomal membranes (Foot et al., 1982). The cause-and-effect nature of change in composition of brain structural constituents remains to be resolved, but the brain is clearly sensitive to alteration of dietary lipid intake, even in a nutritionally complete diet. It is, therefore, logical to postulate that the dietary lipids fed during the early postnatal period are important determinants for structural and functional parameters of developing brain tissues. During this time, diet may also produce changes important for the response of neuronal tissue to challenges encountered later in the life cycle (i.e., aging, disease). In this regard, developmental processes involving membrane phospholipid biosynthesis and turnover, as well as the interaction of membrane lipids with lipid-dependent enzymes controlling metabolic events in the brain, may be influenced.

Brain Structure

Transport of nutrients into the brain is regulated by specific carrier systems located at the blood brain barrier and involving interdigitation of astrocytes, neurons, pericytes, and endothelial cells to form the functional blood brain barrier characterized by tight junctions and few fenestrations of the endothelial cell. The presence of this blood brain barrier implies a division between the effects of neuronal and extraneuronal factors on the central nervous system. Although this area has not been studied, it is likely that dietary lipid affects endothelial cell function by influencing receptor and transporter sites located at the membrane to alter the balance of neuronal and extraneuronal effects.

Membrane Lipid Composition

Research in our laboratory has focused on the composition of dietary fat and its effects on membrane structure and phospholipid metabolism in young, rapidly growing animals. When weanling rats were fed diets high in $\omega6$ fatty acids with or without $\omega3$ fatty acids, a strong relationship between $22:5\omega6$ content was demonstrated for brain microsomal membrane and synaptic plasma membrane phosphatidylethanolamine (Fig. 1). Dietary treatment had no effect on the synaptosomal membrane content of phosphatidylethanolamine, ethanolamine plasmalogen, or sphingomyelin (Foot et al., 1982) (Table 1). However, levels of phosphatidylcholine and cholesterol were altered by diet. Dietary modulation of synaptosomal phosphatidylcholine levels appeared to be mediated through the relative $\omega6$, $\omega3$, and monounsaturated fatty acid content of the diet. As in synaptosomal membranes, phosphatidylcholine levels in microsomal membranes were altered by diet. The overall effect of dietary fat on synaptosomal and microsomal membrane lipid composition was similar, with minor variations likely due to differences in rates of synthesis and turnover of lipids within each membrane. A highly significant correlation was observed between membrane phosphatidylcholine and cholesterol levels in both synaptosomal and microsomal membranes. From these observations several questions arise. i) Is dietary fat exerting an extrinsic irrepressible

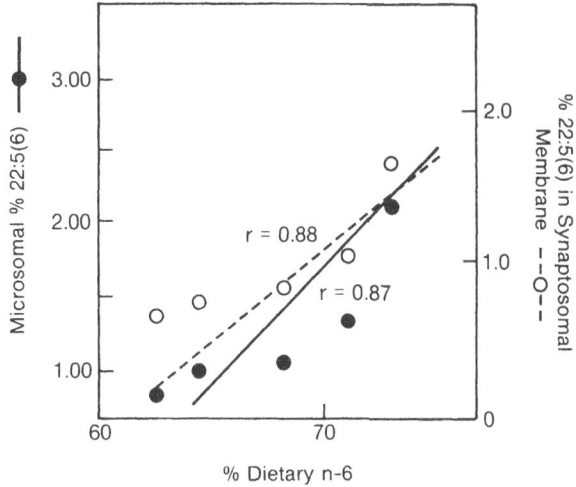

Figure 1. Relationship between diet fat composition and fatty acid content of membrane phosphatidylethanolamine. Animals were fed diets containing mixtures of soybean and sunflower oil (20% w/w fat) for 24 days. Each point represents the mean (n=4).

influence over phospholipid biosynthesis and turnover or is alteration in membrane lipid composition resulting from some form of feedback control within the membrane biosynthetic pathway? ii) Do changes in the fatty acids available for membrane phospholipid synthesis, and thus changes in phospholipid fatty acid composition, alter some specific physical characteristic of the membrane, resulting in concomitant alteration in polar head group composition necessary to maintain membrane physical properties? If either of these mechanisms applies, then it would appear that dietary fatty acids affect phosphatidylcholine biosynthesis *in vivo*.

Table 1. Effect of feeding dietary fat for 21 days on composition of brain synaptosomal membranes

	Content (nmol/mg protein)				
Diet	Phosphatidyl-ethanolamine	Ethanolamine plasmalogen	Phosphatidyl-choline	Sphingo-myelin	Cholesterol
0 days	198±14.6	168±10.8	284±26.5[ac]	43.8± 2.9	480±32.3[a]
Soybean oil	205±17.4	185±17.3	305±13.2[ae]	41.6± 7.3	509±39.5[a]
Sunflower oil	207± 6.2	204± 0.5	266± 9.5[ce]	32.1± 5.1	414± 0.0[a]
Chow	230±14.6	187±29.2	294±19.5[ade]	41.6±13.9	526±91.0[a]
LER	211±12.9	216±29.2	249±15.1[bc]	48.2± 6.6	414±25.8[a]
Significant comparisons	NS	NS	$p<0.01$	NS	$p<0.05$

Values given are mean ± SD. 0 days indicates rats killed before diet treatment. Analysis of variance and the effect of diet are indicated. Three replicates of four rats were used for each treatment. Values without a common superscript are significantly different. LER, low erucic acid rapeseed oil; NS, not significant.

Biosynthesis of phosphatidylcholine *de novo* involves two pathways, the phosphatidylethanolamine methyltransferase pathway and the CDP-choline pathway. It is possible that altering dietary fat could affect the synthesis of phosphatidylcholine through changes in cellular pools of metabolites that activate or inhibit the controlling enzyme, cytidyltransferase (Vance and Choy, 1979). Dietary fat could also modulate phosphatidylcholine biosynthesis through the methylation pathway (Bremer and Greenberg, 1961). Low levels of methyltransferase activity in brain microsomes have been demonstrated (Bremer and Greenberg, 1961). More recently, this methylation pathway in synaptosomal preparations appears to be mediated through levels of substrate or products of the reactions (Hoffman and Cornatzer, 1981; Hoffman et al., 1981). It has also been suggested that methyltransferase shows a preference for unsaturated fatty acyl species (Le Kim et al., 1973), with the major products of this pathway being highly unsaturated phosphatidylcholines (Trewhella and Collins, 1973; Salerno and Beeler, 1973). Thus a mechanism is suggested whereby dietary fat could alter the fatty acid of phosphatidylethanolamine, which in turn could determine the extent of methylation of phosphatidylethanolamine to phosphatidylcholine.

In subsequent studies in which varying levels of dietary ω6 fatty acids were fed, a strong relationship between dietary ω6 fatty acid intake and membrane 22:5ω6 content was demonstrated for brain microsomal membrane and synaptic plasma membrane phosphatidylethanolamine (Fig. 1). Brain, like other excitable tissue, has characteristically high levels of long-chain polyunsaturated fatty acids, particularly 22:6ω3. The importance of polyunsaturates in brain membranes is emphasized by the fact that changes occurring in essential fatty acid deficiency maintain the overall membrane content of polyunsaturates, despite reductions in the proportion of fatty acids of the linoleic and linolenic acid series (Alling et al., 1972; Karlsson, 1975). Feeding a diet containing soybean oil providing ω6 and ω3 fatty acids versus a sunflower oil diet rich in ω6 fatty acids and providing 0.2% ω3 fatty acids produces higher levels of ω3 fatty acids in phosphatidylcholine of microsomal and synaptic plasma membranes and a lower ω6 to ω3 ratio in constituent fatty acids (Fig. 2). Phosphatidylcholine ω3 fatty acid levels increased in microsomal membranes of animals fed diets containing 20% (w/w) fat for 4 weeks, compared with weanling animals. Levels of ω3 fatty acids in synaptic plasma membranes were lower for these animals when compared with weanling animals. These observations demonstrate a membrane-specific response to nutritional factors during development. Conflicting reports exist in the literature concerning the rate of turnover of brain lipids *in vivo*. It is evident that there is considerably greater turnover than previously conceived, even in so-called stable myelin fractions (Gould and Dawson, 1976). Estimated half-lives are less than 20 days for phosphatidylethanolamine, ethanolamine plasmalogen, and phosphatidylcholine (Jungalwala and Dawson, 1971). Diet has also been shown to alter phospholipid fatty acid composition in brain membranes within 24 days (Foot et al., 1982). Studies of brain *in vivo* involving labeled precursors indicate rapid labeling of subcellular membrane lipid fractions in the order of microsomes > mitochondria > myelin. The extent of transfer of lipid from the endoplasmic reticulum to other subcellular fractions, transport of phospholipids down the axon to the nerve terminal, and/or intracellular exchange of lipid is not known. The control mechanisms regulating dynamic changes in levels and composition of phospholipids during brain development are also not understood, but presumably relate to precursor availability and levels and modulation of enzyme activity.

The CDP-choline pathway for phosphatidylcholine biosynthesis is the quantitatively major route for phosphatidylcholine synthesis in brain, with the phosphatidylethanolamine methyltransferase pathway contributing to the total amount of phosphatidylcholine synthesized *de novo*. Phosphocholinetransferase is a lipid-dependent membrane-

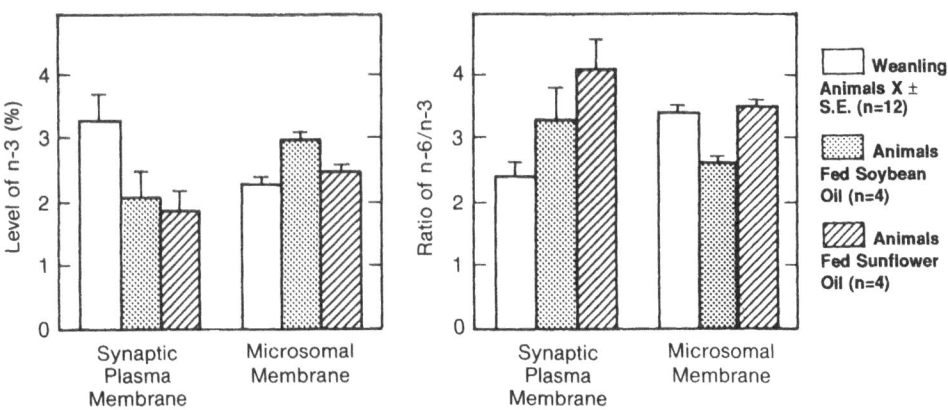

Figure 2. Effect of diet on brain membrane phosphatidylcholine fatty acid during development.

bound enzyme catalyzing synthesis of phosphatidylcholine from CDP-choline and a diglyceride. Activity of phosphocholinetransferase in the microsomal membrane of weanling rat brain approaches that of adult rat liver, demonstrating the great importance of neuronal phospholipid biosynthesis in the weanling animal (Hargreaves and Clandinin, 1987a). A similar observation is made for phosphatidylethanolamine methyltransferase activity in the synaptic plasma membrane of weanling animals (Hargreaves and Clandinin, 1987b). Higher levels of phosphatidylethanolamine methyltransferase activity observed in synaptic plasma membranes versus microsomal membranes suggests a specific function for phosphatidylcholine produced via this pathway at this subcellular site. The pool of phosphatidylcholine synthesized via the phosphatidylethanolamine methyltransferase pathway also has a fatty acid profile that is distinct from the bulk membrane phosphatidylcholine pool (Trewhella and Collins, 1973; Strittmatter et al., 1979) and has a rapid metabolic rate of turnover (Mogelson and Sobel, 1981). The phosphatidylethanolamine methyltransferase pathway has been implicated in specific regulatory processes and in processes of neurotransmitter synthesis and release (Blustajn and Wurtman, 1984; Mozzi et al., 1982). Substrate preference by methyltransferases of the phosphatidylethanolamine methyltransferase pathway for polyunsaturated species of phosphatidylethanolamine may contribute to the polyunsaturated pool of phosphatidylcholine produced via this pathway (Hargreaves and Clandinin, 1987a,b). Thus, it is conceivable that nutritional regulation of distinct pools of phosphatidylcholine within the membrane may have significant implications for regulation of varied enzyme-linked cellular events.

Phospholipid Species

Diet-induced alterations in microsomal membrane phosphatidylethanolamine have been compared with phosphatidylethanolamine composition found in weanling animals (Hargreaves and Clandinin, 1987b) and indicate that feeding a diet rich in $\omega 3$ fatty acids elevates membrane levels of $\omega 3$ fatty acids compared with weanling animals or animals fed diets containing sunflower oil. Diets containing fish oil (28% 20:5$\omega 3$ and 9% 22:6$\omega 3$) or linseed oil (37% 18:3$\omega 3$) produce the highest levels of membrane $\omega 3$ fatty acids, while maintaining membrane levels of $\omega 6$ fatty acids.

The composition of membrane phosphatidylethanolamines has been evaluated by examining phospholipid species. Separation of microsomal membrane phosphatidylethanolamine species by argentation-TLC produces four major molecular species (Table 2). Species containing six double bonds comprised primarily 18:0 plus 22:6ω3. Species containing four double bonds combined 18:0 with 20:4ω6 or with 22:4ω6. Monoenoic species contained 16:0 or 18:0 plus 18:1. A fraction containing four to six double bonds contained the long-chain polyunsaturated fatty acids 20:4ω6, 22:4ω6, 22:5ω6, and 22:6ω3. The distribution and composition of these molecular species differed for weanling animals and for animals fed different dietary treatments (Table 3). Weanling animals exhibited the highest levels of 22:6ω3 and 20:4ω6 in species containing six and four double bonds, respectively. Feeding animals diets containing 20% w/w sunflower oil produced very high levels of 22:5ω6 in the four-to-six double bond species, and feeding diets containing linseed oil or fish oil failed to increase ω3 levels in the major species isolated. The unusual nature of triglycerides in these diets may have resulted in minor phosphatidylethanolamine species rich in long-chain ω3 fatty acids, such as 22:6ω3. It may be significant that addition of cholesterol increased levels of 22:6ω3 in the fraction containing six double bonds and 18:1 in the monoenoic fraction. A possible explanation for this observation is the fact that cholesterol is known to affect fatty acid desaturase activity (Garg et al., 1988).

Feeding animals a diet high in ω6 fatty acids (sunflower oil) produced species of microsomal phosphatidylethanolamine particularly rich in 22:5ω6, 22:4ω6, and 20:4ω6. This observation was related to measurements of the rate of phosphatidylcholine biosynthesis (Fig. 3). Diets containing different ratios of soybean oil and sunflower oil were mixed to produce a range of dietary ω6 to ω3 fatty acid ratios. Resulting levels of 22:5ω6 found in microsomal membrane phosphatidylethanolamine correlated with production of phosphatidylmethylethanolamine via the phosphatidylethanolamine methyltransferase pathway, but was negatively correlated with phosphocholinetransferase activity. The fatty acid composition of the dimethyl intermediate was also shown to be high in 20:4ω6, 22:4ω6, and 22:5ω6 fatty acids (Hargreaves and Clandinin, 1989).

Other Membrane Functions

The interaction between diet, membrane polar head group content, cholesterol concentration, phospholipid fatty acid composition, and function of integral membrane proteins is likely to be extremely complex. In brain, as many enzymes of neurotransmitter metabolism are lipid dependent, it is logical to postulate a possible interaction

Table 2. Fatty acid composition of rat brain microsomal phosphatidylethanolamine species

| Fatty acid | Number of double bonds | | | |
	6	4-6	4	1
16:0	12.9	15.1	8.1	18.1
18:0	38.8	24.4	42.8	22.2
18:1		6.8		41.5
18:2(ω6)		0.6		
20:4(ω6)		14.6	30.6	
22:4(ω6)		5.8	7.2	
22:5(ω6)		9.2		
22:6(ω3)	34.6	6.4		

Results given as % (w/w). Data abstracted from Hargreaves and Clandinin, 1990.

Table 3. Effect of diet on composition of rat brain microsomal membrane phosphatidyl-ethanolamine species

| | Number of double bonds | | | | |
| | 6 | 4-6 | | 4 | 1 |
	22:6(ω3)	22:5(ω6)	22:4(ω6)	20:4(ω6)	18:1
Weanling	35.3	5.6	6.8	30.5	40.8
Dietary oil treatment:					
Soybean oil	31.7	2.6	7.0	18.0	33.4
Sunflower oil	31.8	15.9	8.8	24.6	41.9
Linseed oil	23.1	1.3	2.4	12.4	30.1
Fish oil	18.6	1.5	2.9	21.7	24.8
Pooled SD (n=3)	4.6	1.1	0.7	2.4	5.1

Results given as % (w/w). Data abstracted from Hargreaves and Clandinin, 1990.

between dietary lipid and brain neurotransmitter metabolism. In our laboratory, these changes in brain membrane composition have been shown to alter the thermotropic behavior of acetylcholinesterase activity (Foot et al., 1983) (Fig. 4).

From these studies it may be concluded that for brain synaptosomal membranes, diet-induced changes in phospholipid fatty acid composition and polar lipid content significantly alter the functional properties of a lipid-dependent membrane protein within its membrane microenvironment. The interaction between diet, polar head group distribution, cholesterol content, and fatty acyl-tail composition of phospholipids is complex and likely more specific to the environment of the integral protein than can

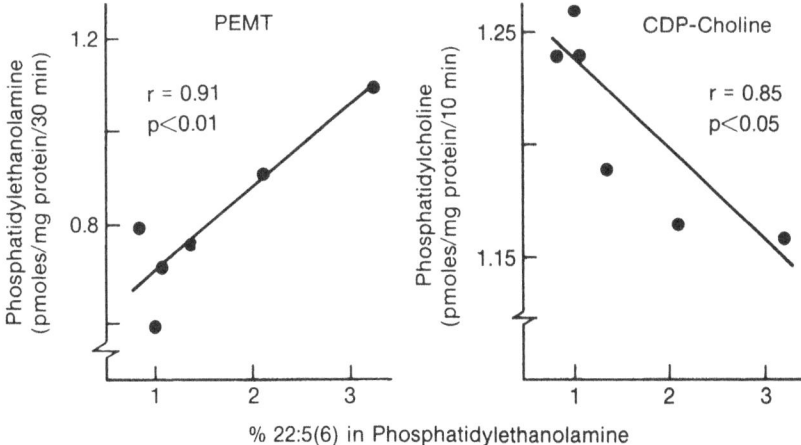

Figure 3. Relationship between microsomal membrane phosphatidylethanolamine 22:5(ω6) content and phosphatidylcholine synthesis via phosphatidylethanolamine methyltransferase (PEMT) and CDP-choline pathways. Animals were fed diets containing mixtures of soybean and sunflower oil (20% w/w fat) for 24 days. Each point is the mean (n=4). Data abstracted from Hargreaves and Clandinin, 1990.

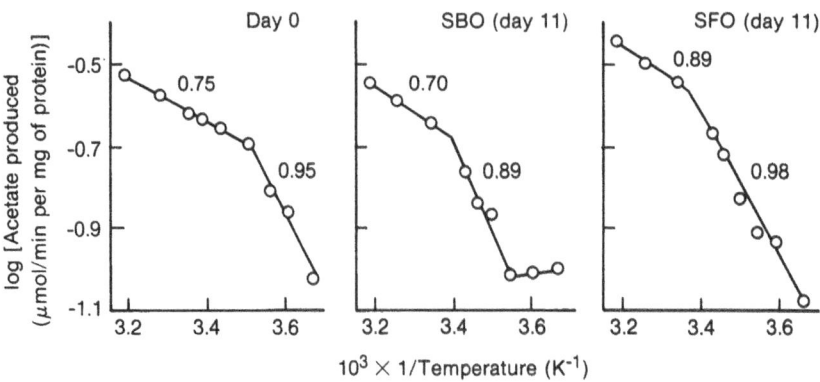

Figure 4. Arrhenius plots showing changes in thermotropic behavior of acetylcholinesterase activity induced by dietary fat. Synaptosomal membranes were prepared from brains of rats fed diets containing soybean oil (SBO) or sunflower oil (SFO). Each point represents the mean of four groups of four rats per diet treatment. Two separate assays were performed on each group. The data were analyzed by the method of least squares for determination of break points. Straight lines were fitted by regression analysis. Regression coefficients are given for each line. Energies of activation calculated from these slopes are below transition temperature (J/mol per °C): SBO, 240 and SFO, 200; $p < 0.05$.

be assessed by electrospin resonance (esr) probes. Most significantly, it is apparent that membrane-dependent enzymes of neurotransmitter metabolism, such as acetylcholinesterase, may be sensitive to alterations in balance of dietary fatty acids consumed, even in a nutritionally complete diet.

IMPACT OF DIETARY FAT ON RETINAL PHOSPHOLIPID COMPOSITION

Excellent studies by Bazan and co-workers have examined the synthesis and incorporation of 20:4ω6 and 22:6ω3 into phospholipids in the developing retina and the rod outer segments, as well as during degenerative conditions of the retina (Aveldaño and Bazan, 1983; Aveldaño et al., 1983; Bazan et al., 1985; Reddy and Bazan, 1985; Bazan et al., 1986; Reddy et al., 1986; Scott et al., 1987). Phospholipid species containing C20-C36 ω6 and ω3 fatty acids are abundant in the rod outer segments of the retina and are apparently associated with rhodopsin of the photoreceptor membrane (Aveldaño, 1988). Moreover, the metabolism of 20:4ω6 and 22:6ω3 is altered during impaired photoreceptor cell differentiation (Scott et al., 1988). Exposure of the retina to light stimulates metabolism of 20:4ω6 in the rod outer segment of the retina, resulting in accelerated formation of leukotriene B_4, hydroxyeicosatetraenoic acid, and prostaglandin D_2 (Birkle and Bazan, 1989). Thus, it is clear that the metabolism of 20:4ω6 and 22:6ω3 is in some way crucial to visual functions in the retina.

The impact of dietary fat on the metabolism of polyunsaturated fatty acids in the retina has not been exhaustively examined. In this regard we have conducted studies in normal and diabetic rats to examine the effect of diets high or low in ω3 fatty acids on the metabolism of 20:4ω6 and 22:6ω3 in the retina in conjunction with the levels of C26-C34 ω6 and ω3 fatty acids found in individual phospholipid species. Feeding weanling rats nutritionally adequate diets containing 0.2% of dietary fat as 18:3ω3 versus up to 5% of dietary fatty acids as 20:5ω3 and 22:6ω3 for 6 weeks altered the phospholipid fatty acid composition in the retina (Table 4). Feeding a diet containing increased levels of ω3 fatty acids increased the ω3 fatty acid level in each phospholipid

Table 4. Effect of high and low dietary ω3 fatty acid intake on very long-chain polyenoic fatty acid content of phospholipid in the retina

	PE		PC		PS	
Fatty acid	High	Low	High	Low	High	Low
ω6						
20:4	8.7	10.2	10.3	11.1	4.7	7.3
22:4	1.2	2.1	0.5	1.0	1.1	4.0
22:5	0.1	1.9	0.2	1.5	0.2	2.9
LCPE x1	0.0	0.0	0.2	0.1	0.1	0.1
LCPE x2	0.0	0.1	0.1	0.2	0.2	1.5
LCPE x3	0.0	0.1	0.0	0.2	-	0.2
LCPE x4	0.0	0.1	0.1	0.1	0.1	0.1
ω3						
20:5	0.5	0.1	0.4	0.0	0.3	0.1
22:5	1.1	0.5	0.1	0.4	2.1	0.9
22:6	33.4	27.6	31.7	21.2	37.4	31.7
LCPE x5	0.1	0.1	0.5	0.1	1.5	0.5
LCPE x6	0.2	0.1	0.5	0.3	1.5	1.0

Weanling rats were fed semipurified diets containing 10% (High) or 0.2% (Low) ω3 fatty acids for 42 days. The retinas were removed and phospholipids were isolated. Results are given as % (w/w). Values represent the mean of four samples (three for phosphatidylcholine) each prepared from six retinas pooled from three animals. LCPE, fatty acids of chain length of C24 to C36. Analysis of fatty acid constituents by gas chromatography using BP-20 and BP-1 fused silica columns, in conjunction with argentation TLC, indicated the presence of the following LCPE fatty acids: 24:4, 26:4, 30:4, 32:4, 24:5, 26:5, 28:5, 24:6, 26:6, 28:6. PE, phosphatidylethanolamine; PC, phosphatidylcholine; PS, phosphatidylserine.

fraction while decreasing the level of long-chain ω6 fatty acids. Phosphatidylethanolamine, phosphatidylserine, and phosphatidylcholine in retina also contain highly unsaturated ω6 and ω3 fatty acids of C24-C32 chain length. The content of these constituents in the membrane is also influenced by dietary intake of ω3 and ω6 fatty acids (Table 4).

At present it is not possible to speculate on the functional role of these highly unsaturated fatty acids in the retina, nor the degree of change occurring in the content of these components in retina during development or in relation to change in photoreceptor functions in disease states. It is clear that this new area of research has considerable potential to reveal important mechanisms through which highly unsaturated fatty acids play functional roles in excitable membranes.

CONCLUSION

These observations illustrate that in neural tissues dietary fat affects a complex metabolic pathway for phosphatidylcholine biosynthesis in a coordinated fashion. This may be important to aspects of development concerning choline metabolism, or regulatory processes dependent on signals from a changing milieu in the microenvironment of the plasma membrane. In this regard, the stimulatory effect of longer-chain polyunsaturated essential fatty acid homologues in membrane phosphatidylethanolamine increases phosphatidylethanolamine methyltransferase and decreases CDP-choline activity. It is also concluded that dietary fat influences phosphatidylethanolamine composition in microsomal and synaptic plasma membrane of the brain. Increasing the dietary fat $\omega6/\omega3$ ratio increases microsomal and synaptic plasma membrane $\omega6/\omega3$ fatty acid ratio in phosphatidylethanolamine fractions. Membrane phosphatidylethanolamine species containing one, four, or six double bonds and the ratio of $22:5\omega6/22:6\omega3$ in phosphatidylethanolamine containing four to six double bonds are also affected by the fatty acid composition of the diet. Increase in membrane $\omega6$ fatty acid content is associated with increased phosphatidylethanolamine methyltransferase activity and decreased phosphocholine transferase activity, thus indicating a mechanism by which change in an exogenous factor (i.e., dietary fat intake) may alter brain phospholipid biosynthesis. It is provocative to ponder whether diet could be used to induce formation of membrane structures more resistant to specific insults that cause degeneration of brain structural material and/or degeneration of cholinergic neurons or to reverse degenerative changes occurring in neural membrane structure and function.

ACKNOWLEDGMENTS

This work was supported by a grant from the Natural Sciences and Engineering Research Council of Canada, the Medical Research Council of Canada, and the Canadian Diabetes Association. K.M. Hargreaves was a recipient of an Alberta Heritage Foundation for Medical Research Studentship and is currently a Canadian Heart and Stroke Foundation Postdoctoral Fellow. M.T. Clandinin is a Scholar of the Alberta Heritage Foundation for Medical Research.

REFERENCES

Alling C, Bruce A, Karlsson I, Svennerholm L (1972) The effect of different dietary levels of essential fatty acids on growth of the rat. Nutr Metab 16:38.

Aloia RC and Raison JK (1989) Membrane function in mammalian hibernation. Biochim Biophys Acta 988:123.

Aveldaño MI (1988) Phospholipid species containing long and very long polyenoic fatty acids remain with rhodopsin after hexane extraction of photoreceptor membranes. Biochemistry 27:1229.

Aveldaño MI and Bazan NG (1983) Molecular species of phosphatidylcholine, -ethanolamine, -serine, and -inositol in microsomal and photoreceptor membranes of bovine retina. J Lipid Res 24:620.

Aveldaño MI, Pasquare de Garcia SJ, Bazan NG (1983) Biosynthesis of molecular species of inositol, choline, serine and ethanolamine glycerophospholipids in the bovine retina. J Lipid Res 24:628.

Bazan NG, Reddy TS, Bazan HEP, Birkle DL (1986) Metabolism of arachidonic and docosahexaenoic acids in the retina. Prog Lipid Res 25:595.

Bazan NG, Reddy TS, Redmond TM, Wiggert B, Chader GJ (1985) Endogenous fatty acids are covalently and noncovalently bound to interphotoreceptor retinoid-binding protein in the monkey retina. J Biol Chem 260:13677.

Birkle DL and Bazan NG (1989) Light exposure stimulates arachidonic acid metabolism in intact rat retina and isolated rod outer segments. Neurochem Res 14:185.

Blustajn JK and Wurtman RJ (1984) Alzheimer's disease: Advances in basic research and therapies (Wurtman RJ, Corkin SH, Growdon JH, eds) pp 183-198. Center for Brain Sciences and Metabolism Charitable Trust.

Bremer J and Greenberg DM (1961) Methyl transferring enzyme system of microsomes in the biosynthesis of lecithin (phosphatidylcholine). Biochim Biophys Acta 46:205.

Capaldi RA, ed (1977) Membrane proteins and their interaction with lipids. Vol 1, New York: Marcel Dekker.

Clandinin MT, Cheema S, Field CJ, Garg ML, Venkatraman J, Clandinin TR (1991) Dietary fat: Exogenous determination of membrane structure and cell function. FASEB J 5:2761.

Clandinin MT, Field CJ, Hargreaves K, Morson L, Zsigmond E (1985) Role of diet fat in subcellular structure and function. Can J Physiol Pharmacol 63:546.

Cohen EL and Wurtman RJ (1976) Brain acetylcholine: Control by dietary choline. Science 191:561.

Foot M, Cruz T, Clandinin MT (1983) Effect of dietary lipids on synaptosomal acetylcholinesterase activity. Biochem J 211:507.

Foot M, Cruz TF, Clandinin MT (1982) Influence of dietary fat on the lipid composition of rat brain synaptosomal and microsomal membranes. Biochem J 208:631.

Garg ML, Sebokova E, Thomson ABR, Clandinin MT (1988) Delta6-desaturase activity in liver microsomes of rats fed diets enriched with cholesterol and/or omega-3 fatty acids. Biochem J 249:351.

Gould RM and Dawson RMC (1976) Incorporation of newly formed lecithin into peripheral nerve myelin. J Cell Biol 68:480.

Hannun YA and Bell RM (1989) Functions of sphingolipids and sphingolipid breakdown products in cellular regulation. Science 243:500.

Hargreaves K and Clandinin MT (1987a) Phosphocholinetransferase activity in plasma membrane: Effect of diet. Biochem Biophys Res Commun 145:309.

Hargreaves K and Clandinin MT (1987b) Phosphatidylethanolamine methyltransferase: Evidence for influence of diet fat on selectivity of substrate for methylation in rat brain synaptic plasma membranes. Biochim Biophys Acta 918:97.

Hargreaves KM and Clandinin MT (1989) Coordinate control of CDP-choline and phosphatidylethanolamine methyltransferase pathways for phosphatidylcholine biosynthesis occurs in response to change in diet fat. Biochim Biophys Acta 1001:262.

Hargreaves K and Clandinin MT (1990) Dietary lipids in relation to postnatal development of the brain. Upsala J Med Sci Suppl 48:79.

Hoffman DR and Cornatzer WE (1981) Microsomal phosphatidylethanolamine methyltransferase: Some physical and kinetic properties. Lipids 16:533.

Hoffman DR, Haning JA, Cornatzer WE (1981) Microsomal phosphatidylethanolamine methyltransferase: Inhibition by S-adenosylhomocysteine. Lipids 16:561.

Jope RS and Jenden DJ (1979) Choline and phospholipid metabolism and the synthesis of acetylcholine in rat brain. J Neurosci Res 4:69.

Jungalwala FB and Dawson RMC (1971) The turnover of myelin phospholipids in the adult and developing rat brain. Biochem J 123:683.

Karlsson I (1975) Effects of different dietary levels of essential fatty acids on the fatty acid composition of ethanolamine phosphoglycerides in myelin and synaptosomal plasma membranes. J Neurochem 25:101.

Kikkawa U, Kishimoto A, Nishizuka Y (1989) The protein kinase C family: Heterogeneity and its implications. Annu Rev Biochem 58:31.

Le Kim D, Betzing H, Stoffel W (1973) Studies *in vitro* and *in vivo* on methylation of phosphatidyl-N-N-dimethylethanolamine to phosphatidylcholine in rat liver. Hoppe Seylers Z Physiol Chem 354:437.

Lee RE (1985) Membrane engineering to rejuvenate the aging brain. Can Med Assoc J 132:325.

Marinetti GV and Cattieu K (1982) Tightly (covalently) bound fatty acids in cell membrane proteins. Biochim Biophys Acta 685:109.

McMurchie EJ (1988) Physiological regulation of membrane fluidity. In: Advances in membrane fluidity (Aloia RC, Curtain CC, Gordon LM, eds) Vol 3, pp 189-237. New York: Alan R. Liss.

Mogelson S and Sobel BE (1981) Ethanolamine plasmalogen methylation by rabbit myocardial membranes. Biochim Biophys Acta 666:205.

Mozzi R, Siepi D, Adreoli V, Porcellati G (1982) Biochemistry of SAM and related compounds (Usdin E, Borchardt RT, Creveling CR, eds) pp 129-138. New York: MacMillan Press.

Reddy TS and Bazan NG (1985) Synthesis of docosahexaenoyl-, arachidonoyl- and palmitoyl-coenzyme A in ocular tissues. Exp Eye Res 41:87.

Reddy TS, Birkle DL, Packer AJ, Dobard P, Bazan NG (1986) Fatty acid composition and arachidonic acid metabolism in vitreous lipids from canine and human eyes. Curr Eye Res 5:441.

Salerno DM and Beeler DA (1973) The biosynthesis of phospholipids and their precursors in rat liver involving *de novo* methylation and base-exchange pathways, *in vivo*. Biochim Biophys Acta 326:325.

Scott BL, Racz E, Lolley RN, Bazan NG (1988) Developing rod photoreceptors from normal and mutant *rd* mouse retinas: Altered fatty acid composition early in development of the mutant. J Neurosci Res 20:202.

Scott BL, Reddy TS, Bazan NG (1987) Docosahexaenoate metabolism and fatty acid composition in developing retinas of normal and *rd* mutant mice. Exp Eye Res 44:101.

Singer SJ and Nicolson GL (1972) The fluid mosaic model of the structure of cell membranes. Science 175:720.

Spector AA and Yorek MA (1985) Membrane lipid composition and cellular function. J Lipid Res 26:1015.

Strittmatter WJ, Hirata F, Axelrod J (1979) Phospholipid methylation unmasks cryptic beta-adrenergic receptors in rat reticulocytes. Science 204:1205.

Stubbs CD and Smith AD (1984) The modification of mammalian membrane fluidity and function. Biochim Biophys Acta 779:89.

Tanford C (1978) Hydrophobic effect and organization of living matter. Science 200:1012.

Trewhella MA and Collins FD (1973) Pathways of phosphatidylcholine biosynthesis in rat liver. Biochim Biophys Acta 296:51.

Vance DE and Choy PC (1979) How is phosphatidylcholine biosynthesis regulated? Trends Biochem Sci 4:145.

Wurtman RJ, Hefti F, Melamed E (1980) Precursor control of neurotransmitter synthesis. Pharmacol Rev 32:315.

Yeagle PL (1989) Lipid regulation of cell membrane structure. FASEB J 3:1833.

STRUCTURAL AND FUNCTIONAL IMPORTANCE OF DIETARY

POLYUNSATURATED FATTY ACIDS IN THE NERVOUS SYSTEM

Jean-Marie Bourre,[1] Michelle Bonneil,[2]
Jean Chaudière,[1] Michel Clément,[1]
Odile Dumont,[1] Georges Durand,[3]
Huguette Lafont,[2] Gilles Nalbone,[2]
Gérard Pascal,[3] and Michèle Piciotti[1]

[1]INSERM Unité 26
 Hôpital Fernand Widal
 75475 Paris cedex 10

[2]INSERM Unité 130
 18 avenue Mozart
 13009 Marseille

[3]INRA
 CNRZ Jouy-en-Josas, France

ABSTRACT

The nervous system is the organ with the second greatest concentration of lipids. These lipids participate directly in membrane functioning. Brain development is genetically programmed. It is therefore necessary to ensure that nerve cells receive an adequate supply of nutrients, especially of lipids, during their differentiation and multiplication, and throughout their lives. The effects of polyunsaturated fatty acid deficiency have been extensively studied; prolonged deficiency leads to death in animals. Linoleic acid is now universally recognized to be an essential nutrient. Until recently, however, α-linolenic acid was considered non-essential.

Feeding animals with oils that have a low α-linolenic content results in all brain cells and organelles and various organs having reduced amounts of 22:6n-3, which is compensated for by an increase in 22:5n-6. The speed of recuperation from these anomalies is extremely slow for brain cells, organelles, and microvessels, in contrast to other organs. A decrease in α-linolenic series acids in the membranes results in a 40% reduction in the Na^+-K^+-ATPase of nerve terminals and a 20% reduction in 5′-nucleotidase. Some other enzymatic activities are not affected, although membrane fluidity is altered. A diet low in α-linolenic acid induces alterations in the electroretinogram which disappear with age; motor function and activity are little affected, but learning behavior is markedly altered. The presence of α-linolenic acid in the diet confers a greater resistance to certain neurotoxic agents (triethyl-lead).

During the period of cerebral development, there is a linear relationship between brain content of n-3 acids and the n-3 content of the diet up to the point where α-linolenic levels reach 200 mg for 100 g of food intake. Beyond that level there is a plateau. For other organs, such as the liver, the relationship is also linear up to 200 mg/100 g, but then there is merely an abrupt change in slope and not a plateau. When dietary 18:2n-6 content was varied, it was noted that 20:4n-6 optimum values were obtained at 150 mg/100 g for all nerve structures, 300 mg for testicle and muscle, 800 mg for kidney, and 1200 mg for liver, lung and heart. A deficiency in α-linolenic acid and an excess of linoleic acid have the same main effect: an increase in 22:5n-6 levels.

Bearing in mind the relative metabolisms of human and rat, their rates of development, the difference between their brain/body weight ratios, and the similarity in fatty acid composition of their nerve membranes, it is possible to affirm that results obtained in the rat are necessarily, and at the very least, valid for man. For the brain and the other organs, the requirement in α-linolenic acid is 200 mg/100 g food intake (0.4% of calories), provided that linoleic acid requirements of 1200 mg/100 g food intake (2.4% of calories) are met.

During pre- and postnatal development, Δ6 desaturase in brain decreases dramatically (12-fold) up to postnatal day 21 and remains nearly constant thereafter. In liver, the activity increases approximately 9-fold between day 3 before birth and day 7 after birth. Thereafter, it decreases slightly up to weaning and is approximately constant up to 4 months of age. From then on Δ6-desaturase decreases with age (40% between 4 and 17 months). The question remains whether the residual Δ6-desaturase activity after day 21 is sufficient to support the turnover of brain membranes. If it is not, the very long-chain fatty acids would have to be synthesized by the liver. Since liver synthesis decreases during aging, this source may be insufficient. It should be noted that cultured nerve cells differentiate, multiply, and take up and release neurotransmitters only if the medium contains 20:4n-6 and 22:6n-3, but not if it contains 18:2n-6 and 18:3n-3. Thus, the fatty acids that are essential for the brain could be those with very long chains. They are probably synthesized in the liver from α-linolenic and linoleic acids. They can also be furnished directly by the diet.

However, a dietary excess of fish oil can prove to be toxic due to perturbation of the composition of cerebral membranes. Pharmacological doses of fish oil do not alter fatty acid composition of liver and brain, and do not change protection against peroxidation. Increasing dietary fish oil in rat had the following effects on brain lipids: 20:4n-6 regularly decreased and cervonic acid was increased by 30% at high fish oil concentration. In contrast, in the liver nearly all fatty acids (saturated, monounsaturated, and polyunsaturated) were affected by high dietary content of fish oil, but liver function was normal and serum triacylglycerols were reduced.

As we have shown that DHA and not α-linolenic acid is essential for brain cell culture and that, at least in liver, desaturating activity decreases with certain pathologies and aging, it is fundamental to consider the addition of very long chains in the diet.

Polyunsaturated fatty acids in membranes are protected against peroxidation, mainly by vitamin E. In the peripheral nervous system during development and aging, a highly significant correlation between vitamin E and n-6 PUFA was observed, but there was not a significant correlation between n-3 PUFA and vitamin E.

INTRODUCTION

Brain is one of the tissues containing the highest amount of lipids which, in turn, play a role in modulating the structure, fluidity, and function of brain membranes (Farias, 1980; Brenner, 1984; Mead, 1984). Brain lipids contain polyunsaturated fatty acids derived from dietary essential linoleic and α-linolenic acids. More than one-third of the brain fatty acids are polyunsaturated with a prevalence of acids containing very long chains (mainly arachidonic acid, 20:4n-6, and cervonic acid, 22:6n-3) (Paoletti and Galli, 1972; Galli, 1973; Crawford et al., 1977; Lamptey and Walker, 1978a; McKenna and Campagnoni, 1979; Bourre et al., 1984; Greenwood and Craig, 1987; Connor and Neuringer, 1988; Holman, 1988; Yamamoto et al., 1988; Bazan, 1990). In fact, brain cells and organelles contain only trace amounts of linoleic acid and α-linolenic acid (Bourre et al., 1984). Essential fatty acid deficiency is known to have dramatic effects on various organs.

The polyunsaturated fatty acids in membranes are not the same as the dietary precursors (linoleic and α-linolenic acids), but have longer and more highly unsaturated chains (mainly arachidonic and cervonic acids). These acids, in particular arachidonic acid, are the precursors of important hormonal substances (prostaglandins and leukotrienes), but their structural role is also important (Mead, 1984) since they play a major role in the structure, enzymatic activities, and function of the membrane (Paoletti and Galli, 1972; Galli, 1973; Crawford et al., 1977; Lamptey and Walker, 1978a; McKenna and Campagnoni, 1979; Farias, 1980; Menon and Dhopeshwarkar, 1982; Bourre et al., 1984; Brenner, 1984; Mead, 1984; Stubbs and Smith, 1984; Holman, 1986; Greenwood and Craig, 1987; Sanders and Rana, 1987; Connor and Neuringer, 1988; Holman, 1988; Yamamoto, 1988; Bazan, 1990). Thus, either these fatty acids are rapidly and completely transformed into the longer chain fatty acids after crossing the blood brain barrier, or the fatty acids essential for the brain are in fact the very long chain fatty acids which are either synthesized in the liver or are provided with the diet.

Cerebral development is genetically programmed, the renewal of neurons and oligodendrocytes is nil, and that of nervous membranes is often very slow. Therefore, during differentiation and multiplication, cells require adequate supplies of nutrients, especially lipids and particularly polyunsaturated fatty acids (PUFA). A lipid abnormality leads to an alteration in the function of membranes.

Saturated and monounsaturated fatty acids are mainly synthesized by nerve tissue itself, via complex mechanisms that differ according to cell type and organelle (Bourre et al., 1984). In the nervous system, on average, one fatty acid out of three is polyunsaturated. In fact, the polyunsaturated fatty acids present in the membranes are not the dietary precursors (linoleic and α-linolenic acids) but longer and more desaturated chains (mainly 20:4 n-6 and 22:6 n-3). These control the composition of membranes and hence their fluidity, and, as a result, their enzymatic activity, the binding between molecules and their receptors, cellular interactions, and the transport of nutrients. As far as the nervous system is concerned, these fatty acids can also influence certain electrophysiological parameters as well as learning functions. Dietary polyunsaturated fatty acids to a great extent determine membrane levels of these fatty acids (Mead, 1984; Holman, 1986) and are particularly important for ensuring harmonious cerebral development (Menon and Dhopeshwarkar, 1982). There are many reports on the influence of polyunsaturated fatty acids on the structure and function of the nervous system (Paoletti and Galli, 1972; Svennerholm et al., 1972; Galli, 1973; Sun and Sun, 1974; Crawford et al., 1977; Cook, 1978; Lamptey and Walker, 1978a, 1978 b; McKenna and Campagnoni, 1979; Clandinin et al., 1980a, 1980b; Sprecher, 1981; Bazan et al., 1982; Holman et al., 1986; Berkow and Campagnoni, 1983; Bourre et al., 1984; Kaare and Drevon, 1986; Crawford et al., 1987; Greenwood and Craig, 1987; Philbrick et al., 1987; Sanders and Rana, 1987; Connor and Neuringer, 1988; Holman, 1988; Yamamoto et al., 1988; Bourre et al., 1989a; Bazan, 1990). However, polyunsaturated fatty acids of the n-3 series play a very special role in membranes, especially in the nervous system: all cerebral cells and organelles are extremely rich in these fatty acids (Table 1). It is therefore extremely important to know precisely what quantity should be supplied by the diet.

Table 1. Occurrence of polyunsaturated fatty acids in nervous cells and organelles

	Total PUFAs	20:4n-6	22:6n-3
Neurons	32	15	8
Synaptosomes	33	18	12
Astrocytes	29	10	11
Oligodendrocytes	20	9	5
Myelin	5	9	5
Capillaries	35	16	10
Mitochondria	30	16	12
Microsomes	29	11	12
Retina	45	5	35
Photoreceptor membranes	65	4	56
Peripheral nerve	10	7	2
Schwann cells	22	11	5

Results are expressed as percentage of total fatty acids (mg %). Animals were fed standard lab chow containing both n-3 and n-6 fatty acids.

A-LINOLENIC ACID CONTROLS THE COMPOSITION AND FUNCTION OF NERVE MEMBRANES

The effects on the nerve cells of a specific deficiency in n-3 acids (with a normal supply of n-6 fatty acids) are interesting (Bourre et al., 1984). We have previously compared animals fed a diet containing a normal amount of n-6 fatty acids but lacking n-3 fatty acids (a diet containing sunflower oil) with animals fed a diet containing both types of fatty acids (a diet containing soybean oil). For the diet deficient in n-3, two groups of Wistar rats were fed for several generations with a semi-synthetic diet containing either sunflower oil or peanut oil. For the normal diet containing n-3, two groups of rats were fed diets containing either soybean oil or rapeseed oil. Rapeseed oil-fed animals were compared with peanut oil-fed animals; soybean oil-fed animals were compared with sunflower oil-fed animals. The brain cells and the intracellular organelles conserve a normal total quantity of polyunsaturated fatty acids, but the various cell types and organelles show a considerable deficit in cervonic acid (22:6n-3), which is compensated for by an excess of docosapentaenoic acid (22:5n-6) as measured for whole brain and other organs (Galli et al., 1971; Alling et al., 1972) (Table 2). Comparison of animals fed for 60 days on either the sunflower oil diet or the soybean oil diet showed n-3/n-6 ratios of 1/20 in the diet, 1/16 in the oligodendrocytes, 1/12 in the myelin, 1/2 in the neurons, 1/6 in the synaptosomes, and 1/3 in the astrocytes (Bourre et al., 1989a). The importance of n-3 fatty acids has also been shown by a study of phosphatidylethanolamine in animals fed a peanut or rapeseed oil diet (Nouvelot, 1986).

The Rate of Recovery of Anomalies is Extremely Slow

Rats switched from a diet deficient in n-3 diet to a diet containing n-3 (Youyou et al., 1986; Bourre et al., 1989b) required several months for brain cells and organelles

Table 2. Quantities of 22:6n-3 and 22:5n-6 in the nervous system of animals fed an n-3 deficient diet, expressed as percentages of quantities in animals fed a diet without deficiencies

	22:6n-3	22:5n-6
Neurons	28	214
Synaptosomes	27	1,088
Oligodendrocytes	10	240
Myelin	14	1,200
Astrocytes	47	344
Mitochondria	25	917
Microsomes	28	592
Retina	36	1,280
Sciatic nerve	28	1,000

Data from Bourre et al., 1984 and Bourre et al., 1989a

to recover normal levels of 22:6n-3 and lose excess 22:5n-6 (Fig. 1).

This slow recovery was the same whatever the cell or organelle. It would be expected that recuperation would not be rapid in myelin, which has a slow turnover. But it is very surprising that nerve terminals also have a very slow recovery although

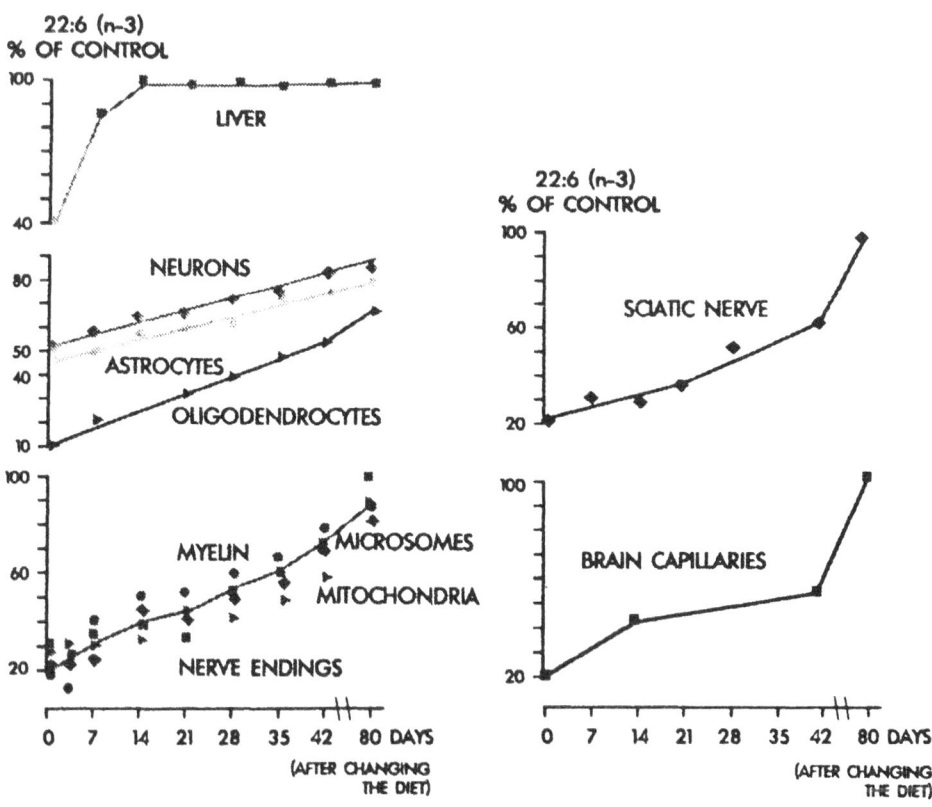

Figure 1. Rate of recovery of 22:6n-3 in rats fed n-3 deficient diet followed by resumption of normal diet.

Table 3. Deficiency in α-linolenic acid and brain membrane enzyme activities

	Brain	Myelin	Nerve endings
5′-nucleotidase	0.70	0.74	1.20
Na$^+$-K$^+$-ATPase	0.95	1.10	0.55
CNPase	0.95	0.78	0.00

Values represent enzyme activities obtained with animals fed an n-3 deficient diet divided by those obtained with nondeficient animals. Data from Bourre et al., 1989a. 5′-nucleotidase: EC 3.1.3.5; Na$^+$-K$^+$-ATPase: EC 3.6.1.3; CNPase (2′-3′ cyclic nucleotide-3′-phosphodiesterase): EC 3.1.4.37.

turnover of their membrane molecules is supposed to be rapid. It may be that regulation of recuperation occurs at the level of either synthesis of chain ends in the liver (22:6n-3 and 20:4n-6), transport across the blood-brain barrier, or the enzymatic activities of desaturation and elongation. It is interesting to note that cerebral microvessels and capillaries (Homayoun et al., 1988) also have a very slow rate of recuperation, even though they are in contact with plasma lipoproteins of normal composition, since the liver recuperates very rapidly (2 weeks).

Effects of α-Linolenic Acid Deficiency on Enzymatic Activities

The activity of 5′-nucleotidase is decreased by 30% in whole brain, but not in myelin or in nerve terminals, signifying that its activity is probably altered considerably in cell membranes (Table 3) (Bourre et al., 1989a). These results are in agreement with those of Bernshohn and Spitz (1974), who have shown that a decrease in the activity of this enzyme produced by simultaneous deficiency in linoleic and α-linolenic acids is corrected only by the addition of α-linolenic acid to the diet.

Na$^+$-K$^+$-ATPase is reduced nearly by half in the nerve terminals of animals fed an n-3 deficient diet compared with those fed the n-3 containing diet. On the other hand, simultaneous deficiencies in both linoleic and α-linolenic acid lead to an increase in Na$^+$-K$^+$-ATPase activity (Sun and Sun, 1974). This enzyme controls ion transport produced by nerve transmission. It consumes half the energy used by the brain.

It is interesting to note that CNPase (2′-3′-cyclic nucleotide-3′-phosphodiesterase), which is specific for myelin, decreases as a result of α-linolenic acid deficiency, even though the myelin membrane is considered to be very rigid and not very metabolically active. The activity of another enzyme, acetylcholine-esterase, is also modulated by dietary lipids (Beaugé et al., 1988).

Membrane Fluidity

In nerve-ending membranes, fluidity is affected by the diet, depending on the membrane region. Feeding the sunflower oil diet compared to the soybean oil diet results in less fluidity in the surface polar part of the membranes probed by TMA- or PROP-DPH but greater fluidity in the apolar part of the membranes (probed by DPH) (Table 4) (Beaugé et al., 1988).

The fluidizing effect of ethanol shown with diphenylhexatriene (DPH) is also decreased significantly in animals fed sunflower oil (Beaugé et al., 1988). Concomitantly,

Table 4. Effect of diet on membrane organization in terms of basal degree of fluorescence polarization of DPH, TMA-DPH, and PROP-DPH in rat synaptic membranes

Probe	Diets		Percentage change
	Soya	Sunflower	
DPH	0.333±0.004	0.320±0.001*	– 4%
TMA-DPH	0.352±0.001	0.359±0.002*	+ 2%
PROP-DPH	0.360±0.001	0.370±0.003*	+ 3%

Data are means ± SD. Statistical significance: *p < 0.01.
Data from Beaugé et al., 1988.

rats fed sunflower oil are more sensitive to ethanol-induced hypothermia, illustrating the importance of diet to membrane sensitivity and animal response to ethanol, regardless of the exact mechanisms (Table 5).

EFFECT OF α-LINOLENIC ACID ON ELECTROPHYSIOLOGICAL, BEHAVIORAL, AND TOXICOLOGICAL PARAMETERS

Electroretinogram

In the retina, 22:6 n-3 level is high (Bazan et al., 1982). Prolonged deficiency in PUFAs induces changes in the distribution of membrane fatty acids in the retina which are associated with changes in the electroretinogram (Neuringer and Connor, 1986; Connor and Neuringer, 1988). In 4-week-old animals fed an n-3 deficient diet, the threshold of detection (10 Mv) of wave A required a light stimulation 10 times stronger than that of the nondeficient animals (Fig. 2). In 6-week-old deficient animals, electroretinogram changes were less marked; in adult deficient animals, only the A wave remained abnormal (Bourre et al., 1989a).

Learning Tests

A simultaneous deficiency in linoleic and α-linolenic acids affects the learning capacities of animals (Lamptey and Walker, 1978b), as does a selective deficiency in α-linolenic acid (Lamptey and Walker, 1976). Although motor activity and open field tests were nearly normal in animals fed the n-3 deficient diet, learning capacities were

Table 5. Effect of diet on ethanol membrane tolerance in terms of fluidizing efficacy of ethanol in rat synaptic membranes

Probe	Diets		Percentage change
	Soya	Sunflower	
DPH	0.0185±0.0040	0.0152±0.0012*	– 18%
TMA-DPH	0.0088±0.0002	0.0080±0.0012	
PROP-DPH	0.0087±0.0003	0.0070±0.0009	

Data are means ± SD. Statistical significance: *p < 0.01.

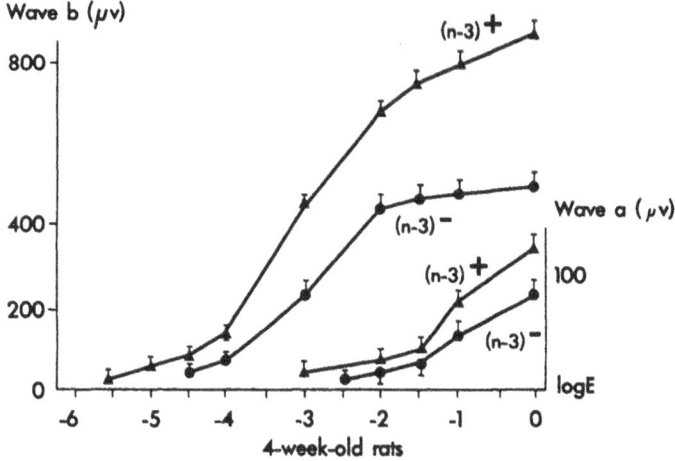

Figure 2. Effect of dietary n-3 deficiency on a- and b-waves in electroretinograms of 4-week-old rats.

severely perturbed, as shown by the shuttle box test. In the first session, animals fed the diet containing n-3 made a more rapid association between the light stimulus and the electric shock, since they avoided on average seven shocks out of 30, whereas n-3 deficient diet animals avoided only two. These differences diminished with further conditioning and disappeared at the fourth session (Fig. 3) (Bourre et al., 1989a). Interestingly, the extinction of learning capacities was significantly longer in animals deficient in dietary α-linolenic acid (Yamamoto, et al., 1987; Togashi and Tamai, 1988).

Mortality in Animals Tested with the Neurotoxic Agent Triethyltin

The LD_{50} of animals fed the soybean oil or sunflower oil diets did not differ significantly (6.18 vs 6.02 ml/kg, respectively), but the animals fed the sunflower oil diet died earlier than those fed the soybean oil diet (Bourre et al., 1989a).

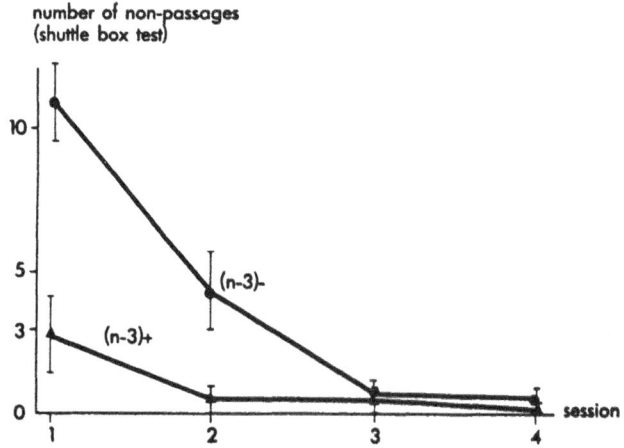

Figure 3. Effect of dietary n-3 deficiency on learning capabilities in rats.

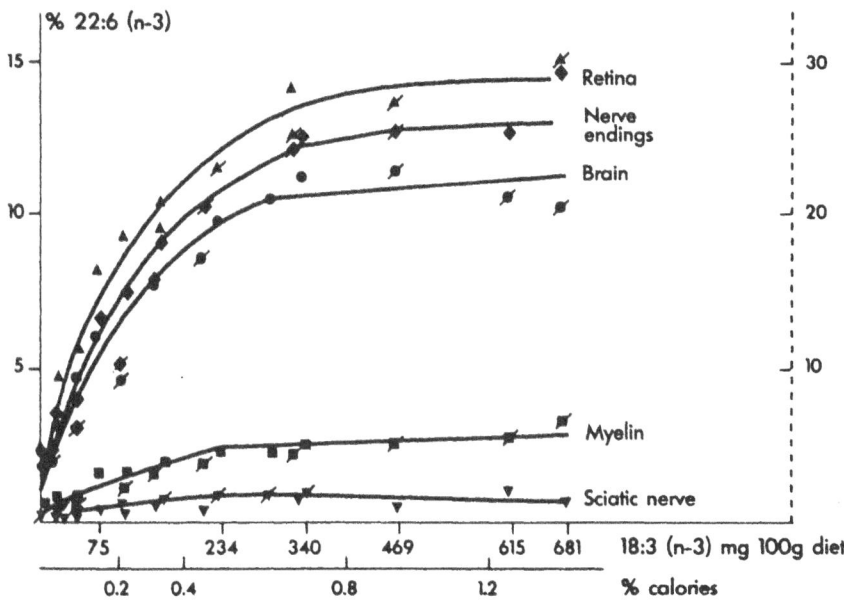

Figure 4. Relationship between dietary α-linolenic acid and cervonic acid level in nervous tissue.

MINIMUM REQUIREMENTS

Minimum Requirement of α-Linolenic Acid in Cerebral Membranes

When 12 groups of rats were fed diets with intermediate levels of α-linolenic acid for three weeks before mating, increased amounts of 18:3 n-3 led to an overall increase in 22:6 n-3 and a decrease in 22:5 n-6 in the 21-day-old pups (Fig. 4). In brain, levels of 22:6 n-3 increased linearly with intakes of 18:3 n-3 from 0 to 200 mg/100 g diet and then reached a plateau; the opposite was observed for 22:5n-6 (Bourre et al., 1989a). In liver, kidney, lung, heart and muscle the same threshold was found but the plateau was less clear.

Minimum Requirement of Linoleic Acid

For three weeks before mating, 12 groups of female rats were fed different amounts of linoleic acid (18:2n-6) (Bourre et al., 1990a) (Fig. 5). The resulting male pups were killed at 21 days of age. Dietary 18:2n-6 contents between 150 and 6200 mg/100 g food intake produced the following results. Linoleic acid levels remained very low in brain, myelin, synaptosomes, and retina. In contrast, 18:2 n-6 levels increased in sciatic nerve. In heart, linoleic acid levels were high, but were not related to dietary linoleic acid intake. Levels of 18:2n-6 were significantly increased in liver, lung, kidney, and testicle and were even higher in muscle. On the other hand, in heart a constant amount of 18:2n-6 was found at a low level of dietary 18:2n-6. Constant levels of arachidonic acid (20:4n-6) were reached at 150 mg/100 g diet in all nerve structures, at 300 mg/100 g diet in testicle and muscle, at 800 mg/100 g diet in kidney, and at 1200 mg/100 g diet in liver, lung, and heart. Constant adrenic acid (22:4n-6) levels were obtained at 150, 900, and 1200 mg/100 g diet in myelin, sciatic nerve, and brain, respectively. Minimal levels

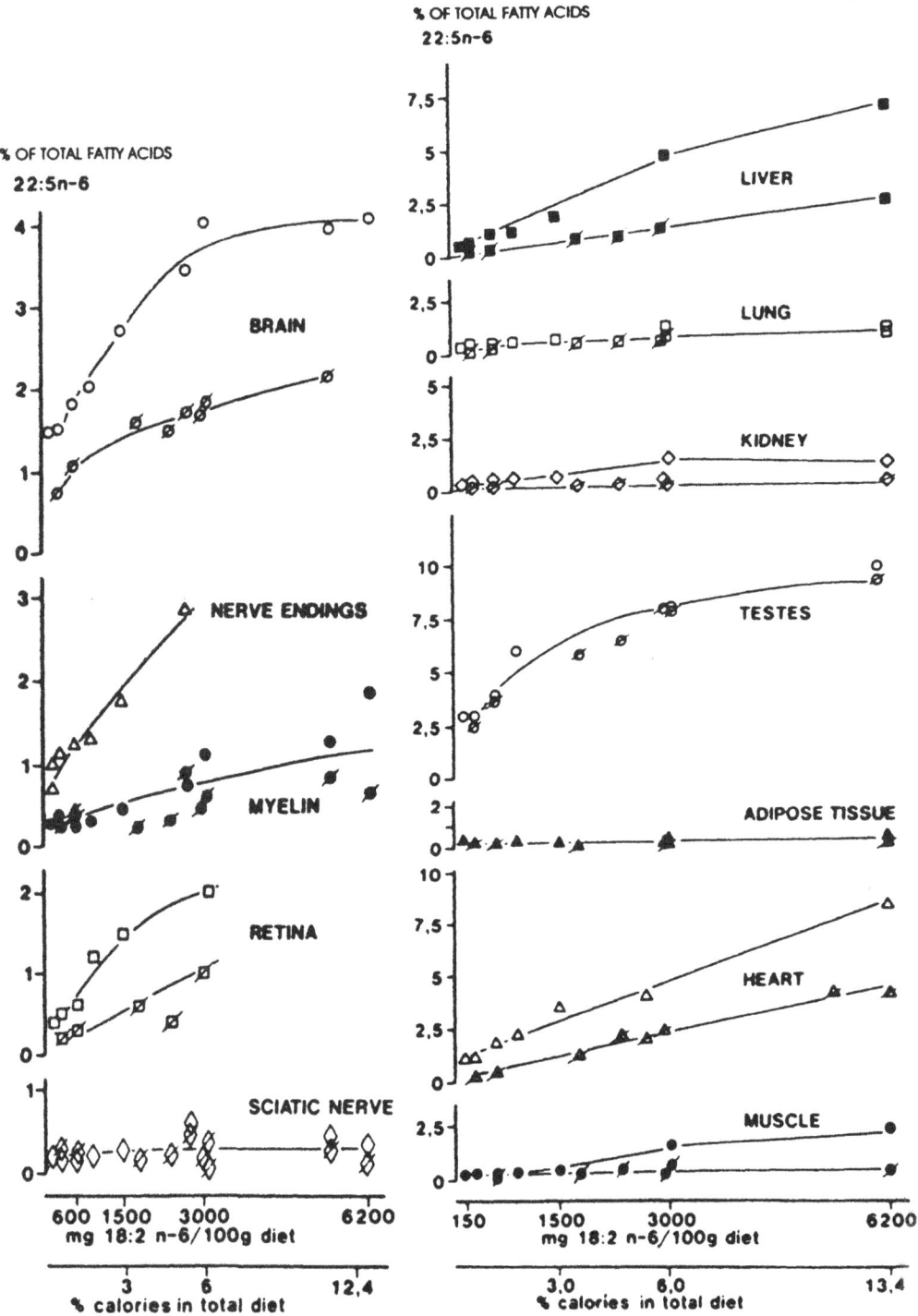

Figure 5. Accumulation of docosapentaenoic acid (22:5n-6) various tissue from 21-day-old male pups of rats fed various levels of linoleic acid (18:2n-6) for three weeks prior to mating.

were difficult to determine. In all fractions examined accumulation of docosapentaenoic acid (22:5n-6) was the most direct and specific consequence of excess amounts of dietary 18:2n-6. Tissue eicosapentaenoic acid (20:5n-3) and 22:5n-3 levels were relatively independent of dietary 18:2n-6 intake, except in lung, liver, and kidney. In several organs (muscle, lung, kidney, liver, heart), as well as in myelin, very low levels of dietary linoleic acid led to an increase in 20:5n-3. Dietary requirements for 18:2n-6 varied from 150 to 1200 mg/100 g food intake, depending on the organ and the nature of the tissue fatty acid. Therefore, the minimum dietary requirement is estimated to be about 1200 mg/100 g (i.e., the level that ensures stable and constant amounts of arachidonic acid).

These precursors, linoleic and α-linolenic acids, are elongated and desaturated by the liver into longer chains, which are in fact the essential fatty acids for the brain, as cell cultures seem to have demonstrated (Bourre et al., 1983). Nerve cells in culture differentiate, multiply, and capture and liberate neurotransmitters only if the medium contains 20:4n-6 and 22:6n-3, but not in the presence of 18:2n-6 and 18:3n-3 (Bourre et al., 1983, Loudes et al., 1983).

Δ6 DESATURASE IN BRAIN AND LIVER
DURING DEVELOPMENT AND AGING

The amount of Δ6 desaturase was measured in the mouse brain and liver using linoleic acid as substrate. During pre- and postnatal development, Δ6 desaturase in brain decreased dramatically (12-fold) up to postnatal day 21 (Strouve-Vallet and Pascaud, 1971; Bordoni et al., 1988; Bourre et al., 1990b; Ulmann et al., 1991) and remained nearly constant thereafter. In liver, the activity increased approximately 9-fold between day 3 before birth and day 7 after birth. Thereafter, it decreased slightly up to weaning and was approximately constant up to 4 months. From then on, Δ6 desaturase decreased with age (40% between 4 and 17 months).

The mouse was chosen as model because it is known that desaturating activity in mice is lower than that in rats (Cunnane et al., 1984; Horrobin et al., 1984) and thus is closer to that in humans, taking into account the ratio of very long-chain polyunsaturated fatty acids to their precursors in the blood and liver.

In brain, Δ6 activity is very high during early development up to 7 days after birth. This corresponds to the period of neuronal and glial multiplication, the latter event reaching a peak 3-5 days after birth in the mouse. Early brain development requires large quantities of polyunsaturated fatty acids for membrane synthesis. Interestingly, Δ6 activity does not peak during myelination, although myelin contains large amounts of polyunsaturated fatty acids. The same pattern was found for Δ9 desaturase activity (Carreau et al., 1979). Thus, polyunsaturated fatty acids required for myelination are either accumulated in the oligodendrocytes before myelination or possibly are supplied through the bloodstream. This is in contrast to the synthesis of saturated and monounsaturated long-chain and very long-chain fatty acids which peaks during myelination and is impaired in neurological dysmyelinating mutants (Bourre et al., 1977). Interestingly, chain lengthening of eicosapentaenoic acid (EPA) is less affected in these mutants than elongation of erucic and arachidic acids (Bourre et al., 1976).

The question remains whether the residual Δ6 desaturase activity after 21 days is sufficient for synaptogenesis and, later on, to support the turnover of brain membranes. If it is not, the very long-chain fatty acids would have to be synthesized by the liver. Since liver synthesis decreases during aging, this source may be insufficient. It has been hypothesized that one aspect of aging could be the reduced activity of Δ6 desaturase, which would impede membrane renewal (Horrobin, 1981).

RELATIONSHIP BETWEEN POLYUNSATURATED FATTY ACID IN MEMBRANES AND VITAMIN E

Persistence of High Correlations with Total and Specific n-6 Polyunsaturated Fatty Acids

Vitamin E, as an integral part of membranes, is seen as a biological antioxidant which, by sequestering free radicals, functions to terminate the propagation of auto_oxidation processes such as polyunsaturated fatty acid peroxidation. In addition, specific effects of α-tocopherol that do not involve its antioxidant function and that act on the architecture of membranes by controlling the profiles of their unsaturated phospholipid and cholesterol components have been suggested. It will be interesting to determine whether there is a correlation between alteration of vitamin E and polyunsaturated fatty acids.

A highly significant correlation between vitamin E and n-6 PUFA (18:2n-6, 20:4n-6, and total n-6) was observed but not between n-3 PUFA and vitamin E (Clément and Bourre, 1990). It is suggested that there may be a relationship between vitamin E and n-6 PUFA in the peripheral nervous system membranes during development and aging.

In contrast to brain, the sciatic nerve concentration of vitamin E in rats increased rapidly during the postnatal period (approximately 5-fold between days 1 and 8), then decreased dramatically (about 2-fold between days 8 and 30), and further decreased slowly between days 30 and 60 and remained constant up to 2 years. While the sciatic nerve concentration of vitamin E decreased by 58% between days 8 and 30, the concentration of vitamin E in serum presented a marked decrease (75%). The age-related changes in fatty acid concentration of the endoneurial fraction of the sciatic nerve were characterized by a large increase in content of saturated and monounsaturated fatty acids up to 6 months (2-fold for saturated and 4-fold for monounsaturated fatty acids). Then, up to 24 months, the amount of these fatty acids decreased very slowly. The content of n-6 polyunsaturated fatty acids decreased rapidly up to 1 year and slowly thereafter. In contrast, during development the amount of n-3 PUFA was relatively stable and decreased during aging.

EFFECTS OF FISH OIL

Effect of Pharmacological Doses

Feeding rats a diet enriched in n-3 PUFAs (200 ml/day menhaden oil) increased the content of eicosapentaenoic acid (20:5n-3) in brain phospholipids. Conversely, 22:4n-6 was reduced. These changes were not associated with alterations in either vitamin E concentration or glutathione peroxidase and catalase activities in cerebrum and cerebellum. No increase in peroxidative damage was found. Interestingly, the major very long-chain fatty acids (22:6n-3 and 22:5n-3) were not affected (Claudière et al., 1987).

Significant differences were observed in total fatty acids with the percentage of 22:4n-6 being nine times lower in the menhaden group, while the percentage of 20:5n-3 was six times higher. In the phosphatidylethanolamine fraction, which contains 45-50% PUFAs, the percentage of 20:5n-3 was nine times higher in the menhaden group, while the percentages of 22:6n-3 were not significant in different groups. In the phosphatidylcholine fraction which only contains 11% PUFAs, the percentage of 20:5n-3 and 22:6n-3 was higher in the menhaden group, but these differences were minor when compared to those observed in the phosphatidylethanolamine fraction.

In total lipids from menhaden oil-fed animals, 22:4n-6 was decreased but no significant alterations were found in either phosphatidylethanolamine or phosphatidylcholine. It is speculated that large alterations probably exist in phosphatidylserine.

The main differences observed between groups in both total lipids and the phosphatidylethanolamine fraction did not change the percentage of total PUFAs, which remained close to 29% of the total lipids in each group.

The vitamin E content, as well as the activity of glutathione peroxidase and catalase in the brain and cerebellum, was unchanged in both the menhaden and olive oil groups.

Effects of Increasing the Amount of Fish Oil

Increasing dietary fish oil in rat had the following effect on brain lipids (Table 6): arachidonic acid decreased progressively; eicosapentaenoic acid, normally nearly undetectable, was present; 22:5n-3 dramatically increased but remained below 1% of total fatty acids; and cervonic acid increased by 30% at high fish oil concentration. Saturated and monounsaturated fatty acids were not affected regardless of chain length. In contrast, in the liver nearly all fatty acids (saturated, monounsaturated and polyunsaturated) were affected by high dietary content of fish oil, but liver function was normal: serum vitamin A and E, glutathione peroxidase, alkaline phosphatase and transaminases were not affected. Total serum cholesterol, unesterified cholesterol and phosphatidylcholine were slightly affected. In contrast, triacylglycerols were dramatically reduced in proportion to the fish oil content of the diet (Bourre et al., 1990c).

For brain, Table 6 shows that 20:4n-6 decreased proportionately with increasing dietary fish oil content (and decreasing corn oil). We have previously shown that 20:4n-6 content is independent of dietary linoleic acid content in excess of the minimal level (0.3% of the calories); this was largely true in all diets tested. Thus the decrease in 20:4n-6 in brain membranes is due only to the increase of fish oil in the diet. While 22:4n-6 was less affected than arachidonic acid, 22:5n-6 was reduced by about 60%. The high amount of dietary corn oil could increase 22:5n-6, since its level in membranes parallels dietary excess of linoleic acid (Bourre et al., 1990a) as well as the deficiency in α-linolenic acid (Holman et al., 1982; Bourre et al., 1989a).

Table 6. Total lipid fatty acid composition (weight %) of rat brain after various diets

		Composition of the diet			
Salmon oil, g/100 g diet	-	1.5	4.0	7.0	10.0
Corn oil, g/100 g diet	10	8.5	6.0	3.0	-
Ratio of n-6/n-3	75	10.0	3.0	1.0	0.1
		Brain fatty acids, weight %			
18:2n-6	0.9	1	0.9	0.7	0.3
20:4n-6	10.2	9.1	8.5	7.7	7.8
22:4n-6	3.5	2.7	2.7	3.0	2.7
22:5n-6	0.8	0.4	0.3	0.3	0.3
20:5n-6	-	-	0.1	0.2	0.3
22:5n-3	0.1	0.2	0.4	0.6	0.7
22:6n-3	12.1	12.3	15.4	14.6	15.6

Five groups of 12 male Wistar rats (IFFA-Credo, l'Arbresle, France) weighing 190-200 g were housed two per cage. For 8 weeks, all groups received the same semi-synthetic diet having the same total amount of lipids, but varying in fish oil (increasing salmon oil was compensated for by decreasing corn oil). Data from Bourre et al., 1990c.

Table 7. Fatty acid composition of the forebrain and liver from rats fed various oil-supplemented diets

Fatty acid	Brain				Liver		
	Control	Corn	Cod liver	Salmon	Control	Corn	Cod liver
C18:2n-6	0.62	1.3	0.82	0.85	14.0	30.6	7.8
C20:4n-6	8.8	9.1	7.2	7.7	20.36	18.34	3.2
C22:4n-6	3.15	3.5	2.1	2.2	0.52	1.16	-
C20:5n-3	0.2	0.2	0.35	0.27	0.13	0.1	1.0
C22:5n-3	0.22	0.27	0.98	0.65	0.5	0.32	0.3
C22:6n-3	10.70	10.8	13.0	13.9	3.72	2.9	1.6

Control: diet containing 4.4% (w/w) fat consisting of a lard (2.2%) and corn oil (2.2%) mixture. All experimental animals were fed a 17% lipid diet. Corn: 2% corn oil with 15% cod liver oil; Cod liver: same diet as group 2 but supplemented with cod liver oil; Salmon: salmon oil-enriched diet (12.5% w/w) supplemented with 4.5% corn oil. Corn oil, as supplied, contained 45 mg α-tocopherol per 100 g oil. Salmon oil was supplemented with 100 mg α-tocopherol/100 g oil. Therefore, the amount of vitamin E supplied by the salmon oil-enriched diet was 295 mg/kg of diet. Data from Bourre et al., 1988.

Eicosapentaenoic acid was nearly undetectable in the diet containing low amounts of fish oil and increased in the diet containing 4% or more fish oil, but the content was still extremely low, even in the diet containing a very high amount of fish oil. In 10% the fish oil diet, 22:5n-3 was increased 7-fold but brain content was always below 1% of the total fatty acids and 22:6n-3 was increased by about 30%.

Effect of Large Excess of Very Long-Chain PUFA

In agreement with others (Philbrick et al., 1987), we (Bourre et al., 1988) have shown (Table 7) that cod liver oil as well as salmon oil (15 and 12.5% in the diet, respectively) supplemented with α-tocopherol (100 mg/100 g) induce similar alterations in forebrain PUFA levels (but saturated and monounsaturated fatty acids are little affected if at all). Although the brain is considered to be well protected, an 8-week fish oil diet in 60-day-old animals increased the n-3 series and decreased the n-6 series. When cod liver oil- and salmon oil-fed animals were compared with those fed corn oil, the brain contents of 20:4n-6, 22:4n-6, and 22:5n-6 were decreased by 16-19, 37-40, and 64-79%, respectively, whereas brain 20:5n-3 and 22:6n-3 were increased by 375, 141-22, and 20-29%, respectively. In the cod liver oil-fed animals all liver fatty acids were significantly affected. Brain is not protected against a large excess of very long-chain n-3 PUFA; such an excess increases the n-3/n-6 ratio, could lead to abnormal function, and might be difficult to reverse.

ESSENTIAL FATTY ACIDS IN CULTURED BRAIN CELLS

The biochemical and morphological effects of polyunsaturated fatty acids on fetal brain cells grown in a chemically defined medium were studied. Fetal brain cells were dissociated from mouse cerebral hemispheres taken on the 16th day of gestation. After cells had grown in chemically defined medium for 8 days, the proportion of polyunsaturated fatty acids in cultured cells was only one-half of that observed at day 0 and about 1.5 times less than that of cells grown in serum-supplemented medium. The fatty acid 20:3n-9 was present in cultured cells grown in either chemically defined or serum-supplemented medium, demonstrating the deficiency of essential fatty acids. The reduced amount of polyunsaturated fatty acids in cells grown in the chemically defined medium was balanced by an increase in monounsaturated fatty acids. The saturated fatty acids were not affected. When added at seeding time, linoleic, α-linolenic, arachidonic, or docosahexaenoic acid stimulated the proliferation of small dense cells (Bourre et al., 1983). In addition, we demonstrated that each of the four fatty acids studied was incorporated into phospholipids. Adding fatty acids of the n-6 series increased the content of n-6 fatty acids in the cells, but also provoked an increase in the n-3 fatty acids. Among several combinations of fatty acids, only 20:4 and 22:6, when added to the culture in a ratio of 2/1, restored a fatty acid profile similar to that of controls (i.e., *in vivo* tissue taken at postnatal day 5).

As noted above, cultured nerve cells differentiate, multiply, take up and release neurotransmitters only if the medium contains 20:4n-6 and 22:6n-3, but not if it contains 18:2n-6 and 18:3n-3 (Loudes et al., 1983; Bourre et al., 1983). Hepatic desaturase must be functional for transformation of dietary precursors into longer chains.

DISCUSSION

Essentiality of α-Linolenic Acid

A diet deficient in α-linolenic acid caused marked alterations in the fatty acid composition of all cellular and subcellular fractions examined. The total content (number of moles) of PUFAs was not altered, the marked decrease in 22:6n-3 being compensated for by an increase in 22:5n-6. This compensation is quantitative, but total unsaturation remains in deficit. It is evidence that PUFAs control the fluidity of biological membranes, hence many of their activities. A specific deficiency in n-3 fatty acids perturbs the activities of membrane enzymes, alters some electrophysiological activities as shown by the electroretinogram, and disturbs learning abilities. After switching from a deficient to a normal diet, the rate of recovery is remarkably slow: it is several months before brain cells and organelles recover normal levels of cervonic acid. This rate is the same in all other organelles. It is therefore crucial to supply the fatty acids necessary for cerebral structures at the developmental stage. A deficiency is difficult to correct.

Mean fatty acid levels in human brain differ little from those of similar regions in rat brain. Human development involves a greater daily increase in brain mass over a longer period, and the ratio of brain weight:total body weight is greater in man, even taking the 2/3 coefficient into account. Consequently, the minimal requirements in rat are *a fortiori* those in man. In any case, for obvious ethical reasons, it is not possible to determine the effects of increasing dietary fatty acid levels on the composition of human cerebral membranes. Our study is the first to measure simultaneously the variations of all the polyunsaturated fatty acid levels in several organs as a function of variations in dietary linoleic acid content, minimal α-linolenic acid requirements being satisfied. In contrast to α-linolenic acid requirements, which are the same for all organs (200 mg/100 g food intake) (Bourre et al., 1989a), the linoleic acid requirements differ according to organ (Bourre et al., 1990a). The minimal requirements in man may, therefore, be taken as 1200 mg/100 g food intake (2.4% of calories) for linoleic acid and 200 mg/100 g food intake (0.4% of calories) for α-linolenic acid.

A pathogenesis of α-linolenic acid deficiency has been described in the monkey (Fiennes et al., 1986) and in humans (Holman et al., 1982; Anonymous, 1986; Bjerve et al., 1987a; Bjerve et al., 1987b). A deficiency in n-3 fatty acids has been proposed as a syndrome of modern society (Rudin, 1982). It is, therefore, very important to verify the precise amount of n-3 acids in the diet. The results of this study indicate that, in order to avoid deficiency, α-linolenic acid should be present at 0.4% of the total dietary energy, in agreement with studies in animals (Pudelkewicz et al., 1968) and in humans (Holman, et al., 1982; Lasserre et al., 1985; Kinsella, 1986).

Advantages (and Possible Adverse Effects) of Very Long-Chain Precursors Derived from Fish Oil

Since cerebral structures contain very long-chain PUFAs, it might seem wise to provide these acids directly in the diet, especially since the ability of the organism to transform linoleic and α-linolenic precursors diminishes rapidly during development (Strouve-Vallet and Pascaud, 1971; Cook, 1978; Bordoni et al., 1988; Bourre et al., 1990b; Ulmann et al., 1991). If the diet of rats is supplemented with menhaden oil (1% by weight added to the normal diet), our results indicate that the profile of cerebral fatty acids is little altered, peroxidized derivatives do not appear, and there is no change in the activity of enzymes that protect against peroxidation: cytosolic superoxide dismutase containing Cu and Zn, and mitochondrial superoxide dismutase containing Mn, glutathione peroxidase, glutathione reductase, and catalase (Claudière et al., 1987).

On the other hand, large quantities (up to 12%) of dietary fish oil, even supplemented with vitamin E, perturb the fatty acid profile of the liver as well as that of the brain. In brain, there is a deficiency of arachidonic acid and a marked decrease in 22:4n-6 and 22:5n-6, associated with excess 22:6n-3 and 22:5n-3.

As subtle changes in brain membrane PUFAs determined by dietary alterations in α-linolenic acid provoke alterations in brain membrane PUFAs, membrane fluidity, enzymatic activities, electrophysiological parameters, learning tests, and resistance to poisons (Bourre et al., 1989a), the question can be raised whether increased fish oil intake leads to functional alterations in the nervous system.

Although blood parameters, except triacylglycerol, were normal in all animals, alterations in membrane PUFAs could alter some as yet undetermined nature liver or brain function.

It is clear that consumption of fish oils containing n-3 PUFA may have beneficial effects on ischemic heart disease and thrombosis (Salem et al., 1986; Simopoulos, 1986; Ackman, 1989). However, as the ingestion of large amounts of n-3 PUFA in experimental animals gives rise to adverse effects, it is possible that a diet abundant in fish oil may be harmful in man. Not much is known about human susceptibility to n-3 PUFA with respect to disturbances in vitamin E metabolism. Interestingly, during development and aging, n-3 fatty acid content and vitamin E content in rat peripheral nerve are not correlated (Clément and Bourre, 1990). The PUFA composition of the diet regulates the fatty acid composition of the liver endoplasmic reticulum (Tahin et al., 1981), and this in turn is an important factor controlling the rate and extent of lipid peroxidation *in vitro* and possibly *in vivo* (Hammer and Wills, 1978; Mounié et al., 1986). The replacement of cell membrane n-6 fatty acids by dietary n-3 fatty acids and subsequent alterations of membrane composition remain to be elucidated. Maintenance of membrane fluidity within narrow limits is presumably a prerequisite for proper functioning of the cell. Lipids play a key role in determining membrane fluidity, and changes in lipid and fatty acid composition have been reported to alter important cellular functions (Farias, 1980; Brenner, 1984; Stubbs and Smith, 1984; Greenwood and Craig, 1987). Therefore, dietary modification of membrane phospholipids by fish oil supplements may have significant effects. The n-3 fatty acids are being promoted in pharmacological doses for the prevention of coronary artery disease. However, the use of fish-oil supplements in patients should be considered equivalent to drug therapy, and further studies of their long-term efficacy, toxicity, and the possibility of overdosage must be conducted before recommendations can be made about their general use.

Brain contains high amounts of n-3 PUFA, and it is well known that fish oil alters the PUFA composition of various organs, especially liver and heart (Swanson and Kinsella, 1986; Huang et al., 1986), thus special attention must be paid to the brain (Bourre et al., 1988).

Requirements for n-3 acids are very high in humans during the neonatal period (Clandinin et al, 1980a, Clandinin et al., 1980b; Martinez, 1991) and must be supplied to the mother during gestation and then to the newborn. Human milk contains α-linolenic acid and also cervonic acid, which are often absent from infant formulas. Human newborns receiving formula milk have red blood cells that are deficient in cervonic acid (Putnam et al., 1982). The fatty acid composition of red blood cells can serve as an index of cerebral membrane composition (Carlson et al., 1986). In addition, there is undoubtedly a relationship between dietary lipids, serum fatty acids (Guesnet et al., 1988), and the properties of red blood cells (Popp-Snijders et al., 1984) and their structure (Durand et al., 1986).

The origin of brain saturated and unsaturated fatty acids (*in situ* synthesis and dietary origin) is well documented (Bourre, 1980); in contrast, polyunsaturated fatty acid metabolism in nerve tissue needs further study.

REFERENCES

Ackman RG (1989) Nutrition 4:251-253.
Alling C, Bruce A, Karlsson I, Sapia O, Svennerholm I (1972) J Nutr 102:773.
Anonymous (1986) Nutr Rev 44:301.
Bazan NG (1990) In: Nutrition and the brain, Vol. 8 (Wurtman RJ, Wurtman JJ, eds) pp.1-24. New York: Raven Press.
Bazan N, Di Fazio De Escalante S, Careaga M, et al. (1982) Biochim Biophys Acta 712:702.
Beaugé F, Zerouga M, Niel E, et al. (1988) In: Biomedical and social aspects of alcohol and alcoholism. (Kuriyama K, Takada A, Ishich Y, eds), pp. 291-294. Amsterdam: Elsevier Biomedical Press.
Berkow SE and Campagnoni AT (1983) J Nutr 113:582.
Bernshohn J and Spitz FJ (1974) Biochem Biophys Res Commun 57:293.
Bjerve KS, Fisher S, Alme L (1987a) Am J Clin Nutr 46:570.
Bjerve KS, Mostad IL, Thoresen L (1987b) Am J Clin Nutr 45:66.
Bordoni A, Biagi PL, Turchetto E, Hrelia S (1988) Biochem Int 17:1001.
Bourre JM (1980) Origin of aliphatic chains in brain. In: Neurological mutations affecting myelination. INSERM Symposium 14 (Baumann N, ed), pp 187-206. Amsterdam: Elsevier Biomedical Press.
Bourre JM, Bonneil M, Dumont O, et al. (1988) Biochim Biophys Acta 960:458.
Bourre JM, Bonneil M, Dumont O, et al. (1990c) Biochim Biophys Acta 1043:149.
Bourre JM, Daudu O, Baumann N (1976) Biochim Biophys Acta 1976, 424:1.
Bourre JM, Durand G, Pascal G, Youyou A (1989b) J Nutr 119:15.
Bourre JM, Faivre A, Dumont O, et al. (1983) J Neurochem 41:1234.
Bourre JM, François M, Weidner C, et al. (1989a) J Nutr 119:1880.
Bourre JM, Pascal G, Durand G, Masson M, Dumont O, Piciotti M (1984) J Neurochem 43:342.
Bourre JM, Paturneau-Jouas M, Daudu O, Baumann N. Eur J Biochem 72:41.
Bourre JM, Piciotti M, Dumont O, Pascal G, Durand G (1990a) Lipids 25:465.

Bourre JM, Piciotti M, Dumont O (1990b) Lipids 25:354.

Brenner RR (1984) Prog Lipid Res 23:69.

Carlson SE, Carver JD, House SG (1986) J Nutr 116:718.

Carreau JP, Daudu O, Mazliak P, Bourre JM (1979) J Neurochem 32:659.

Chaudière J, Clément M, Driss F, Bourre JM (1987) Neurosci Lett 82:233.

Clandinin MT, Chappell JE, Leong S, Heim T, Swyer PR, Chance GW (1980a) Early Hum Devel 2:121.

Clandinin MT, Chappell JE, Leong S, Heim T, Swyer PR, Chance GW (1980) Early Hum Devel 2:131.

Clément M and Bourre JM (1990) J Neurochem 54:2110.

Cook HW (1978) J Neurochem 30:1327.

Connor WE and Neuringer M (1988) In: Biological membranes: aberrations in membrane structure and function. (Karnovsky ML, Leaf A, Bolls LC, eds), pp 275-294. New York: Alan R. Liss.

Crawford MA, Hassam AG, Stevens PA (1981) Prog Lipid Res 20:31.

Crawford MA, Hassam AG, Williams G, Whitehouse W (1977) In: Function and biosynthesis of lipids. Advances in experimental medicine and biology. (Bazan NG, Brenner RR, Giusto NM, eds), pp 135-144. New York: Plenum Press.

Cunnane SC, Napoleon Keeling PW, Thompson RP, Crawford MA (1984) Br J Nutr 51:209.

Drevon CA (1986) Arteriosclerosis 6:352.

Durand G, Guesnet P, Desnoyer F, Pascal G (1986) Prog Lipid Res 25:395.

Farias RN (1980) J Lipid Res 17:251.

Fiennes RNT, Sinclair AJ, Crawford MA (1973) J Med Prim 2:155.

Foot M, Cruz TF, Clandinin MT (1982) Biochem J 208:631-640.

Galli C (1973) In: Dietary lipids and postnatal development (Galli C, Jacini G, Pecile A, eds), pp 191-202. New York: Raven Press.

Galli C, Trzeciak HI, Paoletti R (1971) Biochim Biophys Acta 248:449.

Greenwood CE and Craig REA (1987) In: Current topics in nutrition and disease. Basic and clinical aspects of nutrition and brain development. pp 159-216. New York: A. R. Liss.

Guesnet P, Pascal G, Durand G (1988) Rep Nutr Develop 28:275.

Hammer CT and Wills ED (1978) Biochem J 174:585.

Holman RT (1986) J Am Coll Nutr 5:183.

Holman RT (1988) Progr Chem Fats and Other Lipids. 279.

Holman RT, Johnson SB, Hatch TF (1982) Am J Clin Nutr 35:617.

Horrobin D (1981) Med Hyp 7:1211.

Horrobin DF, Huang YS, Cunnane SC, Manku MS (1984) Lipids 19:806.

Huang YS, Nassar BA, Horrobin DF (1986) Biochim Biophys Acta 879:22.

Kaare RN, Homayoun P, Durand G, Pascal G, Bourre JM (1988) J Neurochem 51:45.

Kinsella JE (1986) Food Technology 146:89-97.

Lamptey M and Walker BL (1976) J Nutr 106:86-93.

Lamptey MS and Walker BL (1978a) J Nutr 108:351-357.

Lamptey MS and Walker BL (1978b) J Nutr 108:358.

Lands WE (1986) Nutr Rev 44:189.

Lasserre M, Mendy F, Spielmann D, Jacotot B (1985) Lipids 20:227-233.

Loudes D, Faivre A, Barret A, et al. (1983) Dev Brain Res 9:231.

Martinez M (1991) In: Health effects of ω3 polyunsaturated fatty acids in seafoods, World rev nutr diet, Vol. 66. (Simopoulos, AP, Kifur RR, Maretin RE, Barlow SM, eds), pp 87-102. Basel: Karger.

McKenna MC and Campagnoni AT (1979) J Nutr 109:1195-1204.

Mead JF (1984) J Lipid Res 25:1517-1521.

Menon NK and Dhopeshwarkar GA (1982) Prog Lipid Res 21:309.

Mounié J, Faye B, Magdalou J, Goudonnet H, Truchot R, Siest G (1986) J Nutr 116:2034.

Neuringer M and Connor WE (1986) Nutr Rev 44:289.

Nouvelot A, Delbart C, Bourre JM (1986) Ann Nutr Metab 30:316.

Paoletti R and Galli C (1972) In: Lipid malnutrition and the developing brain (Ciba Foundation Symposium), pp 121-140. Amsterdam: Elsevier Biomedical Press. 1972.

Philbrick DJ, Mahadevappa VG, Ackman G, Holub J (1987) J Nutr 117:1663.

Popp-Snijders C, Schouten JA, De Jong AP, Van Der Veen EA (1984) Scand J Clin Lab Invest 44:39.

Pudelkewicz C, Seufert J, Holman RT (1968) J Nutr 94:138-146.

Putnam JC, Carlson SE, De Voe PW, Barnes IA (1982) Am J Clin Nutr 36:106.

Rudin DO (1982) Med Hypotheses 8:17.

Salem N, Kim HY, Yergey JA (1986) In: ω3 Fatty acids and nutrition. (Simopoulos AP, ed), pp. 263-317. Academic Press.

Sanders JA and Rana SC (1987) Ann Nutr Metab 81:349.

Simopoulos AP (1986) J Nutr 116:2350.

Sprecher H (1981) Prog Lipid Res 20:13.

Strouve-Vallet C and Pascaud M (1971) Biochimie 53.

Stubbs CD and Smith AD (1984) Biochim Biophys Acta 779:89.

Sun GY and Sun AY (1974) J Neurochem 22:15.

Svennerholm L, Alling C, Bruce A, Karlsson I, Sapia O (1972) In: Lipids, malnutrition and the developing brain. Ciba Foundation Symposium, pp. 141-157. Amsterdam: Elsevier Biomedical Press.

Swanson JE and Kinsella JE (1986) J Nutr 116:514.

Tahin QS, Blum M, Carafoli E (1981) Eur J Biochem 121:5.

Togashi T and Tamai Y (1988) J Lipid Res 29:1013-1021.

Ulmann L, Blond JP, Maniongui C, Poisson JP, Durand G, Bezard J, Pascal G (1991) Lipids 26:127.

Yamamoto N, Hashimoto A, Takemoto Y, Okuama H, Nomura M, Kitajima R, Togashi R, Tamal Y (1988) J Lipid Res 29:1013-1021.

Yamamoto N, Sarton M, Moriuchi A, Nomura M, Okuyama H (1987) J Lipid Res 28:144-151.

Youyou A, Durand G, Pascal G, Piciotti M, Dumont O, Bourre JM (1986) J Neurochem 46:224.

LONG AND VERY LONG POLYUNSATURATED FATTY ACIDS OF RETINA AND SPERMATOZOA: THE WHOLE COMPLEMENT OF POLYENOIC FATTY ACID SERIES

Marta I. Aveldaño

Instituto de Investigaciones Bioquímicas
Universidad Nacional del Sur-CONICET
8000 Bahía Blanca, Argentina

Photoreceptor cells and spermatozoa have long been known to be the richest sources of polyunsaturated fatty acids (PUFA) in vertebrates. Docosahexaenoate (22:6n-3) and docosapentaenoate (22:5n-6), according to species and in diverse proportions, are common major acyl chains of the glycerophospholipids of these highly specialized cells. Both 22:6n-3 and 22:5n-6 are the products of a reaction that involves the introduction of a double bond in the Δ4 position of the chain in 22:5n-3 and 22:4n-6 respectively. Even though this reaction has yet to be fully characterized, the most highly unsaturated fatty acids that occur in vertebrate membranes are known to result from such desaturation. The four fatty acids just mentioned and their predecessors are grouped into two "lineages," defined by the position of the first double bond counting from the methyl end, as the n-3 and the n-6 families. However, each of the four can be elongated, thus producing four lines of descendants, each line comprising various n-3 hexaenoic, n-6 pentaenoic, n-3 pentaenoic, and n-6 tetraenoic PUFA respectively. This chapter summarizes our findings on retina, joining them with more recent ones on spermatozoa, PUFA components, and the peculiar glycerophospholipids in which they occur. It is shown that, just as the glycerophospholipids of retina contribute to extend our view of existing PUFA by disclosing a whole variety of chain lengths within each of the known PUFA "lineages," the glycerophospholipids of spermatozoa do so by disclosing new lineages for each of the known PUFA lengths. Intricate as this may sound, this overview expands, but at the same time simplifies, our picture of PUFA structural and metabolic relationships.

VERY LONG-CHAIN POLYUNSATURATED FATTY ACIDS (VLCPUFA) OF RETINA

Glycerophospholipid molecular species characterized by having long-chain PUFA esterified to both the sn-1 and the sn-2 positions of the glycerol backbone were described some time ago in retina (Aveldaño and Bazan, 1977; Akino and Tsuda, 1979; Miljanich et al., 1979). Such "supraenoic" or "dipolyunsaturated" species were shown to be prominent components of the major glycerophospholipids of the outer segments of the rods of bovine retina, accounting for nearly 30%, 20%, 50%, and 10%, respectively, of the total molecular species of phosphatidylcholine, -ethanolamine, -serine, and -inositol in those membranes (Aveldaño and Bazan, 1983). The group of dipolyunsatur-

Neurobiology of Essential Fatty Acids
Edited by N.G. Bazan *et al.*, Plenum Press, New York, 1992

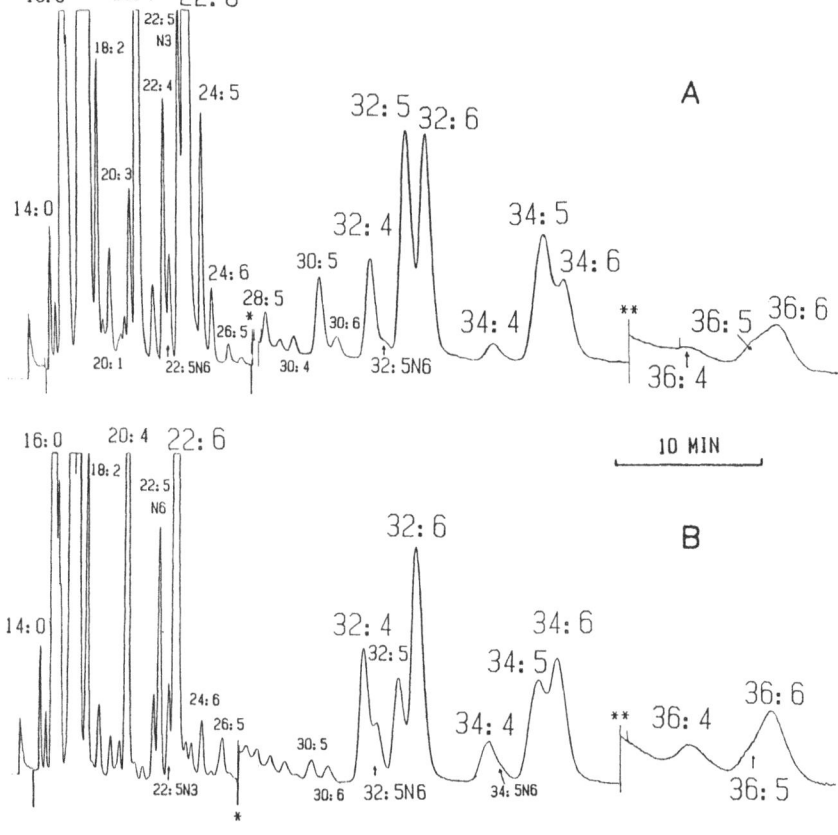

Figure 1. Fatty acids of cattle (A) and rabbit (B) retina phosphatidylcholine (PC). A standard set of two GLC packed columns (temperature-programmed at 5°C/min up to the elution of 22:6 approximately, then run isothermally) and a flame ionization detector were used. One and two asterisks denote a 10- and a 100-fold increase in detector sensitivity respectively. Cattle retina PC was characterized by the highest proportion of VLC pentaenes of the n-3 series of all species examined. Rabbit retina had a relatively larger proportion of VLC pentaenes of the n-6 series than cattle (consistent also with a higher 22:5n-6/22:5n-3 ratio). Both species contained VLC tetraenes (all of the n-6 series) and VLC hexaenes (all of the n-3 series). The major VLCPUFA of all unsaturation groups were 32 and 34 carbons long.

ated species of cattle rod phospholipids was highly heterogeneous in turn, being composed of species with two hexaenoic fatty acids per molecule, a hexaenoic and a pentaenoic acid, pentaenoic-pentaenoic, and so on, each bound to one of the positions of the glycerol backbone (dodecaenoic, undecaenoic, decaenoic, etc. molecular species of phospholipids).

After being separated by standard chromatographic techniques, the dodecaenoic, undecaenoic, and decaenoic groups of molecular species of phosphatidylcholine (PC) also proved to be very diverse. There were several dodecaenoic PCs that contained 22:6n-3 as one of their acyl groups, the other one being another hexaenoic fatty acid of an interesting group whose lengths varied from 22 to 36 carbon atoms (22:6 to 36:6). Undecaenoic PCs contained in turn 22:6 in one of the positions, the other one being occupied by a pentaenoic fatty acid (20:5 to 36:5), while most decaenoic species of retina PC were made up of 22:6 in one of the positions and one of several tetraenoic

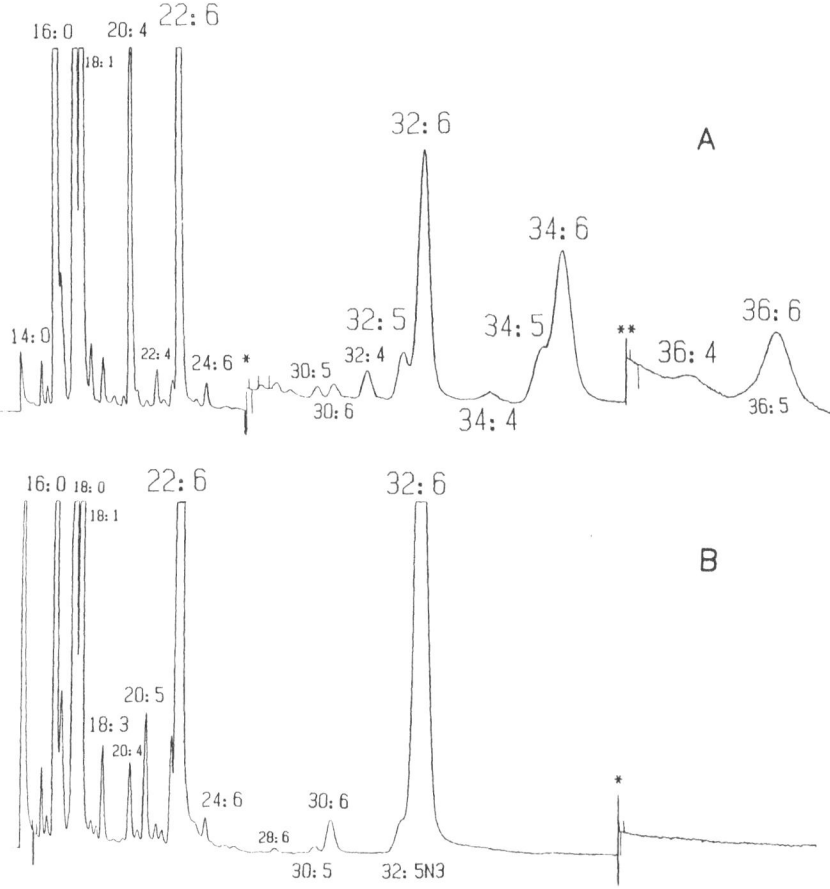

Figure 2. Fatty acids of rat (A) and cod (B) retina phosphatidylcholine (PC). In contrast to the diversity of VLCPUFA profiles of Figure 1, VLC-hexaenoic fatty acids of the n-3 series predominated in rat retina PC (A) with low proportions of n-3 pentaenes and negligible n-6 pentaenes including 22:5n-6. (It is noteworthy that these retinas were from rats similar to those whose spermatozoa virtually lacked 22:6n-3, being instead exceedingly rich in 22:5n-6 and n-9 polyenes, see Figure 4 and Table 1.) The fatty acid profiles of cod retina PC (B) were even simpler than those of rat; the bulk of all polyenes belonged to the n-3 series and, of these, the majority were n-3 hexaenes (22:6n-3 + 32:6n-3: more than 80% of the fatty acids of PC).

fatty acids (20:4 to 36:4) in the other (Aveldaño, 1987).

Positional analysis of the double bonds (Aveldaño and Sprecher, 1987) in the novel polyenes showed that all very long hexaenes belonged to the n-3 series, that all tetraenes belonged to the n-6 series, and that there were pentaenes of both the n-3 and the n-6 series in retina PC. Analysis of the distribution of fatty acids between the sn-1 and the sn-2 positions in dipolyunsaturated PCs (Aveldaño, 1988) showed a considerable specificity: the VLCPUFA were predominantly located at the sn-1 position of the glycerol backbone, 22:6n-3 being the major acyl chain at sn-2.

Most of the features described for the PCs of cattle retina were also observed in other vertebrates, and it is also interesting to note that there were characteristic predominances of particular groups of VLCPUFA in certain species (Figs. 1-3, slightly modified from Aveldaño, 1987, with permission). Thus, cattle retina contained the largest proportions of pentaenoic VLCPUFA of the n-3 series, cod retina contained the

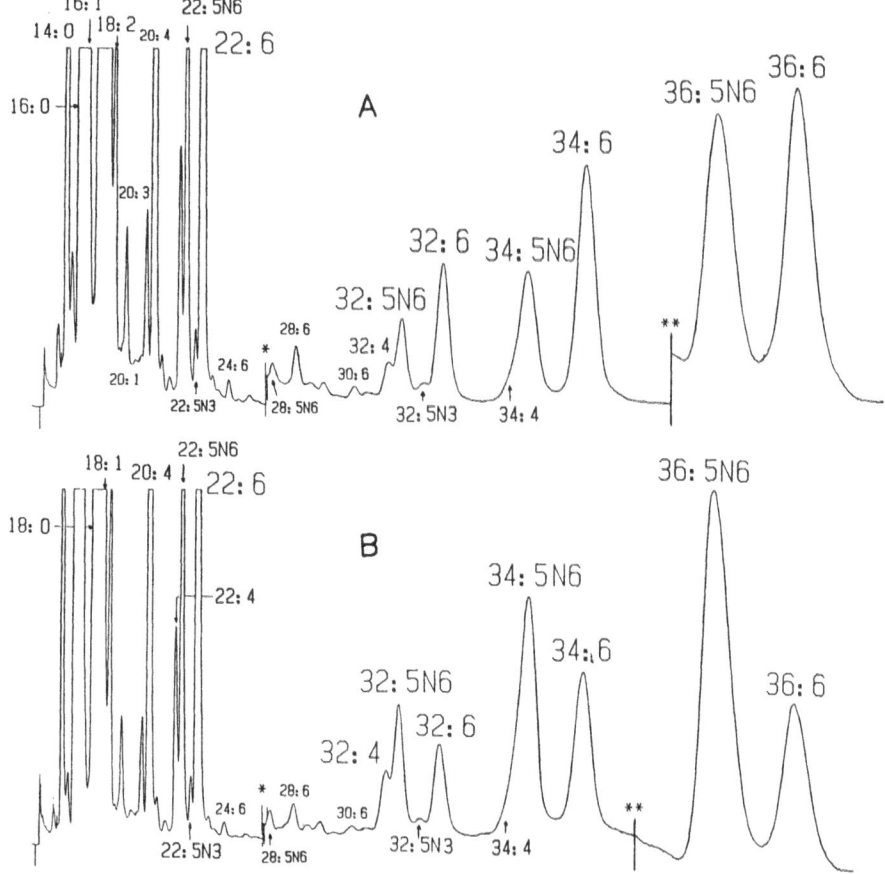

Figure 3. Fatty acids of retina PC from young chickens raised by two different breeders. In contrast to the cattle retina (Fig. 1A), the bulk of the VLC pentaenes of chicken retina belonged to the n-6 series, n-3 pentaenes being minor components. This was consistent with the fact that the proportions of 22:5n-6 in comparison with 22:5n-3 were considerably higher in chicken (and opposite to those of cattle) retina PC. Panels A and B show retinas from chickens raised on diets rich in n-3 and n-6 polyenoic fatty acids (fish byproducts and corn) respectively. The 22:5n-6/22:6n-3 ratio, as well as the ratios between the n-6 pentaenoic and n-3 hexaenoic fatty acids with very long chains, were larger in B than in A, suggesting a correlation between the type of dietary fatty acids and the type of retinal VLCPUFA.

highest levels of hexaenoic VLCPUFA of the n-3 series, and chicken retina the largest of pentaenoic VLCPUFA of the n-6 series among the animals analyzed. Cattle, rabbit and chicken also had abundant tetraenes of the n-6 series. For all four series of VLCPUFA, the chain length appearing in largest proportions was that of 32 carbon atoms.

The synthesis of VLCPUFA was investigated in bovine retinas and in subcellular fractions isolated therefrom after incubations with $(1\text{-}^{14}C)$acetate (Rotstein and Aveldaño, 1988). Phosphatidylcholine was the phospholipid class most highly labeled, all its radioactivity being recovered in its acyl moieties. Of these, argentation thin-layer chromatography (TLC) showed that mostly saturated and polyunsaturated fatty acids were labeled, each containing nearly half the radioactivity incorporated from acetate. Further separation of the radiolabeled fractions by high pressure liquid chromatography (reverse phase) showed that most of the radioactivity incorporated in saturates was in palmitate, while most of that incorporated in polyenes was not in 20:4n-6, 20:5n-3, 22:5n-6 or 22:6n-3, in which labeling was negligible, but in longer polyenes of each of the series. To synthesize 20:4n-6 and 20:5n-3 a $\Delta 5$ desaturation, and to produce 22:5n-6 and 22:6n-3 a $\Delta 4$ desaturation, are the required reactions. For labels in these fatty acids to have been detected, their precursors should also have had incorporated acetate, which was not the case in the described conditions. In contrast, the label from acetate was significant in VLCPUFA, indicating that they were actively synthesized within the retina from preexisting tetra-, penta-, and hexaenoic fatty acids by sequential additions of two-carbon units, in complete analogy with established mechanisms by which other fatty acids are elongated.

Most of the VLCPUFA synthesized from acetate were recovered as acyl chains of PC, even at relatively short intervals of incubation, in various subcellular fractions isolated from retinas after incubations with the precursor (including cytosol, microsomes, mitochondria, and photoreceptor membranes from the outer segments of rods). A similar result had been obtained in previous experiments using $(1-^{14}C)$-labeled 20:4n-6, 22:5n-3, and 22:6n-3 (Rotstein and Aveldaño, 1987a,b). In both cases a high percentage of the PUFA incorporated in lipids (whether endogenously synthesized or exogenously provided) was in PC, and a high percentage of the label in PC was in its dipolyunsaturated species.

Since VLCPUFA predominantly occur in dipolyunsaturated molecular species of PC, and since such PCs are highly characteristic components of photoreceptor membranes, i.e., the outer segments of the rods and cones, the bulk of the VLCPUFA formed *de novo* from labeled acetate in isolated retinas can be expected to reflect the synthesis that takes place in the inner segment of photoreceptor cells, even in such a complex tissue as the entire retina. This is not the case for the labeled palmitate also synthesized, since this fatty acid can be formed in any or at least in many retinal cells. By the same token, the (^{14}C)-labeled dipolyunsaturated PCs that are formed from (^{14}C)-labeled 20:4n-6, 22:5n-3, and 22:6n-3 can be expected to be produced mainly by photoreceptor cell enzymes, in contrast with the corresponding tetraenoic, pentaenoic, and hexaenoic species, which are much more ubiquitous and less specific. In other words, the lipid products are so peculiar that they may serve as "markers" of photoreceptor membranes in a similar sense as diphosphatidylglycerol or galactosylceramide may be good "markers" of mitochondrial and myelin membranes respectively.

Concerning lipids, and especially with this kind of experimental approach, there is always a certain degree of apparent overlap in functions between subcellular organelles (e.g., mitochondria and endoplasmic reticulum in fatty acid elongation, or mitochondria and peroxisomes in fatty acid beta oxidation, or endoplasmic reticulum and mitochondria in the synthesis of certain lipids). However, the subcellular distribution of the dipolyunsaturated PCs containing the VLCPUFA synthesized *in situ* from labeled acetate, or incorporating the exogenously provided labeled polyenes, showed that the highest specific radioactivities were attained by PCs recovered from cytosolic fractions, followed by PCs of microsomes, mitochondria, and photoreceptor membranes. The results were consistent with the idea that the most active synthesis of these peculiar PCs should take place in the endoplasmic reticulum, and disclosed that an apparently very intense intracellular traffic of newly synthesized PCs exists among subcellular membranes. The transport of such PCs through cytosol (protein-mediated) is one of the important mechanisms by which the highly specialized photoreceptor membranes may acquire and/or replace the specific phospholipid molecules they cannot synthesize *de novo* themselves. Such a molecular replacement of intact phospholipid does not exclude, and in fact coexists with, other forms of phospholipid turnover *in situ* in photoreceptor membranes (Giusto et al., 1986 and references therein).

The fact that the synthesis of the bulk of retina VLCPUFA takes place simply by successive elongations of existing polyenes within the photoreceptor cells is perfectly compatible with the possibility that, in a given animal, conditions that affect the availability of any of the 22-carbon PUFA eventually affect the levels of the corresponding VLCPUFA in retina. A notable effect on retina VLCPUFA correlating with the ratio between 22:5n-6 and 22:6, in turn dependent on the types of essential fatty acids predominating in the diet, was observed in young chickens of the same age and origin raised on diets rich in n-6 or n-3 polyenes (Fig. 3). Thus, the ratios between VLC (n-6) pentaenes and VLC (n-3) hexaenes in retina PC were consistently larger in the animals in which the 22:5n-6/22:6n-3 ratio was larger (Aveldaño, 1987), namely in those fed corn. The types of changes shown in Figure 3 are of a reversible, compensatory nature, suggesting there must be a delicate balance such that, if the enzyme or enzymes performing the Δ4 desaturation in the retina photoreceptor cells find themselves with more 22:4n-6 than 22:5n-3, they will synthesize more 22:5n-6 than 22:6n-3 (and vice versa), the very long polyenes eventually predominating being secondary to the relative abundances of these "precursors."

The above was not the case in aging rats, in whose retinas both (the minor) 22:5n-6 and (the major) 22:6n-3 decreased significantly, with no compensatory changes in other polyenes (Rotstein et al, 1987). As a consistent consequence, the corresponding VLC (n-6) pentaenes and VLC (n-3) hexaenes also decreased. Since these alterations could not be attributed to any change in the diet of the animals (the same throughout their life span), and since the retinas from aged rats *de novo* incorporated (^{14}C)22:6n-3 even more avidly in PC than those of young animals, the results suggested the possibility that, among the diverse effects of aging, an alteration of the properties of the enzyme(s) performing the Δ4 desaturation may be a primary event, the observed decrease in VLCPUFA being secondary to such alteration.

The compositional and metabolic observations described suggest that it might be worth reexamining the VLCPUFA of retina lipids as a means of gaining a new insight into some problems of vision research. Studies of photoreceptor membrane biogenesis, development, renewal, aging, deterioration caused by chemicals, radiation, and diets, and especially the study of the diversity of human diseases, both congenital and acquired, that affect photoreceptor membranes and impair vision, might benefit from the investigation of these specific polyenes. In addition to the impact of diverse factors on the VLCPUFA and the phospholipid species in which they occur, there are still many basic avenues open for research that clarifies their *raison d'etre*, their physiological role in the photoreceptor membrane, their physical properties, their mode of interaction with the visual pigments, how the elongases regulate their synthesis in the rods and cones, and how their catabolism runs—probably within the pigment epithelial cells—as well as many other questions.

Abnormal increases in VLCPUFA of the type described here, with 26 to 38 carbon atoms, were observed by Poulos et al. (1986b) to occur in the brains of patients with Zellweger syndrome (ZS). This is a rapidly progressive, fatal genetic disorder characterized by abnormal biogenesis of peroxisomes that affects the CNS and the retina in addition to liver, kidney, adrenals and many other tissues (Lazarow and Moser, 1989). Should the enzymes that degrade VLCPUFA be located in peroxisomes, such defective peroxisomes would be expected to result in increased levels and impaired capacity to degrade the VLCPUFA of retina photoreceptor membranes. Even though retina and/or pigment epithelium VLCPUFA have not yet been measured in these diseases, there is in fact a dramatic visual impairment in disorders in which peroxisomes either are absent or are present but defective in peroxisomal fatty acid oxidation enzymes. Typical examples of these diseases are, respectively, the classic Zellweger syndrome and neonatal adrenoleukodystrophy, in both of which changes resembling those of retinitis pigmentosa occur in the retina and pigment epithelium (Cohen et al., 1984). More recently, evidence that children affected by these disorders show a dramatic decrease in the levels of both 22:6n-3 and 22:5n-6 in the brain (and retina) was presented by Martinez (1989, 1990). Her work raised the exciting possibility that Δ4 desaturation of PUFA could also be a peroxisomal activity.

THE SPERMATOZOA CONNECTION

Our interest in the fatty acids of rat spermatozoa started as an extension of our observations on the effects of aging on 22:6n-3 and 22:5 n-6 levels in retina, which had suggested an impairment of the Δ4 desaturation activity (Giusto et al, 1986). Since spermatozoa had long been known to contain large amounts of 22:5n-6, a minor component of rat retina, spermatozoa were deemed a good model in which to test whether or not aging affected Δ4 desaturase products. (We did not find any significant effects of aging on spermatozoa PUFA until animal ages that can be considered far advanced into senility). Another reason for interest in these peculiar cells was that VLCPUFA had been detected in rat testes (Grogan, 1984) and in spermatozoa of several mammals (Poulos et al., 1986a). As in retina, the VLCPUFA Poulos described in sperm had up to 36 carbon atoms and belonged to the n-6 and/or the n-3 series (n-6 polyenes occurring in boar, n-6 and n-3 polyenes being present in bull and ram, and n-3 VLCPUFA predominating in man). Shortly afterwards the VLCPUFA of sperm were reported by the same group to occur mainly as components of sphingomyelin (Poulos et al., 1987).

Studying the epididymal spermatozoa of boar, we confirmed the findings of Poulos, i.e., that boar sperm sphingomyelin was rich in very long (n-6) tetraenoic fatty acids. In epididymal spermatozoa from rats, however, the proportion of VLCPUFA in sphingo-myelin was smaller, and the proportion of sphingomyelin itself was much smaller than in boar sperm (less than 3% of the phospholipids versus more than 15%, respectively). Apart from this quantitative difference, looking at other lipids we found ourselves in the presence of an unexpected group of polyenes, which were far from minor compo-nents of rat sperm (Fig. 4).

THE UNCOMMON GROUP OF PUFA OF RAT SPERMATOZOA

The analysis of the fatty acids of rat spermatozoa isolated from epididymal regions showed abundant proportions of unfamiliar long-chain PUFA identified by several criteria as polyenes of the n-9 series (Aveldaño et al., in press/a). The most abundant component of this novel group was 22:4n-9, followed by 22:3n-9, 20:3n-9, and 24:3n-9. The unusual polyenes were highly concentrated in the major phospholipid class of spermatozoa, the choline glycerophospholipids (CGP). Physiological maturation of spermatozoa within the epididymal tract was found to result in several changes affecting the lipid composition (Aveldaño et al., in press/b) and the fatty acid composition of major lipid classes (Fig. 4). Of the changes in phospholipids, the most prominent was a decrease in the levels of diacylglycerophospholipid classes and a concomitant relative increase in the ether-linked subclasses of both choline and ethanolamine glycerophos-pholipids. Of the changes in fatty acids, the most significant was a decrease in the levels of 18:1n-9 in most lipid classes, and a concomitant increase in the proportion of long-chain PUFA of the n-9 and the n-6 series.

In spermatozoa from cauda epididymidis, the peculiar long-chain polyenes of the n-9 series made up nearly one-fourth of the fatty acids of CGP, and about 15% of the fatty acids of the total phospholipids (Fig. 4). When the subclasses of CGP were resolved, it became clear that plasmenylcholine was the major phospholipid component of mature rat spermatozoa, precisely the lipid in which the n-9 PUFA were highly concentrated (Table 1). The fatty acids in question were undetectable in the lipids of epididymal spermatozoa of boar, analyzed for comparison. In this species the alkyl-acyl subclasses (plasmanyl) were larger than the plasmenyl- and phosphatidyl- subclasses in both CGP and ethanolamine glycerophospholipids (EGP). The major plasmanyl subclasses were exceedingly rich in 22:6n-3 and 22:5n-6 (Table 1).

RETINA AND SPERMATOZOA: THE WHOLE COMPLEMENT OF PUFA

The fatty acids characterized in rat spermatozoa as long and very long-chain polyenes of the n-9 series—of which 22:4n-9 was a major one—occurred concomitantly with large amounts of 22:5n-6 (Fig. 4, Table 1). Both PUFA occurred in rat sperm to the virtual exclusion of 22:6n-3, a significant component, in turn, of lipids of other tissues such as brain and retina (e.g., see Fig. 2). A Δ4 desaturation is the key enzymatic activity required for the synthesis of the three 22-carbon polyenes, namely 22:4n-9, 22:5n-6, and 22:6n-3 (Fig. 5).

It has long been established that a severe deficiency of essential fatty acids of both the n-3 and the n-6 series in the diet is necessary to promote a compensatory increase in PUFA of the n-9 series in the lipids of mammalian tissues. The most familiar of these polyenes is 20:3n-9, known to be made from 18:1n-9 when there is not enough 18:2n-6 in the diet to synthesize 20:4n-6 (see Fig. 5). The described findings constitute an unexpected departure from the idea that long-chain n-9 PUFA can only be observed in lipids of tissues or cells from animals with chronic dietary deficiencies of essential n-3

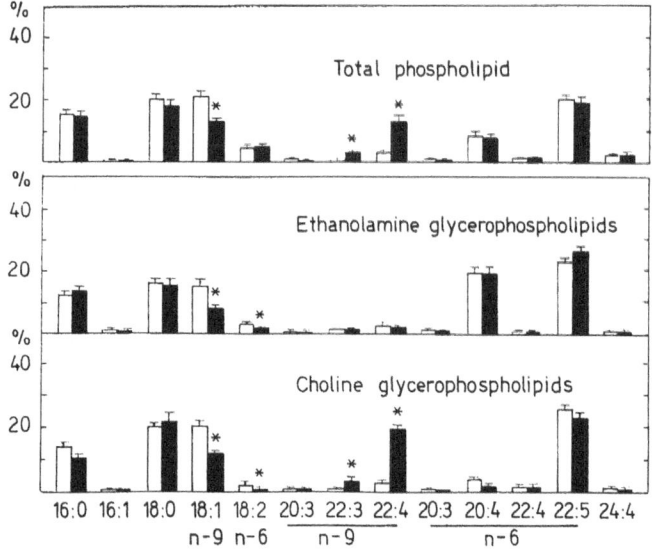

Figure 4. Fatty acid composition of the total phospholipid and of the two major phospholipid classes (CGP>EGP) of rat spermatozoa isolated from the epididymis (white bars, caput; black bars, cauda). There were significant changes in fatty acid composition due to the physiological maturation of spermatozoa within the epididymal tract, with obvious decreases in 18:1n-9 and accretion of 22:3 and 22:4n-9. Other important components of sperm, the n-6 PUFA, were present in similar proportions in mature and immature sperm cells. Decreases in the phosphatidyl and relative increases in the plasmenyl subclasses of CGP, plus changes in fatty acids within each subclass, contributed to the modifications observed in CGP and total phospholipid. Other lipids contributing to the "total phospholipid" fatty acid profile were diphosphatidylglycerol and phosphatidylinositol (Aveldaño et al., in press/b).

Table 1. Fatty acid composition of choline and ethanolamine glycerophospholipids from epididymal spermatozoa

| | RAT | | | | | | BOAR | | | |
| | CGP | | | EGP | | | CGP | | EGP | |
	Phosphatidyl	Plasmanyl	Plasmenyl	Phosphatidyl	Plasmanyl	Plasmenyl	Phosphatidyl	Plasmanyl	Phosphatidyl	Plasmanyl
14:0	0.1 ± 0.02	0.3 ± 0.1	0.1 ± 0.04	0.6 ± 0.02	0.2	0.3 ± 0.02	13.9	0.3	1.1	0.2
15:0	0.04 ± 0.01	0.2 ± 0.02	0.1 ± 0.1	0.3 ± 0.01	0.03	0.3 ± 0.1	1.3	0.2	0.6	1.3
16:0	15.4 ± 0.8	5.0 ± 0.7	1.4 ± 0.2	23.2 ± 1.5	3.6	3.2 ± 0.3	28.0	1.3	11.3	1.5
16:1	1.0 ± 0.1	1.6 ± 0.3	0.4 ± 0.1	-	1.5	1.0 ± 0.1	1.2	-	-	-
17:0	0.4 ± 0.04	-	0.3 ± 0.1	1.3 ± 0.4	-	-	0.6	0.4	0.8	-
18:0	26.0 ± 0.4	4.9 ± 2.0	1.4 ± 0.2	16.2 ± 0.3	2.9	2.2 ± 0.1	9.4	1.1	31.1	0.5
18:1	18.3 ± 1.2	15.7 ± 1.5	3.3 ± 0.6	6.2 ± 0.3	7.7	7.1 ± 0.4	10.5	0.2	7.3	1.3
18:2	1.0 ± 0.04	1.3 ± 0.1	0.2 ± 0.1	0.8 ± 0.3	1.0	0.8 ± 0.1	3.1	0.3	†7.7	0.1
18:3	0.2 ± 0.03	0.2 ± 0.03	-	0.1 ± 0.05	0.3	0.02	0.9	-	0.4	0.1
20:1	0.2 ± 0.01	0.1 ± 0.1	0.1 ± 0.02	0.1	0.3	-	0.6	-	0.1	-
20:2	0.03	0.06	-	-	-	-	2.0	0.5	0.1	-
20:3n-9	0.5 ± 0.04	0.4 ± 0.2	0.1 ± 0.1	0.8 ± 0.3	0.6	3.4 ± 0.7	-	-	-	-
20:3n-6	0.1 ± 0.05	0.04	0.06 ± 0.02	0.1	-	-	7.0	0.2	3.6	0.1
20:4n-6	1.9 ± 0.1	4.1 ± 0.2	1.2 ± 0.3	27.3 ± 2.2	5.9	8.5 ± 1.0	3.9	0.7	26.0	0.3
22:2n-9	0.7 ± 0.2	0.9 ± 0.2	1.2 ± 0.1	-	1.2	-	-	-	-	-
22:3n-9/20:5*	1.5 ± 0.1	4.6 ± 0.4	8.3 ± 0.3	0.7 ± 0.3	1.4	3.0 ± 0.5	0.3*	0.3*	0.7*	0.1*
22:4n-9	8.5 ± 0.3	12.0 ± 0.1	50.0 ± 1.0	1.0 ± 0.5	2.3	7.5 ± 0.6	-	-	-	-
22:4n-6	0.5 ± 0.1	3.2 ± 0.3	2.1 ± 0.1	0.7 ± 0.2	3.2	3.0 ± 0.2	0.8	3.4	0.7	4.1
22:5n-6	23.1 ± 0.4	40.5 ± 1.5	27.3 ± 0.8	19.3 ± 0.4	59.5	55.1 ± 3.1	12.1	65.0	7.5	66.0
24:4/22:6†	0.3 ± 0.1	3.0 ± 0.7	1.6 ± 0.2	0.9 ± 0.1	5.6	3.4 ± 0.2	4.3†	26.0†	2.6†	21.3†
24:5n-6	0.2 ± 0.1	1.9 ± 0.3	0.9 ± 0.1	0.4	2.8	1.2 ± 0.1	0.1	0.1	4.4	3.1

The lipids from cauda epididymal spermatozoa were isolated by TLC, and the corresponding subclasses of choline and ethanolamine glycerophospholipids (CGP, EGP) were resolved by another TLC after conversion to acetyldiglycerides. Fatty acids were analyzed by GLC. Results are given as percentage (wt%). *In rat, this is mostly 22:3n-9; in boar, 20:5n-3. †In rat, this is mostly 24:4n-9; in boar, 22:6n-3. Plasmenyl- and plasmanylcholine, respectively, were the major phospholipid components of mature rat and boar spermatozoa. In boar sperm lipids, the predominant 22-carbon PUFA were 22:5n-6 and 22:6n-3, with negligible 22:4n-9, while in rat 22:5n-6 and 22:4n-9 predominated, with negligible 22:6n-3.

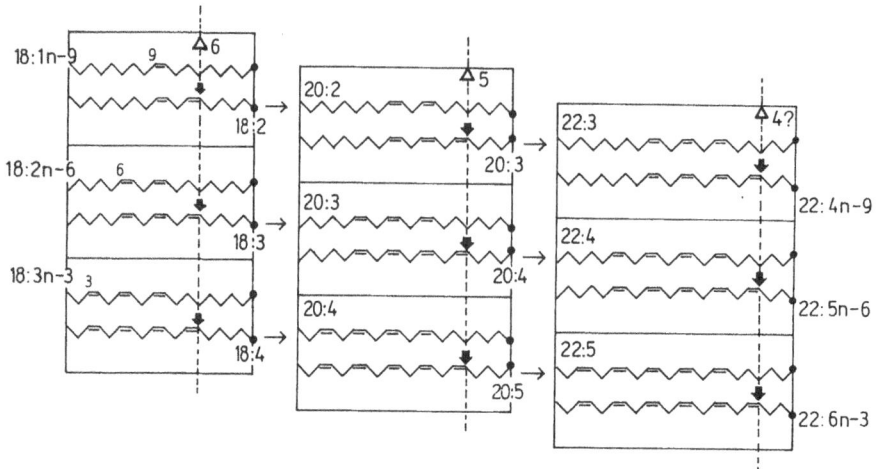

Figure 5. Structural and metabolic relationships between the three major families of polyenoic fatty acids of vertebrate tissues, now including the PUFA of rat spermatozoa. Oleate (18:1n-9), linoleate (18:2n-6), and linoleate (18:3n-3) are the precursors, and a Δ4 desaturation is *the common* final requisite for the biosynthesis of the most highly unsaturated members of each series, 22:4n-9, 22:5n-6, and 22:6n-3, respectively. From the 22-carbon PUFA, by elongation through stepwise additions of 2-carbon units, six homologous series of very long-chain PUFA may be synthesized in certain cells and for specific membranes. The longest n-9 PUFA we observed were 24:3 and 24:4n-9 (in rat sperm). Very long-chain PUFA of the n-3 and the n-6 series were detected in spermatozoa of this and other mammals (Poulos et al., 1986a; 1987) and in retina (Figs. 1-3).

and n-6 PUFA, vindicating the n-9 PUFA as normal physiological products of certain cells.

The n-9 polyenes of rat spermatozoa, which in all likelihood derive from oleate (Fig. 5), i) did not occur in significant amounts in tissues other than the epididymal tract and spermatozoa isolated therefrom; ii) did not occur in large amounts in lipids of spermatozoa other than the choline plasmalogen; and iii) did not result from a deficiency of essential fatty acids of the n-3 or n-6 series in the diet, which had plenty of 18:2n-6 and 18:3 n-3 (Aveldaño et al., in press/a). The n-9 PUFA in question occurred, in fact, in well-fed, healthy, long-lived laboratory rats that were perfectly fertile for generations, rats capable of producing large amounts of 22:5n-6 and 22:6n-3, rats that maintained a remarkably constant lipid and fatty acid composition in their retinas and in their spermatozoa and epididymis from an early (2 months) to a very advanced (2 years) age (Rotstein et al., 1987, and unpublished observations).

Even though 18:2n-6 and 18:3n-3 were available in the diet, a local relative deficiency of such precursors within the epididymal tract could occur since in the rat the process of spermiogenesis is very active and continuous throughout the animal's life. Such a possibility would not, however, satisfactorily explain why long-chain PUFA are so specifically concentrated in the choline plasmalogen of sperm, and why they have not been detected in other cells also characterized by an intense renewal, such as blood, intestinal, or photoreceptor cells.

Figure 5 summarizes the structural and metabolic relationships among all major PUFA of the three families: that from oleate (18:1n-9), that from linoleate (18:2n-6), and that from linolenate (18:3n-3), pointing to the major routes through which 22-carbon PUFA from each one are synthesized. These routes involve a common, simple and logical series of reactions: two chain elongations (→) interspersed with three chain desaturations (↓), which can account for all major PUFA recognized so far in mammalian tissues. They lead to six 22-carbon polyenes which, through further elongations, may give rise to very long-chain polyunsaturated fatty acids in specific cells. In the retina our group, and in brain and spermatozoa the group of Poulos et al., have identified VLCPUFA evolving from the lower two-thirds of the last panel, namely (n-6) tetraenes, (n-6) pentaenes, (n-3) pentaenes, and (n-3) hexaenes. The possibility cannot be excluded that very long polyenes of the n-9 series could also be found in the future, and in fact 24:3 and 24:4 n-9 were identified in rat sperm by us. A key role in the formation of the most highly unsaturated members of each of the three series is played by the last desaturation (the $\Delta 4$), which is performed by the elusive enzymatic step that has not yet been fully characterized but that is crucial because of the biological importance of its fatty acid products.

REFERENCES

Akino T and Tsuda M (1979) Characteristics of phospholipids in microvillar membranes of octopus photoreceptor cells. Biochim Biophys Acta 556:61.

Aveldaño MI (1987) A novel group of very long chain polyenoic fatty acids in dipolyunsaturated phosphatidylcholines from vertebrate retina. J Biol Chem 262:1172.

Aveldaño MI (1988) Phospholipid species containing long and very long chain polyenoic fatty acids remain with rhodopsin after hexane extraction of photoreceptor membranes. Biochemistry 21:1229.

Aveldaño MI and Bazan NG (1977) Acyl groups, molecular species, and labeling by (^{14}C) glycerol and (^{3}H)arachidonic acid of vertebrate retina glycerolipids. Adv Exp Med Biol 83:397.

Aveldaño MI and Bazan NG (1983) Molecular species of phosphatidyl -choline, -ethanolamine, -serine and -inositol in microsomal and photoreceptor membranes of bovine retina. J Lipid Res 24:620.

Aveldaño MI and Sprecher H (1987) Very long chain (C_{24} to C_{26}) polyenoic fatty acids of the n-3 and n-6 series in dipolyunsaturated phosphatidylcholines from bovine retina. J Biol Chem 262:1180.

Aveldaño MI, Rotstein NP, Vermouth NT (in press/a) Occurrence of long and very long polyenoic fatty acids of the n-9 series in rat epididymal spermatozoa. Lipids.

Aveldaño MI, Rotstein NP, Vermouth NT (in press/b) Lipid remodelling during epididymal maturation of rat spermatozoa. Enrichment in plasmenylcholines containing long chain polyenoic fatty acids of the n-9 series. Biochem J.

Cohen SMZ, Brown FR, Martin L, Moser HW, Chen W, Kistenmacher M, Punett H, Grover W, De La Cruz C, Chan NR, Green WR (1984) Ocular histopathological and biochemical studies of the cerebro-hepatorenal (Zellweger) syndrome and its relation to neonatal adrenoleukodystrophy. Am J Ophthalmol 96:488.

Giusto NM, de Boschero MI, Sprecher H, Aveldaño MI (1986) Active labeling of phosphatidylcholines by (1-^{14}C)docosahexaenoate in isolated photoreceptor membranes. Biochim Biophys Acta 860:137.

Grogan WM (1984) Metabolism of arachidonate in rat testes: characterization of 26 - 30 carbon polyenoic fatty acids. Lipids 19:341.

Lazarow PB and Moser HW (1989) Disorders of peroxisomal biogenesis. In: The metabolic basis of inherited disorders, 6th ed (Scriver CR, Beaudet AL, Sly WS, Valle D, eds) Vol 2, p 1479. New York: McGraw-Hill.

Martinez M (1989) Polyunsaturated fatty acid changes suggesting a new enzymatic defect in Zellweger syndrome. Lipids 24:261.

Martinez M (1990) Severe deficiency of docosahexaenoic acid in peroxisomal disorders: A defect of Δ4 desaturation? Neurology 40:1292.

Miljanich GP, Sklar LA, White DL, Dratz EA (1979) Disaturated and dipolyunsaturated phospholipids in the bovine retinal rod outer segment membrane. Biochim Biophys Acta 552:294.

Poulos A, Johnson DW, Beckman K, White IG, Easton C (1987) Occurrence of unusual molecular species of sphingomyelin containing 28-34 carbon polyenoic fatty acids in ram spermatozoa. Biochem J 248: 961.

Poulos A, Sharp P, Johnson D, White I, Fellenberg AJ (1986a) The occurrence of polyenoic fatty acids with greater than 22 carbons in mammalian spermatozoa. Biochem J 246:891.

Poulos A, Sharp P, Singh J, Johnson D, Fellenberg AJ, Pollard A (1986b) Detection of a homologous series of C_{26}-C_{38} fatty acids in the brain of patients without peroxisomes (Zellweger syndrome). Biochem J 235:607.

Rotstein NP and Aveldaño MI (1987a) Labeling of lipids of retina subcellular fractions by [1-^{14}C]eicosatetraenoate (20:4n-6), docosapentaenoate (22:5n-3), and docosahexaenoate (22:6n-3). Biochim Biophys Acta 921:221.

Rotstein NP and Aveldaño MI (1987b) Labeling of phosphatidylcholines of retina subcellular fractions by [1-^{14}C]eicosatetraenoate (20:4n-6), docosapentaenoate (22:5n-3), and docosahexaenoate (22:6n-3). Biochim Biophys Acta 921:235.

Rotstein NP and Aveldaño MI (1988) Synthesis of very long chain (up to 36 carbon) tetra, penta, and hexaenoic fatty acids in retina. Biochem J 249:191.

Rotstein NP, de Boschero MI, Giusto NM, Aveldaño MI (1987) Effects of aging on the composition and metabolism of docosahexaenoate-containing lipids of retina. Lipids 22:253.

PHOSPHOLIPID METABOLISM IN RAT INTESTINAL MUCOSA AFTER

ORAL ADMINISTRATION OF LYSOPHOSPHOLIPIDS

Alessandro Bruni,[1] Paolo Orlando,[2]
Lucia Mietto,[3] and
Giampietro Viola[3]

[1]Department of Pharmacology
University of Padova
Padova, Italy

[2]Servizio Radioisotopi
Universita Cattolica Sacro Cuore
Roma, Italy

[3]Fidia Research Laboratories
Abano Terme, Italy

INTRODUCTION

Appropriate changes in dietary fat may produce systemic effects beneficial in certain pathological states. Extensive investigations have been performed with polyunsaturated fatty acids of the ω-3, ω-9, and ω-6 series. As competitors for arachidonate in prostanoid biosynthesis (Garg et al., 1990), ω-3 polyunsaturated fatty acids have attracted considerable interest for their antithrombotic effect in man (Herold and Kinsella, 1986). Furthermore, they may help to prevent atherosclerosis (Von Schacky, 1987), an effect possibly due to inhibition of very low density lipoprotein assembly and secretion (Lang and Davis, 1990). In a different approach, anionic phospholipids such as phosphatidyl-serine (PS) have been tested in aged rats as a supplement in drinking water (Nunzi et al., 1987; Zanotti et al., 1989). The administration of PS has been effective in improving spatial memory and passive avoidance retention. Impressive morphological correlates to this action of PS have been obtained by examining the dendritic spine density of pyramidal cells in the hippocampus. While aged rats exhibit dendritic atrophy and a loss of synapses and cells, these features are not seen in animals treated with PS. Since the dose of phospholipid is low (50 mg/kg/day) in comparison to an ordinary dietary supplement, the action of PS may result from a pharmacological (regulatory) effect, rather than an altered food intake.

Based on these findings, we have begun a series of experiments designed to study the action of relatively low doses of PS on the phospholipid metabolism of intestinal mucosal cells, the first cells to come in contact with the orally administered phospholipid. The general aim of the study is to investigate the possibility that a

Neurobiology of Essential Fatty Acids
Edited by N.G. Bazan *et al.*, Plenum Press, New York, 1992

PS-induced modification of intestinal function may extend its influence to more distant sites. In this latter context, improved learning ability in rats has also been observed with a dietary supplement of ω-3 polyunsaturated fatty acids (Yamamoto et al., 1987).

Metabolism of PS Given Orally

Although PS is present in the diet and can be synthesized in liver and intestinal cells, only trace amounts or none at all are recovered in plasma lipoproteins (Table 1). A large amount of PS is instead found in blood cells; however, its contact with the plasma is prevented by a distribution of this phospholipid in the inner layer of the plasma membrane. The main barriers preventing contact of PS with plasma constituents are intestinal PS decarboxylase and the recently discovered PS translocase, found in erythrocyte plasma membranes (Seigneuret and Devaux, 1984). Early studies on PS decarboxylase of mucosal intestinal cells demonstrated its high activity, thereby explaining the absence of dietary PS in lipoproteins of intestinal origin (Wise and Elwyn, 1965). The small amounts of lysoPS able to enter mucosal cells are promptly decarboxylated after reacylation into PS. On the other hand, when PS is exposed to the plasma surface of erythrocytes, PS translocase quickly promotes an inward flip-flop movement to restore the natural position of this phospholipid in the inner layer of the plasma membrane.

In this study the fate of PS in the rat was followed after oral administration of 50 mg/kg of this phospholipid labeled with [^{3}H]-glycerol (Fig. 1). PS was given by duodenal infusion as a suspension of multilamellar vesicles in saline buffer. The metabolites were then analyzed in mesenteric lymph. Extraction with chloroform-methanol showed that the liposoluble radioactivity recovered in lymph over 5 hr accounted for 19% of the given dose. Triglycerides formed 80% of this fraction, suggesting that part of the PS was completely degraded, with the labeled glycerol used for the *de novo* synthesis of phospholipids and acylglycerols. Phospholipids constituted 11% of liposoluble radioactivity, whereas a minor part was found in diacylglycerol (9%). Analysis of the phospholipid fraction revealed that phosphatidylethanolamine (PE) was the major com-

Table 1. Blood and diet phospholipids in man

	Diet[a] (mmol/day)	Plasma[b] (μmol/ml)	Erythrocytes[c] (μmol/ml of packed cells)
Total phospholipids	2.6	2.5	3.95
Phosphatidylcholine	1.2	1.5	1.12
Phosphatidylethanolamine	0.7	0.08	1.0
Sphingomyelin	0.1	0.4	1.0
Phosphatidylinositol	0.15	0.05	0.07
Phosphatidylserine	0.1	absent	0.5
Lysophosphatidylcholine	-	0.14	0.04

[a]Bile phosphatidylcholine (9-27 mmol/day) not included. The amount of phospholipids is calculated on the basis of a daily intake of 200 g bovine skeletal muscle whose composition is given by White, 1973. [b]Punzi et al., 1986. [c]Broekhuyse, 1969. Reproduced with permission from Bruni et al., 1989.

ponent (66%), followed by phosphatidylcholine (PC, 8%), PS (6%), and lysoPE (3%). These experiments confirmed the extensive metabolism of PS given orally. It is interesting to note that although most of the labeled phospholipid was hydrolyzed or converted into PE in the intestinal tract, a small fraction reached the systemic circulation as a component of the phospholipid pool.

Appearance of Unsaturated Fatty Acids in Plasma

Since PE, the final product of PS metabolism, is a minor component of plasma lipoproteins, the extensive lysoPS reacylation occurring in the intestinal cells might influence the transport of fatty acids in plasma lipids. We therefore tested polyunsaturated fatty acids, as these compounds are mainly added to position 2 of glycerol where lysoPS is acylated. The lysoderivative of PS was administered orally to rats together with radiolabeled arachidonate. After 1 hr, liposoluble radioactivity was measured in plasma, liver, and the mucosa of the proximal segment of the small intestine. LysoPC was also administered for comparative purposes. As shown in Table 2, the two lysophospholipids differed in their ability to increase arachidonate transfer in plasma, lysoPS being completely ineffective. A similar conclusion was indicated by the analysis of liver radioactivity. The inability of lysoPS to promote the transfer of arachidonate into plasma lipids was not due to inhibition of fatty acid absorption, as the appearance of radioactivity in the intestinal mucosa was increased by either lysophospholipid. When arachidonate distribution in mucosal cell phospholipids was examined, lysoPS was seen to increase the fraction incorporated into PE and PS at the expense of the PC fraction.

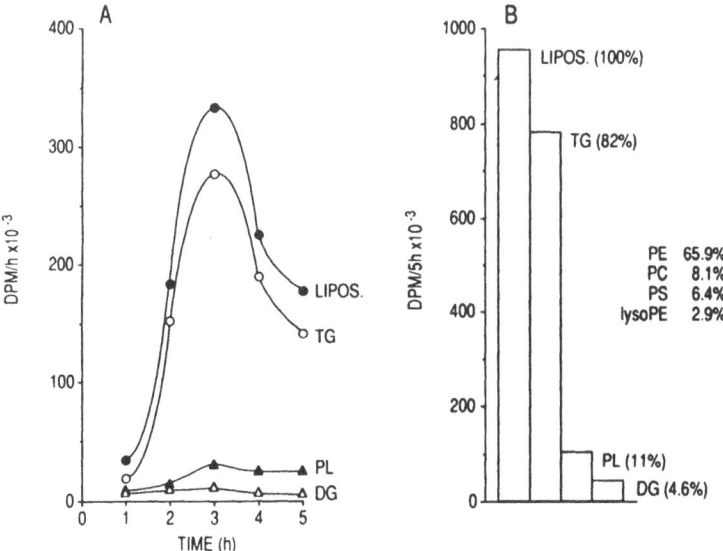

Figure 1. Metabolism of PS given by intraduodenal infusion to rats. [³H]Glycerol-PS (50 mg/kg; 21x10⁶ DPM/kg) was given by intraduodenal infusion as a dispersion in saline buffer. The mesenteric lymph was then collected for 5 hr and extracted with chloroform-methanol. (A) Radioactivity in the total liposoluble extract (LIPOS) and in the triglyceride (TG), phospholipid (PL), and diacylglycerol (DG) fractions. (B) Percent composition of liposoluble fraction in the total 5 hr period. The main labeled components of the phospholipid fraction are also indicated. The data were obtained from one of two rats which showed the same pattern.

Table 2. Effect of lysoPC and lysoPS given by oral route on arachidonate absorption and distribution

	% of arachidonate dose in the liposoluble fraction		
	None	LysoPC	LysoPS
Total plasma	1.1±0.3	2.3±0.7	1.1±0.2
Whole liver	2.8±0.6	8.0±1.5	2.9±0.2
Intestinal mucosa (1 g wt w)	17.3±4.8	32.9±3.5	28.9±3.2

Phospholipid labeling in the intestinal mucosa (% of total radioactivity found in liposoluble fraction)

	None	LysoPC	LysoPS
Total phospholipids	86.2±0.8	84.8±2.0	87.5±2.7
PC	55.4±1.1	63.9±2.3	42.1±2.5
PE	15.6±0.7	7.6±0.4	24.1±1.4
PS	1.3±0.0	1.0±0.2	7.8±1.2
PI	6.7±0.4	5.9±0.3	6.0±0.8

[^{3}H]Arachidonate (8×10^6 DPM/kg) was given to rats (260 g) by the oral route in 0.5 ml of saline buffer containing 10 mg/ml of bovine serum albumin. LysoPC or lysoPS (30 mg/kg) was added where indicated. Plasma (4% of body weight), liver (5% of body weight) and the mucosa of the proximal segment of small intestine were extracted with chloroform-methanol after 1 hr. Values are means ± SEM from four rats.

Taking into account the increase in arachidonate incorporation in the mucosa due to lysophospholipid administration (line 3 of Table 2) and expressing the results as a percent of arachidonate found in each phospholipid, it was calculated that lysoPC increased PC labeling 2.2-fold, leaving the other phospholipids unaffected. In contrast, lysoPS increased PC labeling only 1.3-fold, producing instead 2.6- and 11-fold increases in PE and PS labeling, respectively ($P<0.01$ for the difference between the two lysophospholipids).

To test whether lysoPS was unable to transfer arachidonate into plasma lipoproteins under conditions allowing maximal synthesis of lipid constituents, the lysophospholipid was given as a gastric bolus together with a mixture of lysoPC and triolein. As reported in Table 3, lysoPS did not change the effectiveness of lysoPC when the two lysophospholipids were given in the absence of triolein. On the contrary, lysoPS further promoted the absorption of arachidonate into the intestinal mucosa. However, when triolein was also added to the mixture, lysoPS did not enhance the transfer of arachidonate, but rather induced a significant inhibition of fatty acid appearance in plasma and liver. The amount found in the intestinal mucosa was not decreased. Similar results were obtained when linoleate was substituted for arachidonate. When the distribution of radioactivity in plasma lipids was measured, lysoPS was found to favor the incorporation of arachidonate into the triglyceride fraction at the expense of the phospholipid fraction (the respective values were 13.0% ± 1.1% and 54.6% ± 0.9% without lysoPS versus 19.7% ± 1.4% and 49.6% ± 1.5% in its presence; n = 22, $P<0.01$). Mass determinations of phospholipid and triglycerides in plasma and of phospholipid in the intestinal mucosa did not show lysoPS-induced changes.

Two hypotheses were considered to explain the lysoPS effect in the presence of triolein (Fig. 2). In the first, transient increases of PS and diacylglycerol in the absorptive cells cause protein kinase C translocation and activation. Activated protein

Table 3. Triacylglycerol-dependent inhibition by lysoPS of arachidonate absorption and distribution

	Triolein added	Percent of arachidonate dose in the liposoluble fraction	
		LysoPC	LysoPC + LysoPS
Total plasma	No	3.8±0.6	3.6±0.2
	Yes	5.5±0.5	4.0±0.4 (27%)[a]
Whole liver	No	11.2±1.1	9.9±0.9
	Yes	13.9±1.1	10.4±1.0 (25%)[a]
Intestinal mucosa	No	25.3±2.0	41.2±4.9[a]
(1 g wt w)	Yes	22.9±2.9	29.2±2.1

[^{3}H]Arachidonate (9×10^6 DPM/kg) was given to rats (280 ± 23 g) orally in 0.5 ml of saline buffer containing 10 mg/ml of serum albumin, lysoPC (10 mg/kg), and triolein (10 mg/kg) where indicated. In a parallel group of rats, lysoPS (30 mg/kg) was also added to the lipid mixture. Organs were extracted after 2 hr. Values are means ± SEM from five rats (without triolein) or 12 rats (with triolein). The percent inhibition by lysoPS is shown in parentheses; [a]$P<0.02$.

kinase C has been reported to increase the release of arachidonate from phospholipids (Godson et al., 1990) and to increase the CDP-choline pathway of PC biosynthesis (Pelech and Vance, 1984). However, early reports (Johnston et al., 1970) indicate that diacylglycerol originating from monoglycerides forms a separate pool, which may not be available for protein kinase C activation. In the second scheme, PS and lysoPE, whose generation is favored by the conversion of absorbed PS into PE (see also Fig. 1), would activate the cytidylyltransferase in the presence of fatty acids and diacylglycerol (Choy and Vance, 1978; Pelech and Vance, 1984). In contrast, lysoPC is known to induce a negative regulation of the enzyme. In the two hypotheses, the decreased appearance in plasma of labeled fatty acids with lysoPS present may be the consequence of an increased ratio between the PC formed via the CDP-choline pathway (less labeled due to isotopic dilution of arachidonate with fatty acids originated from triglycerides) and that formed by lysoPC reacylation (more labeled due to prefer-

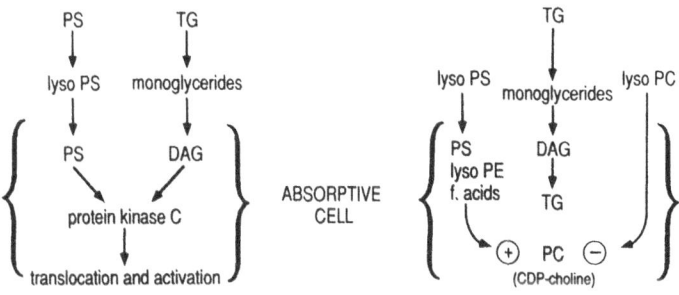

Figure 2. Possible mechanisms of action of lysoPS (see text for details).

tial acylation with arachidonate). In agreement with this possibility it was found that the administration of lysoPC together with triglycerides and radiolabeled glycerol induced a low level of radioactive PC, whereas the inclusion of glycerol in phosphatidylinositol (PI) was increased, indicating a preferential utilization of diacylglycerol through the diacylglycerol kinase pathway. Enhanced labeling of PI by lysoPC was also observed when radiolabeled inositol was substituted for glycerol. In contrast, lysoPS did not reproduce this pattern, but rather induced an increased incorporation of glycerol in PC (Viola G, Mietto L, Ping L, Bruni A, unpublished data). Activation by exogenous PS of *de novo* synthesis of PC has been reported during long-term incubation of this phospholipid with cultured neuroblastoma cells (Slack et al., 1989).

SUMMARY

The present results indicate that PS, a phospholipid contained in small amounts in the human diet and not included in plasma lipoproteins, may be used to influence phospholipid metabolism in intestinal mucosal cells. Since PS influx into absorptive cells occurs after its hydrolysis to lysoPS, this metabolite may be used to increase the absorption of this phospholipid. These data show that lysoPS, after diffusion into intestinal cells, is sequentially converted into PS and PE (which make up a minor fraction of the lipids present in lipoproteins). As expected, lysoPS given together with radiolabeled unsaturated fatty acids was unable to promote their transfer into plasma lipoproteins. In this respect lysoPS differed from lysoPC, the latter increasing the appearance of dietary fatty acids in plasma. When given together, lysoPS decreased the lysoPC-induced transfer of unsaturated fatty acids into plasma. This effect required addition of triglycerides to the lipid mixture. In attempting to explain this triacylglycerol-dependent inhibition by lysoPS, we found that this phospholipid increased the incorporation of glycerol into mucosal cell PC. In contrast, lysoPC was inhibitory. Furthermore, in the presence of labeled inositol, lysoPC (but not lysoPS) promoted the appearance of labeled phosphatidylinositol. The data thus suggest that the two lysophospholipids differ in promoting the two main pathways of PC synthesis in the intestinal cells. While lysoPC favors PC synthesis by reacylation, lysoPS enhances the CDP-choline pathway of PC synthesis.

REFERENCES

Broekhuyse RM (1969) Quantitative two-dimensional thin-layer chromatography of blood phospholipids. Clin Chim Acta 23:457.

Bruni A, Mietto L, Bellini F, Boarato E, Toffano G (1989) Pharmacological and autopharmacological action of phosphatidylserine. In: Phospholipids in the nervous system: Biochemical and molecular pathology (Bazan NG, Horrocks LA, Toffano G, eds), p 217. Padova: Liviana Press.

Choy PC and Vance DE (1978) Lipid requirements for activation of CTP:phosphocholine cytidylyltransferase from rat liver. J Biol Chem 253:5163.

Garg ML, Thomson ABR, Clandinin MT (1990) Interactions of saturated, n-6 and n-3 polyunsaturated fatty acids to modulate arachidonic acid metabolism. J Lipid Res 31:271.

Godson C, Weiss BA, Inset PA (1990) Differential activation of protein kinase C alpha is associated with arachidonate release in Madin-Darby canine kidney cells. J Biol Chem 265:8369.

Herold PM and Kinsella JE (1986) Fish oil consumption and decreased risk of cardiovascular disease: a comparison of findings from animal and human feeding trials. Am J Clin Nutr 43:566.

Johnston JM, Paultauf F, Schiller CM, Schultz LD (1970) The utilization of the alpha-glycerophosphate and monoglyceride pathway for phosphatidylcholine biosynthesis in the intestine. Biochim Biophys Acta 218:124.

Lang CA and Davis RA (1990) Fish oil fatty acids impair VLDL assembly and/or secretion by cultured rat hepatocytes. J Lipid Res 31:2079.

Nunzi MG, Milan F, Guidolin D, Toffano G (1987) Dendritic spine loss in hippocampus of aged rats. Effect of brain phosphatidylserine administration. Neurobiol Aging 8:501.

Pelech SL and Vance DE (1984) Regulation of phosphatidylcholine biosynthesis. Biochim Biophys Acta 779:217.

Punzi L, Todesco S, Toffano G, Catena R, Bigon E, Bruni A (1986) Phospholipids in inflammatory synovial effusions. Rheumatol Int 6:7.

Seigneuret M and Devaux PF (1984) ATP-dependent asymmetric distribution of spinlabeled phospholipids in the erythrocyte membrane: Relation to shape changes. Proc Natl Acad Sci USA 81:3751.

Slack BE, Liscovitch M, Blusztajn JK, Wurtman RJ (1989) Uptake of exogenous phosphatidylserine by human neuroblastoma cells stimulates the incorporation of [methyl-^{14}C]choline into phosphatidylcholine. J Neurochem 53:472.

Von Schacky C (1987) Prophylaxis of atherosclerosis with marine ω-3 fatty acids: A comprehensive strategy. Ann Int Med 107:890.

White DA (1973) The phospholipid composition of mammalian tissues. In: Form and function of phospholipids. BBA Library 3 (Ansell GB, Dawson RMC, Hawtorne JN, eds) p 441. Amsterdam: Elsevier Scientific Publishing Company.

Wise EM and Elwyn D (1965) Rates of reactions involved in phosphatide synthesis in liver and small intestine of intact rats. J Biol Chem 240:1537.

Yamamoto N, Saitoh M, Moriuchi A, Nomura M, Okuyama H (1987) Effect of dietary alpha-linolenate/linoleate balance on brain lipid compositions and learning ability of rats. J Lipid Res 28:144.

Zanotti A, Valzelli L, Toffano G (1989) Chronic phosphatidylserine treatment improves spatial memory and passive avoidance in aged rats. Psychopharmacology 99:316.

CARBACHOL-STIMULATED RELEASE OF ARACHIDONIC ACID AND EICOSANOIDS FROM BRAIN CORTEX SYNAPTONEUROSOME LIPIDS OF ADULT AND AGED RATS

Joanna Strosznajder and Marek Samochocki

Department of Neurochemistry
Medical Research Centre
Polish Academy of Sciences
3 Dworkowa str.
00-784 Warsaw, Poland

SUMMARY

Synaptoneurosomes from the brain cortex of adult rats (4 months old) and aged rats (27 months old), prelabeled with [^{14}C]arachidonic acid (AA), were used as the source of enzyme(s) and substrates to study the effect of a cholinergic agonist on the release of AA and eicosanoids. In synaptoneurosomes from adult brains, carbachol, the nonhydrolyzable analog of acetylcholine, increased AA release by 16% in the presence of 2 mM calcium. This agonist-mediated AA release occurred specifically from phosphatidylinositol (PI). Concomitantly, carbachol in the presence of 2 mM Ca^{2+} significantly activated the formation of 15-HETE and $PGF_{2\alpha}$. This effect of carbachol on the level of eicosanoids was also observed in the presence of endogenous calcium. In synaptoneurosomes from aged brains, carbachol had no effect on the release of AA and eicosanoids. The results of studies involving inhibitors of phospholipase A_2 (PLA_2) and phospholipase C (PLC) suggested that PLA_2 is almost completely responsible for the Ca^{2+}-dependent, carbachol-mediated AA liberation. The distribution of labeled AA in the lipids after incubation of synaptoneurosomes in the presence of 2 Mm Ca^{2+} and carbachol indicated that in aged synaptoneurosomes, the muscarinic receptor-mediated degradation of phosphoinositides through phospholipase C is preserved, but the turnover of the phosphoinositide cycle is probably suppressed. These results indicate that aging significantly affects the population of cholinergic-muscarinic receptors coupled to PLA_2.

INTRODUCTION

Brain aging is associated with significant disturbances of memory and membrane deterioration (Boruts et al., 1982; Wood and Schroeder, 1988). The synaptic membrane changes are probably a primary cause of age-related losses of function in brain (Sun and Sun, 1979). Boehme et al. (1979) and Sun and Sun (1979) found that aged brain, and particularly nerve endings isolated as synaptosomes, contain increased proportions of lysoglycerophospholipids and free fatty acids and that the ability to repair damage and to renew the phospholipids in membranes is decreased. However, it seems that in the aging brain the essential signal circuitry remains intact. Studies of the metabolic events involved in signal transmission may provide an important clue. In the central nervous system there are several adaptive mechanisms used to maintain synaptic transmission. Among these mechanisms are agonist-mediated arachidonic acid and eicosanoids release. This release seems to be critical for normal function as well as for certain mechanisms of plasticity such as long-term potentiation in which N-methyl-D-aspartate (NMDA) and cholinergic receptors appear to be necessary (Ascher and Nowak, 1986; Williams and Johnston, 1988; Schaad et al., 1991). In the central nervous system acetylcholine is quantitatively a minor neurotransmitter; however, it is thought to play a key role in intellectual activities, including memory. In the brain, deficits in various cholinergic markers suggest a decline in cholinergic synaptic transmission during aging (Bartus and Dean, 1983).

Until now little has been known about the effect of aging on the cholinergic receptor-mediated release of arachidonic acid and eicosanoids. The aim of our studies was to characterize activity of the calcium-dependent and cholinergic agonist-mediated AA liberation from prelabeled lipids of adult and aged brain cortex synaptoneurosomes.

MATERIALS AND METHODS

Materials

[1-^{14}C]Arachidonic acid (55 mCi/mmol) was purchased from Amersham, UK. EGTA, GTPγS, quinacrine, neomycin sulphate, carbachol, dithiothreitol, and sucrose were purchased from Sigma, St. Louis, MO, USA. TLC plates were from Merck, Germany. Wistar rats (4 and 27 months old) were used.

Preparation of Synaptoneurosomal Fractions from Rat Brain Cortex

The synaptoneurosomal fraction was obtained from rat cerebral cortex, as described by Hollingsworth et al. (1985). The slices of cerebral cortex were prepared manually with a cooled razor blade. Then they were homogenized by hand (5 strokes) in 7 ml of Ca^{2+} free Krebs-Henseleit bicarbonate buffer (KRBS), pH 7.4, using a Dounce-type glass homogenizer. After dilution to 35 ml with KRBS buffer and centrifugation at 1100 × g for 15 min, the pellet was resuspended in KRBS buffer, incubated at 37°C for 30 min under 95% O_2: 5% CO_2 atmosphere, and subsequently used for the assay of AA incorporation. This synaptoneurosomal fraction was examined by electron microscopy. The preparation was highly enriched in synaptoneurosomes as described previously by Strosznajder and Samochocki (1991).

Assay of Arachidonic Acid Incorporation

Arachidonic acid incorporation was assayed in the incubation system, which contained 2 μCi [^{14}C]AA, 2.5 mM ATP, 10 mM MgCl$_2$, 0.1 mM CoA, 0.3 mM dithiothreitol, and 15 mg of synaptoneurosomal protein in a total volume of 5 ml. Incubation was carried out for 15 min at 37°C in a shaking water bath. After incubation, synaptoneurosomes were washed twice with KRBS buffer containing fatty acid-free bovine serum albumin (BSA) by centrifugation at 1100 × g for 15 min and once with buffer without BSA. The pellet was used for experiments. The details have been described elsewhere (Strosznajder and Samochocki, 1991).

Assay of Arachidonic Acid Release

Synaptoneurosomes prelabeled with [^{14}C]AA were preincubated in KRBS buffer containing 1 mg protein and 20 μg/ml of saponin for 15 min at 37°C. Then, depending on the experimental conditions, other reagents were added to a final volume of 1 ml, and incubation was carried out for 60 min. The reaction was terminated with chloroform:methanol (1:2 by vol) and the lipids were extracted according to Bligh and Dyer (1959). To improve extraction of polyphosphoinositides, an acidified chloroform:methanol mixture was used, as described previously by Strosznajder et al. (1987).

Assay of Prostaglandins (PGE$_2$ and PGF$_{2\alpha}$) and 15-HETE in Synaptoneurosomes

Synaptoneurosomes were incubated at 37°C for 60 min in the presence of endogenous and 2 mM added $CaCl_2$ with or without 1 mM carbachol. The reaction was terminated by the addition of 3 ml of absolute ethanol and eicosanoids were extracted as described by Powell (1980). Separation of eicosanoids was performed on SEP-PAK minicolumns (C$_{18}$ type) using a sequential elution system (H$_2$O:ethanol, 85:15 by vol; petroleum ether; and methyl formate). The final fractions were collected and evaporated to dryness, then resuspended in assay buffer. Eicosanoids were determined using commercial RIA kits from Amersham in accordance with the manufacturer's protocol.

Lipid Analysis

Brain lipid extract was applied to silica gel TLC plates, which were then subjected to the separation-reaction-separation procedure, according to Horrocks and Sun (1972). The second solvent system was modified to contain chloroform:methanol:acetone:acetic acid:0.1 M ammonium acetate, 130:60:55:3.5:10 by volume. Individual lipid spots were scraped from the plates into scintillation vials with 10 ml of scintillation fluid. Radioactivity of samples was measured in a Beckman LS9000 scintillation counter.

Polyphosphoinositides: PIP and PIP$_2$ were separated on high performance thin layer chromatography (HPTLC) plates using the solvent system chloroform:methanol:NH$_4$OH, 9:7:2 by volume, as described by Strosznajder et al. (1987). Protein was estimated according to Lowry et al. (1951).

RESULTS

Uptake of arachidonic acid into aged brain cortex synaptoneurosome lipids and the pattern of AA incorporation into particular lipids of the brain cortex synaptoneurosomes were not significantly changed by aging (Table 1). Prelabeled synaptoneurosomes from adult and aged brain cortex were used for the investigation of AA release.

In the presence of 2 mM EGTA, the activity of AA liberation after 1 hr of incubation at 37°C was the same in adult and aged brain. In both adult and aged synaptoneurosomes, 2 mM Ca^{2+} significantly activated AA release. In adult synaptoneurosomes, carbachol in the presence of Ca^{2+} enhanced this release by 16%. In synaptoneurosomes from aged brain, however, the effect of carbachol on AA release was significantly less. The addition of atropine (10 μM) completely eliminated the effect of carbachol (Fig. 1). Quinacrine almost completely suppressed Ca^{2+}-dependent carbachol-mediated AA release. Neomycin, which also had an inhibitory effect, suppressed this process by about 30% (Fig. 1).

Table 1. Pattern of arachidonic acid incorporation into particular lipids of adult and aged rat brain cortex synaptoneurosomes

	Adult (%)	Aged (%)
Phosphatidylcholine	25.6 ± 2.26	29.0 ± 2.56
Phosphatidylethanolamine	9.3 ± 1.32	8.5 ± 0.42
Phosphatidylinositol	22.5 ± 3.52	23.4 ± 2.32
Phosphatidylserine	3.3 ± 0.73	3.8 ± 0.88
Phosphatidic acid	9.7 ± 1.64	8.2 ± 1.26
Neutral lipids	29.1 ± 3.45	26.1 ± 4.06

The values express percentage distribution of particular lipid radioactivity in total lipid radioactivity. The results are means ± SD from three experiments carried out in triplicate.

The effect of carbachol on Ca^{2+}-dependent AA release in adult brain cortex synaptoneurosomes was time dependent (Fig. 2). Carbachol-mediated AA release was absolutely dependent on the presence of a high concentration of exogenous Ca^{2+} ions (2 mM). In synaptoneurosomes from aged brain cortex this agonist had no effect. Analysis of the AA radioactivity distribution among different lipids after incubation of synaptoneurosomes in the presence of carbachol indicated that carbachol activates degradation of the phosphoinositides in both adult and aged synaptoneurosomes. The carbachol-mediated AA release occurred mainly from phosphatidylinositol (PI). Simul-

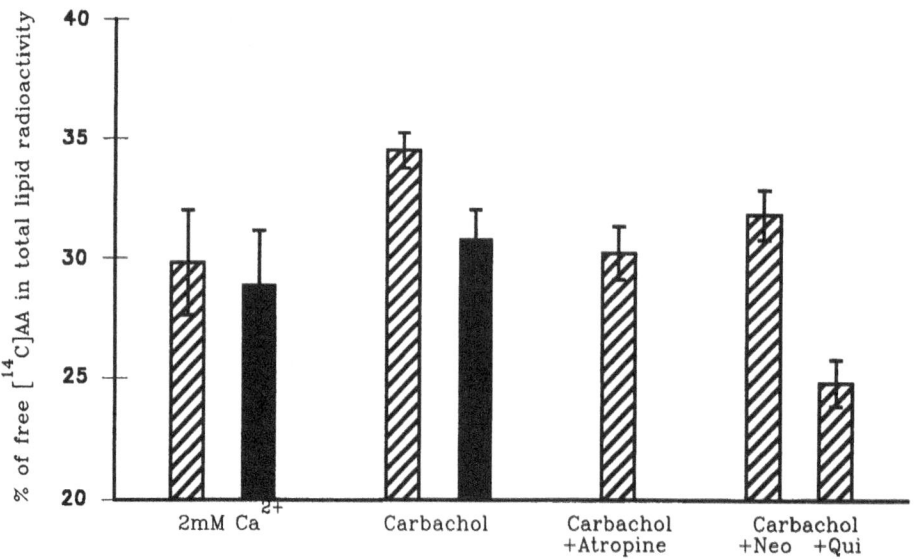

Figure 1. Calcium-dependent and carbachol-mediated arachidonic acid release in cortex synaptoneurosomes from adult and aged brain. Effects of atropine and inhibitors of phospholipases, neomycin (Neo) and quinacrine (Qui), on carbachol-stimulated AA release in adult brain are shown. Concentration of atropine 10 μM, quinacrine 150 μM, and neomycin 1 mM. The synaptoneurosomes were incubated 60 min at 37°C. Results are means ± SD from three to six experiments carried out in triplicate. Striped bar, adult brain; solid bar, aged brain.

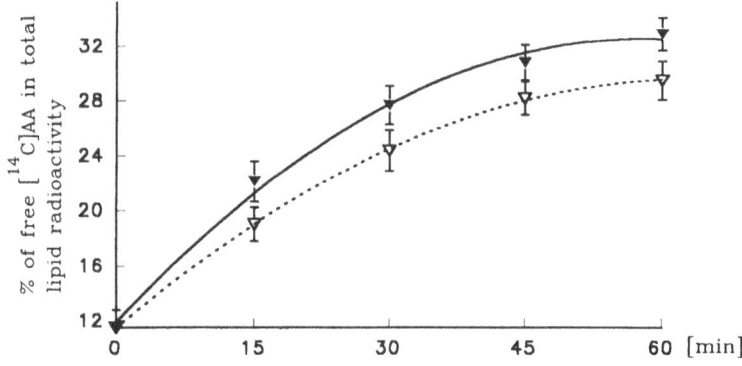

Figure 2. The time-dependent effect of Ca^{2+} and carbachol on [^{14}C]AA release from prelabeled lipid of adult brain cortex synaptoneurosomes. The results represent the percentage changes of free [^{14}C]AA in total lipid radioactivity. The results are means ±SD of triplicate determinations from a typical experiment. Open triangles, 2 mM $CaCl_2$; filled triangles, 2 mM $CaCl_2$ + carbachol.

taneously, carbachol also significantly enhanced the level of AA radioactivity in phosphatidic acid (PA) in both aged and adult brains. The formation of PA, however, occurred less actively in aged as compared to adult synaptoneurosomes (Fig. 3).

The effect of carbachol on the formation of eicosanoids in adult and aged synaptoneurosomes was also investigated. The endogenous level of eicosanoids was significantly higher in aged brain as compared to adult (data not shown). In adult brains, but

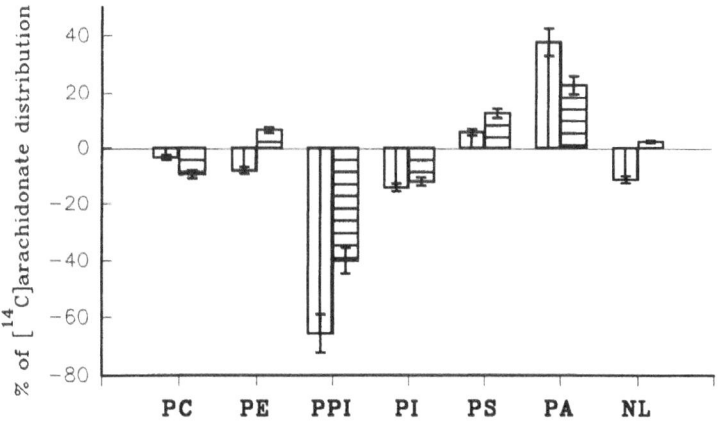

Figure 3. Effect of carbachol on [^{14}C]AA distribution in lipids of adult and aged brain cortex synaptoneurosomes. The results represent the percentage changes of particular lipid radioactivity compared with lipid radioactivity in the presence of 2 mM $CaCl_2$. The results represent the means ± SD of six experiments carried out in triplicate. In each group of two bars, left bar indicates adult, right bar indicates aged. PC, phosphatidylcholine; PE, phosphatidylethanolamine; PPI, polyphosphoinositides; PI, phosphatidylinositol; PS, phosphatidylserine; PA, phosphatidic acid; NL, neutral lipids.

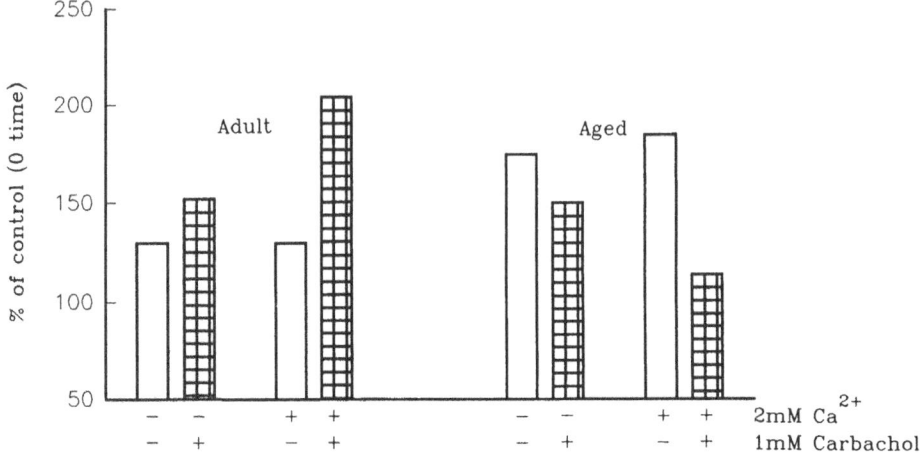

Figure 4. Effect of carbachol on $PGF_{2\alpha}$ formation in adult and aged brain cortex synaptoneurosomes. The synaptoneurosomes were incubated at 37°C for 60 min in the presence and absence of 2 mM $CaCl_2$ and 1 mM carbachol. The results represent the mean value of a triplicate determination from a typical experiment.

not in aged brains, carbachol enhanced the level of $PGF_{2\alpha}$ significantly in the presence of 2 mM Ca^{2+} and slightly in the presence of endogenous $CaCl_2$ (Fig. 4).

A small effect of carbachol on 15-HETE formation was also observed in adult brain in the presence of endogenous and 2 mM $CaCl_2$. Carbachol had no effect on 15-HETE formation in aged brain (Fig. 5).

DISCUSSION

Our results indicate that aging decreases Ca^{2+}-dependent, carbachol-mediated AA

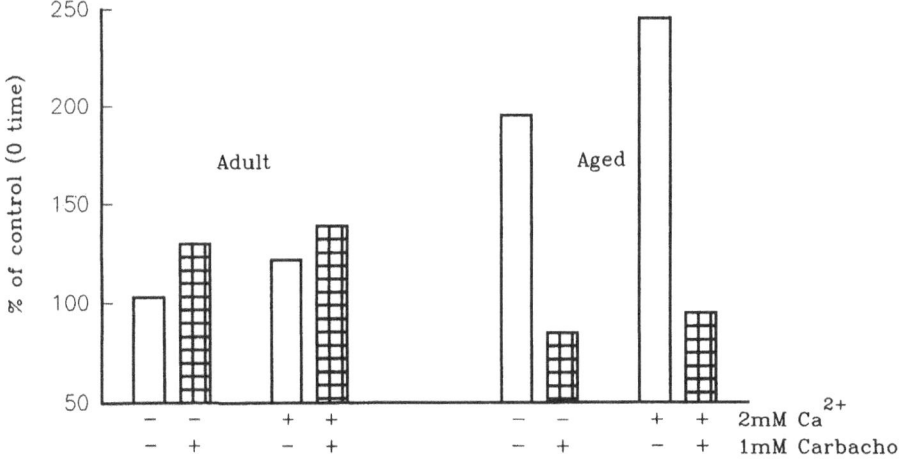

Figure 5. Effect of carbachol on 15-HETE formation in adult and aged brain cortex synaptoneurosomes. The synaptoneurosomes were incubated at 37°C for 60 min in the presence and absence of 2 mM $CaCl_2$ and 1 mM carbachol. The results represent the mean value of a triplicate determination from a typical experiment.

release. It seems that aging suppresses the muscarinic receptor-mediated release of AA and eicosanoids but does not alter the cholinergic receptor-mediated degradation of phosphoinositides. Our results agree with the observations of Surichamorn et al. (1989), who also found that aging affects neither muscarinic receptor-mediated hydrolysis of phosphoinositides nor its inhibition by phorbol esters and tetrodotoxin. These results demonstrate that aging affects Ca^{2+}-dependent AA liberation through the action of PLA_2, which is coupled to the muscarinic receptor.

Biegon et al. (1989) found that both M1 and M2 muscarinic receptor subtypes show an anatomically selective pattern of decreased density during aging, specifically a small but significant reduction in the cortical region and in the striatum. Moreover, the studies of Pietrzak et al. (1989) indicate that the plasticity of muscarinic receptors is affected by aging. A decrease in the density of the receptors after antagonist treatment was marked in the frontal cortex of young rats, less marked in adult rats, and absent in senescent rats (Pedigo and Polk, 1985).

Our studies indicate that aging suppresses Ca^{2+}-dependent muscarinic cholinergic receptor-mediated AA release by PLA_2. Concomitantly, it seems that the receptor response coupled to phosphoinositide degradation is preserved in aged brain synapto-neurosomes.

The significant changes induced by aging in cholinergic receptor-mediated arachidonic acid release and eicosanoids formation may have essential implications for the alterations of synaptic transmission and intellectual activity observed with aging.

ACKNOWLEDGMENTS

This work has been supported by Polish Academy of Sciences no 0602.09. The authors thank Mrs. M. Skorupka and St. Kuciak for their excellent technical assistance. We are also grateful to Mrs. M. Izak for her assistance in preparing the manuscript.

REFERENCES

Ascher P and Nowak L (1986) Calcium permeability of the channels activated by N-methyl-D-aspartate (NMDA) in mouse central neurones. (Abstr) J Physiol (Lond) 377:35P.

Bartus R and Dean RL (1983) Developing and utilizing animal models in the search for an effective treatment for aged-related memory disturbances. In: Physiological aging and dementia (Gottfries C, ed) pp 231-267. Basal: S Karger Press.

Biegon A, Hanau M, Greenberger V, Segal M (1989) Aging and brain cholinergic muscarinic receptor subtypes: An autoradiographic study in the rat. Neurobiol Aging 10:305.

Bligh EG and Dyer WJ (1959) A rapid method of total lipid extraction and purification. Can J Biochem Physiol 37:911.

Boehme DH, Kosecki R, Corson S, Stern F, Marks N (1979) Lipoperoxidation in human and rat brain tissue. Developmental and regional studies. Brain Res 136:11.

Boruts RT, Dean RL, Beer B, Lippa AS (1982) The cholinergic hypothesis of geriatric memory dysfunction. Science (Wash DC) 217:408.

Hollingsworth EB, McCeal ET, Burton JL, Williams RJ, Daly JW, Creveling CR (1985) Biochemical characterization of a filtered synaptoneurosome preparation from guinea pig cerebral cortex: Cyclic adenosine 3':5'-monophosphate-generating systems, receptors and enzymes. J Neurosci 5:2240.

Horrocks LA and Sun GY (1972) Ethanolamine plasmalogen. In: Research methods in neurochemistry, Vol 1 (Marks N, Rodnight R, eds) p 223. Totowa, NJ: Plenum Press.

Lowry OH, Rosebrough NJ, Farr AL, Randall RJ (1951) Protein measurement with Folin phenol reagent. J Biol Chem 193:265.

Pedigo NW and Polk DM (1985) Reduced muscarinic receptor plasticity in frontal cortex of aged rats after chronic administration of cholinergic drugs. Life Sci 37:1443.

Pietrzak ER, Wilce PA, Shanley BC (1989) Plasticity of brain muscarinic receptors in aging rats: The adaptive response to scopolamine and ethanol treatment. Neurosci Lett 104:331.

Powell WS (1980) Rapid extraction of oxygenated metabolites of arachidonic acid from biological samples using octadecylsilyl silica. Prostaglandins 20:947.

Schaad NC, Magistretti PJ, Schorderet M (1991) Prostanoids and their role in cell-cell interactions in the central nervous system. Neurochem Int 18:303.

Strosznajder J and Samochocki M (1991) Ca^{2+}-independent, Ca^{2+}-dependent and carbachol-mediated arachidonic acid release from rat brain cortex membrane. J Neurochem 57:1198.

Strosznajder J, Wikiel H, Sun GY (1987) Effect of cerebral ischemia on [^{3}H]inositol lipids and [^{3}H]inositol phosphates of gerbil brain and subcellular fractions. J Neurochem 48:943.

Sun GY and Sun AY (1979) Neurochemical aspects of the membrane hypothesis of aging. In: Interdisciplinary topics in gerontology, Vol 15 (von Hahn HP, ed) pp 34-53. New York: S Karger.

Surichamorn W, Abdallah EAM, El-Fakahany EE (1989) Aging does not alter brain muscarine receptor-mediated phosphoinositide hydrolysis and its inhibition by phorbol esters, tetrodotoxin and receptor desensitization. J Pharmacol Exp Ther 251:543.

Williams S and Johnston D (1988) Muscarinic depression of long-term potentiation in CA3 hippocampal neurones. Science 242:84.

Wood WG and Schroeder F (1988) Membrane structure in aged humans and animals. In: Central nervous system disorders of aging. Clinical intervention and research, Vol 33 (Strong R, Gibson W, Wood WG, Burke WJ, eds) pp 199-209. New York: Raven Press.

CARBACHOL-INDUCED STIMULATION OF INOSITOL PHOSPHATES,

ARACHIDONIC ACID AND PROSTAGLANDIN $F_{2\alpha}$ IN RABBIT RETINA

Neville N. Osborne

Nuffield Laboratory of Ophthalmology,
Oxford University,
Walton Street, Oxford OX2 6AW, U.K.

INTRODUCTION

Acetylcholine is a mammalian retinal transmitter (Masland and Mills, 1979). Muscarinic receptors linked to the second messengers inositol phosphates (InsPs)/diacylglycerol (Cuttcliffe and Osborne, 1978; Osborne, 1988) and cAMP (Ghazi and Osborne, 1989) have been located in the retina. Furthermore, binding studies have shown that at least two types of muscarinic receptors exist in the retina (Vanderheyden et al., 1988).

Receptor-mediated release of arachidonic acid (AA) has recently been demonstrated in various tissues of the nervous system. N-Methyl-D-aspartate stimulates the release of AA from striatal neurons and cerebellar granule cells (Dumuis et al., 1988). The release of AA in rat cerebral cortical astrocytes in cultures has been shown to be induced by bradykinin (Burch and Kniss, 1988), substance P (Hartung et al., 1988) and ATP (Gebicke-Haerter et al., 1988). Acetylcholine causes release of AA and prostaglandins in *Torpedo* electric organ and brain slices (Reichmann et al., 1987) and stimulation of α_1-adrenergic receptors in spinal neurons also mediates release of AA (Kanterman et al., 1990). As far as the mammalian retina is concerned, receptor-mediated release of AA has not been reported, although it has been shown by Birkle and Bazan (1984) that potassium chloride-induced stimulation of the rat retina increases levels of AA and eicosanoids.

There are at least two pathways that can lead to the generation of AA. Receptors can be coupled to the activation of phospholipase A_2, which acts directly on phospholipids to release AA (Farago and Nishizuka, 1990). Alternatively, activation of phospholipase C produces InsPs/diacylglycerol from phosphatidylinositol 4,5-bisphosphate; diacylglycerol can then be further metabolized to form AA and prostaglandins (Farago and Nishizuka, 1990).

The aim of our study was to discover whether acetylcholine can stimulate the formation of AA and prostaglandins in retinal tissues and whether such a process may be mediated via phospholipase C or phospholipase A_2. In addition, since there is a lot of evidence for "crosstalk" between secondary messengers (Kanterman et al., 1990), studies were undertaken to see whether AA and/or prostaglandins can influence either InsPs formation or protein kinase C (PKC) activation. Some preliminary data on the effect of AA on GABA release in the rabbit retina are also reported. This study was begun because an association between PKC-γ and GABA appears to occur in the rabbit retina, as has been reported for Purkinje cells (Houslay, 1991). Further studies on Purkinje cells have shown that AA activates PKC-γ and thus enhances the release of GABA from these cells (Taniyama et al., 1990).

METHODS AND RESULTS

Primary rabbit retinal cultures on 13-mm diameter glass cover slips were used (Ghazi and Osborne, 1988). Cells were incubated with ^{3}H-AA or ^{3}H-inositol for 18-24 hr in serum-free medium. The effect of carbachol on the release of ^{3}H-InsPs was then investigated as described elsewhere (Ghazi and Osborne, 1988).

Cells incubated in the presence of ^{3}H-AA were first washed in three changes of serum-free medium (as used for InsPs assay, see Ghazi and Osborne, 1988) containing 0.2% fatty acid-free bovine serum albumin. The bovine serum albumin was used to trap free AA. Thereafter, the cells were placed in serum-free medium, experimental agents were added (final volume 0.5 ml), and the reaction was allowed to proceed for 10 min at 37°C. Aliquots of the medium (500 μl) were placed directly in vials on ice, scintillant was added, and samples were counted in a scintillation spectrophotometer.

The dose-response curve for carbachol-stimulated ^{3}H-AA and ^{3}H-InsPs release is shown in Figure 1. Although the curves appear to be similar, the EC_{50} values are different, the EC_{50} value for AA being lower (0.7 μM) than that for InsPs (8 μM). Atropine antagonized the stimulation of AA and InsPs produced by carbachol (Fig. 2). Neither process was affected by pertussis toxin (PTX), but the phorbol ester PDbut (phorbol 12,13 dibutyrate) specifically reduced the carbachol-induced stimulation of InsPs without influencing the production of AA. The carbachol-induced stimulation of InsPs was unaffected by either AA or $PGF_{2\alpha}$; neither AA nor $PGF_{2\alpha}$ influenced basal levels of InsPs.

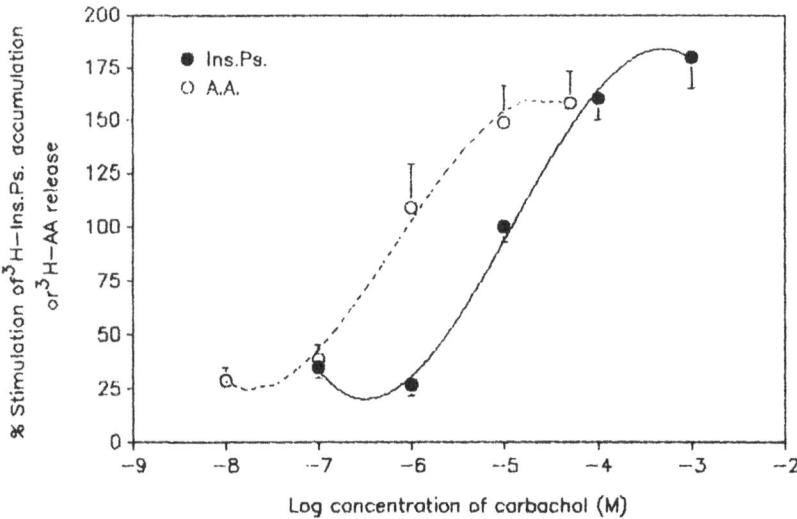

Figure 1. Dose response curves for carbachol-induced stimulation of AA and InsPs in primary rabbit retinal cultures. Results are means ± SEM for three separate experiments.

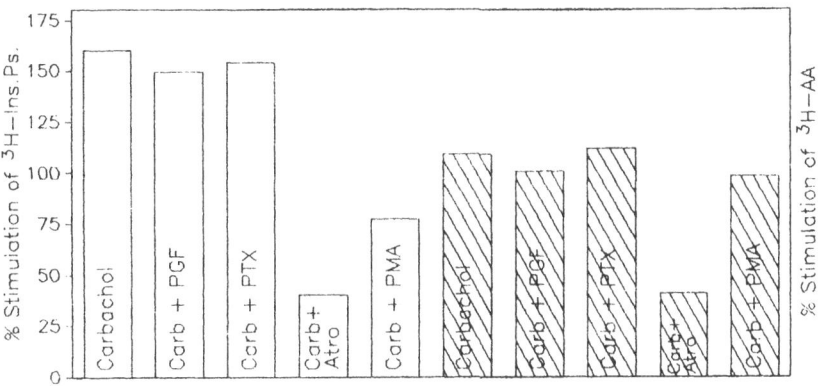

Figure 2. Effect of PDbut (1 μM), atropine (atro, 1 μM), PTX (200 μg/ml) and PGF$_{2\alpha}$ (PGF, 50 μM) on the carbachol (carb, 100 μM for InsPs, 1 μM for AA)-induced stimulation of ^{3}H-InsPs and ^{3}H-AA. Results are expressed as % stimulation and are the mean values of three separate experiments.

Carbachol-induced Stimulation of PGF$_{2\alpha}$

In these experiments rabbit retinal slices were pre-incubated in physiological solution containing glucose (10 mM) and 1% bovine serum albumin (BSA) for 30 min at 37°C. The slices were then washed in fresh solution containing BSA and finally allowed to settle. Aliquots of slices (500 μl) were transferred to 500 μl of physiological solution containing added experimental agents and incubated for 30 min at 37°C. The samples were then centrifuged briefly and 500 μl of the supernatant was transferred to another tube, lyophilized, and assayed for PGF$_{2\alpha}$ content using an assay kit system from Amersham International.

Figure 3 shows that carbachol induces a release of PGF$_{2\alpha}$ from retinal slices in a way that appears to be concentration dependent. Although fairly high concentrations

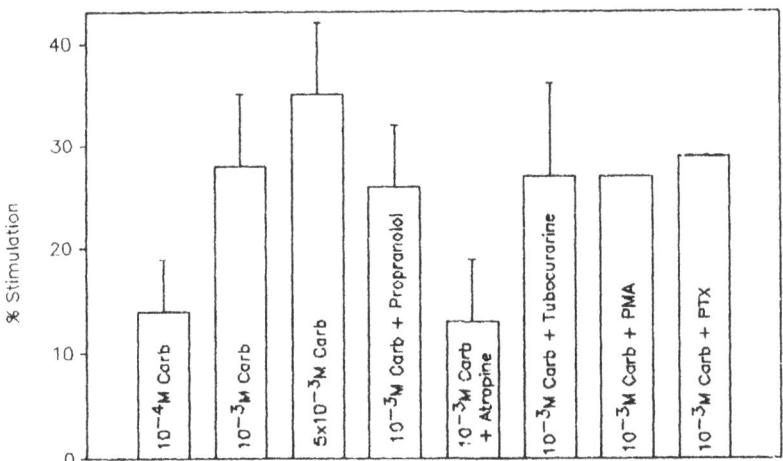

Figure 3. Effect of various concentrations of carbachol (Carb) on the stimulation of PGF$_{2\alpha}$ in rabbit retina. Atropine (5 μM) nullified the carbachol (1 mM) effect, while all other substances were without effect. Results are mean values of four experiments $\pm$ SEM, or the mean of two experiments.

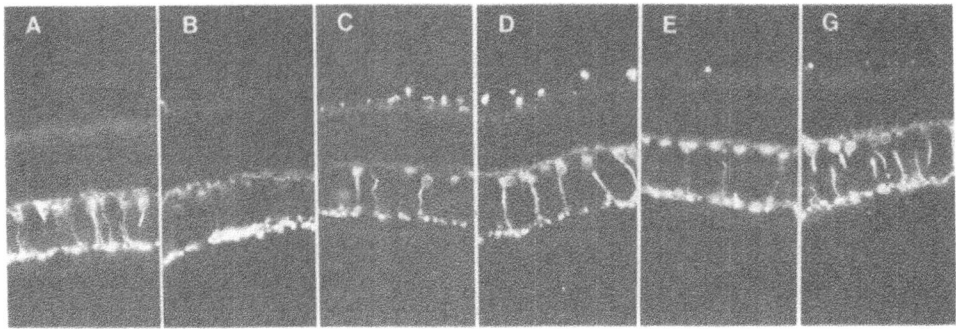

Figure 4. Effect of various substances on PKC-α immunoreactivity in rabbit retina. PKC-α is associated with bipolar cells (A). The effect of 1 μM PDbut is to cause the PKC-α to appear mostly in the terminals of the bipolar cells (B). In contrast, 50 μM PGF$_{2\alpha}$ (C), 10 μM AA plus 10 μM indomethacin (D), 10 μM AA (E), and 10 μM indomethacin (G) were without effect.

of carbachol are required to produce a small but significant release of PGF$_{2\alpha}$, it appears that the process is receptor mediated as atropine blocked the response while propranolol and tubocurarine (respectively ß-receptor and nicotine-receptor antago-nists) were without effect. Moreover, pre-incubation with PDbut or PTX had no effect on the carbachol-induced stimulation of PGF$_{2\alpha}$.

Effect of AA on PKC-α and PKC-γ

In these experiments rabbit retinal tissues were incubated in physiological solution containing PDbut (1 μM), AA (10 μM), or PGF$_{2\alpha}$ (50 μM) for 30 min at 37°C. Thereafter the tissues were fixed for immunohistochemical localization of PKC isoenzymes or the cytosolic and membrane fractions were isolated and subjected to SDS PAGE and Western blotting for the determination of PKC-α and PKC-γ (see Osborne et al., in press).

As shown in Figure 4, PKC-α immunoreactivity is primarily present in bipolar cells in the rabbit retina. The effect of PDbut is to cause a movement of PKC-α from the cell bodies to the terminals of the bipolar cells, as we reported elsewhere (Osborne et al., 1991), while AA and PGF$_{2\alpha}$ did not influence PKC-α immunoreactivity.

PKC-γ immunoreactivity is primarily associated with amacrine cells in the rabbit retina. As can be seen in Figure 5, there was no obvious effect on PKC-γ immunoreact-ivity from exposure of the retina to PDbut, AA, or PGF$_{2\alpha}$. The effect of PDbut and AA

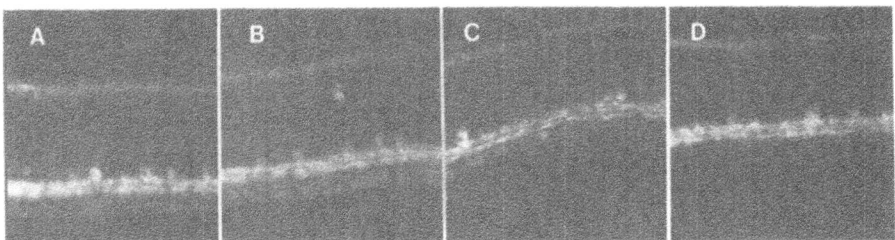

Figure 5. Effect of various substances on PKC-γ immunoreactivity in rabbit retina. PKC-γ is almost exclusively associated with amacrine cells (A). PDbut (B), PGF$_{2\alpha}$ (C), and AA (D) were without effect on PKC-γ immunoreactivity.

Figure 6. Effect of various substances on the translocation of PKC-α (A) and PKC-γ (B) from cytosolic to membrane fractions. Lanes 1-3, 7 and 8 are cytosolic fractions; lanes 4-6, 9 and 10 are membrane fractions. It can be seen that PDbut causes a translocation of PKC-α from cytosolic (3) to membrane (6) compartments. In the case of control (1 and 4) and AA-treated retinas (2 and 5) no apparent translocation of PKC-α has taken place. Neither does AA cause a translocation of PKC-γ (8 and 10) when compared with controls (7 and 9).

on the translocation of PKC-α is shown in Figure 6; only PDbut causes a translocation of the enzyme from the cytosolic to the membrane compartments, while AA has no effect. AA did not cause a translocation of PKC-γ (Fig. 6), which was confirmed by densitometric scanning of the blots.

Effect of AA on the Release of GABA

In these experiments rabbit retinas in physiological saline were exposed to AA (10 μM), PGF$_{2\alpha}$ (50 μM), or PDbut (1 μM) for 30 min at 37°C. Thereafter the tissues were fixed and processed for localization of GABA immunoreactivity. As shown in Figure 7, none of the substances caused an obvious change in GABA immunoreactivity.

In another series of experiments, ^{3}H-GABA was injected intravitreally in the rabbit eye. One hour later the retinas were removed and exposed under *in vitro* conditions to AA or PDbut for 30 min at 37°C as already described. Tissues were fixed and processed for the autoradiographic localization of ^{3}H-GABA. Initial results suggest that PDbut causes a release of accumulated ^{3}H-GABA, but no obvious release could be attributed to AA.

GENERAL CONCLUSIONS

The present results, although preliminary in nature, show that carbachol stimulates the production of AA, PGF$_{2\alpha}$, and InsPs in rabbit retina. When atropine is added, the

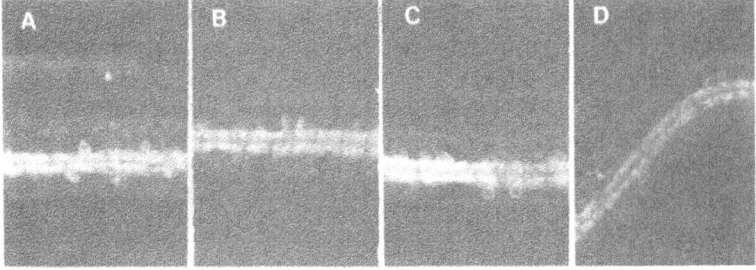

Figure 7. Effect of various substances on GABA immunoreactivity in rabbit retina. In the normal retina (A), GABA is associated with amacrine cells. PDbut (B), PGF$_{2\alpha}$ (C), and AA (D) did not influence normal GABA immunoreactivity.

stimulation of all three substances by carbachol is nullified, showing that muscarinic receptors are involved. Moreover, PTX did not influence the production of AA, $PGF_{2\alpha}$, or InsPs. Some authors have shown that PTX influences G proteins associated with phospholipase C (Casey and Gilman, 1988) and phospholipase $A_{2\alpha}$ (Axelrod et al., 1988), but this is not a universal observation. Variations in data result from the difficulty of treating living cells effectively or appropriately with PTX.

It is clearly necessary to extend the present studies to resolve the issue as to whether AA (and subsequently $PGF_{2\alpha}$) stimulation is accounted for by phospholipids using phospholipase A_2 or by pathways involving phospholipase C. A phospholipase C pathway clearly exists and is documented by the carbachol-induced stimulation of InsPs. It is also possible that the stimulation of AA occurs via receptors linked to phosphatidylcholine-specific phospholipase C and phospholipase D (Exton, 1990).

The present studies show that while carbachol simultaneously stimulates the production of AA, InsPs, and $PGF_{2\alpha}$, neither AA nor $PGF_{2\alpha}$ modulated InsPs production, either by influencing basal levels of InsPs or by acting on the carbachol-induced effect. However, the stimulation of InsPs by carbachol is inhibited by the activation of PKC with a phorbol ester. Thus there is limited "crosstalk" between products formed by carbachol stimulation of muscarinic receptors: PKC activation negatively inhibits InsPs production while produced AA and $PGF_{2\alpha}$ neither potentiate nor inhibit InsPs production.

The present data also provide no evidence that AA and $PGF_{2\alpha}$ have an effect on PKC-α or PKC-γ. It is clear from the immunoblot analysis (and can be inferred from the immunohistochemistry data in the case of PKC-α) that AA does not cause a translocation (activation) of the PKC isoenzymes. Although it has been reported that high concentrations (100 μM) of AA activate PKC (Taniyama et al., 1990), only micromolar concentrations have been shown to influence PKC-γ specifically. We used 10 μM AA in our studies, but could not demonstrate this effect on the enzyme in retinal tissue. Furthermore, while a close relationship exists between PKC-γ and GABAergic cells in the retina, we could not definitely show that AA can cause a release of GABA, as has been described for Purkinje cells (Taniyama et al., 1990).

ACKNOWLEDGMENTS

This work has been supported by a grant from the Wellcome Trust.

REFERENCES

Axelrod J, Burch RN, Jelsema CL (1988) Receptor mediated activation of phospholipase A_2 via GTP-binding proteins: arachidonic acid and its metabolites as second messengers. TINS 11:117.

Birkle DL and Bazan NG (1984), Effect of K^+ depolarization on synthesis of prostaglandins and hydroxyeicosatetra(5,8,11,14)enoic acids (HETE) in rat retina. Biochim Biophys Acta 785:564.

Burch RM and Kniss DA (1988) Modulation of receptor-mediated signal transduction by diacylglycerol mimetics in astrocytes. Cell Mol Neurobiol 8:251.

Casey PJ and Gilman AG (1988) G-protein involvement in receptor-effector coupling. J Biol Chem 263:2477.

Cuttcliffe N and Osborne NN (1978) Serotonergic and cholinergic stimulation of inositol phosphates formation in the rabbit retina. Evidence for the presence of 5-HT_2 and muscarinic receptors. Brain Res 421:95.

Dumuis S, Sebben M, Haves L, Pin JP, Bockaert J (1988) NMDA receptors activate the arachidonic acid cascade system in striatal neurons. Nature 236:68.

Exton JH (1990) Signalling through phosphatidylcholine breakdown. J Biol Chem 265:1.

Farago A and Nishizuka Y (1990) Protein kinase C in transmembrane signalling. FEBS Lett 264:350.

Gebicke-Haerter PJ, Wurster S, Schubert A, Hertting G (1988) P_2-purinergic induced prostaglandin synthesis in primary rat astrocyte cultures. Naunyn Schmiedebergs Arch Pharmacol 338:704.

Ghazi H and Osborne NN (1988) Agonist-induced stimulation of inositol phosphates in primary rabbit retinal cultures. J Neurochem 50:1851.

Ghazi H and Osborne NN (1989) Muscarinic inhibition of adenylate cyclase activity by rabbit retinal cells. In: Neurobiology of the inner retina (Welter R, Osborne NN, eds), pp 437-444. Heidelberg: Springer Verlag.

Hartung H, Heininger J, Schafter B, Toyka K (1988) Substance P and astrocytes: stimulation of cyclooxygenase pathway of arachidonic acid metabolism. FASEB J 2:48.

Houslay MD (1991) "Crosstalk": a pivotal role for protein kinase C in modulating relationships between signal transduction pathways. Eur J Biochem 195:9.

Kanterman RY, Felder CC, Brenneman DE, Ma AL, Fitzgerald D, Axelrod J (1990) α_1-Adrenergic receptor mediates arachidonic acid release in spinal cord neurons independent of inositol phosphate turnover. J Neurochem 54:1225.

Masland RH and Mills JW (1979) Autoradiographic identification of acetylcholine in rabbit retina. J Cell Biol 83:159.

Osborne NN (1988) Muscarinic stimulation of inositol phosphates formation in the rat retina: Developmental changes. Vision Res 8:875.

Osborne NN, Broyden NJ, Barnett NL, Morris NJ (1991) Protein kinase C (α and ß) immunoreactivity in rabbit and rat retina. Effect of phorbol esters and transmitter agonists on immunoreactivity and the translocation of the enzyme from cytosolic to membrane compartments. J Neurochem 57:594-604.

Osborne NN, Barnett NL, Morris NJ, Huang FL (in press) The occurrence of three isoenzymes of protein kinase C (α, ß and γ) in retinas of different species. Brain Res.

Reichmann M, Nen W, Hokin LE (1987) Acetylcholine releases prostaglandins from brain slices incubated *in vitro*. J Neurochem 49:1216.

Taniyama K, Saito N, Kose K, Matsuyama S, Nakayama S, Tanaka C (1990) Involvement of the γ-subtypes of protein kinase C in GABA release from the cerebellum. In: The biology and medicine of signal transduction (Nishizuka Y, Endo M, Tanaka C, eds), pp 399-404. New York: Raven Press.

Vanderheyden P, Ehinger G, Vanquelin G (1988) Characterization of M1 and M2 muscarinic receptors in calf retina membranes. Vision Res 8:247.

INDUCED AND SPONTANEOUS SEIZURES IN MAN PRODUCE INCREASES IN REGIONAL BRAIN LIPID DETECTED BY *IN VIVO* PROTON MAGNETIC RESONANCE SPECTROSCOPY

Bryan T. Woods and Tak-Ming Chiu

Neurology Department
Harvard Medical School, and
Neurology Department and Brain Imaging Center
McLean Hospital
Belmont, MA 02178

ABSTRACT

Elevations of brain concentrations of arachidonic acid and other free fatty acids (FFAs) by seizures induced in animals were demonstrated some years ago. Similarly, large shifts of potassium (K^+) from intra- to extracellular space during seizure activity have been documented in numerous studies. More recent studies of cell membrane function demonstrated a direct effect of FFAs on membrane K^+ conductance, suggesting that FFAs may play a primary role in seizure evolution in brain tissue. Using electroconvulsive therapy (ECT), in which generalized seizures are induced in patients by passage of electrical current, as a controlled human model of seizures, we studied the *in vivo* biochemical effects of single generalized seizures with localized proton magnetic resonance spectroscopy (^1H MRS). We found that ECT reliably induces an elevation in the lipid signal that resonates at approximately 1.2 ppm. We observed a similar increase in brain lipids in a patient with temporal lobe epilepsy temporarily off medication; the signal disappeared after re-medication. Similar observations were noted for a subject with focal gliosis bordering a resected brain tumor. Finally, acute alcohol effects seem also to induce observable lipid changes. The ^1H MRS technique does not yet permit direct identification of the specific lipids involved but analysis of cerebrospinal fluid obtained by lumbar puncture before and immediately after ECT may permit more precise characterization of the observed lipid increases. Theoretical and clinical implications of these results for the study of brain FFAs and epilepsy will be discussed.

INTRODUCTION

The role of brain phospholipids in a number of cellular functions was first suspected more than 30 years ago (Hokin and Hokin, 1958), but has been the subject of major (and still increasing) research attention in the last decade. However, a clear linkage of the results of so many elegant insights at the basic biochemical level to disorders of human nervous system function has lagged behind. Perhaps the most important single

Neurobiology of Essential Fatty Acids
Edited by N.G. Bazan *et al.*, Plenum Press, New York, 1992

reason for the disparity is methodological; the tools to study phospholipid activity and turnover directly *in vivo* in human subjects have not been available. It has been possible to detect certain metabolic products of phosphatidylinositol breakdown in cerebrospinal fluid (CSF) in several clinical conditions (White, 1989), but this approach is indirect, non-localizable, and difficult to characterize temporally.

The recent availability of *in vivo* magnetic resonance spectroscopy (MRS) in clinical whole body magnetic resonance imagers is a powerful addition to the arsenal of chemical analysis (Weiner and Hetherington, 1989). Recent technical improvements in magnetic resonance software and hardware have made proton (^{1}H) spectroscopy on currently available high-field-strength clinical magnetic resonance scanners practical in terms of sensitivity, chemical specificity, localization, and brevity of acquisition times. Spectroscopic methods are available for selectively suppressing the proton signal from water by a factor of 10^2-10^3 while leaving other signals undisturbed, and for studying the resonances of interest in volumes as small as 1 ml. This permits the localized detection of a number of biochemically important compounds in concentrations of about 1 mM (Prichard and Shulman, 1986). It has emerged, however, that under standard acquisition conditions brain lipids produce almost no ^{1}H MRS signal, presumably because the echo times used are too long for membrane-bound lipids. Thus there has been as yet little interest in using ^{1}H MRS to study *in vivo* brain lipid metabolism.

However, in the course of a proton MRS (^{1}H MRS) study to investigate changes in brain lactate concentration following electroconvulsive therapy (ECT) in human patients with major psychiatric illness, the authors observed an unexpectedly large resonance overlapping the lactate resonance region at short echo times (TE = 30 ms), but not at the longer echo times (TE = 270 msec) that are believed to be optimal for detecting lactate (Woods and Chiu, 1990). The observed broad signal is typical for the methylene protons of long-chain fatty acids. The first concern was that this was contamination from scalp lipids, but the repeated observation of this signal at about 1.2 ppm in the voxel of interest (VOI) only after ECT and not before, convinced the authors that it was genuine and that it was lipid (Woods and Chiu, 1991a). Since it had been observed by Bazan (1970, 1971) that seizures in experimental animals result in large, relatively long-lasting increases in brain diacylglycerol (DAG) and arachidonic acid (AA) as a result of a neurotransmitter-mediated breakdown of phosphatidylinositol (PI), the human ^{1}H MRS results are tentatively interpreted as arising from the same process. Although direct confirmation that PI breakdown is the major source of these observed lipid signals is lacking, we have subsequently obtained indirect evidence supporting this hypothesis from several different "model systems."

METHODS

Proton (^{1}H) spectroscopy was performed on a 1.5T Signa whole body imager (General Electric Medical Systems, Milwaukee, Wisconsin) using the STEAM (stimulated echo acquisition mode) protocol. The Signa system is equipped with a quadrature head coil and shielded gradient coils. The STEAM sequence uses three slice-selective radio frequency (RF) pulses; water suppression is achieved by means of an initial presaturation pulse prior to the first selective RF pulse and a second saturation pulse between the second and third RF pulse.

Informed consents are obtained from all participating subjects before the spectroscopy session. Subjects are supine on the scanner table and a reference position (the intersection of the axial and sagittal light beams) is located at the glabella. From a series of T1 weighted coronal and sagittal scout images (echo time TE = 30 msec, repetition time TR = 600 msec), one or more voxel of interest (VOI) are chosen for study. Good spectral resolution is achieved through shimming, which results in a typical water line width of 5-8 Hz (full width at half-maximum). Typically, 64 scans are collected at a TE = 30 msec and TR = 2000 msec with mixing time TM = 50 msec. The free induction decays (FID) are processed on the GE 1280 spectroscopy data station with exponential line broadening (1 Hz), zero-filled to 2K points, and Fourier transformed.

RESULTS

The first "model system" for study of free brain lipids is ECT. The basic observations are that at 1-1.5 hours after treatment a large broad spectral peak can be observed at about 1.2 ppm, and this peak diminishes (but may not return to baseline) in 32 hours (Woods and Chiu, 1990). If ECT is unilateral (i.e., if the stimulating current is primarily restricted to one cerebral hemisphere), the lipid peak is larger on the stimulated side.

The second example of increased lipid signals detectable with ^{1}H MRS is provided by poorly controlled unilateral temporal lobe seizures. In this case the relationship to time of seizure of the increased lipid signal is uncertain, but treatment with anticonvulsant is associated with a marked reduction of the lipid signal (Chiu T-M and Woods BT, unpublished data).

The third source of an observable focal lipid signal comes from a patient with gliosis bordering an area of prior resection of a benign tumor. In this patient, anticonvulsant medications also suppressed the lipid signal which, however, returned following medication withdrawal (Chiu T-M and Woods BT, unpublished data). The final example of an unexpected lipid increase comes from the ^{1}H MRS study of subjects ingesting alcohol (Mendelson JH, Chiu T-M, Woods BT, Mello NK, Teoh Sk, unpublished data). Not only can one observe a brain alcohol signal, but there appears to be an increased lipid signal (Fig. 1) that precedes any detectable alcohol signal.

DISCUSSION

Taken alone, each of these examples permits alternative interpretations as to the origin of the lipid signal. For example, ECT and seizures could produce localized increases in blood flow and leakage of serum triglycerides into brain parenchyma. On the other hand, the patient with gliosis actually had a study with single photon emission computed tomography that showed reduced regional blood flow. The changes with ECT could be due to an electrical current effect on membrane fluidity that results in increased membrane lipid "visibility" to ^{1}H MRS. Membrane fluidity could also explain the alcohol results (Chin and Goldstein, 1977), but this mechanism is irrelevant to the seizure and gliosis models. The simplest unifying explanation for all the results would seem to be that, in each case, breakdown of PI (in the phosphatidylinositol bisphosphate [PIP_2] form) is stimulated and releases free DAG and FFA in quantities that are "visible" to ^{1}H MRS.

If this proposed explanation based on mutual exclusion of alternative hypotheses is correct, then it might be concluded that certain specific properties of the PI second messenger system, namely responsiveness to many different effectors, marked signal amplification due to an enzymatic cascade, and the high MRS "visibility" of long-chain fatty acid molecules, all combine to make the PI system a sensitive *in vivo* marker for regional brain activity in man. If this were all that could be established, then the potential for clinically useful MRS would already be significantly advanced.

It is, however, plausible that observation with ^{1}H MRS of PI-system involvement in the above-described model systems may offer not only a marker of cell activity but also

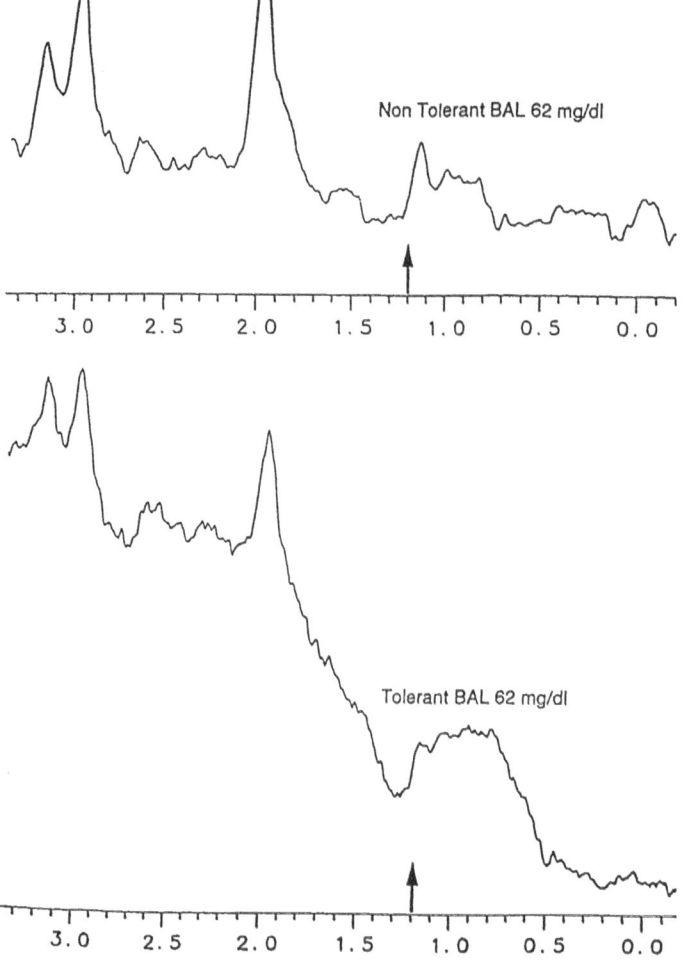

Figure 1. Magnetic resonance spectra of an ethanol peak (arrow) at 1.2 ppm for a non-tolerant subject (top panel) and dominance of a lipid peak at 1.2 ppm for a tolerant subject (bottom panel). Blood alcohol level (BAL) was identical for both subjects, 62 mg/dl.

insight into the mechanisms that explain a number of hitherto apparently unconnected clinical phenomena. Thus, the immediate and longer-term consequences of PI-system stimulation may offer partial explanations for 1) the efficacy of ECT in both mania and depression; 2) the propagation and evolution of seizures in the brain; 3) the deleterious long-term consequences of poorly controlled focal seizures; and 4) some aspects of alcohol tolerance and alcohol dependence.

Figure 2 shows a schematic diagram indicating the interaction of PI stimulation induced by ECT with one receptor putatively overactive in depression, and another overactive in mania. It is suggested that in both conditions the high level stimulation of the PI system results in an activation of protein kinase C (PKC) which, in turn, directly or indirectly downregulates membrane receptors (Chuang, 1989), including not only the

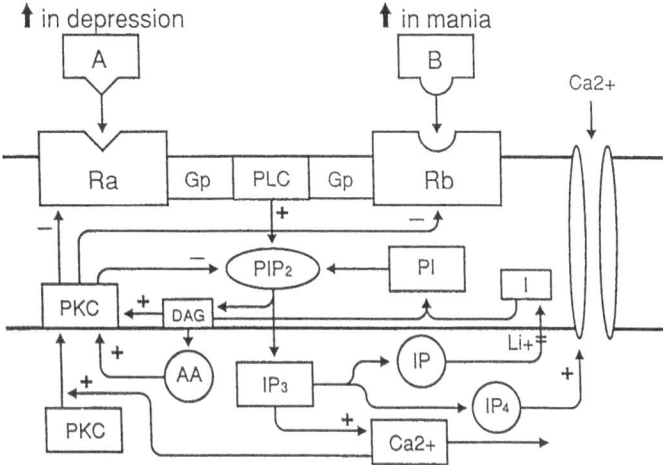

Figure 2. Schematic diagram of the phosphatidylinositol second messenger system and its hypothesized overactivity in both depression and mania. Abbreviations: AA - arachidonic acid; A, B = neurotransmitters; DAG = diacylglycerol; Gp = G protein; I = inositol; IP_3 = inositoltrisphosphate; IP_4 = inositoltetrakisphosphate; PI = phosphatidylinositol; PIP_2 = phosphatidylinositol bisphosphate; PKC = protein kinase C; PLC = phospholipase C; Ra, Rb = receptors. Arrows with + or — signs indicate stimulation or inhibition, respectively; unsigned arrows indicate reactions or actual movement.

overactive one but also the one mediating the inverse effect as well. The consequence is stabilization rather than a switch to the opposite condition. A similar mechanism has been proposed to explain the effectiveness of lithium (Berridge et al., 1982) but in the case of lithium the ion has to be present to maintain an inhibitory block of the PI system, whereas the PKC effect would be expected to last longer.

A second clinically significant condition in which PI-system metabolites may play a major short-term and long-term role is epilepsy. The short-term effects may come about because increased levels of free fatty acids increase membrane potassium (K^+) channel conductances (Ordway et al., 1991). Such changes in K^+ conductance may be of considerable relevance to both the propagation and evolution of seizure activity (Woods and Chiu, 1991b). First of all, an increase in K^+ outflow increases the local concentration of $[K^+]_o$, which in turn has a depolarizing effect on other local area presynaptic terminals, leading to further excitatory neurotransmitter release (Hablitz and Lundervold, 1981). Subsequently, however, the increase in local $[K^+]_o$ reaches a ceiling level (Heinemann et al., 1986), probably because of influx of K^+ into surrounding glia (Barres et al., 1990), which have large numbers of inward-rectifying K^+ channels (Barres et al., 1990). Finally, continued K^+ outflow from neurons has a hyperpolarizing effect, tending to terminate firing and thus local seizure activity (Gutnick et al., 1979).

The first, most rapid effect of increased fatty acid levels would be the increase in $[K^+]_o$ due to efflux from activated cells, and this would serve to spread the paroxysmal activity through depolarizing presynaptic terminals. The next, slower effect would be the efflux of arachidonate itself and/or its lipoxygenase metabolites out of activated cells (Lynch et al., 1989; Williams et al., 1989) to affect surrounding glia and stimulate inward-rectifying K^+ influx, leading to decreased $[K^+]_o$. Finally, the persistence of both effects together would lead to enhancement of neuronal hyperpolarization and termination of paroxysmal discharges.

As to long-term seizure effects, it is well known clinically that poorly controlled focal seizures tend to result in both local enlargement of the region of epileptic tissue and the development of distal foci in regions connected to the original area. It has also been shown in experimental models that excessive quantities of arachidonate may be associated with free radical formation that may exceed the capacity of naturally occurring anti-oxidants, resulting in cell injury or even death (Seisjö, 1981; Chan and Fishman, 1984). Moreover, since arachidonate and its metabolites can diffuse out of affected cells into the surrounding extracellular space (Lynch et al., 1989; Williams et al., 1989), other cells not directly stimulated may also be injured. Our findings of an increased lipid signal in a gliotic border region, even in the absence of observable seizures or ictal abnormalities on the electroencephalogram, might explain why gliotic areas can develop into active seizure foci long after the original trauma.

Finally, in the case of alcohol, there are several different aspects of its effect on brain that one might explain by invoking a major role for PI and free fatty acid metabolism. Acutely, alcohol produces a large number of responses at the cellular level, several of which taken together have the physiological effect of decreasing neuronal excitation and repetitive firing. For example, alcohol increases GABA-stimulated chloride (Cl^-) influx (Allan and Harris, 1987) and Ca^{2+}-dependent K^+ efflux (Carlen et al., 1982), while mobilizing intracellular Ca^{2+} ($[Ca^{2+}]_i$) and decreasing the influx of extracellular Ca^{2+} ($[Ca^{2+}]_o$) (Carlen et al., 1982). Since the increase in $[Ca^{2+}]_i$ with acute alcohol exposure also enhances the hydrolysis of PI leading to release of DAG and AA as well as cytosolic Ca^{2+} (Eberhard and Holz, 1988) and, as noted above, increases in AA and metabolites directly increase K^+ conductance, all these physiological effects may be closely interlinked.

Furthermore, acute alcohol results in increased levels of cyclic-AMP, via enhancement of the effects of adenosine at A_2 receptors (Diamond, 1990) and norepinephrine at ß-adrenergic receptors (Saito et al., 1987). Since PI hydrolysis and AA release cause increased PKC activity (Fig. 2) and increased PKC increases receptor-initiated generation of cAMP (Berridge et al., 1987), these responses to alcohol may also be closely linked.

Continued exposure to alcohol results in decreases in all of these acute effects (i.e., tolerance). In the case of decreased Ca^{2+} influx, tolerance is actually accompanied by increased numbers of voltage-sensitive calcium channels (VSCCs), and this increase can be prevented by substances blocking PKC effects (Gonzales and Hoffman, 1991). Once this and other changes take place, alcohol withdrawal is often accompanied by seizures (dependence state). If tolerance is also accompanied by decreased stimulation of PI hydrolysis and arachidonate release by alcohol, then one can infer that the processes by which this aspect of tolerance occurs may also play a role in alcohol dependence, including withdrawal seizures.

CONCLUSION

Proton spectroscopic observation of several *in vivo* human examples of seizure activity and of the effects of alcohol on brain tissue indicate that in each case significant transient elevations of brain lipids occur. Although there is as yet no direct evidence that these lipids are, in fact, free fatty acids, the large body of literature based on *in vivo* animal models and *in vitro* model systems makes that assumption highly plausible.

REFERENCES

Allan A and Harris R (1987) Acute and chronic ethanol treatments alter GABA receptor-operated chloride channels. Pharmacol Biochem Behav 27:665-670.

Barres BA, Chun LLY, Corey DP (1990) Ion channels in vertebrate glia. Annu Rev Neurosci 13:441-474.

Bazan NG (1970) Effects of ischemia and electroconvulsive shock on free fatty acid pool in the brain. Biochim Biophys Acta 218:1-10.

Bazan NG (1971) Changes in free fatty acids of brain by drug-induced convulsions, electroshock and anesthesia. J Neurochem 18:1379-1385.

Berridge MJ (1987) Inositol trisphosphate and diacylglycerol: two interacting second messengers. Annu Rev Biochem 56:159-193.

Berridge MJ, Downes CP, Hanley MR (1982) Lithium amplifies agonist-dependent phosphatidylinositol responses in brain and salivary glands. Biochem J 206:587-595.

Carlen PL, Gurevich N, Durand D (1982) Ethanol in low doses augments calcium-mediated mechanisms measured intracellularly in hippocampal neurons. Science 215:306-309.

Chan PH and Fishman RA (1984) The role of arachidonic acid in vasogenic brain edema. Fed Proc 43:210-213.

Chin JH and Goldstein DB (1977) Effects of low concentrations of ethanol on the fluidity of spin-labeled erythrocyte and brain membranes. Mol Pharmacol 13:435-441.

Chuang D-M (1989) Neurotransmitter receptors and phosphoinositide turnover. Annu Rev Pharmacol Toxicol 29:71-110.

Diamond I (1990) Models for the investigation of alcohol withdrawal seizures. In: Alcohol and seizures (Porter RK, Hattson RH, Cramer JA, Diamond I, eds), Philadelphia: F.A. David Co.

Eberhard DA and Holz RW (1988) Intracellular Ca^{++} activates phospholipase C. Trends Neurosci 11:517-520.

Gonzales RA and Hoffman PA (1991) Receptor-gated ion channels may be selective CNS targets for ethanol. Trends Pharmacol Sci 12:1-3.

Gutnick MJ, Heinemann U, Lux HD (1979) Stimulus induced and seizure related changes in extracellular potassium concentration in cat thalamus (VPL). Electroencephalogr Clin Neurophysiol 47:329-344.

Hablitz JJ and Lundervold A (1981) Hippocampal excitability and changes in extracellular potassium. Exp Neurol 71:410-420.

Heinemann U, Konnerth A, Pumain R, Wadman WJ (1986) Extracellular calcium and potassium concentration changes in chronic epileptic brain tissue. In: Basic mechanisms of the epilepsies (Delgado-Escueta AV, Ward AA, Woodbury DM, Porter RJ, eds), V. 44. New York: Raven Press.

Hokin LE and Hokin MR (1958) Acetylcholine and the exchange of inositol and phosphate in brain phosphoinositide. J Biol Chem 233:818-821.

Lynch MA, Errington ML, Bliss TVP (1989) Nordihydroguaiaretic acid blocks the synaptic component of long-term potentiation and the associated increases in release of glutamate and arachidonate: an *in vivo* study in the dentate gyrus of the rat. Neuroscience 30:693-701.

Ordway RW, Singer JJ, Walsh JV Jr (1991) Direct regulation of ion channels by fatty acids. Trends Neurosci 14:96-100.

Prichard JW and Shulman RG (1986) NMR spectroscopy of brain metabolism *in vivo*. Annu Rev Neurosci 9:61-85.

Saito T, Lee JM, Hoffman PL, Tabakoff B (1987) Effects of chronic ethanol treatment on the beta-adrenergic receptor-coupled adenylate cyclase system of mouse cerebral cortex. J Neurochem 48:1817-1822.

Seisjö BK (1981) Cell damage in the brain: A speculative synthesis. J Cereb Blood Flow Metab 1:155-185.

Weiner MW and Hetherington HP (1989) The power of the proton. Radiology 172:318-320.

White RP (1989) Cerebrospinal fluid eicosanoids as an index of cerebrovascular status. In: Acid metabolism in the nervous system (Barkai AI, Bazan NG, eds), New York: New York Academy of Sciences.

Williams JH, Errington ML, Lynch MA, Bliss TVP (1989) Arachidonic acid induces a long-term activity-dependent enhancement of synaptic transmission in the hippocampus. Nature 341:739-742.

Woods BT and Chiu T-M (1990) *In vivo* ^{1}H spectroscopy of the human brain following electroconvulsive therapy. Ann Neurol 28:745-749.

Woods BT and Chiu T-M (1991a) Reply to comment on "Electroconvulsive Therapy and Brain Lipids." Ann Neurol 30:429-430.

Woods BT and Chiu T-M (1991b) Effects of fatty acid elevation on ictal activity (letter). Trends Neurosci 14:405.

NETWORK OF SIGNAL TRANSDUCTION PATHWAYS INVOLVING LIPIDS:

PROTEIN KINASE C-DEPENDENT AND -INDEPENDENT PATHWAYS

R. Bell, D. Burns,
T. Okazaki, and Y. Hannun

Departments of Biochemistry and Medicine
Duke University Medical Center
Durham, NC 27710

INTRODUCTION

The cellular glycerol- and sphingolipids are turning out to play not only fundamental roles in cellular and organellar membranes but also new and exciting roles in cellular regulation. The great diversity and complexity of the glycerolipids and sphingolipids present in mammalian cells had long portended important cellular functions, yet few molecular clues were uncovered until protein kinase C was discovered by Nishizuka (1986) to be regulated by sn-1,2-diacylglycerol (DAG), a second messenger derived by cleavage of phosphatidylinositol 4,5-bisphosphate (PIP_2). Glycerolipids turned out to play two major roles in protein kinase C regulation: i) as essential phospholipid cofactors and ii) as lipid second messenger activators, DAG (Bell, 1986). In 1985, detailed structure-activity studies on protein kinase C using advanced mixed micellar methods to investigate specificity and stoichiometry of phospholipid cofactor and DAG activator function produced molecular insight into the mechanism of regulation (Hannun et al., 1985); these studies also led to the discovery that sphingosine inhibits protein kinase C activity and blocks [3H]phorbol 12,13-dibutyrate ([3H]PDBu) binding (Hannun et al., 1986). The work opened up a growing field of research involving the role of sphingolipid breakdown products in normal cellular regulation and perturbation of these normal pathways in disease and pathobiology (Hannun and Bell, 1987; 1989).

The objective of this article is to focus on the structure, function, and regulation of protein kinase C, with special emphasis on the role of the cellular glycerol- and sphingolipids in these processes. The recent discovery of a sphingomyelin cycle operating in human cells (Hannun and Bell, 1989) has opened research into new lipid second messengers/lipid mediators that clearly are independent of protein kinase C. The roles of glycerolipids and sphingolipids in signal transduction processes are summarized in Tables 1 and 2 along with references to key articles. The present data permit some speculation about the functions of and interrelationships between these lipid-dependent pathways.

Table 1. Role of glycerolipids in signal transduction

Lipid	Lipid second messenger/ mediator	Key enzyme-producing effector	Target of effector lipid	References
PIP$_2$	DAG	PIP$_2$	PKC	(1)(2)(3)(4)
PC	PA	PLD	GAP or other	(5)
	DAG	PLD and PA phosphatase	PKC	(6)
	PAF	PLA$_2$	PAF receptor	(7)
PC	C$_{20:4}$ →	PLA$_2$	Multiple receptors	(8)
PI				
PE	Eicosanoids			(9)

References: (1) Nishizuka, 1986; (2) Bell, 1986; (3) Bell and Burns, 1991; (4) Nishizuka, 1989; (5) Tsai et al., 1990; (6) Exton, 1990; (7) Prescott et al., 1990; (8) Laychock, 1982; (9) Moncada, 1982. PIP$_2$, phosphatidylinositol 4,5-bisphosphate; DAG, sn-1,2-diacylglycerol; PKC, protein kinase C; PC, phosphatidylcholine; PA, phosphatidic acid; PLD, phospholipase D; GAP, GTPase activating protein; PAF, platelet-activating factor; PLA$_2$, phospholipase A$_2$; PI, phosphatidylinositol; PE, phosphatidylethanolamine.

Table 2. Emerging roles of sphingolipid in signal transduction processes

Lipid	Lipid second messenger/ mediator	Key enzyme	Target of effector lipid	References
SM and other SL	Sphingosine	SMase-ceramide	PKC	(1)
			Cam kinase	(2)
			PLD	(3)
			EGF receptor kinase	(4)
			Others	(1)
SM	Ceramide	SMase	Unknown	(5)
GM3	GM3		EGF receptor	(6)
			PDGF receptor	(7)
SM	Lysosphingo-myelin	??	Ca^{2+} mobilization	(8)

References: (1) Hannun and Bell, 1989; (2) Jefferson and Schulman, 1988; (3) Lavie et al., 1990; (4) Davis et al., 1988; (5) Okazaki et al., 1990; (6) Bremer et al., 1986; (7) Bremer et al., 1984; (8) Ghosh et al., 1990. SM, sphingomyelin; SL, sphingolipids; EGF, epidermal growth factor; PDGF, platelet-derived growth factor; GM3, ganglioside M3.

MATERIALS AND METHODS

Protein kinase C was purified from rat brain or from the baculovirus/Sf9 insect cell expression system (Burns et al., 1990). Protein kinase C was assayed using mixed micellar methods (Hannun et al., 1985). [3H]PDBu (New England Nuclear) binding was determined as described by Hannun et al. (1986).

HL60 cells were cultured and treated with vitamin D_3 as described by Okazaki et al. (1989). The metabolism of sphingomyelin in HL60 cells, as well as the synthesis and use of cell-permeable ceramides, has been described previously (Okazaki et al., 1989; 1990).

SUMMARY OF RESULTS

The Family Members — Protein Kinase C

The phospholipid- and diacylglycerol (phorbol ester)-regulated serine/threonine protein kinases belonging to the protein kinase C family are now known to number nine (Bell and Burns, 1991). While sharing some common characteristics, these enzymes should be considered as closely related family members having distinct and distinguishable functions. The exact physiological role of the individual protein kinase C family members is under intense investigation now that appropriate oligonucleotide and antibody tools are becoming available. Links between these protein kinase C members and human disease are also being sought.

Protein Kinase C — Domain Structure and Function

The primary structures of the members of the protein kinase C family have been inferred from cDNA sequences. These are compared in Figure 1. The distinguishing structural features are a pseudosubstrate domain, and one or two cysteine-rich regions which resemble zinc binuclear clusters (Pan and Coleman, 1990) (Fig. 2). The cysteine-rich regions of protein kinase C also contain the high-affinity phorbol ester binding sites (Ono et al., 1989a; Burns and Bell, 1991). The C2 region confers calcium sensitivity on the protein kinase C family members that possess this region (Ono et al., 1989b). The catalytic domain possesses ATP and protein substrate binding sites. The regulatory and catalytic domains are separated by a variable hinge region.

Model of Protein Kinase C Regulation

In Figure 3, a current model (Bell and Burns, 1991) of protein kinase C regulation is depicted. In this model, the pseudosubstrate site is shown bound to the protein substrate binding site when protein kinase C is inactive. This inactive state of the enzyme can occur when the enzyme is bound to the membrane or in the cytoplasm. Activation is believed to occur when the membrane-bound form of the enzyme binds diacylglycerol or phorbol ester. Binding causes a conformational change that dislocates the pseudosubstrate site from its protein substrate binding site and renders the enzyme active. The hinge region allows for the flexibility between these two domains and also accounts, in part, for the increased sensitivity of the enzyme to proteolytic degradation when it is activated.

Role of Lipids In Activation

Since the early observation that membrane factors were required for the activation

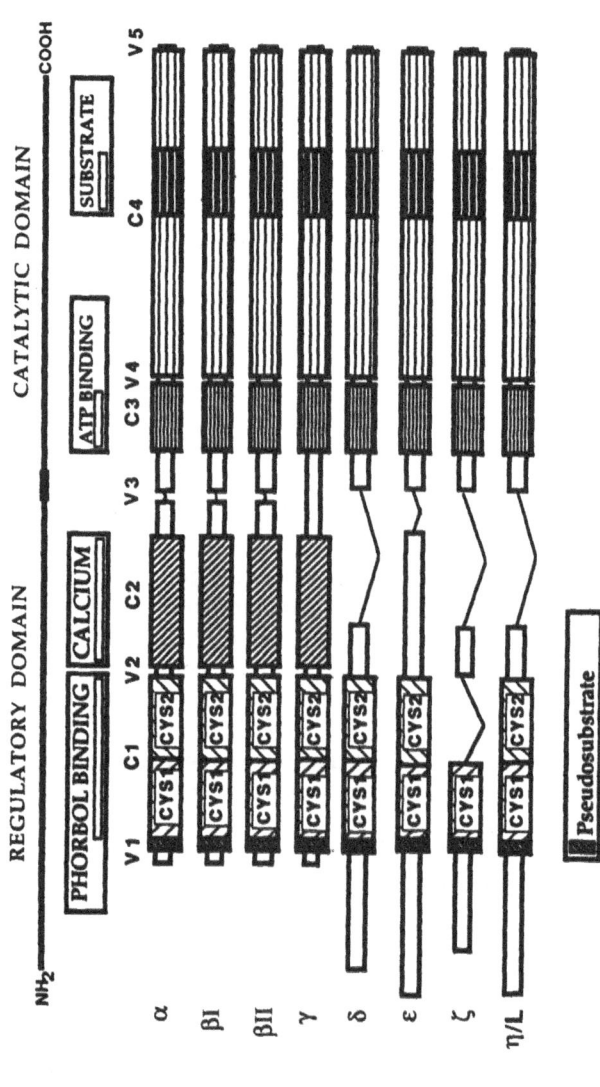

Figure 1. The protein kinase C family members. The nine members of the protein kinase C family are depicted. Important domains are highlighted. V, variable regions of the PKC molecule; C, conserved regions of the PKC molecule (Bell and Burns, 1991).

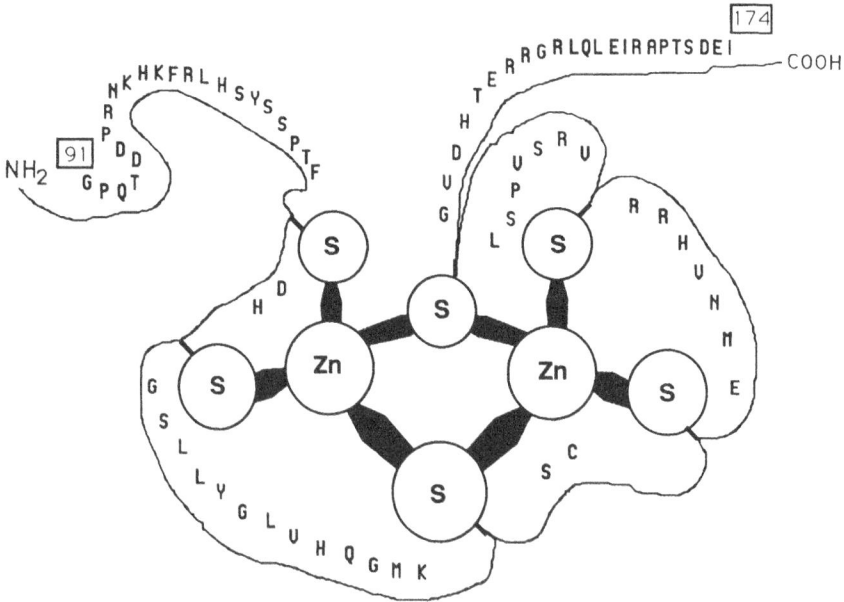

Figure 2. The cysteine-rich region of protein kinase C resembles a zinc binuclear cluster. The second cysteine-rich region of a protein kinase C molecule (amino acids 91-174) is depicted as a potential $Zn(II)C_6$ binuclear cluster similar to the *GAL4* transcription factor (Pan and Coleman, 1990; Bell and Burns, 1991).

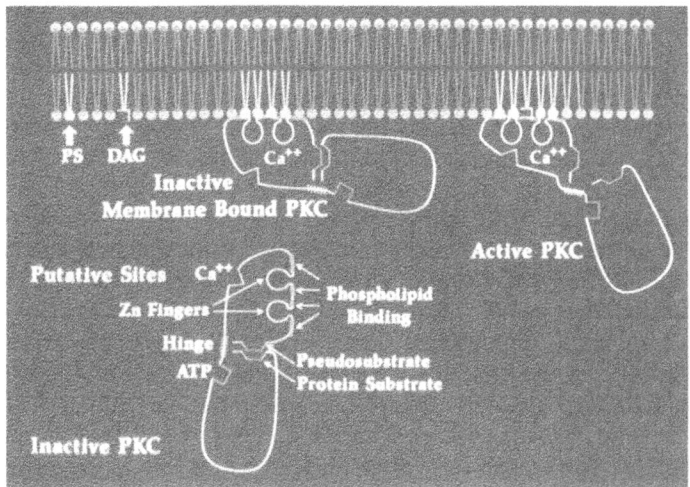

Figure 3. Activation of protein kinase C. Putative sites on a Ca^{2+}-dependent member of the PKC family are marked with arrows. In this model, protein kinase C becomes active when the membrane-bound form of the enzyme binds diacylglycerol or phorbol ester. This binding induces a conformational change in the enzyme causing the dislocation of the pseudosubstrate site from the substrate binding site (Bell and Burns, 1991).

279

of protein kinase C (Takai et al., 1979) and the discovery that these membrane factors were, in fact, phospholipids and diacylglycerols, investigators have been interested in the molecular detail of these critical lipid-protein interactions. At present, a multiplicity of protein kinase C lipid interactions has been described. The principal interactions involve stereospecific interactions with phosphatidyl-L-serine (Lee and Bell, 1989) and stereospecific interactions with *sn*-1,2-diacylglycerol (Nishizuka, 1986). There also appear to be non-specific interactions with anionic phospholipids (Lee and Bell, unpublished data). Protein kinase C-lipid interactions are either calcium dependent or calcium independent depending on the family member. Critical activation events occur upon binding of diacylglycerol to the regulatory domain. Fatty acids (Sekiguchi et al., 1987) and PIP$_2$ also activate protein kinase C *in vitro* (Chauhan et al., 1989; Lee and Bell, 1991).

Procedures to Investigate Lipid — Protein Interactions

Since phospholipids spontaneously self-aggregate to form vesicles or non-physically defined aggregates, mixed micellar methods were developed to provide a physically defined aggregate to investigate lipid-protein interactions (Hannun et al., 1985). Systematic and independent variation of the number of phospholipid cofactors and of diacylglycerol activator molecules is possible in mixed micelles. Such studies provided useful mechanistic insights; it was inferred that four or more phosphatidylserine interaction sites existed and that a single molecule of diacylglycerol per micelle was capable of activating monomeric protein kinase C (Bell and Burns, 1991).

Protein Kinase C Domains Possessing Significant Phospholipid Protein Interactions

A series of truncations, deletions, and fusions of cDNAs encoding protein kinase C were constructed to locate regions of significant protein-lipid interactions. These regions were confined to a subdomain of the regulatory domain of the enzyme. It was demonstrated that fusions or deletions expressing both the first and second cysteine-rich regions (cys1 and cys2) bound phorbol esters with high-affinity. Surprisingly, it was found that when either the cys1 or the cys2 regions were expressed by themselves as fusion peptides with ß-galactosidase or as deletions with protein kinase C, high affinity phorbol ester binding persisted (Burns and Bell, 1991). This clearly indicates that there are two phorbol ester binding regions within the cysteine-rich regions of protein kinase C.

Polypeptide segments of 83 amino acids were capable of high-affinity phorbol ester binding. These contained six cysteine molecules which most closely resemble the zinc binuclear cluster of the *GAL4* transcription factor (Pan and Coleman, 1990). Studies on zinc binding to native protein kinase C were undertaken. In preliminary experiments (Bell and Bloomenthal, unpublished data), it became clear that protein kinase C preparations contain zinc. While the preliminary stoichiometry of 4 moles of zinc per mole of protein kinase C is consistent with the zinc binuclear cluster, present data are inadequate to allow firm conclusions as to the number of zinc molecules per mole of protein kinase C. These studies have not advanced far enough to describe the function of zinc within the protein kinase C molecule. Clearly, the role of zinc within protein kinase C will be a topic for future investigations.

Activation of Protein Kinase C by PIP$_2$

Protein kinase C has been reported to be activated by PIP$_2$ (Chauhan et al., 1989). Phosphatidylserine and calcium were required for the PIP$_2$-dependent activation. In detailed studies on the mechanism of this activation, Lee and Bell (1991) found that the activation resembled, in part, the activation by diacylglycerol. However, unlike earlier studies by Brockeroff and colleagues (Chauhan et al., 1989), PIP_2 was found not to displace phorbol esters. This observation has been confirmed by Huang and coworkers (Huang and Huang, 1991). These data suggest that protein kinase C contains a PIP_2 site distinct from the phorbol ester (DAG) binding site. In that homologies between the C2 region of protein kinase C and PIP_2-specific phospholipase C were noted (Baker, 1989), the C2 region is a likely site of PIP_2 interaction. Detailed molecular biological constructs are likely to define the site of PIP_2 interaction. Interestingly, the non-calcium-dependent members of the protein kinase C family do not contain the C2 region. A clear prediction of this is that the calcium-independent members of the protein kinase C family will not be activated *in vitro* by PIP_2, because they do not contain a C2 region. The physiological significance of PIP_2 activation of protein kinase C is not clear at the present time. The multiplicity and specificity of significant lipid-protein interactions with protein kinase C are just beginning to define a set of potential interactions for physiological regulation.

Sphingosine as an Inhibitor of Protein Kinase C

Sphingosine, the basic building block of the cellular sphingolipids, was discovered to inhibit protein kinase C in a competitive manner with respect to diacylglycerol (Hannun and Bell, 1987; 1989). In addition, sphingosine blocked the binding of phorbol ester to protein kinase C/Triton X-100/phosphatidylserine (PS) mixed micelles, but did not block the interaction of protein kinase C with the PS/Triton X-100 mixed micelles in the presence of calcium. This indicates that sphingosine inhibits the significant interactions with phorbol esters without blocking the interactions between protein kinase C and the phospholipid cofactors. In addition, Hannun and Bell were able to show that any deacylated sphingolipid (lysosphingolipids) would also inhibit protein kinase C activity *in vitro* (Hannun and Bell, 1987). This raised the possibility of physiological regulation of protein kinase C by sphingosine/lysosphingolipids; the latter was of particular interest because lysosphingolipids accumulate in certain of the sphingolipidoses. The hypothesis that sphingosine and lysosphingolipid inhibition of protein kinase C signal transduction processes represents the missing pathobiological link in the sphingolipidoses was put forth (Hannun and Bell, 1987).

Sphingomyelin Cycle

These hypotheses of physiological regulation of protein kinase C and signal transduction processes involving the breakdown of the cellular sphingolipids led to the search for regulated sphingolipid cycles within cells. The first such cycle was discovered in a human promyelocytic leukemic cell line, HL60 cells, in response to vitamin D_3. Vitamin D_3 induces the differentiation of these cells toward the macrophage lineage. Treatment of these cells with physiological levels of vitamin D_3 (1,25 dihydroxyvitamin D_3) led to a transient loss of sphingomyelin. The production of ceramide and phospho-choline reached a maximum when the cellular sphingomyelin content reached a minimum at 2 hr; at 5-6 hr, the level of sphingomyelin had returned to normal. Thus, vitamin D_3 caused a transient degradation of sphingomyelin and defined the first sphingolipid cycle, the "sphingomyelin cycle" in cells (Okazaki et al., 1989).

Studies were undertaken to discern whether the turnover of sphingomyelin is actually involved in the differentiation of these cells, and not just an epi-phenomenon to the addition of vitamin D_3. The definitive studies were done by showing that cell-permeable ceramides could, in the absence of vitamin D_3, cause differentiation of these cells (Okazaki et al., 1990). This implies that ceramide is biologically active. Ceramide could form a new class of lipid mediators or second messengers. Further, it was shown that cell-permeable ceramides were not hydrolyzed to sphingosine. Sphingosine by itself could not cause differentiation. Moreover, the cell-permeable ceramide had to be present in the cells for only 2-4 hr, the period that ceramide was present in HL60 cells in response to vitamin D_3. All of these observations support the physiological significance of the "sphingomyelin cycle" (Okazaki et al., 1990).

Vitamin D_3-Inducible Sphingomyelinase

Since the metabolic studies had definitively established that the turnover of sphingo-myelin involved a sphingomyelinase of the C type, studies were initiated to define whether there were significant changes in neutral sphingomyelinase activity in HL60 cells; indeed, a neutral vitamin D_3-inducible sphingomyelinase activity was discovered (Okazaki et al., 1989). The activity of this sphingomyelinase mirrored the formation of ceramide in the turnover of sphingomyelin that had been observed in response to vitamin D_3 In crude extracts, a change in this activity of up to 60-80% was noted; following DEAE column chromatography, larger changes in the vitamin D_3-inducible sphingomyelinase activity were observed—up to 3- to 5-fold (Okazaki, Bell, and Hannun, unpublished data). Furthermore, recent studies have shown that other effectors, including tumor necrosis factor (TNF) and interferon gamma (Kim et al., 1991), are able to effect sphingomyelin turnover in HL60 cells. This indicates that a multiplicity of effector systems can stimulate the cycle.

Structure-Activity Relationship on Cell-Permeable Ceramides

A number of ceramide analogues have been synthesized with physical properties similar to the cell-permeable ceramides shown to be effective in the differentiation of HL60 cells. These molecules are beginning to describe structure-activity relationships that are critical for ceramide to exert its biological actions (Hannun and Bell, unpublished data). The biological effects of ceramides on cells raise the question as to how ceramides elicit their pharmacologic activities. What is the target for ceramide within the cell? Of course, protein kinases may be involved. However, at the moment there is no clear evidence in support of this hypothesis. A search for the ceramide-sensitive component of the signal transduction pathway is underway.

CONCLUSIONS AND PERSPECTIVES

Clearly an abundance of data implicates the cellular glycerol- and sphingolipids in the critical regulation of protein kinase C and other intracellular signal transduction components. The recent findings of the role of sphingolipids and sphingolipid break-down products in cellular regulation are particularly intriguing given the extreme complexity of this class of lipids. The high abundance of sphingolipids in the central nervous system (CNS) may suggest a key role of sphingolipids and their metabolites in the higher-order signaling processes of the CNS. These, of course, may be intimately linked to the pathology seen in the sphingolipidoses. In addition, there are a couple of interesting new links to toxins, sphingolipid metabolism, and signal transduction processes. Merrill and coworkers discovered that the fumonisins, which are sphingosine analogues, inhibit ceramide synthase (Wang et al., in press). This inhibition leads to the accumulation of dihydrosphingosine in cells. This work has created an opportunity to investigate the mechanism of the pathogenesis of these molecules. Fumonisins have been linked to diseases such as porcine pulmonary edema (Harrison et al., 1990), rat liver cancer (Gelderblom et al., 1988), and esophageal cancer (Marasas, 1982), as well as other disease states (Marasas et al., 1984). In addition, the dermatonecrotic factor of the brown recluse spider venom has been unequivocally shown by King and coworkers (Rees et al., 1988) to be nothing more than a sphingomyelinase D. This implicates one of its products, ceramide phosphate, in pathogenesis. However, the role of ceramide phosphate in signal transduction processes is not clear at the present time. There are other tentative, interesting links to sphingosine metabolites and signal transduction processes. These have stemmed from the work of Ghosh and Gill, who showed that lysosphingomyelin may cause calcium mobilization in smooth muscle cells (Ghosh et al., 1990). In addition other sphingolipid metabolites studied by Speigel and coworkers may be involved in the mitogenic process in 3T3 cells (Zhang et al., 1990). All of these are, of course, tantalizing links to the role of sphingolipids in cellular regulation. Over the next five years, we are likely to see a number of these and other tantalizing leads to biology tied to molecular mechanisms involving sphingolipid metabolites in signal transduction processes and pathogenesis. The accumulation of sphingolipids as prominent tumor antigens may be caused by defects in the production or metabolism of these molecules. In the tumorigenic process, it has been hypothesized that the sphingolipids produce negative regulatory molecules and the progression of tumors involves the loss of these negative regulatory molecules through the loss of anti-oncogenes or tumor suppressor genes. All of these are, of course, interesting speculations awaiting clear and insightful experiments.

Protein kinase C has served and continues to serve to unify once-divergent fields of research. Clearly the union of tumor promotion, calcium mobilization, the phosphatidyl-inositol cycle, and the now clear connection to the roles of sphingolipids in cellular regulation continues to demonstrate that protein kinase C exists as a pivotal cellular regulator. Further detailed molecular analysis of the role of lipids as cofactors and lipids as regulators of protein kinase C is a promising area of research. Penetrating insights into cellular regulation by sphingolipid metabolism may occur because of the production of significant tools; these include radiolabeled sphingosine as well as cell-permeable ceramides and inhibitors to perturb these metabolites in living cells. Also, quantitative methods for sphingosine and ceramide have been developed. It appears that this area of research is poised for rapid progress.

REFERENCES

Baker M (1989) Mol Cell Endocrinol 61:129.
Bell RM (1986) Cell 45:631.
Bell RM and Burns DJ (1991) J Biol Chem 266:4461.
Bremer E, Hakomori S, Bowen-Pope D, Raines E, Ross R (1984) J Biol Chem 259:6818.
Bremer E, Schlessinger J, Hakomori S (1986) J Biol Chem 261:2434.
Burns DJ and Bell RM (1991) J Biol Chem 266:18330.
Burns DJ, Bloomenthal J, Lee M, Bell RM (1990) J Biol Chem 265:12044.
Chauhan A, Chauhan V, Deshmukh D, Brockerhoff H (1989) Biochemistry 28:4952.
Davis R, Girones N, Faucher M (1988) J Biol Chem 263:5373.
Exton J (1990) J Biol Chem 265:1.
Gelderblom W, Jaskiewicz K, Marasas W, Thiel P, Horak R, Uleggar R, Kriek N (1988) Appl Environ Microbiol 54:1806.
Ghosh T, Bian J, Gill D (1990) Science 248:1653.
Hannun Y and Bell RM (1987) Science 235:670.
Hannun Y and Bell RM (1989) Science 243:500.
Hannun Y, Loomis C, Bell RM (1985) J Biol Chem 260:10039.
Hannun Y, Loomis C, Bell RM (1986) J Biol Chem 261:12604.
Harrison L, Colvin B, Greene J, Newman L, Cole J (1990) J Vet Diagn Invest 2:217.

Huang F and Haung KP (1991) J Biol Chem 266:8727.

Jefferson A and Schulman H (1988) J Biol Chem 263:15241.

Kim M, Kinardic C, Obeid L, Hannun Y (1991) J Biol Chem 266:484.

Lavie Y, Piterman O, Liscovitch M (1990) FEBS Lett 277:7.

Laychock S (1982) Cell Calcium 3:43.

Lee M and Bell RM (1989) J Biol Chem 264:14797.

Lee M and Bell RM (1991) Biochemistry 30:1041.

Marasas W (1982) Products in oesophageal cancer areas in Transkei. In: Cancer of the oesophagus (Pfeidder C, Ed) Vol 1, pp 29. Boca Raton: CRC Press.

Marasas W, Kriek N, Fincham J, van Rensburg S (1984) Int J Cancer 34:383.

Moncada S (1982) Br J Pharmacol 76:3.

Nishizuka Y (1986) Science 233:305.

Nishizuka Y (1989) JAMA 262:1826.

Okazaki T, Bell RM, Hannun Y (1989) J Biol Chem 264:19076.

Okazaki T, Bielawaska A, Bell RM, Hannun Y (1990) J Biol Chem 265:15823.

Ono Y, Fujii T, Igarashi K, Kono T, Tanaka C, Kikkawa U, Nishizuka Y (1989a) Proc Natl Acad Sci USA 86:4868.

Ono Y, Fujii T, Ogita K, Kikkawa U, Nishizuka Y (1989b) Proc Natl Acad Sci USA 86:3099.

Pan T and Coleman J (1990) Proc Natl Acad Sci USA 87:2077.

Prescott S, Zimmerman G, McIntyre T (1990) J Biol Chem 265:17381.

Rees R, Nanney L, Yates R, King Jr L (1988) Toxicon 26:1035.

Sekiguchi K, Tsukuda M, Ogita K, Kikkawa U, Nishizuka Y (1987) Biochem Biophys Res Commun 145:797.

Takai Y, Kishimoto A, Iwasa Y, Kawshara Y, Mori T, Nishizuka Y (1979) J Biol Chem 254:3692.

Tsai M, Yu C, Wei F, Stacey D (1990) Science 250:982.

Wang E, Norred W, Bacon C, Riley R, Merrill A (in press) J Biol Chem.

Zhang H, Desai N, Murphey J, Spiegel S (1990) J Biol Chem 265:21309.

CONSERVATION OF DOCOSAHEXAENOIC ACID IN THE RETINA

R.E. Anderson,[1,2] P.J. O'Brien,[3] R.D. Wiegand,[1]
C.A. Koutz,[1] and A.M. Stinson[2]

Departments of [1]Ophthalmology and [2]Biochemistry
Baylor College of Medicine
Houston, TX 77030

[3]Health Research Associates
Rockville, MD 20850

Docosahexaenoic acid (22:6n-3), a polyunsaturated fatty acids (PUFA) of the n-3 family, makes up about half of the total fatty acids of vertebrate rod outer segment (ROS) phospholipids (see review by Fliesler and Anderson, 1983). This level of 22:6n-3 is among the highest reported for any membrane. The reason for the large amount of 22:6n-3 in ROS is not known, although it has been observed that membranes that are active metabolically, as in ROS, mitochondria, sperm, and synaptic vesicles, have high levels of PUFAs, while relatively inactive membranes such as myelin have low PUFA levels (Dratz and Deese, 1986; see books edited by Simopoulos et al., 1986 and Lands, 1987). Over the past 15-20 years, evidence has accumulated suggesting that 22:6n-3 is important in maintaining the normal structure and function of the retina.

DYNAMICS OF 22:6n-3 IN THE RETINA

ROS are dynamic structures whose lipid and protein components are constantly being renewed; mammalian retinas may replace as much as 10% of their ROS mass each day (Young, 1976). Since 22:6n-3 and its shorter chain precursors are essential, animals must be provided with a dietary source of these polyunsaturates in order to maintain the high levels found in retinal membranes. Thus, it should be possible to alter the composition of the ROS phospholipids by feeding diets deficient in n-3 and/or n-6 fatty acids, because of the large amount of membrane material renewed each day in the vertebrate retina. However, retinas (Futterman et al., 1971; Anderson et al., 1977; Tinoco, 1982) and ROS (Anderson and Maude, 1972; Anderson et al., 1977) of rats raised on fat-free diets did not lose their n-3 PUFA to the same extent as non-neural tissues. Deprivation of n-3 fatty acids in rats, chicks, and monkeys resulted in partial replacement of 22:6n-3 in the retina (Benolken et al., 1973; Anderson et al., 1977; Neuringer et al., 1986; Anderson et al., 1989; 1990) and brain (Galli et al., 1971; Neuringer and Conner, 1986; Connor et al., 1990; Anderson et al., 1989; 1990; Lin et al., 1990) by 22:5n-6. However, returning some of these animals to diets containing n-3 fatty acids resulted in replacement of 22:5n-6 with 22:6n-3 in both tissues (Benolken et al., 1973; Connor et al., 1990; Anderson et al., 1989; 1990). Thus, the retina and brain have a preference for 22:6n-3 over n-6 PUFA and each has a mechanism for conservation of 22:6n-3 during essential fatty acid (EFA) deficiency.

CONTROL OF 22:6n-3 IN THE RETINA

Several hypotheses have been proposed to explain the conservation of 22:6n-3 in the retina during n-3 deprivation (Wiegand et al., 1991; Stinson et al., in press).

1) The retina is capable of *de novo* synthesis of 22:6n-3.
2) The retina can reduce the requirements for 22:6n-3 by either shortening the lengths of the ROS or reducing the number of photoreceptor cells.
3) The rate of renewal of rod outer segment membranes is decreased in the absence of a dietary source of 22:6n-3.
4) There is a reduction in the turnover rate of molecular species containing 22:6 n-3 in the ROS during EFA deficiency, which is independent of turnover of integral proteins.
5) There is a reduction in the turnover rate of 22:6n-3 esterified in phospholipids of ROS membranes.
6) The retina can selectively sequester the small amounts of 22:6n-3 that are present in the plasma, even during prolonged EFA deficiency.
7) There is a recycling of 22:6n-3
   a) between the retina, the retinal pigment epithelium (RPE), and the blood,
   b) between the retina and RPE, or
   c) within the retina.

TESTS OF THE HYPOTHESES

Hypothesis 1

The possibility that the retina is unique among animal tissues and could synthesize 22:6n-3 *de novo* was tested in our laboratory (Dudley, 1976). Rats raised for 10 weeks on fat-free or lab chow diets were injected with ^{3}H-acetate and the incorporation into retina fatty acids was determined at various points. Those fed lab chow incorporated label into all fatty acids, including 22:6n-3. However, those fed the fat-free diet incorporated label into mainly saturated and monoenoic acids, with only small amounts (less than 5% of controls) incorporated into n-3 and n-6 fatty acids. Thus, we reject the hypothesis of *de novo* biosynthesis of long-chain PUFA in the retina.

Hypothesis 2

Loss of photoreceptor cells would result in a decrease in the area of the outer nuclear layer (ONL), which contains the nuclei of these cells. To test this hypothesis, the ONL area was determined in retinas from each diet group described above (Wiegand et al., 1991). There were no significant differences in ONL area measured along the superior and inferior meridians extending from the optic nerve to the ora serrata. The average length of ROS in the central superior retina was greater than in the central inferior retina (Battelle and LaVail, 1978; Rapp et al., 1982). However, n-3 dietary status had no effect on ROS length. These results allow us to reject the second hypothesis.

Hypothesis 3

The renewal of ROS integral proteins was studied in weanling rats fed a semi-synthetic diet (AIN-76A basal diet) supplemented with either 10% by weight hydrogenated coconut oil (COC), linseed oil (LIN), or safflower oil (SAF) (Wiegand, et al., 1991). The COC diet contained neither n-3 nor n-6 fatty acids, the SAF diet contained predominately n-6 fatty acids (n-6/n-3 = 830), and the LIN diet contained primarily n-3 fatty acids (n-6/n-3 = 0.4). The rate of renewal of ROS membranes was measured by autoradiography. The apical displacement of labeled integral ROS proteins was determined as a function of time following an intravitreal injection of ^{3}H-leucine. The percent displacement per day was calculated from linear regression analysis and ranged from 9.2% per day for the SAF group to 10.2% per day for the LIN group. These values are not significantly different. Therefore, there was no effect of n-3 deprivation on the renewal rate of ROS integral proteins.

Hypothesis 4

To test the hypothesis that n-3 deprivation leads to a reduction in the turnover of 22:6n-3-containing phospholipids in ROS, weanling rats were fed COC, LIN, or SAF diets for 15 weeks and injected intravitreally with [2-^{3}H]-glycerol (Stinson et al., in press). Animals were killed at various times and ROS phospholipids were isolated, derivatized, and fractionated into individual molecular species by HPLC. Peaks were quantified, collected, and counted for radioactivity. The turnover rates of 22:6n-3-containing molecular species of ROS phospholipids were determined from plots of log specific activity versus time and compared using analysis of covariance. The P-values resulting from the pairwise comparisons (COC vs LIN, COC vs SAF, and LIN vs SAF) showed no significant differences between dietary groups at the 0.05 level. Thus, the level of 22:6n-3 in ROS phospholipids does not affect the turnover rate of these lipids as measured by the disappearance of radioactive glycerol. We therefore reject the hypothesis that the retina conserves 22:6n-3 by lowering the turnover rates of 22:6n-3-containing phospholipid molecular species of ROS membranes during n-3 deficiency.

Hypothesis 5

The turnover of docosahexaenoic acid in ROS phospholipids from rats fed COC, SAF, or LIN diets was determined by following the fate of [4,5-^{3}H]-22:6n-3 injected intravitreally (Stinson et al., in press). The half-life of 22:6n-3 was calculated from the slope of the linear regression line of log specific activity versus time. The LIN group had a half-life of 19 days, while the COC group had a half-life of 54 days. No half-life could be calculated for the SAF group, since the slope of the regression line was zero. Thus, during n-3 deficiency, 22:6n-3 is retained within the retina and used for glycerolipid synthesis. Since the turnover of ROS integral proteins (Hypothesis 3) and glycerolipids (Hypothesis 4) is not altered, the retinas of n-3 deprived rats must recirculate 22:6n-3 within the retina or between the retina and the RPE.

An alternative explanation, that the injected 22:6n-3 rapidly equilibrates throughout the entire animal, was refuted by the following experiment (Stinson et al., in press). Rats were injected intravitreally in the right eye with [4,5-^{3}H]-22:6n-3 and the appearance of radioactivity in the contralateral eye, plasma, and liver was determined

6 and 22 days post injection. At both time points, the specific radioactivity of 22:6n-3 in ROS phospholipids in the injected eye was several hundred times greater than in the uninjected eye, the plasma, or the liver.

Hypothesis 6

This hypothesis was tested indirectly in two experiments. In the first (Wiegand et al., 1991), rats fed COC, SAF, and LIN diets were found to have elevated plasma levels of 20:3n-9, 20:4n-6, and 20:5n-3, respectively, which were greater than the levels of 22:5n-6 and 22:6n-3. However, these C-20 fatty acids were not incorporated in ROS phospholipids, except for relatively small amounts of 20:4n-6, which were the same in all three groups. In the SAF and to a lesser extent in the COC groups, 22:5n-6 replaced 22:6n-3 in ROS phospholipids. Thus, C-22 PUFAs appear to be selectively taken up from the blood by the retina, whereas C-20 PUFAs are not. However, this experiment did not establish the minimum levels of C-22 PUFA in the plasma that could be taken up by the retina.

In the second study (Stinson et al., in press), rats were fed the SAF, LIN, or COC diets and injected intravitreally with [4,5-^{3}H]-22:6n-3; the turnover in ROS phospholipids was determined as a function of time. The LIN and COC groups had half-lives of 19 and 54 days, respectively. No half-life could be calculated for the SAF group. The decrease with time in the specific activity of 22:6n-3 in the LIN group could be the result of exchange of labeled 22:6n-3 from ROS for unlabeled 22:6n-3 in the blood. This would be expected, since blood levels of 22:6n-3 are highest in these animals (Wiegand et al., 1991). The specific activity of 22:6n-3 in the SAF group was constant over a 60-day period, indicating that there was no uptake from the blood of unlabeled 22:6n-3, which would dilute the pool of the radiolabeled 22:6n-3 in the retina. The slow but significant turnover of 22:6n-3 in ROS phospholipids from the COC group indicated that there was still some dilution of labeled 22:6n-3 in the ROS with unlabeled 22:6n-3 from the blood. Since the SAF group showed no decrease in specific activity of 22:6n-3 over the 60 days of the experiment, and since 22:3n-3 is not in isotopic equilibrium throughout the animal (see Hypothesis 5), it seems reasonable to conclude that no 22:6n-3 from the blood was taken up by the retinas of this group during the 60-day period. The reduction in uptake may be due in part to competition between 22:5n-6 and 22:6n-3 for uptake, since the level of the former is higher in the blood of the COC and the SAF animals (Wiegand et al., 1991). Thus, during the early phase of n-3 fatty acid deprivation, the retina responds by sequestering 22:6n-3 from the blood. However, during prolonged n-3 deficiency, the uptake of 22:6n-3 from the blood is drastically reduced and the retina maintains its high levels of 22:6n-3 through recycling within the retina or between the retina and the RPE.

Hypothesis 7

Part **a** of this hypothesis was tested (see Hypothesis 5) and rejected. Parts **b** and **c** state that 22:6n-3 is recycled between the retina and RPE or within the retina, respectively. While neither has been tested rigorously, some evidence exists that 22:6n-3 is recycled between the retina and the RPE. Gordon and Bazan (1990) injected frogs systemically with [^{3}H]-22:6n-3 and followed by autoradiography the incorporation of the label into the ROS. Shed tips of ROS phagocytized by the RPE contained radioactivity, presumably in 22:6n-3, and the density of silver grains in these phagosomes was similar to the density in the ROS. Chen et al. (1991) presented preliminary results showing that the RPE of frogs had higher levels of 22:6n-3 and 22:6n-3-containing phospholipid molecular species than RPE from retinas where ROS shedding had not occurred.

Many membrane activities depend on the intimate interaction between the hydrophobic domains of integral proteins and fatty acid chains. This was found to be true for 22:6n-3 in ROS discs. Minor changes in the levels of 22:6n-3 in ROS phospholipids lead to significant changes in retinal function. Dietary deprivation of n-3 fatty acids leads to a reversible decrease in amplitudes (at saturation) of the a- and b-waves of the electroretinogram (ERG) (Wheeler et al., 1975; Benolken et al., 1973; Watanabe et al., 1987; Yamamoto et al., 1988; Bourre et al., 1989). When rats were fed semipurified diets supplemented with 18:1n-9 (a nonessential fatty acid), 18:2n-6, or 18:3n-3, those receiving the 18:2n-6 had the largest a- and b-wave amplitudes, followed by those receiving 18:2n-6 and 18:1n-9 (Wheeler et al., 1975). There was no difference between the ERG amplitudes in the animals fed 18:1n-9 or fat-free diets. Therefore, n-3 fatty acids in ROS phospholipids may in some way be involved in the conductance changes that generate the ERG signals in these animals. Neuringer, Connor, and their colleagues (Neuringer and Conner, 1986; 1987; Neuringer et al., 1984, 1986; 1991) have established the essential nature of n-3 PUFAs in the primate retina. Female rhesus monkeys were fed semipurified diets containing very low levels of 18:3n-3 throughout their pregnancies, and their infants were continued on an n-3-deficient diet. Control mothers and offspring were fed the semi-purified diet containing 18:3n-3. Significant reduction in the brain and retinal content of 22:6n-3 was found in the n-3 deficient group. By 4-12 weeks of age, the deficient group had decreased visual acuity and a prolonged recovery time of the dark-adapted electroretinogram measured after a saturating flash of light (Neuringer et al., 1984; 1986; 1991). More recent studies by this group have shown that peak latencies of both the cone and rod ERGs were delayed in an n-3 deficient group, compared to controls (Neuringer and Conner, 1987). These ERG changes could not be reversed by feeding 20:5n-3 or 22:6n-3 concentrates. The importance of n-3 PUFAs in human retinal development was recently demonstrated by Uauy et al. (1990), who maintained premature infants (30.5 ± 1.5 week gestational age) on diets containing mother's milk or one of three milk formulas. The formulas contained mainly 18:2n-6, 18:2n-6, and 18:3n-3, or both fatty acids plus fish oil. The blood lipids reflected the infant's diet. Cone function was not affected in these infants. However, the infants in the n-3-deficient group had significantly lower rod amplitudes and higher rod thresholds than the infants receiving mother's milk.

Thus, the function of the retina, as measured by the ERG, can be altered by dietary deprivation of the essential precursors of the major fatty acid present in the lipid bilayer. The mechanism(s) by which 22:6n-3 affects (and to some extent controls) the functional activities of retinal membrane is not known, but may be related to its intimate association with the visual pigment rhodopsin. The series of conformational changes that rhodopsin undergoes following photon capture, one of which leads to binding of transducin, may best be carried out in a bilayer rich in 22:6n-3-containing phospholipids (Dratz and Deese, 1986).

ROLE OF 22:6n-3 IN RETINAL STRUCTURE

Polyunsaturated fatty acids are substrates for free radical-induced lipid peroxidation. The levels of PUFAs in biological membranes determine to a major extent the susceptibility of these membranes to oxidative stress. Garrido et al. (1989) recently reported that erythrocyte and hepatic microsomal membranes of rats fed diets supplemented with sardine oil, which contains high levels of 20:5n-3 and 22:6n-3, were more easily oxidized than membranes from animals fed the conventional lab chow.

The retina, because of its high level of oxidative metabolism and exposure to light, provides an ideal environment for lipid peroxidation. By virtue of its six double bonds, 22:6n-3 is especially susceptible to oxidative destruction. When Noell et al. (1966) first reported light damage in albino rats, they suggested that light caused release of some toxic factor(s) that destroyed the photoreceptor membranes. In the case of lipid peroxidation, the toxic factors would be aldehydes generated from lipid hydroperoxides derived from 22:6n-3. There is evidence that lipid peroxidation is involved in degenerations due to constant illumination. We have reported that the levels of 22:6n-3 are specifically reduced (compared to other fatty acids) in ROS of albino rats exposed to constant light (Weigand and Anderson, 1982; Wiegand et al., 1983; 1986) and that lipid hydroperoxides are increased in these membranes (Wiegand et al., 1983). Organisciak and his co-workers have also reported a decrease in 22:6n-3 (1985; 1989a,b) and an increase in peroxides (1983) in retinas of constant light-exposed animals. Other studies have shown that intravitreal injection of lipid hydroperoxides (Armstrong et al., 1982), intravitreal implantation of an iron nail (Hiramitsu et al., 1976), or maintenance of animals in high levels of oxygen (Hiramitsu et al., 1977; Yagi et al., 1977) results in retinal degeneration. We have shown that intravitreal injection of iron leads to a rapid degeneration of photoreceptor cells (Rapp et al., 1982). Surprisingly, the rest of the retina was spared. Morphological changes were apparent as early as two hours after injection and were accompanied by a dramatic decrease in 22:6n-3 in ROS and an increase in lipid hydroperoxides. The ERG was completely extinguished 24 hours after injection. Other evidence supporting peroxidation of 22:6n-3 in retinal degeneration comes from several laboratories. When albino rats were raised in dim or bright cyclic light (Penn and Anderson, 1987; Penn et al., 1987), significantly lower levels of ROS 22:6n-3 were seen in the bright cyclic light group. When these animals were tested for their susceptibility to acute light damage (2000 lux for 24 hours), the group raised in bright cyclic light showed no damage, while the group raised in dim light was damaged severely. Organisciak et al. (1987) did a similar experiment in which the diet of the rats was supplemented with linseed oil prior to light exposure. They observed an inverse relationship between ROS 22:6n-3 content and rhodopsin (a measure of photoreceptor viability) in retinas after a two-week dark recovery period. These data, when coupled with our *in vitro* experiment showing that lipid peroxidation in ROS reduces the ability of rhodopsin to regenerate (Anderson et al., 1985), provide evidence that the damaging effects of constant illumination may be mediated through lipid peroxidation.

The plasma levels of 22:6n-3 are reduced in some animals and humans with inherited retinal degenerations. Converse et al. (1983; 1987a,b) were the first to report low plasma 22:6n-3 in X-linked and autosomal dominant retinitis pigmentosa (RP). Bazan et al. (1986) found low levels of 22:6n-3 in the plasma of Usher's patients. Our laboratory confirmed the finding in autosomal dominant RP (Anderson et al., 1987). Studies on two animals models of human RP have also shown lipid abnormalities. Miniature poodles with progressive rod-cone degeneration (*prcd*) (Aguirre et al., 1982) and Abyssinian cats with a similar inherited retinal degeneration (Narfström, 1983) have lower plasma 22:6n-3 levels than controls (Anderson et al., 1991a,b,c). Plasma from Irish setters with an inherited retinal degeneration that is different from the poodle and cat degenerations in terms of temporal and morphological characteristics showed no differences.

SUMMARY AND CONCLUSIONS

Over the last several years, evidence has accumulated that n-3 fatty acids, particularly 22:6n-3, are essential for the development of the structure and function of the visual system. The importance of 22:6n-3 is reflected in the tenacious manner in which the retina conserves this fatty acid during n-3 deficiency. We have shown that conservation is achieved by recycling 22:6n-3 within the retina or between the retina and the pigment epithelium. Within the retina, recycling could be accomplished by deacylation-reacylation reactions (Louie et al., 1991; Zimmerman and Keys, 1988). Recycling between the retina and the RPE may be achieved through specific transport proteins, possibly interphotoreceptor retinoid-binding protein (Bazan et al., 1985) and/or apolipoprotein E (Bazan et al., 1991).

ACKNOWLEDGMENTS

This research was supported by the Retinitis Pigmentosa Foundation Fighting Blindness; NIH grants EY00871, EY04149, EY07001, and EY02520; The Retina Research Foundation; and Research to Prevent Blindness, Inc. R. E. Anderson is a Senior Investigator of Research to Prevent Blindness, Inc.

REFERENCES

Aguirre G, Alligood J, O'Brien P, Buyukmichi N (1982) Pathogenesis of progressive rod-cone degeneration in miniature poodles. Invest Ophthalmol Vis Sci 23:610-630.

Anderson RE and Maude MB (1972) Lipids of ocular tissues: VII. The effects of essential fatty acid deficiency on the phospholipids of the photoreceptor-membranes of rat retina. Arch Biochem Biophys 151:270-276.

Anderson GJ, Connor WE, Corliss JD (1990) Docosahexaenoic acid is the preferred dietary n-3 fatty acid for the development of the brain and retina. Pediatr Res 27:89-97.

Anderson GJ, Connor WE, Corliss JD, Lin DS (1989) Rapid modulation of the n-3 docosahexaenoic acid levels in the brain and retina of the newly hatched chick. J Lipid Res 30:433-441.

Anderson RE, Benolken RM, Jackson MB, Maude MB (1977) The relationship between membrane fatty acids and the development of the rat retina. In: Function and biosynthesis of lipids (Bazan NG, Brenner RR, Giusto NM, eds) pp 547-559. New York: Plenum Press.

Anderson RE, Maude MB, Alvarez RA, Acland GM, Aguirre G (1991a) Plasma lipid abnormalities in the miniature poodle with progressive rod-cone degeneration. Exp Eye Res 52:349-355.

Anderson RE, Maude MB, Alvarez RA, Nilsson SEG, Narfström K, Acland GM, Aguirre G (1991b) Plasma lipid abnormalities in *prcd*-affected miniature poodles and Abyssinian cats. In: Retinal degenerations (Anderson RE, Hollyfield JG, LaVail MM, eds) pp 131-142. Boca Raton: CRC Press.

Anderson RE, Maude MB, Lewis RA, Newsome DA, Fishman GA (1987) Abnormal plasma levels of polyunsaturated fatty acids in autosomal dominant retinitis pigmentosa. Exp Eye Res 44:779-788.

Anderson RE, Maude MB, Nielsen JC (1985) Effect of lipid peroxidation on rhodopsin regeneration. Curr Eye Res 4:65-71.

Anderson RE, Maude MB, Nilsson SEG, Narfström K (1991c) Plasma lipid abnormalities in Abyssinian cat with a hereditary rod-cone degeneration. Exp Eye Res 53:415-417.

Armstrong D, Hiramitsu T, Gutteridge J, Nilsson SG (1982) Studies on experimentally induced retinal degenerations. I. Effect of lipid peroxides on electroretinographic activity in the albino rabbit. Exp Eye Res 35:157-171.

Battelle B-A and LaVail MM (1978) Rhodopsin content and rod outer segment length in albino rat eyes: Modification by dark adaptation. Exp Eye Res 26:487-497.

Bazan NG, Reddy TS, Redmond TM, Wiggert B, Chader GJ (1985) Endogenous fatty acids are covalently and non covalently bound to interphotoreceptorretinoid-binding protein in the monkey retina. J Biol Chem 260:13677-13680.

Bazan NG, Rodriguez de Turco EB, Gordon WC (1991) Docosahexaenoic acid and phospholipid metabolism in photoreceptor cells and in retinal degeneration. In: Retinal degenerations (Anderson RE, Hollyfield JG, LaVail MM, eds) pp 151-165. Boca Raton: CRC Press.

Bazan NG, Scott BL, Reddy TS, Pelias MZ (1986) Decreased content of docosahexanoate and arachidonate in plasma phospholipids in Usher's syndrome. Biochem Biophys Res Commun 141:600-604.

Benolken RM, Anderson RE, Wheeler TG (1973) Membrane fatty acids associated with the electrical response in visual excitation. Science 182:1252-1254.

Bourre J-M, Francois M, Youyou A, Dumont O, Piciotii O, Pascal G, Durand G (1989) The effects of dietary α-linolenic acid on the composition of nerve membranes, enzymatic activity, amplitude of electrophysiological parameters, resistance to poisons and performance of learning tasks in rats. J Nutr 119:1880-1892.

Chen H, Wiegand RD, Anderson RE (1991) Docosahexaenoic acid-containing phospholipid molecular species increase in frog retinal pigment epithelial cells following photoreceptor shedding. Invest Ophthalmol Vis Sci (Suppl) 32:702.

Connor WE, Neuringer M, Lin DS (1990) Dietary effects on brain fatty acid composition; the reversibility of n-3 fatty acid deficiency and turnover of docosahexaenoic acid in the brain, erythrocytes and plasma of rhesus monkeys. J Lipid Res 31:237-247.

Converse CA, Hammer HM, Packard CJ, Shepherd J (1983) Plasma lipid abnormalities in retinitis pigmentosa and related conditions. Trans Ophthalmol Soc UK 103:508-512.

Converse CA, McLachlan T, Bow AC, Packard CJ, Shepherd J (1987a) Lipid metabolism in retinitis pigmentosa. In: Degenerative retinal disorders: clinical and laboratory investigations (Hollyfield JG, Anderson RE, LaVail MM, eds) pp 93-101. New York: Alan R. Liss.

Converse CA, McLachlan T, Bow AC, Packard CJ, Shepherd J (1987b) Lipid metabolism in retinitis pigmentosa. In: Advances in the biosciences, research in retinitis pigmentosa (Zrenner E, Krastel H, Goebel H-H, eds) pp 557-561. Oxford: Pergamon Journals Ltd.

Dratz EA and Deese AJ (1986) The role of docosahexaenoic acid (22:6n-3) in biological membranes: examples from photoreceptors and model membrane bilayers. In: Health effects of polyunsaturated fatty acids in seafood (Lands WEM, ed) pp 319-351. Orlando: Academic Press, Inc.

Dudley PA (1976) Control of photoreceptor membrane synthesis by essential fatty acids. Houston, TX: PhD Dissertation, Baylor College of Medicine.

Fliesler SJ and Anderson RE (1983) Chemistry and metabolism of lipids in the vertebrate retina. Prog Lipid Res 22:79-131.

Futterman S, Downer JL, Hendrickson A (1971) Effect of essential fatty acid deficiency on the fatty acid composition, morphology, and electroretinographic response of the retina. Invest Ophthalmol 10:151-156.

Galli C, Trzeciak HI, Paoletti R (1971) Effects of dietary fatty acids on the fatty acid composition of the brain ethanolamine phosphoglyceride: Reciprocal replacement of n-6 and n-3 polyunsaturated fatty acids. Biochim Biophys Acta 248:449-454.

Garrido A, Garrido F, Guerra R, Valenzuela A (1989) Ingestion of high doses of fish oil increases the susceptibility of cellular membranes to the induction of oxidative stress. Lipids 24:833-835.

Gordon WC and Bazan NG (1990) Docosahexaenoic acid utilization during rod photoreceptor cell renewal. J Neurosci 10:2190-2204.

Hiramitsu TY, Hirato K, Nishigaki I, Yagi K (1977) The formation of lipoperoxide in the retina of rabbits exposed to high concentrations of oxygen. Experientia 32:622-623.

Hiramitsu TY, Hasegawa Y, Hirata K, Nishigaki I, Yagi K (1976) Lipid peroxide formation in the retina in ocular siderosis. Experientia 32:1324-1325.

Lands WEM, ed. (1987) Proceedings of the AOCS short course on polyunsaturated fatty acids and eicosanoids. Champaign, IL: American Oil Chemists' Society.

Lin DS, Connor WE, Anderson GJ, Neuringer M (1990) Effects of dietary n-3 fatty acids on the phospholipid molecular species of monkey brain. J Neurochem 55:1200-1207.

Louie K, Zimmerman WF, Keys S, Anderson RE (1991) Phospholipid molecular species from isolated bovine rod outer segments incorporate exogenous fatty acids at different rates. Exp Eye Res 53:309-316.

Narfström K (1983) Hereditary progressive retinal atrophy in the Abyssinian cat. J Hered 74:273-276.

Neuringer M and Connor WE (1986) N-3 fatty acids in the brain and retina: Evidence for their essentiality. Nutr Rev 44:285-294.

Neuringer M and Connor WE (1987) The importance of dietary n-3 fatty acids in the development of the retina and nervous system. In: Proceedings of the AOCS short course on polyunsaturated fatty acids and eicosanoids (Lands WEM ed) pp 301-311. Champaign, IL: American Oil Chemists' Society.

Neuringer M, Connor WE, Lin DS, Anderson GL, Barstad L (1991) Dietary omega-3 fatty acids: Effects on retinal lipid composition and function in primates. In: Retinal degenerations (Anderson RE, Hollyfield JG, LaVail MM, eds) pp 117-129. Boca Raton: CRC Press.

Neuringer M, Connor WE, Lin DS, Barstad L, Luck S (1986) Biochemical and functional effects of prenatal and postnatal n-3 fatty acid deficiency on retina and brain in rhesus monkeys. Proc Natl Acad Sci USA 83:4021-4025.

Neuringer M, Connor WE, Van Patten C, Barstad L (1984) Dietary omega-3 fatty acid deficiency and visual loss in infant rhesus monkeys. J Clin Invest 73:272-276.

Noell WK, Walker VS, Kang BS, Berman S (1966) Retinal damage by light in rats. Invest Ophthalmol 5:450-473.

Organisciak DT, Favreau P, Wang H-M (1983) The enzymatic estimation of organic hydroperoxides in the rat retina. Exp Eye Res 36:337-349.

Organisciak DT, Jiang Y-L, Wang H-M, Pickford M, Blanks JC (1989b) Retinal light damage in rats exposed to intermittent light. Invest Ophthalmol Vis Sci 30:795-805.

Organisciak DT, Wang H-M, Li A-Y, Tso MOM (1985) The protective effect of ascorbate in retinal light damage of rats. Invest Ophthalmol Vis Sci 26:1580-1588.

Organisciak DT, Wang H-M, Noell WK (1987) Aspects of the ascorbate protective mechanism in retinal light damage of rats with normal and reduced ROS docosahexaenoic acid. In: Degenerative retinal disorders: clinical and laboratory investigations (Hollyfield JG, Anderson RE, LaVail MM, eds) pp 455-468. New York: Alan R. Liss.

Organisciak DT, Wang H-M, Xie A, Reeves DS, Donoso LA (1989a) Intense light-mediated changes in rat rod outer segment lipids and proteins. In: Inherited and environmentally induced retinal degenerations (LaVail MM, Anderson RE, Hollyfield JG, eds) pp 493-512. New York: Alan R. Liss.

Penn JS and Anderson RE (1987) Effect of light history on rod outer-segment membrane composition in the rat. Exp Eye Res 44:767-778.

Penn JS, Naash MI, Anderson RE (1987) Effect of light history on retinal antioxidants and light damage susceptibility in the rat. Exp Eye Res 44:779-788.

Rapp LM, Wiegand RW, Anderson RE (1982) Ferrous ion-mediated retinal degeneration: Role of rod outer segment lipid peroxidation. In: Problems of normal and genetically abnormal retinas (Clayton R, Haywood J, Reading H, Wright A, eds) pp 109-119. New York: Academic Press.

Simopoulos AP, Kifer RR, Martin RE (1986) Health effects of polyunsaturated fatty acids in seafood. Orlando, FL: Academic Press.

Stinson AM, Wiegand RD, Anderson RE (in press) Mechanisms of conservation of docosahexaenoic acid during n-3 fatty acid deficiency. J Lipid Res.

Tinoco J (1982) Dietary requirements and function of α-linolenic acid in animals Prog Lipid Res 21:1-45.

Uauy RD, Birch DG, Birch EE, Tyson JE, Hoffman DR (1990) Effect of dietary omega-3 fatty acids on retinal function of very-low-birth-weight neonates. Pediatr Res 28:485-492.

Watanabe I, Kato M, Aonuma H, Hishimoto A, Naito Y, Moriuchi A, Okuyama H (1987) Effect of dietary alpha-linolenate/linoleate balance on the lipid composition and electroretinographic responses in rats. In: Advances in the biosciences. Research in retinitis pigmentosa (Zrenner E, Krastel H, Goebel HH, eds) pp 563-570. Oxford: Pergamon Journals Ltd.

Wheeler TG, Benolken RM, Anderson RE (1975) Visual membrane: Specificity of fatty acid precursors for the electrical response to illumination. Science 188:1312-1314.

Wiegand RD and Anderson RE (1982) Determination of molecular species of rod outer segment phospholipids. In: Methods of enzymology, Visual pigments and purple membranes (Packer L, ed) pp 297-304. New York: Academic Press.

Wiegand RD, Giusto NM, Rapp LM, Anderson RE (1983) Evidence for rod outer segment lipid peroxidation following constant illumination of the rat retina. Invest Ophthalmol Vis Sci 24:1433-1435.

Wiegand RD, Joel CD, Rapp LM, Nielsen JC, Maude MB, Anderson RE (1986) Polyunsaturated fatty acid and vitamin E in rat rod outer segments during light damage. Invest Ophthalmol Vis Sci 27:727-733.

Wiegand RD, Koutz CA, Stinson AM, Anderson RE (1991) Conservation of docosahexaenoic acid in rod outer segment of rat retinas during n-3 and n-6 fatty acid deficiency. J Neurochem 57:1690-1699.

Yagi K, Matsuoka S, Ohkawa H, Oshihi N, Takeguchi Y, Kakai H (1977) Lipoperoxide level in the retina of chick embryo exposed to high concentration of oxygen. Clin Chim Acta 80:355-360.

Yamamoto N, Hashimoto A, Takemoto Y, Okuyama H, Nomura M, Kitajima R, Tagashi T, Tamai Y (1988) Effect of the dietary α-linolenate/linoleate balance on lipid compositions and learning ability of rats. II. Discrimination process, extinction process, and glycolipid compositions. J Lipid Res 29:1013-1021.

Young RW (1976) Visual cells and the concept of renewal. Invest Ophthalmol 15:700-725.

Zimmerman WF and Keys S (1988) Acylation and deacylation of phospholipids in isolated bovine rod outer segments. Exp Eye Res 47:247-260.

DOCOSAHEXAENOIC ACID UPTAKE AND METABOLISM IN PHOTORECEPTORS: RETINAL CONSERVATION BY AN EFFICIENT RETINAL PIGMENT EPITHELIAL CELL-MEDIATED RECYCLING PROCESS

Nicolas G. Bazan, William C. Gordon,
and Elena B. Rodriguez de Turco

LSU Eye Center and Neuroscience Center
Medical Center School of Medicine in New Orleans
Louisiana State University
2020 Gravier Street, Ste. B
New Orleans, LA 70112, USA

INTRODUCTION

The vertebrate retina is compartmentalized into two chambers, each of which is isolated from the other and from the circulatory system by tight junctions (Rodriguez de Turco, et al., 1991b). These barriers occur between the cells of the retinal pigment epithelium (RPE), the distal tips of the Müller cells and photoreceptors (forming the outer limiting membrane), and the proximal ends of the Müller cells (forming the inner limiting membrane). The inner chamber contains the elements of the neural retina, while the outer compartment, the interphotoreceptor matrix, surrounds the photoreceptor outer and inner segments (Fig. 1). The synthesis and packaging of photoreceptor membrane phospholipids take place within the endoplasmic reticulum and Golgi apparatus of the inner segments; these phospholipids are subsequently targeted for either the synaptic terminals or the light-sensitive outer segments. Following assembly of the photosensitive membranes, the newly formed discs are added basally to the stack of discs already within the outer segment, resulting in continual elongation. This constant biogenesis of disc membranes is offset by shedding and phagocytosis of outer segment tips by the RPE cells. In addition to their involvement in this daily membrane degradative process, RPE cells participate actively in the synthesis of new photoreceptor membranes. They efficiently take nutrients from the choriocapillaris and deliver them to the photoreceptor cells through the interphotoreceptor matrix.

The newly synthesized disc membranes of photoreceptor cells (Aveldaño de Caldironi and Bazan, 1980; Fliesler and Anderson, 1983; Wiegand and Anderson, 1983), as well as synaptic terminals in the central nervous system (Cotman et al., 1969), are built with phospholipids that are highly enriched in docosahexaenoic acid ($22:6\omega3$). Polyunsaturated phospholipid species, with one $22:6\omega3$ moiety esterified at the second position of the glycerol backbone, and supraenoic molecular species, with two $22:6\omega3$ moieties per molecule (Miljanich et al., 1979; Aveldaño de Caldironi and Bazan, 1980; Aveldaño and Bazan, 1983; Aveldaño, 1989), are selectively used for membrane renewal at the base of photoreceptor outer segments. The precursor of $22:6\omega3$ is linolenic

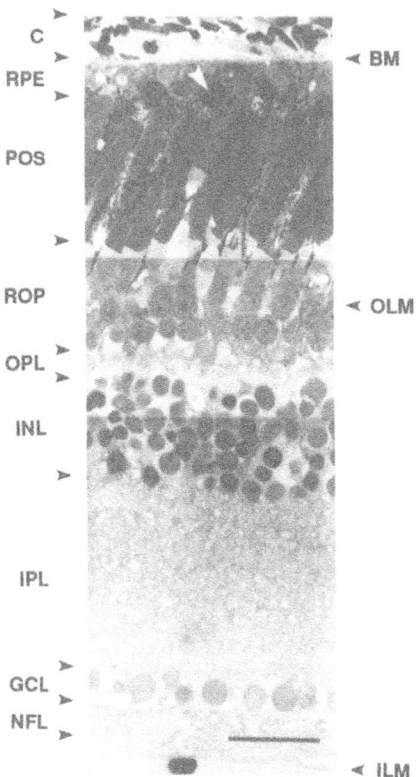

Figure 1. Structure of the frog retina, demonstrating the large rod photoreceptor cells near the top, in contact with the pigment epithelium. The bases of the photoreceptors extend downward to about the middle of the picture, while the neural retina occupies the lower portion. The retina lies just under a richly endowed capillary bed, partially visible at the top. The white arrow indicates the shed tip (phagosome) of a rod photoreceptor within a pigment epithelial cell. The black arrows denote boundaries between the retinal layers. BM, Bruch's membrane; C, choriocapillaris; GCL, ganglion cell layer; ILM, inner limiting membrane; INL, inner nuclear layer; IPL, inner plexiform (synaptic) layer; NFL, nerve fiber layer; OLM, outer limiting membrane; OPL, outer plexiform (synaptic) layer; POS, photoreceptor outer segments; ROP, rest of photoreceptors; RPE, retinal pigment epithelium. Scale: 20 μm.

acid (18:3ω3), an essential fatty acid provided in the diet (Tinoco, 1982; Tinoco et al., 1977; 1978). It undergoes a series of elongation-desaturation steps to produce the most abundant component of the ω3 family, 22:6ω3. While 22:6ω3-rich phospholipids are necessary for excitable membrane functions (Wheeler et al., 1975; Neuringer et al., 1984), the limited ability of the brain and the retina to synthesize 22:6ω3 makes these tissues dependent upon a sustained supply from circulating plasma lipoprotein carriers. Therefore, several very important major questions concerning the 22:6ω3 supply to the brain and retina must be addressed: where is 22:6ω3 synthesized; which is (are) the carrier(s); and how is selective delivery to the brain and retina accomplished? The liver, and to a lesser extent the intestine, are both potential sources of circulating plasma lipoproteins containing 22:6ω3. It appears that acquisition of high levels of 22:6ω3 by neural tissue may depend upon specific organ targeting, high selective uptake, and/or tenacious retention. In fact, a well-developed mechanism for high retention of this essential fatty acid by excitable membranes contributes to 22:6ω3 conservation after long dietary deprivation of all ω3 fatty acids (Tinoco et al., 1977; 1978).

The interorgan trafficking of 22:6ω3, and its uptake by retinas during photoreceptor cell differentiation (postnatal developing mice) and during photoreceptor cell renewal (frog), have been the focus of our recent studies. Moreover, evidence of i) its selective uptake by photoreceptor cells, ii) its active utilization for phospholipid synthesis, iii) the intracellular trafficking toward the base of outer segments and synaptic terminals, iv) its migration in close association with rhodopsin toward the tips of rod outer segments, and v) the recycling of 22:6ω3 from phagocytized disc membrane phospholipids toward photoreceptor cells, are summarized in this overview.

CONVERSION OF 18:3ω3 TO 22:6ω3 IN THE LIVER AND ITS SUPPLY TO BRAIN AND RETINA

The synthesis of 22:6ω3 from its labeled precursor [^{14}C]18:3ω3 and the interorgan transfer of [^{14}C]22:6ω3 from the liver to the brain and retina were followed in 3-day-old mouse pups (Scott and Bazan, 1989). At this early postnatal time, an active and efficient supply of ω3 fatty acids is necessary to support photoreceptor cell differentiation and synaptic membrane biogenesis in the central nervous system. Intraperitoneal injections of [^{14}C]18:3ω3 allowed us to trace, within 2 hr, a rapid uptake of precursor by the liver (Fig. 2). This was followed during the subsequent 70 hr by a sustained decrease of label in the liver, concomitantly with an increase in labeling of brain and retinal lipids. At early times, 18:3ω3 was found in the liver, esterified mainly into triacylglycerols (TAG), and then was gradually released for further desaturation and elongation, followed by esterification of the 22:6ω3 end product into phospholipids (Scott and Bazan, 1989). Less than 5% was metabolized and recovered in other fatty acids (palmitic acid), while 20:5ω3, 22:5ω3, and 22:6ω3 accounted for 18% of total labeled liver lipids at 2 hr and 27% at 6 hr. Despite the large proportion of [^{14}C]18:3ω3 in circulating plasma lipoproteins (48% and 22% of total labeling at 2 and 6 hr, respectively), none was detected in brain or retina at any time up to 72 hr after injection (Figs. 2, 3). In contrast with the liver, a major proportion of 18:3ω3 was metabolized and recovered in other fatty acids (i.e., palmitic acid, 50%) in neural tissue. This reflects the higher efficiency of the liver, as compared with brain and retina, in the handling of 18:3ω3 for 22:6ω3 synthesis. Transient storage of 18:3ω3 as liver TAG preserves these fatty acids from oxidation. This allows the liver gradually to release, elongate, and desaturate 18:3ω3 to 22:6ω3, leading to a sustained availability of 22:6ω3 for lipoprotein synthesis, independent of dietary intake of 18:3ω3. Conversely, brain and retina, due to inadequate transient storage of 18:3ω3 (esterified into TAG), would require a constant supply of this essential fatty acid.

Interestingly, within 2-6 hr postinjection, brain 18:3ω3 was elongated and desaturated to 20:5ω3, 22:5ω3, and then 22:6ω3, while 22:5 was the only ω3 fatty acid labeled in the retina. This strongly suggests that a low activity of Δ4 desaturase, involved in the conversion of 22:5ω3 to 22:6ω3, is the rate-limiting step in retinal synthesis of 22:6ω3. Moreover, it supports the contention that the retina, more than the brain, depends upon an exogenous supply of 22:6ω3 for lipid synthesis.

Of the [^{14}C]-labeled fatty acids that appeared in the liver following [^{14}C]18:3ω3 injection, only [^{14}C]22:6ω3 decreased as a function of time in parallel with an increase in brain and retina (Fig. 4). This observation, and the fact that [^{14}C]22:6ω3 is the main fatty acid labeled in plasma lipoproteins by 24-72 hr, reflect the central role of the liver in elongation and desaturation of 18:3ω3, and in the selective utilization of [^{14}C]22:6ω3 to synthesize lipids secreted into the blood circulation as lipoproteins.

A similar time-dependent profile of labeled lipids in liver, brain, and retina was observed in frogs injected via the dorsal lymph sac with [^{3}H]22:6ω3 (Bazan et al., 1991).

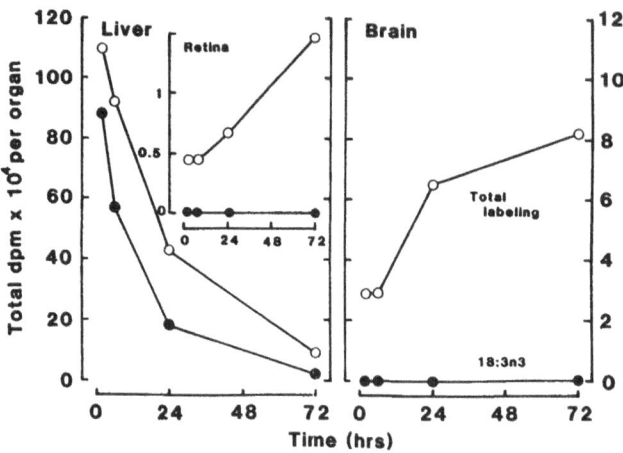

Figure 2. Time-course for labeled lipids in liver, brain, and retina after intraperitoneal injection of [^{14}C]18:3ω3 into 3-day-old mouse pups. Values were replotted from Scott and Bazan, 1989.

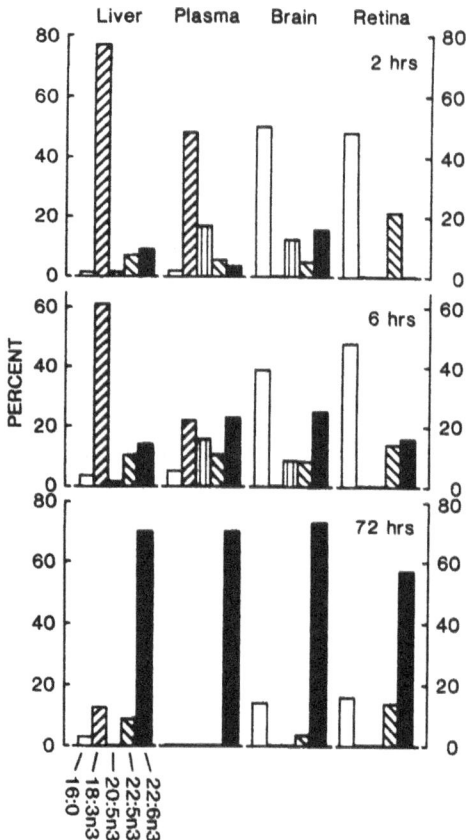

Figure 3. Profile of individual fatty acids labeled in liver, plasma, brain, and retina following intraperitoneal injection of [^{14}C]18:3ω3. Values represent percent distribution with respect to total recovered radioactivity, and were replotted from Scott and Bazan, 1989.

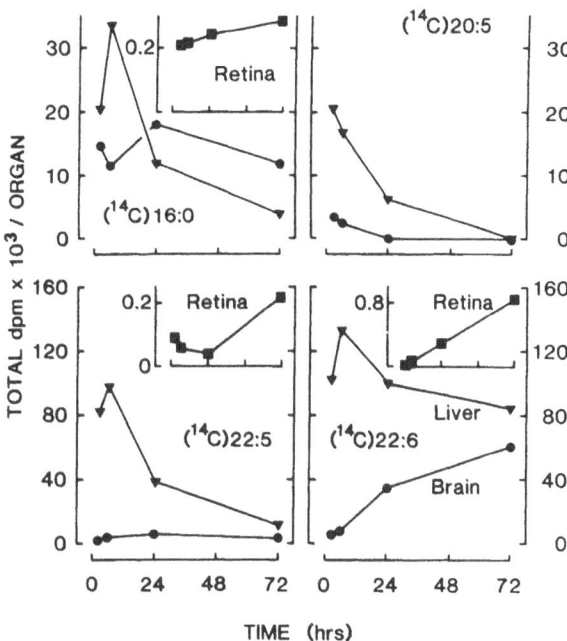

Figure 4. Time course of individually labeled fatty acids in liver, brain, and retina after intraperitoneal injection of [^{14}C]18:3ω3 into 3-day-old mouse pups. Values were recalculated from Scott and Bazan, 1989.

A constant flux of [^{3}H]22:6ω3 from plasma circulating lipoproteins toward brain and retina was evident up to the final time point, 34 days after the injection.

SELECTIVE UPTAKE AND METABOLISM OF [^{3}H]22:6ω3 BY PHOTORECEPTOR CELLS

Labeling of frog retinas *in vitro* with [^{3}H]22:6ω3 (0.11 μM final concentration) for short periods of time up to 6 hr revealed rapid uptake of the precursor by photoreceptor cells and continual esterification into phospholipids (Rodriguez de Turco et al., 1991b). The precursor was not accumulated in the free fatty acid pool but actively transferred to CoA and esterified into lipids. Distribution of labeling among phosphatidylinositol (PI), phosphatidylcholine (PC), and phosphatidylethanolamine (PE) reflected an early preferential labeling of a small pool of PI coupled to [^{3}H]22:6ω3 uptake. Incubation of human and monkey retinas under similar conditions also demonstrated preferential labeling of phospholipids and early extensive labeling of phosphatidic acid (PA) and PI (Rodriguez de Turco et al., 1990; Gordon et al., 1991). The enzymatic pathways involved in [^{3}H]22:6ω3 esterification into lipids (i.e., *de novo* synthesis and/or turnover) are not well defined at present. However, when [^{3}H]glycerol, a marker of *de novo* synthesis, was used as a precursor, an active labeling of retinal polyunsaturated and dipolyunsaturated molecular species of PI and other phospholipids was observed (Aveldaño de Caldironi and Bazan, 1980). Also, an early large uptake of labeled 22:6ω3 into PA, a key intermediate in the *de novo* synthesis of lipids, was observed after both *in vitro* (Bazan et al., 1984; Rodriguez de Turco et al., 1991) and *in vivo* (Bazan et al.,

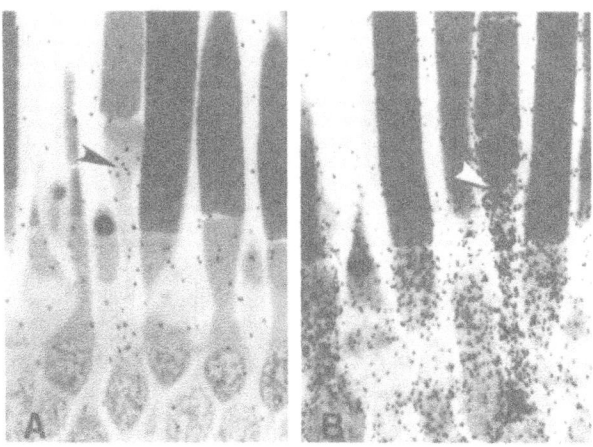

Figure 5. Light microscope-level autoradiograms illustrating the distribution of [³H]22:6ω3 (the black silver grains) within the photoreceptor cells of the frog retina after 30 min (A) and 6 hr (B) of incubation in the presence of this molecule. Label enters the central region (black arrow) of the rod photoreceptor, disperses to other areas, and then accumulates just below the outer segment where new disc membranes will be assembled (white arrow). Notice that the two types of frog rod photoreceptors (the shorter, blue-sensitive 435-rods indicated by the two arrows, and the longer, green-sensitive 502-rods) accumulate label differentially, with the labeling of 502-rods up to three times that of the 435-rods. The cone photoreceptors (apparent here from the presence of their round, dark oil droplets) accumulate only minimal label.

1991) retinal labeling. This observation suggests that a *de novo* pathway may be of major relevance in the building of at least some retinal molecular species of phospholipids containing 22:6ω3.

The profiles of labeled retinal lipids observed following 6 hr of *in vitro* or *in vivo* labeling in frogs injected with [³H]22:6ω3 via the dorsal lymph sac were similar (Rodriguez de Turco et al., 1991a). This suggests that, although [³H]22:6ω3 circulates in lipoproteins, mainly esterified into PC and cholesterol esters (Bazan et al., 1991), after crossing the RPE and the interphotoreceptor matrix it will target retinal cells (in free or esterified form) and follow a metabolic pathway similar to the free precursor offered *in vitro* to the cells. Autoradiographic analysis of [³H]22:6ω3 distribution after both *in vitro* and *in vivo* labeling of frog retinas clearly showed that most of the label (92%) accumulated in the photoreceptor layer. The remainder appeared as diffuse labeling in the neural retina (Rodriguez de Turco et al., 1990; 1991b).

Among the photoreceptor cells, dramatic differences in uptake exist (Fig. 5). In the frog, autoradiographic label in rod cells was as much as 16 times that of cone cells, and similar disparities were observed in both inner and outer segments. One exception involved the highly labeled oil droplets of the red-sensitive, 575-cones. Interestingly, short-term *in vitro* retinal incubations with [³H]22:6ω3 revealed autoradiographic differences between the two rod photoreceptors as well. Although the inner segments of the red rods (green-sensitive, 502-rods) demonstrated extensive labeling, these same regions in the green rods (blue-sensitive, 435-rods) accumulated three times this amount of label over the same period of time (Rodriguez de Turco et al., 1991b).

This preferential uptake of [³H]22:6ω3 was followed autoradiographically as a function of time within the cells of the frog retina. All cell types of the neural retina appeared to label diffusely, reaching a plateau after only 1 hr. No individual morphotype accumulated more label than any other. However, the synaptic areas had measurably more [3H]22:6ω3 than cell body or axonal regions. Conversely, rod photoreceptor cells continued to accumulate [3H]22:6ω3, even after 6 hr of incubation (Rodriguez de Turco et al., 1991b). Within 15-30 min, label appeared in the myoid region, followed by both distal and proximal cytoplasmic dispersion (Fig. 5A). Label gradually increased in the perinuclear cytoplasm and axonal regions, finally forming noticeable concentrations within the synaptic terminals by 2-4 hrs. [3H]22:6ω3 also accumulated within the rod ellipsoids (regions of densely packed mitochondria located near the base of the outer segments), becoming noticeable after 1 hr, and continuing to accumulate for the entire incubation period (Fig. 5B) (Rodriguez de Turco et al., 1991b). The diffuse form of the [3H]22:6ω3 label accumulated evenly throughout the rod outer segments. After 4-6 hr, a dense line was just apparent at the base of some outer segments, suggesting that the newly synthesized, 22:6ω3-rich phospholipids had been assembled into photosensitive membranes. It is the constant basal accumulation of these highly labeled discs that forms the growing dense region of the rod outer segments seen in long-term *in vivo* studies of [3H]22:6ω3 retinal labeling (Gordon and Bazan, 1990).

When both the biochemical and the cytological observations are considered, it becomes apparent that the handling of 22:6ω3 by retinal cells must involve a sequence of well-defined events and pathways. Selective uptake of 22:6ω3 by the photoreceptor cells occurs in a limited area of the plasma membrane surrounding the myoid region, where activation and esterification occur quickly, perhaps almost simultaneously. This is followed by movement of 22:6ω3-phospholipids through the endoplasmic reticulum and, perhaps, the Golgi apparatus, and transport through the ellipsoid to the region of disc morphogenesis at the base of the outer segment.

DISC SYNTHESIS AND COMIGRATION OF [3H]22:6ω3-LABELED PHOSPHOLIPIDS WITH RHODOPSIN DURING MEMBRANE RENEWAL OF ROD OUTER SEGMENTS

Photomembrane discs are assembled at the base of the outer segments from molecular components that are packaged in the inner segment. Proteins such as opsin, complexed with phospholipids, arrive in small vesicles that fuse with the plasma membrane near the base of the connecting cilium (Besharse and Pfenninger, 1980; Deretic and Papermaster, 1991). However, there is strong evidence that other lipids arrive by a separate pathway, independent of the Golgi apparatus (Matheke and Holtzman, 1984; Matheke et al, 1984; Fliesler and Basinger, 1987). As new membrane accumulates, outpocketings at the base of the outer segment develop. Once each of these has expanded to become a flattened, circular extension the diameter of the cell, the membrane of the margin fuses with the outer segment above it. The extracellular space between adjacent layers becomes enclosed as a membrane-bound, free-floating disc within the plasma membrane of the outer segment, and is pushed apically as more discs are formed.

[3H]22:6ω3 enters the outer segment by two pathways. Autoradiographic methods show that all outer segments rapidly label, both *in vitro* and *in vivo*, with an evenly dispersed, diffuse label (Gordon and Bazan, 1990; Rodriguez de Turco et al., 1991b). The outer segment discs rapidly saturate, with little additional label occurring in these [3H]22:6ω3-phospholipids over the next 34 days (Bazan et al., 1991). This first mechanism must involve specific protein carriers, since no labeled vesicles have been seen to enter the outer segments and migrate to the apical discs. Evenly dispersed [3H]22:6ω3 is present *in vitro* within 30 min, and matches the dispersion profiles reported for other fatty acids (palmitic acid, arachidonic acid, and stearic acid; Bibb and Young, 1974), suggesting this to be a general mechanism used by the photoreceptors to replace fatty acids esterified into phospholipids by the acylation/reacylation cycle and/or replacement of phospholipids of the disc membranes.

The second pathway involves the incorporation of [^{3}H]22:6ω3-phospholipids into disc membranes as they are assembled at the base of outer segments. This densely accumulating label reflects membrane replacement, while the diffuse form represents molecular replacement (Bibb and Young, 1974). As new discs are formed, older, densely labeled discs are moved apically, carrying the label with them. In addition, because the liver constantly supplies the retina with [^{3}H]22:6ω3, there is continual basal labeling of outer segments. This dense label uniformly fills the entire length of the outer segments. Parallel experiments with [^{3}H]22:6ω3 and [^{3}H]leucine (an opsin marker) demonstrate that the front of this dense label comigrates with labeled protein. Both labels reach the tip of the outer segment by 28 days, indicating that the dense label does not leave the disc membrane until it is shed into the overlying RPE via membrane replacement. Finally, this growing dense label must be very different from the diffuse form of label that occurs immediately and then remains constant over time. These labeling profiles suggest that the accessibility of free 22:6ω3 to disc phospholipids, its esterification via deacylation-reacylation, and/or the transport of 22:6ω3-phospholipids in replacement reactions is not the predominant pathway involved in rod outer segment labeling. Moreover, this implies that the addition of 22:6ω3-phospholipids at the base of outer segments constitutes the main pathway contributing to the net enrichment of discs with 22:6ω3-phospholipids.

All fatty acids, including 22:6ω3, label outer segments rapidly, evenly, and diffusely. In contrast, 22:6ω3 is the only fatty acid that generates a dense, disc-specific [^{3}H]22:6ω3 label. Selective delivery of 22:6ω3 to the photoreceptors, and subsequent preferential utilization of 22:6ω3-phospholipids for disc membrane synthesis, may account for the region of high density labeling, and, although the 22:6ω3-phospholipids are not covalently associated with protein, i.e., rhodopsin (Gordon and Bazan, 1990), they do not leave the disc membrane as other lipid molecules do. These 22:6ω3-rich phospholipids must, therefore, serve some unique function within the photoreceptor disc membranes, requiring a high number of 22-carbon molecules with six double bonds to interface with other membrane components in the disc membrane.

These findings are corroborated by studies on rats (Anderson and Maude, 1972; Wheeler et al., 1975) and monkeys (Neuringer et al., 1984; 1986) that have been dietarily deprived of 22:6ω3 and all fatty acid precursors. Retinas lacking these ω3 fatty acids substitute 22:5ω6 in the photoreceptor membranes. Under these conditions, electroretinographic investigations demonstrate dramatic changes in light sensitivity and dynamic range (Neuringer et al., 1991), implying that substitution of other, more available fatty acids is not adequate for the support of photic events. All of this evidence suggests a unique role for the 22:6ω3 molecule in photoreceptor cells.

RECYCLING OF 22:6ω3 BY THE RETINAL PIGMENT EPITHELIUM/ PHOTORECEPTOR CELL COMPLEX FOLLOWING SHEDDING AND PHAGOCYTOSIS

One of the most remarkable characteristics of neural tissue (both brain and retina) is the ability to tenaciously retain 22:6ω3, even during long periods of dietary deprivation of ω3 fatty acids. This mechanism has developed to protect excitable membranes from depletion of 22:6ω3 phospholipids essential for normal functioning. The retina, and especially the photoreceptor cells, require a constant supply of 22:6ω3. This input is extremely important since outer segments are continually losing disc membranes to the RPE. Packets of photoreceptor disc membranes are periodically shed from the tips of the outer segments, appearing within the RPE cytoplasm as phagosomes (Young and Bok, 1969).

Careful analysis of autoradiograms obtained from the eyes of frogs that had been injected with [^{3}H]22:6ω3 reveals that labeled 22:6ω3 molecules are contained within these structures (Gordon and Bazan, 1990). Furthermore, biochemical and autoradiographic studies indicate that retention of 22:6ω3 within the outer segments before shedding and phagocytosis is not involved in 22:6ω3 conservation within the retina (Bazan et al., 1991; Gordon et al., in press). Density analyses of diffusely and densely labeled regions of outer segments show that both forms of label remain unchanged throughout the 28-day disc cycle, demonstrating no exchange between the two labeled forms. In addition, the density of label within newly shed phagosomes matches that of the outer segment tip, whether shedding occurred before or after the arrival of the dense label to this region (Fig. 6). This also illustrates that the dense form of [^{3}H]22:6ω3-phospholipids is locked into disc membrane structure until final degradation by the RPE cells. Finally, labeling of the RPE cytoplasm remains low and diffuse throughout a 34-day interval following injection of [^{3}H]22:6ω3. Thus, RPE cells regulate the flow of 22:6ω3 to the interphotoreceptor matrix from both circulating plasma lipoproteins and phagosomes rich in 22:6ω3-phospholipids. In fact, the dense label arriving from the outer segments as phagosomes, contributing up to 12% of total RPE label, is easily managed by the RPE cells and immediately recycled back to the photoreceptors (Gordon et al., in press).

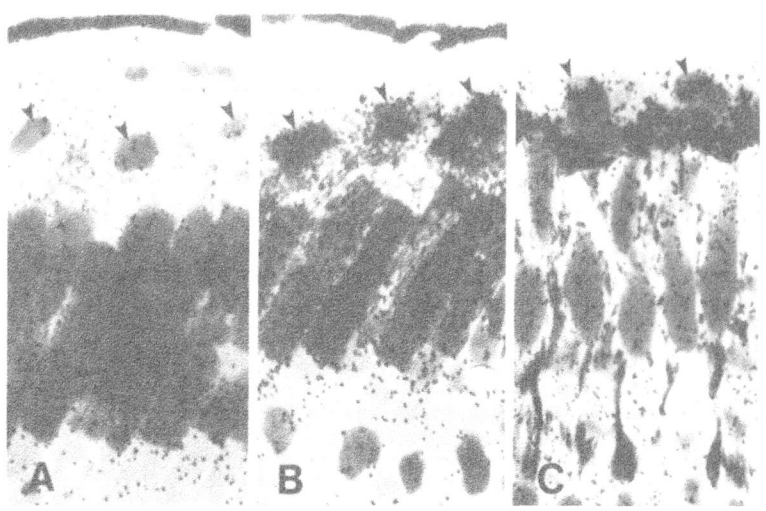

Figure 6. Autoradiograms demonstrating the distribution of [^{3}H]22:6ω3 (A and B) and [^{3}H]leucine (C) in photoreceptor outer segments and retinal pigment epithelial cells of the frog retina. Animals were injected via the dorsal lymph sacs with either the 22:6ω3 or leucine label, and allowed to incorporate these molecules into new photoreceptor disc membranes for a period of 10 days (A) or 28 days (B and C). Because leucine, a protein marker, does not leave the disc membrane, the presence of a dense leucine band (C) within phagosomes indicates the interval of time necessary for a newly formed basal disc to reach the photoreceptor tip. Following a 24-hr interval of constant light, animals were subjected to 1 hr of dark and 1 hr of light to trigger the shedding response. Newly shed phagosomes (arrows) are visible within the pigment epithelium. After only 10 days of labeling (A), the phagosomes contain only diffuse silver grains. However, after the heavily labeled discs reach the photoreceptor tip, phagosomes are densely labeled with [^{3}H]22:6ω3 (B), demonstrating that the form of 22:6ω3 label within the disc membranes is not free to diffuse throughout the outer segment, but remains *in situ* for the 28 days.

SUMMARY AND CONCLUSIONS

After 18:3ω3 is obtained from the diet, it is accumulated by the liver, where it is esterified and temporarily stored as triacylglycerols. As it is required, 18:3ω3 is elongated and desaturated to 22:6ω3, then released into the circulation with lipoprotein carriers. RPE cells remove the 22:6ω3 from the choriocapillaris and subsequently release it to the retina proper. In the frog, all 22:6ω3 input to the photoreceptors occurs by way of the RPE cells. After passing through the interphotoreceptor matrix, it is selectively taken into the myoid region of photoreceptor cells where it is immediately activated and esterified onto position 2 (and sometimes also position 1) of a glycerol molecule. Some phospholipids are passed through the endoplasmic reticulum and Golgi apparatus, while others are not. Generally, transport to the outer segments seems to be independent of the Golgi apparatus. Addition to rod outer segments occurs in two ways: i) a general diffuse pathway, probably common to all fatty acids, which rapidly labels the entire outer segment; and ii) a specific dense pathway, utilized only by 22:6ω3-containing phospholipids, which become locked into the matrix of disc membranes along with opsin. There appears to be no exchange between these two forms of label. Accumulation of newly synthesized basal discs pushes older, 22:6ω3-laden discs apically until the outer segment tips, high in 22:6ω3-phospholipids (the dense form of outer segment label), are shed into the RPE cytoplasm. There, as the 22:6ω3 fatty acids are released from the disc membranes during degradation, a recycling mechanism immediately directs these essential fatty acids back into the interphotoreceptor matrix, thus conserving this molecule in the retina, and permitting it to be again selectively taken up by the photoreceptors for photomembrane synthesis. The process of 22:6ω3 handling and trafficking by the retina is specifically orchestrated around a conservation mechanism that is regulated by the RPE cells and that ensures, through a short feedback loop from the phagosomes to the interphotoreceptor matrix, adequate levels of 22:6ω3 for photoreceptors at all times.

ACKNOWLEDGMENTS

This work was supported in part by US Public Health Service grants EY04428 and EY02377 from the National Eye Institute, National Institutes of Health, Bethesda, Maryland.

REFERENCES

Anderson RE and Maude MB (1972) Lipids of ocular tissues: VIII. The effects of essential fatty acid deficiency on the phospholipids of the photoreceptor membranes of rat retina. Arch Biochem Biophys 151:270-276.

Aveldaño MI (1989) Dipolyunsaturated species of retina phospholipids and their fatty acids. In: Biomembranes and nutrition, Vol 195 (Leger CL, Béréziat G, eds) pp 87-96. Paris: Colleque INSERM.

Aveldaño MI and Bazan NG (1983) Molecular species of phosphatidylcholine, ethanolamine, -serine, and -inositol in microsomal and photoreceptor membranes of bovine retina. J Lipid Res 24:620-627.

Aveldaño de Caldironi MI and Bazan NG (1980) Composition and biosynthesis of molecular species of retina phosphoglycerides. Neurochem Int 1:381-392.

Bazan HEP, Sprecher H, Bazan NG (1984) *De novo* biosynthesis of docosahexaenoil-phosphatidic acid in bovine retinal microsomes. Biochim Biophys Acta 796:11-19.

Bazan NG, Gordon WC, Rodriguez de Turco EB (1991) Delivery of docosahex-aenoic acid (^{3}H-22:6) by the liver to the retina in the frog. Invest Ophthalmol Vis Sci 32(Suppl):701.

Besharse JC and Pfenninger KH (1980) Membrane assembly in retinal photo-receptors. I. Freeze-fracture analysis of cytoplasmic vesicles in relationship to disc assembly. J Cell Biol 87:451-463.

Bibb C and Young RW (1974) Renewal of fatty acids in the membranes of visual cell outer segments. J Cell Biol 61:327-343.

Cotman C, Blank ML, Moehl A, Snyder F (1969) Lipid composition of synaptic plasma membranes isolated from rat brain by zonal ultracentrifugation. Biochemistry 8:4606-4612.

Deretic D and Papermaster DS (1991) Polarized sorting of rhodopsin on post-Golgi membranes in frog retinal photoreceptors. J Cell Biol 113:1281-1293.

Fliesler SJ and Anderson RE (1983) Chemistry and metabolism of lipids in the vertebrate retina. Prog Lipid Res 22:79-131.

Fliesler SJ and Basinger SF (1987) Monensin stimulates glycerol incorporation into rod outer segment membranes. J Biol Chem 262:17516-17523.

Gordon WC and Bazan NG (1990) Docosahexaenoic acid utilization during rod photoreceptor cell renewal. J Neurosci 10:2190-2202.

Gordon WC, Rodriguez de Turco EB, Bazan NG (1992) Retinal pigment epithelial cells play a central role in the conservation of docosahexaenoic acid by photoreceptor cells after shedding and phagocytosis. Curr Eye Res (in press).

Gordon WC, Rodriguez de Turco EB, Peyman GA, Bazan NG (1991) Uptake and distribution of docosahexaenoic acid (^{3}H-22:6,n-3) in detached and attached human retina. Invest Ophthalmol Vis Sci 32(Suppl):702.

Matheke ML and Holtzman E (1984) The effects of monensin and of puromycin on transport of membrane components in the frog retinal photoreceptor. II. Electron microscopic autoradiography of proteins and glycerolipids. J Neurosci 4:1093-1103.

Matheke ML, Fliesler SJ, Basinger SF, Holtzman E (1984) The effects of monensin on transport of membrane components in the frog retinal photoreceptor. I. Light microscopic autoradiography and biochemical analysis. J Neurosci 4:1086-1092.

Miljanich P, Sklar LA, White DL, Dratz EA (1979) Disaturated dipolyunsaturated phospholipids in the bovine rod outer segment disk membrane. Biochim Biophys Acta 55:294-306.

Neuringer M, Connor WE, Lin DS (1991) Altered background adaptation of the ERG in rhesus monkeys postnatally deficient in omega-3 fatty acids. Invest Ophthalmol Vis Sci 32(Suppl):702.

Neuringer M, Connor WE, Lin DS, Barstad L, Luck S (1986) Biochemical and functional effects of prenatal and postnatal ω3 fatty acid deficiency on retina and brain in rhesus monkeys. Proc Natl Acad Sci USA 83:4021-4025.

Neuringer M, Connor WE, VanPetten C, Barstad L (1984) Dietary omega-3 fatty acid deficiency and visual loss in infant rhesus monkeys. J Clin Invest 73:272-276.

Rodriguez de Turco EB, Gordon WC, Bazan NG (1991a) Modulation of uptake and metabolism of ^{3}H-DHA in retina as a function of extracellular concentration of free DHA. Invest Ophthalmol Vis Sci 32(Suppl):702.

Rodriguez de Turco EB, Gordon WC, Bazan NG (1991b) Rapid and selective uptake, metabolism, and cellular distribution of docosahexaenoic acid among rod and cone photoreceptor cells in the frog retina. J Neurosci 11:3667-3678.

Rodriguez de Turco EB, Gordon WC, Peyman GA, Bazan NG (1990) Preferential uptake and metabolism of docosahexaenoic acid in membrane phospholipids from rod and cone photoreceptor cells of human and monkey retinas. J Neurosci Res 27:522-532.

Scott BL and Bazan NG (1989) Membrane docosahexaenoate is supplied to the developing brain and retina by the liver. Proc Natl Acad Sci USA 86:2903-2907.

Tinoco J (1982) Dietary requirements and functions of α-linolenic acid in animals. Prog Lipid Res 21:11-45.

Tinoco J, Babcock R, Hincerbergs I, Medwadowski B, Miljanich P (1978) Linolenic acid deficiency: Changes in fatty acid patterns in female and male rats raised on a linolenic acid-deficient diet for two generations. Lipids 13:6-17.

Tinoco J, Miljanich P, Medwadowski B (1977) Depletion of docosahexaenoic acid in retinal lipids of rats fed a linolenic acid-deficient, linoleic acid-containing diet. Biochim Biophys Acta 486:575-578.

Wheeler TG, Benolkin RM, Anderson RE (1975) Visual membrane: Specificity of fatty acid precursors for the electrical response to illumination. Science 188: 1312-1314.

Wiegand RD and Anderson RE (1983) Phospholipid molecular species of frog rod outer segment membranes. Exp Eye Res 37:159-173.

Young RW and Bok D (1969) Participation of the retinal pigment epithelium in the rod outer segment renewal process. J Cell Biol 42:392-403.

ESSENTIAL FATTY ACIDS AND NEURODEVELOPMENTAL DISORDER

Michael A. Crawford

Institute of Brain Chemistry and Human Nutrition
Hackney Hospital,
Homerton High Street,
London ES 6BE
UK

INTRODUCTION

Although there has been a rapid reduction in the number of babies dying at birth, there has been no corresponding fall in the number of handicapped babies or those who develop mental or visual defects following birth.

Mortality and morbidity of the newborn increase steeply as birth weight falls below 2.5 kg. The incidence of severe neurodevelopmental disorder increases from 6.8/1,000 births in the 3.5-4.5 kg birth weight range to over 200/1,000 below 1.5 kg.

Reports from Sweden and the UK indicate that since 1967, there has been a three-fold increase in the incidence of cerebral palsy among low-birth-weight babies. Other severe neurodevelopmental disorders, including blindness, deafness, mental retardation, and autism, appear with greater frequency in low-birth-weight babies, and often more than one defect occurs in one baby. The risk of repetition is high.

Little is known about the cause of the neurodevelopmental handicaps. Formerly, excess oxygen and lack of vitamin E were thought to be the causes for retinopathy of prematurity. By contrast, lack of oxygen during birth (asphyxia) was considered as the principal reason babies developed mental handicaps such as cerebral palsy. Both assumptions are now being questioned. Many babies are subjected to a sudden lack of oxygen during birth but, in a healthy baby, this doesn't seem to cause much harm. In others damage follows, and it is still difficult to predict which baby will eventually become spastic or develop mental, visual, hearing, or physical handicaps.

The common denominator in these apparently different defects is that they occur during brain development. In humans, the brain holds top priority in fetal development. The major biochemical investment in brain growth is largely in lipids, which are required for the massive membrane area.

Essential fatty acids are known to be determinants of brain development and integrity, and several research groups are currently studying the relevance of docosahex-aenoic acid (DHA) to visual development. We have studied nutritionally induced encephalomalacia in which the cerebellum but not the cerebrum is affected, which results in death following severe hemorrhage and edema. The model uses the chick at a time when growth of the cerebrum is slackening but the cerebellum is undergoing its growth thrust. Hence this model may represent periventricular hemorrhage, a potential precursor to cerebral palsy. In this model, the balance between arachidonic acid (AA) and DHA and tocopherol appears to be causally important.

In our studies of low-birth-weight babies, we have found highly significant correlations between birth weight and AA and DHA in the umbilical endothelium, as well as maternal and cord blood at delivery. Studies of maternal nutrient intakes revealed that reduced nutrient intakes are associated with low birth weight.

ESSENTIAL FATTY ACIDS

The first evidence that the essential fatty acid component of the neural membranes is limiting for neural integrity was derived from studies on allergic encephalomyelitis. Clausen and Moller (1967) reported that rat brain could be made susceptible to an induced autoimmune attack by depleting its membranes through dietary deficiency of essential fatty acids.

We (Crawford and Sinclair, 1972) suggested that both n-3 and n-6 fatty acids are limiting for brain growth, and there is now much evidence (Bourre et al., 1989; Galli et al., 1977; Yamamoto et al., 1987; Bazan, 1989; Neuringer et al., 1988; Budowski et al, 1980) that both neural integrity and function can be permanently disturbed by deficits of n-6 and n-3 essential fatty acids exercised through the mother on fetal and neonatal development.

SPECIFIC FATTY ACIDS REQUIRED FOR NEURAL MEMBRANE GROWTH

Fats are used for two purpose in the body: (i) storage energy reserve and (ii) structural. The structural fat is built from non-essential and essential fatty acids in a manner similar to that by which protein is built from non-essential and essential amino acids.

Interest in structural fat, or lipid, comes from the fact that it is quantitatively the most important structural unit in the brain and nervous system and the second most important in all other soft tissues. The essential fatty acids occur in plants as linoleic (parent n-6) and α-linolenic (parent n-3) acids which are chain elongated and desaturated from 18 carbon chain lengths with two and three double bonds to 20 and 22 carbon chain lengths and much higher degrees of unsaturation with four and six double bonds.

The biosynthetic process responsible is slow, alters with age and, in some species, is almost non-existent. The pre-formed long-chain fatty acids are incorporated into the developing rat brain with a greater than tenfold efficiency when compared to the parent essential fatty acids (Sinclair, 1975). The importance of this point is that *the brain only uses the pre-formed long-chain EFAs and not the parent essential fatty acids*.

In experimental animals, deficits of EFAs during the period of brain growth have been found to lead to retardation of brain cell division, and visual and learning defects. As structural lipid provides the skeleton for cell membranes, those cell systems with specialized architectural requirements for membranes also have a specialized requirement for essential fatty acids. The brain is the prime example of a membrane-rich system as its function depends on a highly active signal transmission across its membranes for which the long-chain, unsaturated fatty acids are employed. It is now believed that both families (n-6 and n-3) are essential (Crawford and Sinclair, 1972; Lamptey and Walker, 1976; FAO/WHO, 1978; Budowski et al., 1987; Yamamoto et al., 1987; Neuringer et al., 1988; Bazan, 1989; Bourre et al., 1989), with the n-3 fatty acids such as docosahexaenoic acid playing a special role in highly active sites such as in the synapse and photoreceptor (Yamamoto et al., 1987; Neuringer et al., 1988; Bazan, 1989; Bourre et al., 1989)

THE VASCULAR SYSTEM

Another membrane-rich system is the endothelial cell layer that lines the arteries. The endothelial cell is flat and thin, so a large part of it is outer membrane. It uses essential fatty acids for its structural integrity and also as precursors for the synthesis of prostacyclin (Moncada et al., 1976), a hormone-like substance that prevents accidental thrombus adhesion during high pressure physical contact between the platelet and the arterial wall. Again, the balance of n-3 and n-6 fatty acids is relevant to the control of vasoconstrictive and thrombogenic activities via the cyclooxygenase products, the prostaglandins, derived from the long-chain EFAs (Moncada et al., 1976; Higgs et al., 1986). Additionally, lipoxygenase products are involved in constriction and dilation, as well as interaction of the white cell population with the endothelium and the immune response; interestingly, these products have activities some 1,000 times more powerful than histamine (Samuelsson, 1986). The synthesis of these physiologically active derivatives of the EFAs is in turn dependent on the balance of EFAs in the diet (Weber, 1987).

Most recently, discovery of the photoreceptor's extensive use of docosahexaenoic acid has led to research that may be relevant to retinopathy of prematurity (Uauy et al., 1990; Carlson et al, 1986).

Understanding the link between the unique involvement of lipids in brain structure is already paying dividends in genetically determined peroxisomal disease (Goldfischer, 1988), as in Zellweger's syndrome, in which very long-chain fatty acids accumulate in the brain causing rapid progressive decline in function and severe handicap. The first studies of preventive treatment for these adreno-leukodystrophies are based on feeding shorter chain fatty acids to suppress the synthesis of the very long-chain components. As with phenylketonuria, it may be possible to prevent the escalation of damage in these genetic disorders with diet, once the fault has been properly identified.

MATERNAL NUTRITION AND PREGNANCY OUTCOME

No nutrient is an island, and the essential fatty acids are only a part of a nutrient complex in which the function of one nutrient usually occurs with, and is dependent on, the availability of another. In this context, linoleic acid occurs with tocopherol and several B vitamins, whereas α-linolenic acid occurs with ß-carotene, ascorbic acid, and tocopherol.

In an attempt to test experimental data in the human context, we studied more than 500 pregnancies and found significantly reduced intakes of several vitamins, minerals, and fatty acids in mothers who produced low-birth-weight babies as compared to those whose babies were in the 3.5-4.5 kg reference range. Of the 44 nutrients we measured in diets of mothers who produced low-birth-weight babies, 43 were taken at significantly lower levels than those taken by mothers who produced babies in the optimal reference range. Additionally, nutrient data tracked with birth weight in a dose-responsive manner up to 3,270 g, above which there was no relationship.

While smoking and alcohol abuse are known to influence the outcome of pregnancy, our studies on diet, which controlled for these factors, indicate that maternal nutrient intake and the habitual diet are independently related to birth weight and infant's head circumference. Smoking and alcohol abuse were not serious issues in our study population, possibly because participants had relatively low disposable incomes, which would have reduced their ability to so indulge. Among those smoking 0-15 cigarettes per day, smoking correlated negatively with nutrient intake.

The data further indicated that maternal nutrition at or before conception was more strongly correlated to birth dimensions than was nutrition during the pregnancy itself. The above data make the hypothesis that maternal nutrition is irrelevant to birth weight and head circumference highly improbable.

The conclusion that maternal nutrition in preparation for conception is the most critical is supported by retrospective studies of the Dutch, German, and Norwegian food shortages (Wynn and Wynn, 1981), the Illinois Hospital study (Orstead et al., 1985), and our own prospective study. It is also supported by the basic biologic principle that the most active period of cell division is within the first few weeks after conception. Indeed, it can be said that the period before a mother knows she is pregnant, sees a health professional, and has pregnancy confirmed is the period of greatest rate of cell division. That means that this sensitive period occurs under conditions that existed before conception.

From the behavior of all other animal systems, there is clear evidence that "nature prepares in advance of conception." The bird's egg is laid with 100% of all nutrients in place that are required to convert the fertilized cell into a chick. Similar advance preparations can be seen in the insect world from bees to ants, and the farming community has learned the same principle empirically. The same principle can be seen intact in humans, with egg secretion being canceled below a certain level of body fat (as in anorexia). Again, fat deposition and placental growth proceed rapidly in the first half of pregnancy, whereas fetal growth is greater in the last half. That is, the mechanism for fetal nourishment is built ahead of the need. Furthermore, the parturition at term in the well-nourished mother leaves her with fat storage that provides for one-third of the energy requirements for the first 100 days of lactation (FAO/WHO, 1978).

The conclusion that nature prepares in advance has important implications for our understanding of (i) the influence of maternal nutrition on fetal growth and (ii) our interpretation of events occurring in the immediate neonatal period.

ESSENTIAL FATTY ACID INDICES OF FETAL UNDERNUTRITION

In an attempt to obtain further objective evidence, the essential fatty acid content of maternal and cord blood phosphoglycerides was studied. It was found that reduced levels of arachidonic acid were associated with low birth weight, as well as with head circumference and placental weight (Crawford et al., 1989). More recent studies on premature infants similarly suggest that arachidonylphosphoglyceryl choline is a closer index of birth weight than the docosahexaenyl glyceride which in its turn may be a better index of degree of prematurity (Leaf et al., in press).

We then studied the endothelium of the umbilical artery as being possibly representative of fetal tissue from a range of low birth weight and normal babies. Strong correlations were obtained with arachidonic and docosahexaenoic acid components of the polar phosphoglycerides in relation to birth weight and head circumference.

We also identified a novel fatty acid, the "Mead Acid," which is a polyunsaturated fatty acid that the body makes from oleic acid if it is deficient in the essential fatty acids needed for membranes. The Mead Acid was present in unusually large amounts. The conventional diagnostic technique for essential fatty acid deficiency is to use the ratio of this unusual fatty acid to arachidonic acid, which it replaces. Applying this test we again found strong correlations with birth weight and head circumference.

The strongest correlations (Pearsons >0.8; $p<0.001$) were found in the endothelial ethanolamine phosphoglycerides (EPG). As this is predominantly an inner membrane phosphoglyceride, its information is likely to be of historical interest relating to the conditions during fetal growth (Crawford et al., 1990).

Examination of twins revealed that these indices were greater in the twin with the lower birth weight. On the other hand the levels of docosahexaenoic acid were identical in the twins, consistent with the suggestion that while arachidonic acid may be related to birth weight, docosahexaenoic acid was more relevant to degree of prematurity.

Further research and development on this technique is much needed as we feel it could be the first technique which specifically defines the extent of fetal malnutrition regardless of a maternal or placental cause. The question is whether it predicts neural deficits or damage and, if so, in what direction. From the model systems of cerebellar (Budowski et al., 1987) and retinal development (Neuringer et al., 1988), deficits during development can specifically affect that part of the brain developing at that time. That is, Dobbing's principle of the period of brain growth being the "vulnerable" period (Dobbing, 1972) is applicable to different parts at different times. The important point derived from this evidence is that the conditions during prenatal growth may be responsible for disorder or susceptibility to disorder appearing postnatally.

The pre-term infant is also exposed to problems associated with incomplete maturation. For example, our data suggest that membrane-bound cupric zinc superoxide dismutase activity is rising in the pre-term infant but does not quite reach the levels at expected date of delivery that are found in term babies. Additionally, babies that are small for date appeared to have lower activities. The oxygen tension in the fetus is significantly below that of the neonate. If the maturation of the defense against free radicals does occur late in fetal growth, then being born early might add the additional stress of a poor defense against free radicals.

PERIVENTRICULAR HEMORRHAGE AND CEREBRAL PALSY

Hemorrhage appears to occur to a varying extent in a large proportion of pre-term infants (Dubowitz, 1991). Hence we need to ask if a reduced supply of the precursors for membrane growth contributes to the fragility of the peri- and intra-ventricular vascular system which is growing rapidly to accommodate the brain growth thrust occurring at that time.

There is now evidence from several laboratories and our own that the premature infant is denied the substantial supply of arachidonic and docosahexaenoic acids which it otherwise would have received if it had remained as fetus fed by the placenta. Within three to six days their levels may fall to less than a fifth of those found in the placental-fetal supply.

Both these fatty acids are key components of neural and vascular membranes. Arachidonic acid and other 20-carbon polyunsaturated fatty acids are also precursors for eicosanoids which regulate blood flow and coagulation. The deficits of these fatty acids induced by prematurity would be expected to lead to loss of the membrane integrity manifested by hemolysis and hemorrhage. The answers to these questions would suggest a route for prevention.

In order to meet the outstanding demands for brain growth, the neonatal cerebrovascular system has to deliver as much as 60% of the total oxygen and nutrient intake by the baby. During the last trimester the brain enjoys its major growth spurt, which includes a growth spurt of the cerebrovascular system. It is known that an inadequate nutrient flow during a period of critical development can induce "susceptibility." Hence the question that needs to be asked is whether the periventricular hemorrhage is due to inadequate nutrient supply to the fetus in the last trimester when the cerebrovascular system is undergoing rapid development.

Winick (1983) has pointed out that fetal growth-retarded babies are born from small placentas with multiple and often massive infarctions (Althabe et al., 1985). He therefore argued on the basis of the pathology that poor placental development is responsible for fetal growth retardation. This implies that the nutritional/biochemical conditions are those that induce inadequate vascular growth and risk of coagulation. The low intake of essential fatty acids we found to be associated with low birth weight would be expected to compromise endothelial growth and function. Indeed, reduced synthesis of prostacyclin has been reported in the umbilical arteries and placentas from low-birth-weight babies in association with increased Mead Acid (Ongari et al., 1984).

The placenta is largely a new and rapidly growing vascular system that develops in the first part of pregnancy to serve the fetal growth thrust of the last trimester. In association with low birth weight, there is evidence of much placental vascular pathology. One must therefore ask whether the conditions that resulted in a poorly developed and infarcted placenta have adversely influenced fetal development.

CONCLUSIONS

From the evidence so far accumulated, the baby born small-for-dates should probably be considered to be a baby that was malnourished as a fetus.

Fetal malnourishment could be due to several causes but placental development and maternal nutrition would appear to be relevant. However, as far as nutrition is concerned, it seems that maternal nutrition prior to conception is more relevant than nutrition during the latter part of pregnancy when it is too late to affect fetal cell division.

The pre-term infant is born at a time when cell membrane development is unprepared for the different conditions of the extra-uterine environment and is likely to be further affected by feeding regimens which currently do not replace the placental provision, especially of the essential fatty acids.

Additionally, if nutritional or other constraints in the supply of nutrients do result in fragile membranes in the pre-term infant, then it would be hardly surprising if the integrity of membrane-rich systems is in jeopardy.

These considerations recommend the thought that the several developmental disorders for which the low-birth-weight infant is at high risk need not be due to obstetric or pediatric mismanagement but may be the result of a poor preparation for life.

REFERENCES

Althabe O, Laberre C, Telenta M (1985) Maternal vascular lesions in placentae of small for gestational age infants. Placenta 6:265-276.

Bazan NG (1989) Lipid-derived metabolites as possible retina messengers: Arachidonic acid, leukotrienes, eicosanoids and platelet activating factor. In: Extracellular and intracellular messengers in the vertebrate retina pp 269-300. New York: Alan R. Liss.

Bourre J-M, Marianne F, Youyou A, Dumont O, Picotti M, Pascal G, Durand G (1989) The effects of dietary alpha-linolenic acid on the composition of nerve membranes, enzymatic activity, amplitude of electrophysiological parameters, resistance to poisons and performance of learning tasks in rats. J Nutr 119:1880-1891.

Budowski P, Hawker CM, Crawford ME (1980) L'effet protecteur de l'acide alpha-linolenique sur l'enceohalomalaciie chez le pouter. Ann Nutr Aliment 34:389-400.

Budowski P, Leighfield MJ, Crawford MA (1987) Nutritional encephalomalacia in the chick: An exposure of the vulnerable period for cerebellar development and the possible need for both ω6 and ω3 fatty acids. Br J Nutr 58:511-520.

Carlson SE, Rhodes PG, Ferguson MG (1986) Docosahexaenoic acid status of pre-term infants at birth and following feeding with human milk formula. Am J Clin Nutr 44:798-804.

Clausen J and Moller D (1967) Allergic encephalomyelitis induced by brain antigen after deficiency in polyunsaturated fatty acids during myelination. Is multiple sclerosis a nutritive disorder? Acta Neurol Scand 43:375-388.

Crawford MA and Sinclair AJ (1972) Nutritional influences on the evolution of the mammalian brain. In: Lipids, malnutrition and the developing brain. Ciba foundation symposium (Elliott K, Knights J, eds) pp 267-287. Amsterdam: Elsevier.

Crawford MA, Costeloe K, Doyle W, Leighfield MJ, Lennon AE, Meadows N (1990) Potential diagnostic value of the umbilical artery as a definition of neural fatty acid status of the fetus during its growth. Biochem Soc Trans 18:761-766.

Crawford MA, Doyle W, Drury P, Lennon A, Costeloe K, Leighfield M (1989) N-6 and n-3 fatty acids during early human development. J Intern Med 225(Suppl 1):159-69.

Dobbing J (1972) Vulnerable periods of brain development. In: Lipids, malnutrition and the developing brain. Ciba Foundation Symposium (Elliott K, Knights J, eds) pp 1-7. Amsterdam: Elsevier.

Dubowitz V (1991) Magnetic resonance studies of the newborn brain in relation to cerebral palsy. Cerebral Palsy Today 1:1-3.

FAO/WHO (1978) Report of an expert consultation: The role of dietary fats and oils in human nutrition. Rome:FAO.

Galli C, Galli G, Spagnuolo C, Bosisio E, Tosi L, Folco CG, Longiave D (1977) Dietary essential fatty acids, brain polyunsaturated fatty acids, and prostaglandin synthesis. In: Function and biosynthesis of lipids. (Bazan NG, Brenner RR, Giusto NM, eds) pp 561-573. New York: Plenum Press.

Goldfischer SL (1988) Peroxisomal disease. In: Biological membranes: aberrations in membrane structure and function (Karnovsky ML, Leaf A, Bollis LC, eds) pp 117-137. New York: Alan R. Liss.

Higgs EA, Moncada S, Vane JR (1986) Prostaglandins and thromboxanes. Progr Lipid Res 25:5-11.

Lamptey MS and Walker BL (1976) A possible essential role for dietary linolenic acid in the development of the young rat. J Nutr 106:86.

Leaf A, Leighfield MG, Costeloe KL, Crawford MA (In press) Long chain polyunsaturated fatty acid composition of plasma choline phosphoglycerides in preterm infants. J Pediatr Gastroenterol Nutr.

Moncada S, Gryglewski R, Bunting S, Vane JR (1976) An enzyme isolated from arteries transforms prostaglandin endoperoxides to an unstable substance that inhibits platelet aggregation. Nature (London) 263:663-665.

Neuringer M, Anderson GJ, Connor WE (1988) The essentiality of n-3 fatty acids for the development & function of the retina & brain. Annu Rev Nutr 8:517-541.

Ongari MA, Ritter JM, Orchard MA, Waddell KA, Blair IA, Lewis PJ (1984) Correlation of prostacyclin synthesis by human umbilical artery with status of essential fatty acid. Am J Obstet Gynecol 149:455-460.

Orstead D, Arrington A, Kamath SK, Olsen R, Kohrs MB (1985) Efficacy of prenatal nutrition counselling: Weight gain, infant birth weight, and cost effectiveness. J Am Diet Assoc 85:40-45.

Samuelsson B (1986) Leukotrienes and other lipoxygenase products. Prog Lipid Res 25:13-18.

Sinclair AJ (1975) The incorporation of radioactive polyunsaturated fatty acids into the liver and brain of the developing rat. Lipids 10:175-184.

Uauy RD, Birch DG, Birch EE, Tyson JE, Hoffman DR (1990) Effect of dietary omega-3 fatty acids on retinol function of very low birth weight neonates. Pediatr Res 28:485-492.

Weber PC (1987) N-3 fatty acids and the eicosanoid system. In: Fat production and consumption - technologies and nutritional implications series A: Life sciences, Vol 131 (Galli C, Fedeli E, eds) pp 123-130. New York: Plenum Press.

Winick M (1983 Nutrition, intrauterine growth retardation and the placenta. Trophoblast Research 1:7-14.

Wynn M and Wynn A (1981) The prevention of handicap of early pregnancy origin. London: The Foundation for Education & Research in Childbearing.

Yamamoto N, Saitoh M, Moriuchi A, Nomura M, Okuyama H (1987) Effect of dietary alpha-linolenate/linoleate balance on brain lipid composition and learning ability in rats. J Lipid Res 28:144-151.

REGULATION OF ARACHIDONIC ACID METABOLISM IN THE PERINATAL

BRAIN DURING DEVELOPMENT AND UNDER ISCHEMIC STRESS

Ephraim Yavin,[1] Baruch Kunievsky,[1]
Nicolas G. Bazan,[2] and Shaul Harel[1]

[1]Department of Neurobiology
Weizmann Institute of Science
Rehovot, Israel

[2]Louisiana State University Medical Center
New Orleans, LA, USA

SUMMARY

Oxygen deprivation following cessation of blood flow to vital organs such as brain, heart, and kidney is a ubiquitous human disease, invariably leading to devastating consequences. Studies in experimental models support the contention that membrane permeability is altered, ion fluxes impaired, and energy stores depleted under these circumstances. Certain lipids such as diglycerides (DG) and arachidonic acid (AA), both of which are important cellular second messengers, appear to increase during ischemia. At this point, the contribution of these and other lipids to cell deregulation, loss of function, and ultimate death has not been clarified because no precise link between lipid alterations as detected in ischemia and subsequent cellular processes has been made.

In this report we examine the origin of lipid-derived second messengers in fetal rat brain prelabeled with [^{3}H]AA and study the fate of various lipids upon obstruction of the fetal-maternal circulation. The data support the possibility of a phospholipase A_2-mediated deacylation of poly-phosphoinositides (poly-PI) to form free AA and a phospholipase C-mediated hydrolysis of PC to form DG during ischemia.

Lipid Pathobiochemistry During Ischemia

Ischemic episodes caused by obstruction of the cerebral blood flow are common brain insults that may lead to permanent damage, depending on duration, degree, and target location. In biochemical terms, the early stages of ischemic sequelae are characterized by immediate ionic and pH changes, depletion of energy stores, and rapid alterations in membrane lipid constituents (Bazan, 1976; Farber et al., 1981; Rehncrona et al., 1982; Siesjö, 1984). Among the proposed changes in membrane lipids, accumulation of free fatty acids (FFA) by a phospholipase A_2 (PLA_2)-mediated hydrolysis, which releases primarily polyunsaturated fatty acids (PUFA), has been well established (Bazan, 1970; Galli and Spagnuolo, 1976). Of equal magnitude is the accumulation of saturated and mono-unsaturated FFA (Gardiner et al., 1981; Abe et al., 1987; Ikeda et al., 1986), which would indicate that additional lipases participate in the process, presumably to further degrade intermediary glyceride and lysophospholipid metabolites. During early ischemia, accumulation of DG by a phospholipase C (PLC)-mediated activity has been noticed (Banschbach and Geison, 1974; Bazan, 1976; Huang and Sun, 1987). In the adult brain, one species of this enzyme catalyzes the breakdown of poly-PI into DG and inositol phosphates (Berridge, 1984).

Ample evidence now exists to indicate that changes in the intracellular Ca^{2+} steady state levels are the main triggering mechanism for these lipid-associated conversions (Siesjö and Wieloch, 1985). Ca^{2+} levels can be elevated either through influx via one or several routes into the cell and/or via an inositol-triphosphate-mediated mobilization from intracellular stores. Ca^{2+} elevation appears to be the driving force for the activation of both PLA_2 and PLC. Each one of these Ca^{2+}-dependent enzymatic activities plays a pivotal role in the dynamic shaping of plasma membrane lipids. They also participate in generating lipid-derived second messengers, such as AA and DG, during normal signal transduction processes (Nishizuka, 1986; 1988). Therefore, elevation of these enzymatic activities in the course of ischemia requires a rigorous examination with respect to the intermediary metabolites that are generated and the possible effects of these metabolites on alterations of cell or tissue function. One example of such an effect is the activation of brain plasmalogenase during the first minute of ischemia and the subsequent production of lyso derivatives (Edgar et al., 1982). Accumulation of the lyso derivatives in plasma membrane as a result of drastic reduction in ATP and phosphocreatine levels may introduce perturbation in the packing of the bilayer. Excess amounts of FFA are yet another example of perturbation of the *cis* bilayer configuration (Klausner et al., 1980; Hill et al., 1983), while the flip-flop properties of DG could interfere with the *trans* lipid bilayer configuration (Allan et al., 1978; Epand and Lester, 1990). Free polyunsaturated fatty acids are believed to serve as substrates for free radical formation and edema induction (Chan and Fishman, 1978). Thus, understanding the molecular mechanisms that cause FFA deacylation under pathological conditions and the metabolic routes that lead to clearance or accumulation of FFA is a prerequisite for the prevention of irreversible tissue damage.

Global Ischemic Episodes in the Developing Fetal Rat Brain During Intrauterine Life

For a number of years, a major effort in our laboratory has been devoted to exploring the molecular basis underlying growth retardation in humans and animal models following intrauterine insults (Harel et al., 1985a,b). Hypoxic-ischemic episodes in perinatal life are considered to be a major cause of neuronal injury and impaired postnatal development (Gottfried, 1973; Frank, 1985). To investigate this important clinical entity we have used two parallel animal models to follow both biochemical and behavioral changes related to brain development.

Using pregnant rabbits in the last trimester of gestation, we have investigated the effects of ischemic intrauterine insults on eicosanoid metabolism (Goldin et al., 1990). We demonstrated that close-to-term rabbit fetuses preinjected intracerebrally with [³H]AA gave rise to radioactively labeled prostaglandin (PG) E_2, thromboxane B_2 (TxB_2), and 6-keto-$PGF_{1\alpha}$ metabolites. TxB_2 content in the restricted fetuses was five- and twofold higher at 3 hr than control fetal brain and placenta tissue values, respectively, and remained significantly high for 24 hr. Also 6-keto-$PGF_{1\alpha}$ levels reached a peak value that was greater by 2.5- and 1.5-fold at 6 hr for the ischemic brain and placenta tissue, respectively, compared with control fetuses. The thromboxane/prostacyclin ratio was maximal in the brain after approximately 3 hr, while that in the placenta continued to rise even after 20 hr. These sustained high levels of thromboxane may be indicative of cerebral vasoconstriction and possible exacerbation of the ischemic insult.

Recently we have established a second experimental model using pregnant rats at 19-20 days gestation to mimic global ischemic episodes (Magal et al., 1988). The selection of the timing and mode of induction of whole body ischemia in the rat model was based on: (a) the similarity to pathophysiological conditions existing in the uteroplacental axis *in vivo*; and (b) the parallelism in the timing of the neuronal growth spurt in rat and man. In both species, stress imposed at critical stages may result in impaired development. Acute sessions of ischemia ranging from 5-45 min were produced during pregnancy by obstructing the entire placental vasculature with clamps; removal of the clamps permitted re-establishment of the blood circulation. Unlike related models of placenta blood vessel ligation (Wigglesworth, 1964), all individual fetuses were made hypoxic such that experimental variability was significantly reduced. Measurement of lactic acid content confirmed the fact that fetuses had been rendered ischemic (Magal et al., 1988).

Protein Kinase C: A Possible Target During Ischemia

The involvement of PKC in transmembrane signaling and its dependence on lipids is now a well-established phenomenon (Nishizuka, 1988). PKC requires phosphatidylserine (PS) and Ca^{2+} and depends on DG for maximal activity. The transmembrane signaling events that elicit DG production during normal cellular activity are now emerging. For the most part a receptor-mediated turnover of PIP_2, coupled via a GTP (G protein)-dependent activation of a specific PLC, appears to be responsible for DG formation (Bell, 1986). AA and oleic acid can activate PKC in a Ca^{2+}-phospholipid-independent manner (Murakami and Routtenberg, 1985; Nishizuka, 1988). Linoleic acid and its methyl ester have been shown in artificial bilayers to partially replace DG or to act synergistically with DG to activate PKC both *in vitro* (Sekiguchi et al., 1987) and in platelet cells (Seifert et al., 1987).

In the fetal ischemic rat brain, we have recently demonstrated a rapid translocation of PKC after maternal-fetal blood flow obstruction (Louis et al., 1988). Prolonged obstruction caused a substantial loss of PKC activity and a concomitant appearance of a Ca^{2+}/lipid-independent enzymatic activity. Therefore a major aim of the current work has been to identify and quantify these lipid-derived second messengers, i.e., DG and FFA, in order to shed some light on the origin of the phospholipids associated with PKC translocation and destruction. Identification of these lipid constituents may provide important clues for understanding ischemic pathophysiology.

Lipid Turnover Studies

To accomplish this aim we have tracer-labeled fetal brain lipids *in vivo* and obstructed fetal-maternal circulation thereafter as detailed elsewhere (Magal et al., 1988). When injected intracerebrally through the uterine wall into the 20-day-old fetal rat, [^{3}H]AA was rapidly incorporated into the majority of brain lipids. After 1 hr of isotope administration, the phospholipid fraction accounted for 90.0±3.1% of the total lipid-associated radioactivity, with the remaining radioactivity found in the neutral lipids fraction (Fig. 1). Most of the label (55.6±8.2%) was found in PC, followed by PI and PE (18.0±3.4% and 13.7±7.3%, respectively). In the non-polar lipids, label was found predominantly in the acylglycerol (mono, di, and tri) species while the levels of free AA remained relatively low (1.5±0.2%). The specific radioactivity of PI was considerably higher (520±78 dpm/mg lipid Pi) compared to that of PC (177±15 dpm/mg lipid Pi) and PE (67±4 dpm/mg lipid Pi) phospholipids.

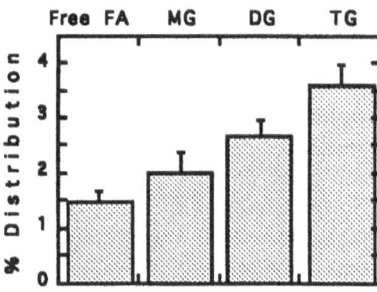 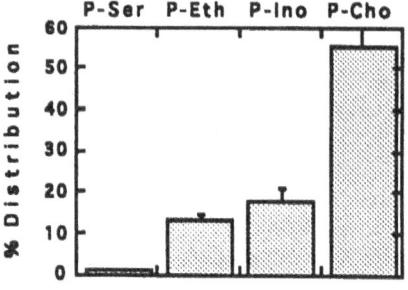

Figure 1. Distribution of [³H]AA into fetal brain lipids. [³H]Arachidonate (Na⁺) (2 μCi/2 μl) was injected intracerebrally into 20-day-old embryos. After 1 hr *in utero*, fetuses were delivered and brains were rapidly excised and immediately homogenized in hexane/isopropanol (3/2 by vol) mixture. Phospholipids and neutral lipids were separated by TLC. Incorporation into phospholipids is expressed as percent of total radioactivity. Values are means ± SE of three experiments.

Incorporation of [³H]AA into cellular lipids after intraventricular administration of the tracer into normal or ischemic fetuses was time dependent and unaffected by blood flow arrest at the placental level. A rapid decrease in radioactivity associated with the FFA pool ($t_{1/2}$=2 min) was noticed in both experimental and control animals. The disappearance of [³H]AA from the FFA pool was accompanied by a parallel increase in labeling of tissue phospholipids, which reached a plateau 5 min after injection. Between 40-70% of the injected radioactivity was recovered in the brain, indicating a very effective acylation process. No significant differences between normal and restricted animals were encountered even after 40 min of ischemia, suggesting that blood flow arrest did not impair reacylation.

Distribution of radioactivity in various fetal brain lipids after maternal-fetal blood arrest was studied. Most significant was a 25% increase in the free AA radioactivity content after 5 min, which nearly doubled when the restriction session was prolonged to 20 min. Equally pronounced was a rapid increase (about 50% compared to control) in the diglyceride radioactivity after 5 min of ischemia, while few or no changes were seen in monoglycerides and triglycerides. Accumulation of AA was also noticed after examining the fatty acid profile by gas chromatography. Surprisingly, there was no significant accumulation of saturated or unsaturated FFA.

The rapid release of AA via a phospholipase A_2 and/or a phospholipase C/diglyceride lipase combined activation following ischemia has been shown in the adult brain (Gardiner et al., 1981; Edgar et al., 1982; Huang and Sun, 1986; Abe et al., 1987; Nakano et al., 1990). These studies emphasized that accumulation of AA is accompanied by a quantitatively greater release of saturated and/or monounsaturated FFA (Yasuda et al., 1985). The significantly low levels of free FFA in the fetal brain may be the result of diminished hydrolysis or rapid reacylation. Evidence for deacylation/reacylation of the 2-position of PI to obtain polyunsaturated derivatives in the nervous system has been observed (Yau and Sun, 1974). The selective release of AA as revealed in our experimental model by both mass and radioactivity data is very consistent and could be part of a signal transduction chain which results in downstream eicosanoid synthesis (Goldin et al., 1990) or PKC activation (Louis et al., 1988).

In conjunction with the possible release of AA, the second question we addressed concerned the origin of DG during ischemia. Examination of DG content in the fetal brain indicated that ischemic animals contained significantly more DG in comparison

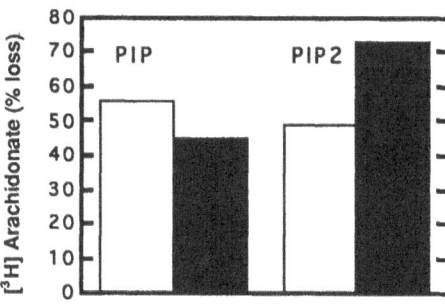

Fig. 2. Losses of [³H]AA in polyphosphoinositides after ischemia. Intracerebral injection of [³H]AA (2 μCi/ml) was performed 1 hr before blood flow arrest. Rats were restricted for 5 min (white bar) or for 20 min (black bar). Values are expressed as percentage of loss in comparison to control values.

to their sham-operated counterparts. There was no preferential increase in the stearate/AA acyl group species, usually abundant in poly-PI lipids, suggesting a limited, PI-specific PLC activity. This is in contrast to the enhanced production of AA-enriched DG species seen in the ischemic adult brain (Aveldaño and Bazan, 1975).

Distribution of radioactivity in fetal brain phospholipids did not vary significantly after various periods of restriction. A gradual reduction in PC became apparent. Within the minor phospholipids, a slight increase in PA levels above the control was found. The most significant change was a reduction of radioactivity in PI after 5 min of ischemia. Losses in poly-PI radioactivity were very prominent after 5 min and amounted to nearly 55% and 45% for PIP and PIP_2, respectively (Fig. 2). After 20-min restriction, the apparent losses in PIP_2 were even greater.

The marked reduction in poly-PI radioactivity and the concomitant appearance of label in the DG pool conformed with the general idea of a phospholipase C-mediated hydrolysis (Berridge, 1984). It remained unclear, however, whether DG was indeed derived from poly-PI largely because, as noted earlier, DG acyl species did not contain excess stearate and arachidonate species.

To obtain a better assessment of the phospholipids mobilized for DG and AA formation during the ischemic onset, [¹⁴C]palmitic acid and [³H]AA were co-injected by the intracerebral route to double label brain lipids. Fetuses were then subjected to a 20-min restriction episode, and restriction was followed by a 30-min reperfusion session. Brain lipids were extracted and purified, and the [³H]/[¹⁴C] ratios in individual lipids were determined.

The poly-PI fraction exhibited the highest [³H]/[¹⁴C] ratio in the nonischemic control animals with apparent values of 2.38, 2.50, and 2.44 for PIP_2, PIP, and PI respectively. The relative ratio in PE was also high (2.2 compared to 1.6 for the FFA pool), most likely due to the presence of vinyl ether bonds in the 1-position which were not accessible to labeling by [¹⁴C]palmitic acid. After 20-min restriction, more than 25% and 40% of ³H label was lost from PIP and PIP_2, respectively, as summarized in Figure 3. At that time, little or no ¹⁴C radioactivity was lost in the poly-PI fraction. PC, PE, and PS exhibited a relative increase in the abundance of ³H label over the control after 20-min ischemia. When reperfusion was continued for 30 min, losses in PC and PE label were notable. This experiment provides strong support for the possibility that PIP and PIP_2 are the major source for AA release via a PLA_2-catalyzed hydrolysis, which

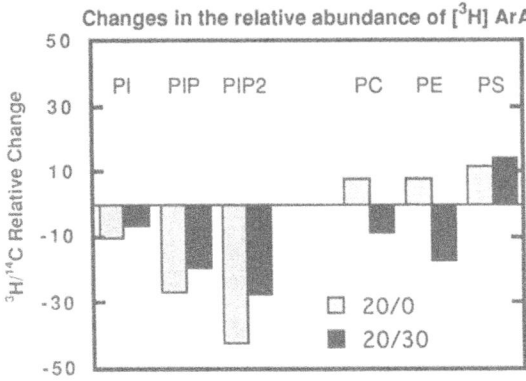

Figure 3. [³H]Arachidonate/[¹⁴C]palmitate relative ratio change in fetal brain lipids after ischemia. [³H]AA (1 µCi) and [¹⁴C]palmitate (1.5 µCi) in 2 ml were co-injected by the intracerebral route to 20-day-old embryos. After 1 hr *in utero* labeling, pups were restricted for 20 min (20/0) or restricted and reperfused (20/30) *in utero*. Brain lipids were extracted with chloroform /methanol/HCl (2/1/0.5% by vol) from designated fetuses for polyphosphoinositides or with hexane/isopropanol (3/2 by vol) for the bulk lipids. Individual lipids were separated by TLC and radioactivity was determined. Values are expressed as change in the relative abundance of [³H]AA over [¹⁴C]palmitate in experimental compared to sham embryos.

is compatible with the selective accumulation of AA. This idea is summarized in the metabolic scheme outlined in Figure 4A. At present, there is no direct evidence for the existence of a poly-PI-specific phospholipase A_2. Neither could we identify lyso-poly-PI intermediates, mainly because they may be rapidly reacylated or further degraded. The diminished ³H radioactivity in poly-PIs after 5 min ischemia (Fig. 2) is not due to conversion of PIP and PIP_2 metabolites into DG as inferred from the relative persistence of the [¹⁴C]palmitate label in these phospholipids.

The possible participation of poly-PI in the generation of brain free fatty acids at the onset of ischemia is not an entirely new concept. Based on the acyl group composi-

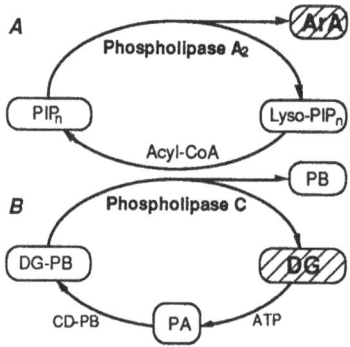

Figure 4. Metabolic routes for AA and DG formation in fetal rat brain. Scheme A depicts a phospholipase A_2-mediated hydrolysis of poly-phosphoinositides (PIP_n) and formation of AA. Scheme B depicts a phospholipase C-mediated hydrolysis of a major phospholipid (DG-PB) and generation of DG and phosphoryl base (PB).

tion of DG, Ikeda et al. (1986) concluded from their studies that AA release was the result of a combined phospholipase C and DG lipase hydrolysis. This is in line with a signal transduction sequence, which involves a PLC-mediated PIP_2 hydrolysis to generate inositol triphosphate second messenger and DG (Berridge, 1984). In contrast we propose herein that the ischemic fetal brain derives its AA from a phospholipase A_2-mediated activation of poly-PIs based on rapid disappearance of PIP and PIP_2, lack of PIP_2-derived DG, and selective accumulation of AA.

Formation of DG during the ischemic onset appears to arise from PC hydrolysis. This assumption (Fig. 4B) is based on two observations: [1] PC appears to change moderately at the onset of ischemia (data not shown); [2] PC contains 22:6 (docosahexaenoic acid), a small but prominent FFA, which is increased twofold in the DG fraction. Some evidence that PC may be involved in certain signal transduction systems has been recently discussed (Exton, 1990).

The molecular basis for participation of highly specific phospholipids and their derivatives in ischemic pathophysiology must be evaluated with respect to the ontogeny of excitatory amino acid receptors and neurotransmitter complexes in the brain in general and the developing brain in particular (Ikonomidou et al., 1989). Substantial evidence now suggests that glutamate and glutamate receptors are actively involved in amplification of the ischemic insult (Meldrum, 1985; Rothman and Olney, 1986). A particular role has been attributed to a PLC-mediated hydrolysis of PI, which releases DG in response to activation/binding of glutamate to quisqualate receptors (Nicoletti et al., 1986; Chen et al., 1988; Sladeczek et al., 1988). The possible linkage between a PLA_2-dependent poly-PI degradation and glutamate receptors in the ischemic perinatal brain remains to be established.

ACKNOWLEDGMENTS

This work was made possible partly through a grant made available by the Gulton Foundation, New York, and by a grant provided by the Revson Foundation of the Israel Academy of Sciences and Humanities, Jerusalem. The authors are most grateful to Jeff Hammer for invaluable assistance with the manuscript.

REFERENCES

Abe K, Kogure K, Yamamoto H, Imazawa M, Miyamoto K (1987) Mechanism of arachidonic acid liberation during ischemia in gerbil cerebral cortex. J Neurochem 48:503-509.

Allan D, Thomas P, Michell RH (1978) Rapid transbilayer diffusion of 1,2-diacyl glycerol and its relevance to control of membrane curvature. Nature 276: 288-290.

Aveldaño MI and Bazan NG (1975) Rapid production of diacylglycerols enriched in arachidonate and stearate during early brain ischemia. J Neurochem 25: 919-920.

Banschbach MW and Geison RL (1974) Post-mortem increase in rat cerebral hemisphere diglyceride pool size. J Neurochem 23:875-877.

Bazan NG (1970) Effects of ischemia and electroconvulsive shock on free fatty acid pool in brain. Biochim Biophys Acta 218:1-10.

Bazan NG (1976) Free arachidonic acid and other lipids in the nervous system during early ischemia and after electroshock. In: Function and metabolism of phospholipids in the central and peripheral nervous system, Vol 72 (Porcellati G, Amaducci L, Galli C, eds) pp 197-205. New York: Plenum Press.

Bell RM (1986) Protein kinase C activation by diacylglycerol second messengers. Cell 45:631-632.

Berridge MJ (1984) Inositol triphosphate and diacylglycerol as second messengers. Biochem J 220:345-360.

Chan PH and Fishman RA (1978) Brain edema: induction in cortical slices by polyunsaturated fatty acids. Science 201:358-360.

Chen CK, Silverstein FS, Fisher SK, Statman D, Johnston MV (1988) Perinatal hypoxic-ischemic brain injury enhances quisqualic acid-stimulated phosphoinositide turnover. J Neurochem 51:353-359.

Edgar AD, Stosznajder J, Horrocks LA (1982) Activation of ethanolamine phospholipase A_2 in brain during ischemia. J Neurochem 39:1111-1116.

Epand RM and Lester DS (1990) The role of membrane biophysical properties in the regulation of protein kinase C activity. TIPS 8:317-320.

Exton JH (1990) Signaling through phosphatidylcholine breakdown. J Biol Chem 265(1):1-4.

Farber JL, Chien KR, Mittnacht S Jr (1981) Myocardial ischemia: The pathogenesis of irreversible cell injury in ischemia. Am J Pathol 102:271-281.

Frank L (1985) Effects of oxygen on the newborn. Fed Proc 44:2328-2334.

Galli C and Spagnuolo C (1976) The release of brain free fatty acids during ischemia in essential fatty acid-deficient rats. J Neurochem 26:401-404.

Gardiner M, Nilsson B, Rehncrona S, Siesjö BK (1981) Free fatty acids in the rat brain in moderate and severe hypoxia. J Neurochem 36:1500-1505.

Goldin E, Harel S, Tomer A, Yavin E (1990) Thromboxane and prostacyclin levels in fetal rabbit brain and placenta after intrauterine partial ischemic episodes. J Neurochem 54:587-91.

Gottfried AW (1973) Intellectual consequences of perinatal anoxia. Psychol Bull 80: 231-242.

Harel S, Tomer A, Barak Y, Binderman I, Yavin E (1985a) The cephalization index: a screening device for brain maturity and vulnerability in normal and intrauterine growth retarded newborns. Brain Dev 7:580-584.

Harel S, Yavin E, Tomer A, Barak Y, Binderman I (1985b) Brain: body ratio and conceptional age in vascular induced intrauterine growth retarded rabbits. Brain Dev 7:575-579.

Hill DJ, Dawidowicz EA, Andrews ML, Karnovsky MJ (1983) Modulation of microsomal glucose-6-phosphate translocase activity by free fatty acids: Implications for lipid domain structure in microsomal membranes. J Cell Physiol 115:1-8.

Huang SF and Sun GY (1986) Cerebral ischemia induced quantitative changes in rat membrane lipids involved in phosphoinositides metabolism. Neurochem Int 9:185-190

Ikeda M, Yoshida S, Busto R, Santiso M, Ginsberg MD (1986) Polyphosphoinositides as a probable source of brain free fatty acids accumulated at the onset of ischemia. J Neurochem 47:123-132.

Ikonomidou C, Mosinger JL, Salles KS, Labruyere J, Olney JW (1989) Sensitivity of the developing rat brain to hypobaric/ischemic damage parallels sensitivity to N-methyl-aspartate neurotoxicity. J Neurosci 9:2809-2818.

Klausner RD, Kleinfeld AM, Hoover RL, Karnovsky MJ (1980) Lipid domains in membranes. Evidence derived from structural perturbations induced by free fatty acids and lifetime heterogeneity analysis. J Biol Chem 255:1286-1295.

Louis JC, Magal E, Yavin E (1988) Protein kinase C alterations in the fetal rat brain after global ischemia. J Biol Chem 263:19282-19285.

Magal E, Goldin E, Harel S, Yavin E (1988) Acute uteroplacental ischemic embryo: lactic acid accumulation and prostaglandin production in the fetal rat brain. J Neurochem 51:75-80.

Meldrum B (1985) Excitatory amino acids and anoxic-ischemic brain damage. Trends Neurosci 8:47-48.

Murakami K and Routtenberg A (1985) Direct activation of purified protein kinase C by unsaturated fatty acids (oleate and arachidonate) in the absence of phospholipid and Ca^{2+}. FEBS Lett 192:189-193.

Nakano S, Kogure K, Abe K, Yae T (1990) Ischemia-induced alterations in lipid metabolism of the gerbil cerebral cortex: I. Changes in free fatty acid liberation. J Neurochem 54:1911-1916.

Nicoletti F, Iadorola MJ, Wroblewski JT, Costa E (1986) Coupling of inositol phospholipid metabolism with excitatory amino acid recognition sites in rat hippocampus. J Neurochem 46:40-46.

Nishizuka Y (1986) Turnover of inositol phospholipids and signal transduction. Science 233:305-312.

Nishizuka Y (1988) The molecular heterogeneity of protein kinase C and its implications for cellular regulation. Nature 334:661-665.

Rehncrona S, Westerberg E, Akeson B (1982) Brain cortical fatty acids and phospholipids during and following complete and severe incomplete ischemia. J Neurochem 38:84-93.

Rothman S and Olney J (1986) Glutamate and pathophysiology of hypoxic-ischemic brain damage. Ann Biochem 19:105-111.

Seifert R, Schachtele C, Schultz G (1987) Activation of protein kinase C by *cis*- and *trans*-octadecadienoic acids in intact human platelets and its potentiation by diacylglycerol. Biochem Biophys Res Commun 149:762-768.

Sekiguchi K, Tsukuda M, Ogita K, Kikkawa U, Nishizuka Y (1987) Three distinct forms of rat brain protein kinase C: Differential response to unsaturated fatty acids. Biochem Biophys Res Commun 145:797-802.

Siesjö BK (1984) Cerebral circulation and metabolism. J Neurosurg 60:883-908.

Siesjö BK and Wieloch T (1985) Molecular mechanisms of ischemic brain damage: Ca^{++} related events in cerebrovascular disease (Plum F, ed) pp 187-200. New York: Raven Press.

Sladeczek F, Recasens M, Bockaert J (1988) A new mechanism for glutamate receptor action: phosphoinositide hydrolysis. Trends Neurosci 11:545-549.

Wigglesworth JS (1964) Experimental growth retardation in the fetal rat. J Pathol Bacteriol 88:1-13.

Yasuda H, Kishiro K, Izumi N, Nakanishi M (1985) Biphasic liberation of arachidonic and stearic acids during cerebral ischemia. J Neurochem 45:168-172.

Yau TM and Sun GY (1974) The metabolism of [1-^{14}C]arachidonic acid in the neutral glycerides and phosphoglycerides of mouse brain. J Neurochem 23:99-104.

Yavin E, Goldin E, Magal E, Tomer A, Harel S (1989) Ischemia stress and arachidonic acid metabolites in the fetal brain. Ann NY Acad Sci 559:248-258.

INTERACTIONS OF PHOSPHOLIPIDS AND FREE FATTY ACIDS WITH ANTIDEPRESSANT RECOGNITION BINDING SITES IN RAT BRAIN

Marta Stockert, Luis M. Zieher, and Jorge H. Medina

Instituto de Biología Celular
Universidad de Buenos Aires
Paraguay 2155, 3° piso
Buenos Aires, Argentina

ABSTRACT

The lipid microenvironment of cell membranes has been shown to regulate both neurotransmitter and hormone receptors. Preincubation of cortical synaptosomal membranes of rat brain with phospholipase A_2 (PLA_2) increases the number of [^{3}H]imipramine ([^{3}H]IMI) high affinity binding sites without altering K_d (B_{max} control: 2.53 ± 0.28 pmol/mg protein vs B_{max} PLA_2: 3.66 ± 0.26 pmol/mg protein). The displacement curves of [^{3}H]IMI binding in synaptosomal membranes with other tricyclic antidepressants are not affected by the presence of PLA_2. The effect of PLA_2 was prevented by incubation with EGTA ($2x10^{-3}$) or bovine serum albumin (BSA; 1:1). In addition, end products of catalytic activity of PLA_2 such as unsaturated fatty acids (arachidonic or oleic acids) mimicked the effect of PLA_2. These effects were entirely prevented by preincubation with BSA. The *in vitro* addition of the acidic phospholipid phosphatidylserine isolated from bovine brain (BC-PS) produced a similar increase in B_{max}. This action was also blocked by addition of BSA. On the other hand, palmitic acid, a saturated fatty acid, and lysophosphatidylserine (lysoPS) or lysophosphatidylethanolamine (lysoPE) failed to modify [^{3}H]IMI binding sites. The chronic administration of tricyclic antidepressant (AD) resulted in a 25% decrease in [^{3}H]IMI binding sites in synaptosomal membranes. Preincubation of these AD-treated membranes with PLA_2 did not alter [^{3}H]IMI binding, whereas the addition of unsaturated free fatty acids (FFA) produced a greater increase in the density of [^{3}H]IMI binding sites in comparison with control membranes. Taken together, these findings suggest that unsaturated free fatty acids could play an important role in the regulation of the number of [^{3}H]IMI high affinity binding sites in the mammalian brain.

INTRODUCTION

Synaptic receptors are intrinsic proteins of neural membranes embedded in the lipid bilayer matrix. This lipid microenvironment has been shown to modulate both

Neurobiology of Essential Fatty Acids
Edited by N.G. Bazan *et al.*, Plenum Press, New York, 1992

neurotransmitter and hormone receptors (see Loh and Law, 1980). Thus, phospholipids and their products may be important for modifying ligand receptor interactions, receptor/second messenger coupling mechanisms, membrane fluidity, ion channel kinetics, and enzyme activities (Schwartz et al., 1988; Levi de Stein et al., 1989; Ordway et al., 1991). Cleavage of membrane phospholipids by endogenous phospholipases releases free fatty acids (FFA) and lysoderivatives having active roles in transmitting signals to target cells (Bruni et al., 1986; Axelrod et al., 1988; Ordway et al., 1991).

A large body of information has recently emerged on the role of phospholipids and endogenous PLA_2 activity in the ligand-receptor interactions for various ligands to acetylcholine, opiate, dopamine, GABA/BDZ receptors, and ß adrenoceptors (for references see Schwartz et al., 1988; Stockert and Medina, 1990). Many of these effects were suggested to be mediated via the generation of unsaturated FFA.

It has been shown that *in vivo* administration of phospholipid liposomes containing phosphatidylserine (PS) potentiates the efficacy of antidepressant treatment in depressed patients (Casacchia et al., 1982) and in an experimental model of depression (Drago et al., 1985). Furthermore, phospholipid liposomes or PS purified from brain (BC-PS) produced biochemical changes similar to those observed after chronic AD administration (Racagni and Brunello 1984; Stockert et al., 1989). Based on these findings, we decided to investigate how changes in the lipid environment affect the binding properties of [³H]imipramine ([³H]IMI) recognition sites. We report here the *in vitro* effects of BC-PS, PLA_2, saturated and unsaturated FFA, and lysophospholipids on cerebral cortical [³H]IMI binding sites in synaptosomal membranes of control and chronic AD-treated rats.

MATERIALS AND METHODS

For these experiments a crude synaptosomal membrane fraction from cerebral cortex of control and chronic AD-treated rats (amitriptyline 12 mg/kg/day. i.p., for 20 days) was used (Stockert et al., 1989; Stockert and Medina, 1990).

The specific binding of [³H]IMI was performed as previously reported (Stockert and Medina, 1990). Briefly, the membranes (0.2 mg/ml) were incubated in triplicate with 0.5 to 8 nM [³H]IMI (47.4 Ci/mmol, NEN) at 0°C for 60 min in 50 mM Tris base, 120 mM NaCl, and 5 mM CLH buffer, pH 7.2. The reaction was stopped by centrifugation at 9000 × g for 10 min. Specific binding was calculated by subtraction of total binding in the presence of 30 μM nortriptyline and represented about 70% of the total.

Membranes of both control and chronic AD-treated rats were preincubated at 37°C for 30 min with different treatments as follows: BC-PS (FIDIA); lysoPS (FIDIA); lysoPE (FIDIA); PLA_2 from snake venom (Sigma); and arachidonic, oleic, palmitic, *cis* 7-10-13-16 docosatetraenoic and *cis* 4-7-10-13-16-19 docosahexaenoic acids (Sigma) in various concentrations (see legends to Figs. 1-3). In some experiments, either lipid-free bovine serum albumin (BSA; 1:1) or EGTA (20 mM) was added before BC-PS, PLA_2, or FFA. In all cases, control samples (no additions) were incubated under identical conditions.

The results were analyzed using an EBDA program for a single type of binding site adapted to an IBM computer. Data are expressed as mean ± SEM of least four independent experiments and differences among groups were determined by one-way ANOVA followed by Student's t test.

RESULTS

Figure 1 summarizes the effects of the different *in vitro* treatments on [³H]IMI bind-

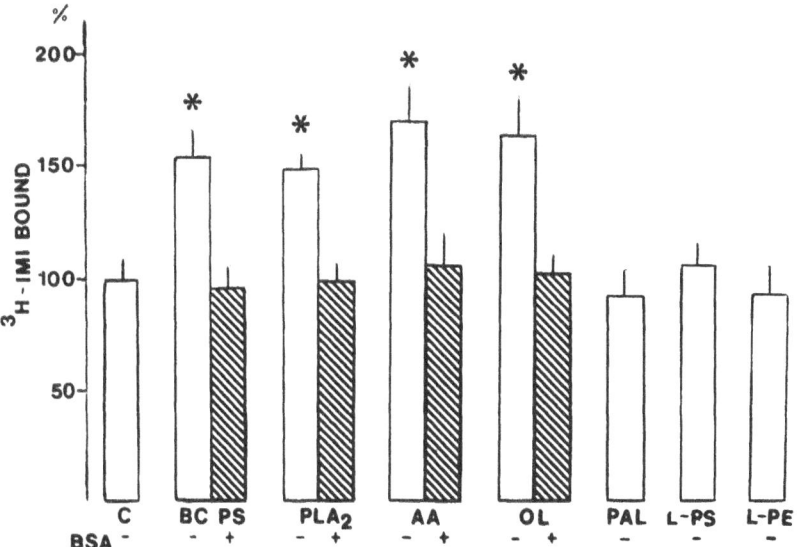

Figure 1. Effects of the addition of BC-PS (100 μM); PLA$_2$ (0.5 U/mg protein); arachidonic acid (AA; 100 μM); oleic acid (OL; 100 μM); palmitic acid (PAL; 200 μM); lysoPS (L-PS; 200 μM); and lysoPE (L-PE; 200 μM) on the B$_{max}$ of [^{3}H]IMI binding to synaptosomal membranes of cerebral cortex from naive (non-AD treated) rats (open bars). Striped bars represent the results after preincubation with lipid-free BSA. C indicates control. Values represent % of mean ± SEM with respect to control of at least seven independent experiments. Asterisk indicates p < 0.01, Student's t test, compared with control.

ing to cerebral cortical synaptosomal membranes of naive rats. Preincubation of these membranes with 50 μM BC-PS produced a remarkable increase in the B$_{max}$ of [^{3}H]IMI binding sites. Similar increments in the density of these sites were observed after incubation with PLA$_2$ (0.5 U/mg protein) and arachidonic and oleic acids (10-100 μM). In all the cases, saturation binding experiments revealed no changes in the apparent dissociation constant (K$_d$ control: 3 ± 0.2 nM; K$_d$ BC-PS: 3.5 ±0.4 nM; K$_d$ PLA$_2$: 2.8 ± 0.2 nM; K$_d$ AA: 2.9 ± 0.3 nM; K$_d$ OL: 3.3 ± 0.5 nM; n=5). The displacement curves of [^{3}H]IMI binding by the triptyline antidepressant amitriptyline were not affected by preincubation of the synaptosomal membranes with PLA$_2$ (Fig. 2).

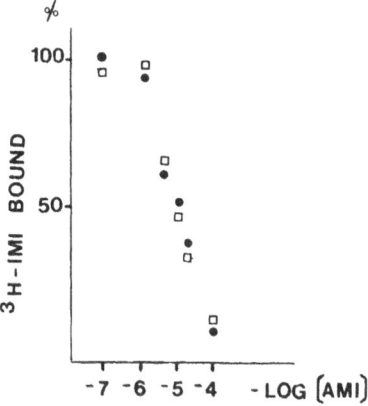

Figure 2. Representative displacement curves of [^{3}H]IMI binding by amitriptyline (AMI) in control (filled circles) and in PLA$_2$-incubated (open squares) membranes.

327

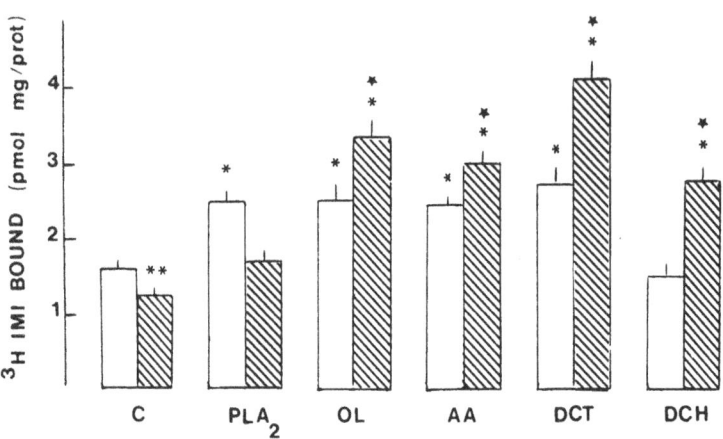

Figure 3. Effects of *in vitro* preincubation of PLA$_2$ (0.5 U/mg protein), and oleic (OL; 100 μM), arachidonic (AA; 30 μM), docosatetraenoic (DCT; 50 μM), and docosahexaenoic (DCH; 50 μM) acids on the density of [^{3}H]IMI binding sites in control membranes (open bars) and membranes from AD-treated rats (striped bars). Data are expressed as mean ± SEM of 4-9 independent experiments. ** p < 0.05; * p < 0.001 with respect to membranes from control rats which were not preincubated; star, p < 0.05 in comparison with the respective incubated membranes from control rats.

The *in vitro* effects of BC-PS, PLA$_2$, and unsaturated FFA on the maximal numbers of [^{3}H]IMI binding sites were eliminated with the use of lipid-free BSA as a scavenger of phospholipase hydrolysis products of membrane phospholipids (Fig. 1). Furthermore, chelation of Ca^{+2}, which is required for PLA$_2$ activity, abolished the effect of PLA$_2$ on [^{3}H]IMI binding.

In contrast, the addition of a saturated FFA, palmitic acid, or the lysoderivatives lysoPS and lysoPE did not modify the binding characteristics of [^{3}H]IMI recognition sites (Fig. 1).

To further characterize the effects of PLA$_2$ and unsaturated FFA on AD binding sites, *in vitro* experiments using synaptosomal membranes from controls and chronic AD-administered rats were carried out. In accord with previous findings (Stockert et al., 1989; Garattini and Samanin, 1988), the chronic administration of AD provoked a downregulation of [^{3}H]IMI binding sites in the cerebral cortical synaptosomal membranes (Fig. 3).

In these membranes, preincubation with PLA$_2$ did not modify binding of [^{3}H]IMI, but the addition of lower unsaturated FFA (e.g., oleic acid) or polyunsaturated FFA (e.g., arachidonic, docosatetraenoic, or docosahexaenoic acid) produced a greater increase in the binding of [^{3}H]IMI to AD-treated membranes in comparison with control membranes (Fig. 3). Furthermore, curves generated with increasing concentrations of arachidonic acid showed increased sensitivity of AD-treated membranes (control EC$_{50}$: 30 μM vs AD-treated EC$_{50}$: 4 μM, n = 3). It is interesting to note that the *in vitro* addition of docosahexaenoic acid to control membranes did not produce changes in the [^{3}H]IMI-specific binding sites.

DISCUSSION

The findings presented here demonstrate that exogenous PS, PLA$_2$, and unsaturated FFA increase the density of [^{3}H]IMI binding sites in cerebral cortical synaptosomal membranes. In contrast, saturated FFA and lysoPS and lysoPE failed to alter these binding sites.

The BC-PS-induced increase in the density of [^{3}H]IMI binding sites seems to be specific because similar experiments performed with [^{3}H]QNB (a muscarinic antagonist), [^{3}H]oxotremorine (a muscarinic agonist), [^{3}H]FNZ (a benzodiazepine receptor agonist), and [^{3}H]prazosin (an α_1-adrenoceptor agonist) revealed no changes in their binding characteristics (Levi de Stein et al., 1989; Rascovsky et al., 1990). However, using a fusion technique we were able to demonstrate that the enrichment of synaptosomal membranes with BC-PS produced a clear-cut increase in the density of benzodiazepine receptors (Levi de Stein et al., 1989).

Based on the findings that the effect of BC-PS is mimicked by the addition of PLA$_2$ or unsaturated FFA, and since all these treatments are blocked by incubating with BSA, which can remove FFA from the membranes, it is tempting to speculate that the action of BC-PS is probably due to the generation of unsaturated fatty acids released by the activity of endogenous PLA$_2$.

The mechanisms by which unsaturated FFA increased the binding of [^{3}H]IMI to synaptosomal membranes is unknown. It is well established that unsaturated fatty acids increase membrane fluidity and modify lipid microviscosity which, in turn, affect the activity of membrane receptors through two factors: the degree of accessibility to ligand binding and the passive rate of diffusion. As a result, membrane "spared" receptors can be exposed and the net effect would be an upregulation of the binding sites.

In experimental and human depression, the biochemical and clinical potentiation of the efficacy of AD treatment induced by phospholipid liposomes containing PS is a well known phenomenon (Casacchia et al., 1982; Drago et al., 1985; Stockert et al., 1989; Racagni and Brunello, 1984). In that context, our *in vitro* experiments suggest that changes in the membrane lipid microenvironment play an important role in regulating [^{3}H]IMI binding sites.

According to a widely accepted hypothesis, deficient cerebral monoaminergic neurotransmission plays a pivotal role in endogenous depression (Langer and Schoemaker, 1988). Many antidepressants are well known for their ability to inhibit the neuronal uptake of serotonin and norepinephrine (Green, 1987). Then, the synaptic cleft concentrations of these neurotransmitters increase. Although antidepressants have been used successfully in the treatment of depression, the exact mechanism of action is still a matter of speculation. In the majority of studies published so far, the density of [^{3}H]IMI binding sites in brain and platelets from untreated depressed patients is lower than that observed in control patients (see Langer and Schoemaker, 1988).

It is important to stress that tricyclic antidepressants enhance incorporation of [^{3}H]arachidonic acid into acidic phospholipids and reduce reacylation of arachidonic acid into neutral phospholipids (Hauser et al., 1986; Barkai and Murthy, 1989). Also, changes have been found in phospholipid composition in rats after chronic AD treatment (see Moor et al., 1988). These findings suggest that antidepressants may be important in controlling the levels of membrane FFA and lysophospholipids. Our *in vitro* results showed a decreased sensitivity of exogenous PLA$_2$ and an increased effect of unsaturated fatty acids on [^{3}H]IMI binding to chronic AD-treated membranes.

It is tempting to suggest that a differential modulation of brain AD binding sites by endogenous unsaturated fatty acids released due to changes in PLA$_2$ activity may occur in membranes chronically exposed to antidepressants.

In conclusion, alterations in membrane phospholipids by phospholipases and the generation of unsaturated FFA may be important factors in the modulation of AD high affinity binding sites. The possible clinical implications of these findings deserve further investigation.

REFERENCES

Axelrod J, Burch RM, Jelsema CL (1988) Receptor-mediated activation of phospholipase A_2 via GTP-binding proteins: Arachidonic acid and its metabolites as second messengers. Trends Neurosci 11:117.

Barkai AI and Murthy LR (1989) Modulation of arachidonate turnover in cerebral phospholipids. In: Arachidonic acid metabolism in the nervous system (Barkai AI and Bazan NG, eds) Ann N Y Acad Sci 559:56.

Bruni A, Nietto L, Battistela A, Boarato E, Palatini P, Toffano G (1986) Serine phospholipids in cell communication. In: Phospholipid research and the nervous system: biochemical and molecular pharmacology (Horrocks LA, Freysz L, Toffano G, eds), p 217. Padova, Italy: Liviana Press.

Casacchia M, Meco G, Pirro R, Di Cesare E, Allegro A, Cusimano G, Manola W (1982) Phospholipid liposomes in depression: a double-blind study versus placebo. Int Pharmacopsychiatry 17:274.

Drago F, Continella G, Mason G, Hernández D, Scapagnini U (1985) Phospholipid liposomes potentiate the inhibitory effect of antidepressant drugs on immobility of rats in a despair test (constrained swim). Eur J Pharmacol 115:179.

Garattini S and Samanin R (1988) Biochemical hypothesis on antidepressant drugs: a guide for clinicians or a toy for pharmacologists? Psychol Med 18:287.

Green AR (1987) Evolving concepts on the interactions between antidepressant treatments and monoamine neurotransmitters. Neuropharmacology 26:815.

Hauser G, Koul O, Lele U (1986) Phospholipid metabolism in nervous tissues: modification of precursor incorporation and enzyme activities by cationic amphiphilic drugs. In: Phospholipid research and the nervous system: biochemical and molecular pharmacology (Horrocks LA, Freysz L, Toffano G, eds), p 93. Padova, Italy:Liviana Press.

Langer SZ and Schoemaker H (1988) Effects of antidepressants on monamine transporters. Prog Neuropsychopharmacol Biol Psychiatry 12:193.

Levi de Stein M, Medina JH, De Robertis E (1989) *In vivo* and *in vitro* modulation of central type benzodiazepine receptors by phosphatidylserine. Mol Brain Res 5:9.

Loh HH and Law PY (1980) The role of membrane lipids in receptor mechanisms. Annu Rev Pharmacol Toxicol 20:210.

Moor M, Honegger UE, Wiesmann UN (1988) Organospecific, qualitative changes in the phospholipid composition of rats after chronic administration of the antidepressant drug desipramine. Biochem Pharmacol 37:2035.

Ordway RW, Singer JJ, Walsh SV (1991) Direct regulation of ion channels by fatty acids. Trends Neurosci 14:96.

Racagni G and Brunello N (1984) Transsynaptic mechanisms in the action of antidepressants drugs. Trends Pharmacol 5:527.

Rascovsky S, Rivas E, Bernik D, Medina JH, Jerusalinsjy D (1990) Modulatory effects of phosphatidylserine on the binding of muscarinic cholinergic receptor ligands: studies *in vitro* and *in vivo*. Mol Chem Neuropathol 13:17.

Schwartz R, Skolnick P, Paul S (1988) Regulation of γ-aminobutyric acid/barbiturate receptor-gated chloride ion flux in brain vesicles by phospholipase A_2: possible role of oxygen radicals. J Neurochem 50:565.

Stockert M and Medina JH (1990) Modulation of cerebral cortical ^{3}H-imipramine binding sites by phospholipase A_2: Possible role of unsaturated free fatty acids. Neurosci Res Commun 6:89.

Stockert M, Buscaglia V, De Robertis E (1989) Action *in vivo* of phosphatidylserine, amitriptyline and stress on the binding of ^{3}H imipramine to membranes of the rat cerebral cortex. Eur J Pharmacol 160:11.

VERY LONG-CHAIN FATTY ACIDS IN PEROXISOMAL DISEASE

A. Poulos, K. Beckman,
D. W. Johnson, B. C. Paton,
B. S. Robinson, P. Sharp,
S. Usher, and H. Singh

Department of Chemical Pathology
Adelaide Medical Centre for Women and Children
North Adelaide
South Australia 5006

ABSTRACT

Fatty acids with from 24 to 28 carbon atoms (very long-chain fatty acids, VLCFA) are present in small amounts in all mammalian tissues. Even longer chain fatty acids with from 30 to 38 carbon atoms (ultra-long-chain fatty acids, ULCFA) are found in certain specialized tissues including retina, brain, and spermatozoa. In patients with inherited defects in peroxisomal structure and/or function, there is an accumulation of VLCFA in most tissues, while VLCFA and ULCFA levels are increased in brain. The most pronounced changes occur in those patients who have defects in peroxisomal assembly (Zellweger syndrome, infantile Refsum's disease, and neonatal adrenoleuko-dystrophy). In the brain of these individuals, ULCFA are distributed largely in molecular species of phosphatidylcholine with penta-, hexa-, and heptaenoic acids. In contrast, patients with X-linked adrenoleukodystrophy have increased levels of phosphatidylcholine with monoenoic rather than polyenoic ULCFA. A defect in a peroxisomal VLCFA CoA synthetase or ligase has been reported for these patients, but assembly of their peroxisomes is apparently normal.

We speculate that ULCFA are normal products of carbon chain elongation. We have confirmed this by demonstrating the elongation of [1-^{14}C]hexacosatetraenoic acid (26:4n-6) by rat brain *in vivo* to a series of longer chain tetraenoic acids with carbon chain lengths up to 34. Elongation to ULCFA can occur as well in non-neural tissues as shown by detection of labeled saturated and monoenoic fatty acids with up to 32 carbon atoms after incubation of normal and Zellweger syndrome fibroblasts with [2-^{14}C] acetate. Increased labeling of VLCFA and ULCFA is observed in cells from patients with peroxisomal disorders.

Our data suggest that ULCFA with up to at least 32 carbon atoms are formed normally, as a part of the elongation process in most mammalian tissues, and that control of carbon chain elongation is a major function of peroxisomes. Impairment of this function as occurs in peroxisomal disease results in the accumulation of VLCFA and ULCFA. The relative enrichment in normal tissues of ULCFA such as 32:6n-3 in ram and bull spermatozoa and 36:4n-6 in human and rat brain suggests a probable physiological role for this class of fatty acids in these tissues.

INTRODUCTION

The study of the metabolic role of peroxisomes has attracted increasing interest over the last few years. One factor which has contributed to this interest is the recognition of patients with disorders of peroxisomal metabolism. These patients can be divided into three groups, including: (i) those with a generalized loss of peroxisomal functions and abnormalities of peroxisomal structure and biogenesis, e.g., Zellweger syndrome (ZS), infantile Refsum's disease (IRD), and neonatal adrenoleukodystrophy (NALD) (Group 1); (ii) those who show a loss of multiple peroxisomal functions, e.g., rhizomelic chondrodysplasia punctata (RCDP) (Group 2); and (iii) those in whom a single biochemical function is affected, e.g., X-linked adrenoleukodystrophy (X-linked ALD) and Refsum's disease (Group 3) (Wanders et al., 1988a). The degradation of saturated very long-chain fatty acids (fatty acids with from 24 to 28 carbon atoms, VLCFA) (Poulos et al., 1986c) and the branched chain phytanic and pristanic acids (Singh et al., 1990) is impaired in ZS, IRD, and NALD, while VLCFA oxidation alone is defective in X-linked ALD (Poulos et al., 1986c; Wanders et al., 1987). Phytanic acid oxidation is also reduced in Refsum's disease and RCDP.

In addition to the changes in saturated VLCFA, Group 1 patients also show evidence of an abnormality in the degradation of monoenoic VLCFA (Poulos et al., 1986c) while the brain from ZS patients contains increased amounts of polyenoic VLCFA and ultra-long-chain fatty acids (30 to 38 carbon atoms, ULCFA) (Poulos et al., 1986b; Sharp et al, 1987). While the accumulation of saturated, monoenoic and polyenoic VLCFA appears to reflect the marked decrease in peroxisomal numbers in ZS patients, the structure of the organelle is unaffected in X-linked ALD and saturated VLCFA accumulation arises as a result of a defect in VLCFA CoA synthetase or ligase (Lazo et al., 1989: Wanders et al., 1988b). As yet the carbon chain length specificity of this peroxisomal synthetase and whether it is capable of activating mono- and polyenoic VLCFA remain uncertain. To understand more fully the role of peroxisomes in the metabolism of fatty acids, we have undertaken an investigation of VLCFA and ULCFA in postmortem brain obtained from patients who show biochemical evidence of a disturbance in oxidation of fatty acids. In addition we have studied the biosynthesis of VLCFA and ULCFA in brain and cultured skin fibroblasts.

MATERIALS AND METHODS

Materials

Reverse phase TLC plates were purchased from Whatman Inc., Clifton, NJ, USA and silica gel 60 plates were obtained from E. Merck, Darmstadt, West Germany. Fatty acid standards were obtained from Nu-Chek Prep, Elysian, Minnesota, USA and from Sigma Chemical Company, St. Louis, MO, USA. Monoenoic VLCFA (30:1n-9 and 34:1n-9) were synthesized as described by Johnson (1990). [1-^{14}C]Hexacosatetraenoic acid (26:4n-6) was synthesized as described by Robinson et al. (1990a). [2-^{14}C]Acetic acid (specific activity 57 mCi/mmol) was obtained from Du Pont Chemical Company, Wilmington, DE, USA and the autoradiography film (Hyperfilm-^{3}H) was purchased from Amersham Australia Pty. Ltd.

Tissues

Brain (mostly frontal lobe containing approximately equal amounts of white and grey matter) was obtained at postmortem from the following patients: Group 1 - a male patient with IRD aged 3 years, a male patient with ZS aged 4 days, and a female patient with NALD aged 3 years; Group 3 - a male patient aged 10 years with X-linked ALD, a male patient aged 13 years with X-linked ALD, and a male patient aged 55 years with a variant form of ALD, adrenomyeloneuropathy (AMN). The diagnosis of peroxisomal disease was based on clinical history, laboratory investigation and, where possible, by ultrastructural analysis.

Controls included a 24-week aborted fetus, a male who died from sudden infant death syndrome aged 9 weeks, a male who died from sudden infant death syndrome aged 20 months, a male who died from a motor vehicle accident aged 4 years, a male who died from a motor vehicle accident aged 7 years, a 55-year-old male who died of a cardiac tamponade, and an 85-year-old female who died from cardio-respiratory failure.

Cell Culture

Skin fibroblasts from normal subjects and from ZS patients were grown in tissue culture (75 cm^2 flasks) in basal Eagle's medium containing 10% fetal calf serum until confluent. One hundred microliters of an ethanolic solution of [2-^{14}C]acetic acid (100 μCi) was added to individual flasks and the flask tops were tightly sealed to prevent the escape of radioactive carbon dioxide into the atmosphere. Incubations were carried out for 3 days prior to the processing of cells. At the end of the incubation period the medium was removed and the cells were washed with Dulbecco's phosphate-buffered saline (calcium- and magnesium-free) prior to harvesting by trypsinization. The harvested cells were washed a further three times with Dulbecco's phosphate-buffered saline and suspended in 450 μl of the same buffer containing 4 mM magnesium chloride. The cells were disrupted by ultrasonication and aliquots were removed for protein analysis and counting.

Lipid Analysis

The extraction and isolation of lipids from brain and the quantitation of VLCFA in individual lipid fractions were carried out as described by Sharp et al. (1991). Gas chromatography-mass spectrometry of VLCFA was performed as outlined earlier (Poulos et al., 1988). The distribution of the double bond in monoenoic VLCFA was determined by mass spectrometric analysis of their picolinyl derivatives as described by Johnson et al. (in press). The intracerebral injection of [1-^{14}C]26:4n-6 fatty acid into rats, the isolation of radiolabeled fatty acids with 0 to 5 double bonds, and reverse phase TLC of the radiolabeled fatty acids (using appropriate fatty acid standards as markers) were carried out as described by Robinson et al. (1990a). HPLC of diglycerides released from molecular species of phosphatidylcholine-containing VLCFA (VLCFA-PC) was carried out as described by Sharp et al. (1991).

Lipids were extracted from the cell suspensions as described by Bligh and Dyer (1959). Butylated hydroxytoluene (BHT) was added at all stages of the extraction and analysis to minimize auto-oxidation of polyunsaturated fatty acids. Aliquots of the lipid extracts were streaked onto silica gel 60 thin-layer plates and chromatograms were developed in hexane-diethyl ether-acetic acid (90:10:1, by volume). The triglyceride and cholesterol ester zones were scraped from the plate, eluted, and the fatty acids released from isolated lipids by methanolysis. Reverse phase TLC of the fatty acids and radioautography were carried out as described by Street et al. (1989).

RESULTS

We had previously shown that normal human and rat brain contain polyenoic n-6 ULCFA and that these fatty acids are distributed mostly in molecular species of phosphatidylcholine (polyenoic ULCFA-PC) (Poulos et al., 1988; Robinson et al, 1990b). The carbon chain length and degree of unsaturation varies significantly in human brain with age (Fig. 1) (Sharp et al., 1991). A similar change with development is also observed in rat brain, with the main features being the predominance of 32 and 34 carbon pentaenoic fatty acids early in development as compared to the predominance of 34 and 36 carbon tetraenoic acids in adults (Robinson et al, 1990b).

Polyenoic ULCFA-PC levels were increased in brain from patients with defects in peroxisomal assembly. Increased levels were found in the three major phenotypic groups, i.e., ZS, IRD, and NALD. The largest proportional increase occurred in the n-6 hexaenoic 34, 36, and 38 carbon fatty acids (Fig. 2) with 13.4, 6.9, and 5.8 µg of total hexaenoic acid per gram wet weight of brain in ZS, NALD, and IRD, respectively. These fatty acids could only be detected in trace amounts in fetal human brain and in the brain of 1-day old rats and were not detected in normal postnatal human brain. While polyenoic VLCFA were present in brain and liver from these patients, polyenoic ULCFA were found only in brain. HPLC separation of the diglycerides released from ULCFA-PC isolated from ZS brain indicated the presence of a large number of molecular species each containing polyenoic ULCFA in the 1-position of the glycerol moiety. The 2-position was occupied by a limited number of fatty acids, with 16:0, 18:1, and 20:4 predominating.

The polyenoic ULCFA composition of ULCFA-PC from X-linked ALD brain was similar to control brain, with the tetra- and pentaenoic fatty acids predominating, although there was a modest increase in the levels of these fatty acids (Fig. 2). However, hexaenoic ULCFA could not be detected. Elevated amounts of 30, 32 and 34 carbon monoenoic fatty acids were detected in this lipid fraction in X-linked ALD patients (1.5, 7.0, and 10.4 µg of total 30, 32, and 34 monoenoic per gram wet weight of brain in the two ALD and one AMN patients vs <0.6 µg in normal 7- to 85-year-old brain). Smaller amounts of the corresponding saturated derivatives were also present. At least 18 separate molecular species of ULCFA-PC were detected in X-linked ALD brain each containing a 1-linked ULCFA and a 2-linked fatty acid, mostly 20:4 and 18:1 (Fig. 3).

At least two isomeric forms of each monoenoic VLCFA and ULCFA were observed in X-linked ALD (Fig. 2). Mass spectrometric analysis of the picolinyl derivatives of the monoenoic fatty acids indicated the presence of n-7 and n-9 fatty acids, the proportion of the two isomeric forms varying according to carbon chain length. Thus, the 24 and 32 carbon fatty acids were mainly n-9 series acids while the 28 carbon acid was mainly an n-7 series acid. Close inspection of the chromatograms of normal brain revealed the presence of monoenoic VLCFA, and monoenoic ULCFA with up to 34 carbon atoms, but these were present in much smaller amounts than in X-linked ALD. Sixteen and 18 carbon n-7 and n-9 fatty acids are components of normal and ALD brains and therefore are probably precursors of the corresponding series even-numbered VLCFA and ULCFA. However, odd-numbered VLCFA may not be formed by elongation of shorter chain fatty acids, because 17 and 19 carbon monoenoic fatty acids occur as either n-8 or n-10 series acids. Presumably, therefore, the monoenoic odd-numbered VLCFA are formed from one carbon longer (even-numbered) fatty acids, by α-oxidation.

In order to investigate the hypothesis that the polyenoic ULCFA are formed by carbon chain elongation of shorter chain polyenoic acids, $[1-^{14}C]26:4$ was synthesized and injected intracerebrally into the brains of 1-day-old and 16-day-old rats. Small amounts of labeled 28 to 34 carbon atom fatty acids were detected 4 hours after injection of the label indicating that the brain can synthesize polyenoic fatty acids with up to at least 34 carbon atoms. Hexacosatetraenoic acid (26:4) is a trace component of normal brain and is probably formed by elongation of arachidonic acid. It is likely therefore that the precursor for the n-6 polyenoic ULCFA is arachidonic acid.

Our data indicated that ULCFA are formed even in tissues such as cultured skin fibroblasts where these fatty acids are not detected by normal gas chromatographic procedures (Poulos et al., 1986c) and only accumulate if there is an abnormality in peroxisomal function. To examine this hypothesis we studied the synthesis of fatty acids from [2-^{14}C] acetate in cultured skin fibroblasts. These cells are known to accumulate both saturated and monoenoic VLCFA in patients with peroxisomal disease (Poulos et al., 1986c). [2-^{14}C]Acetate was incorporated by normal cells in culture into a variety of fatty acids. In a 3-day period approximately 1 to 2% of the total label added was incorporated into lipids (mainly phospholipids and triglycerides) and most of this lipid-incorporated label (>90%) was associated with fatty acids. Most of the label in the saturated and monoenoic fatty acids released from triglycerides (the major labeled lipid product) and cholesterol esters isolated from normal cells was present in the shorter chain fatty acids (mainly 16 and/or 18 carbon) with only minor amounts in VLCFA (Fig.

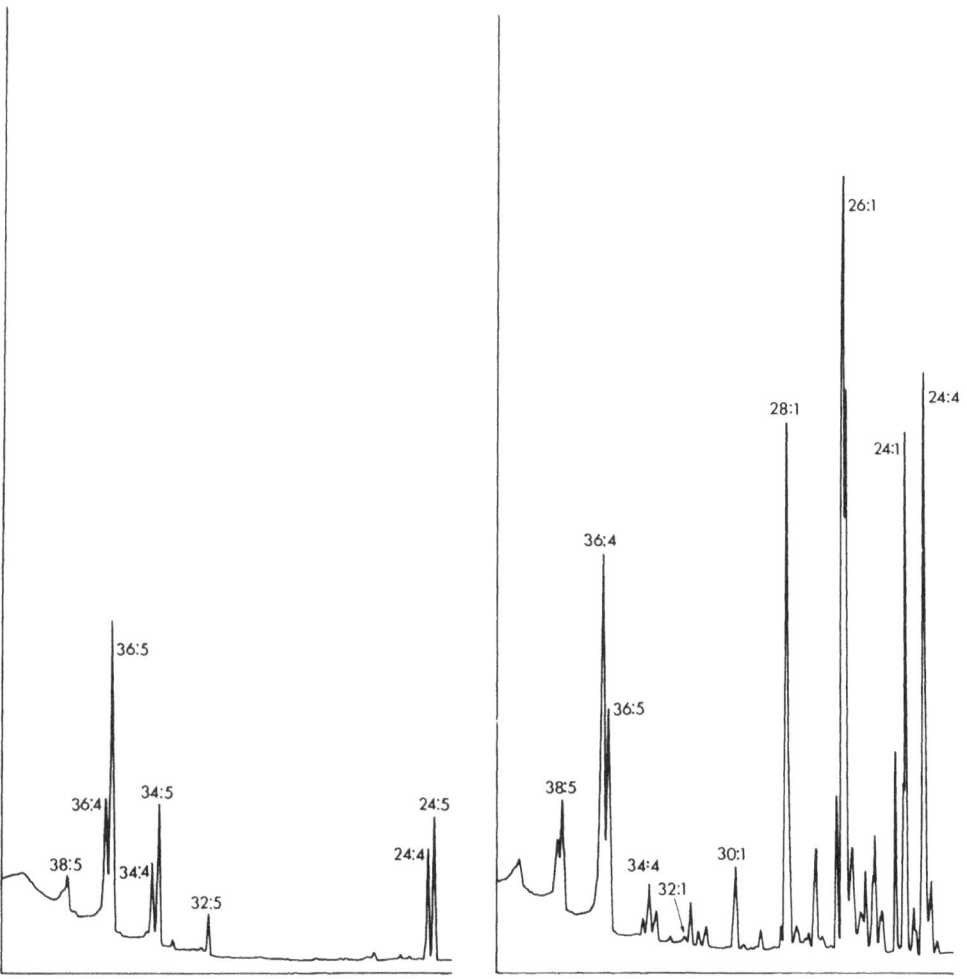

Figure 1. Gas liquid chromatography of fatty acids released from VLCFA-PC isolated from 9-week- (left panel) and 7-year-old (right panel) brain from normal individuals.

4). In Zellweger cells, however, there was a marked increase in the proportion of label in VLCFA and in particular in 26:0 and 26:1. Zellweger cells also contained significant amounts of ULCFA with up to 34 carbon atoms. Radiolabeled ULCFA were also detected in normal cells, but in much smaller amounts.

DISCUSSION

Our data show that polyenoic ULCFA-PC accumulate in patients with inherited defects of peroxisomal assembly. While the accumulation of ULCFA appears to be confined to the brain, VLCFA can be detected in elevated amounts in liver and plasma (Poulos et al, 1989), as well as brain. Molecular species of PC containing polyenoic n-6 ULCFA may therefore be relatively specific for brain, although retina has been shown to have molecular species of PC containing mostly polyenoic n-3 fatty acids with up to 36 carbon atoms (Aveldaño and Sprecher, 1987). The polyenoic ULCFA which accumulate in the brain of peroxisomal disorder patients consist mostly of n-6 hexaenoic

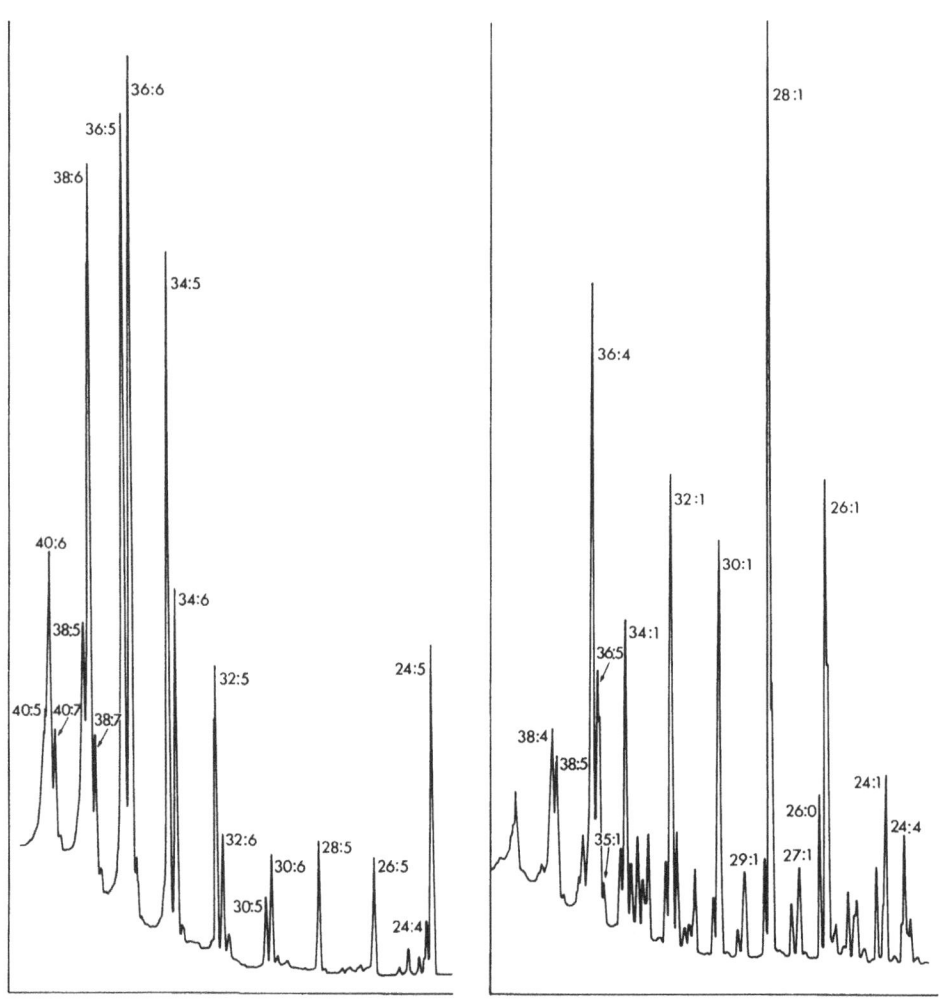

Figure 2. Gas liquid chromatography of fatty acids released from VLCFA-PC isolated from the brains of ZS (left panel) and X-linked ALD (right panel) patients.

acids, derivatives which are normally detected only in fetal tissue. These fatty acids continue to be synthesized even in older normal brain but only accumulate in peroxisomal disease brain, possibly because their degradation is impaired.

The accumulation of monoenoic and saturated, rather than polyenoic, VLCFA-PC and ULCFA-PC in X-linked ALD brain confirms the current view that the defect is in a peroxisomal VLCFA CoA synthetase which acts on saturated and monoenoic VLCFA. This enzyme may also act on saturated ULCFA. However, the accumulation of polyenoic VLCFA and ULCFA in disorders of peroxisomal assembly suggests that peroxisomes are involved in the oxidation of these fatty acids as well and may contain synthetases which are specific for polyenoic VLCFA and ULCFA. These conclusions are supported by our studies which show that suspensions of Zellweger and X-linked ALD fibroblasts have a greatly reduced 24:0 oxidation activity (Singh, Brogan, and Poulos, unpublished data). While we had earlier reported that the oxidation of the corresponding tetraenoic acid (24:4n-6) was also reduced in Zellweger syndrome fibroblasts in culture (Street et al., 1989), in our more recent investigations this has not been substantiated. This would indicate that, if there is an impairment in the degradation of the polyenoic VLCFA in this disorder, it affects fatty acids with more than 24 carbon atoms. However the carbon chain length specificity of the peroxisomal and mitochondrial pathways for the oxidation of polyenoic fatty acids remains unclear.

The biosynthetic pathway for ULCFA is not known. However, it is possible that shorter chain n-6 fatty acids such as 20:4 and 22:4 serve as precursors for the tetraenoic acids while 22:5 is the precursor for the pentaenoic acids. Our investigations with rat brain support this hypothesis. The hexaenoic ULCFA which accumulate in peroxisome-

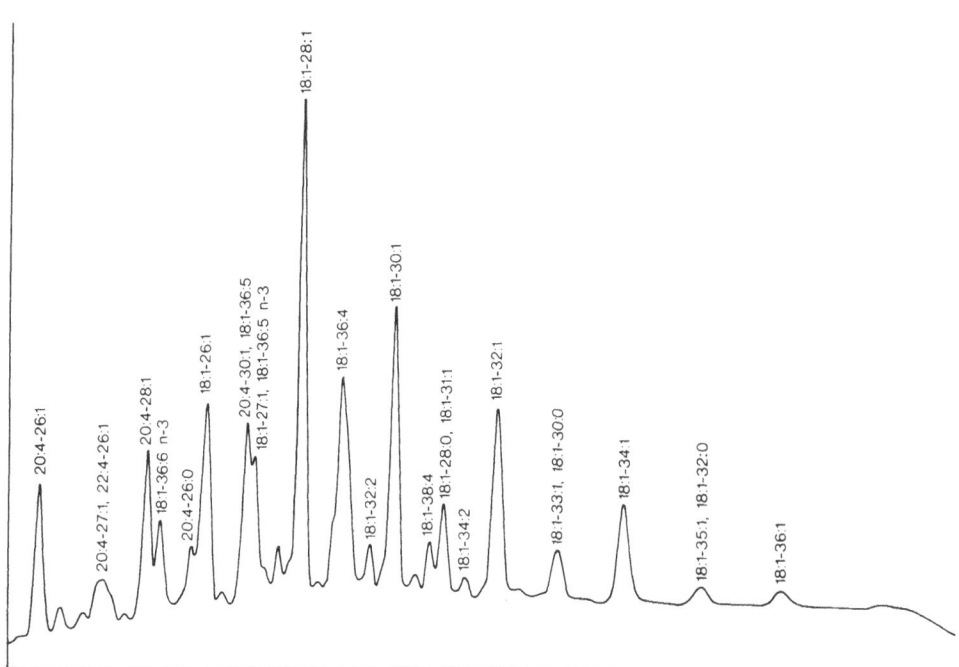

Figure 3. HPLC of p-nitrobenzoyl esters of diglycerides released from VLCFA-PC isolated from X-linked ALD brain. The tentative identification of the fatty acid composition of individual peaks was determined by gas-liquid chromatography.

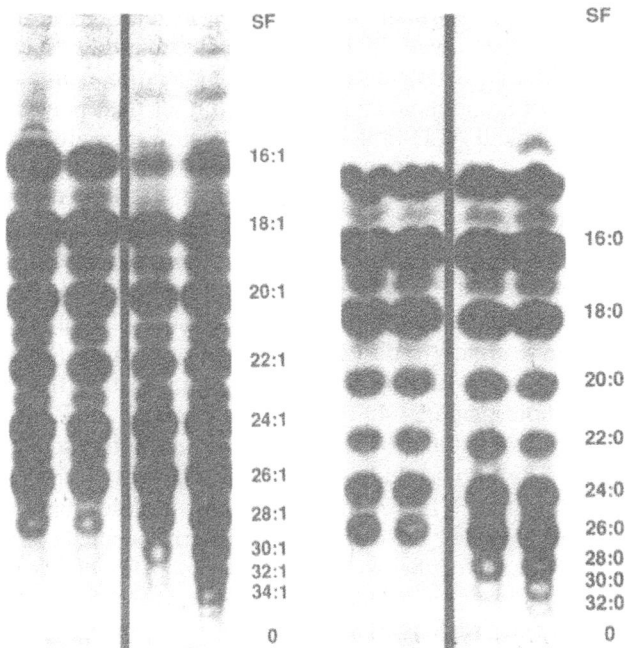

Figure 4. Triglycerides were isolated from normal and ZS fibroblasts after incubation with [2-^{14}C]-acetate and fatty acids were chromatographed and autoradiographs were prepared as described in the text. The numbers represent *carbon chain length:numbers of double bonds*. 0, origin; SF, solvent front. Left panel from left to right: control, control, ZS, and ZS monoenoic fatty acids; right panel from left to right: control, control, ZS, and ZS saturated fatty acids.

deficient patients must arise by desaturation of a pentaenoic VLCFA, but the identities of this fatty acid and the desaturase remain uncertain. We have detected small amounts of 28:6 indicating that desaturation of 28:5 may take place, but if this were the case a Δ7 desaturation would be required. To our knowledge a Δ7 desaturase has never been reported in human brain. The other possibility is desaturation of 26:5, a fatty acid found in significant amounts in Zellweger syndrome brain and plasma. Conversion of 26:5 to the hexaenoic acid would require the more usual -5 desaturase, which has been reported to occur in normal brain (Naughton, 1981).

Our findings indicate that the synthesis of ULCFA can occur in cultured skin fibroblasts. This conclusion is supported by our data showing that normal skin fibroblasts in culture incorporate [2-^{14}C]acetate into saturated and/or monounsaturated 28, 30 and 32 carbon fatty acids. It is clear therefore that pathways for the synthesis of VLCFA and ULCFA occur in tissues as diverse as brain and cultured skin fibroblasts. Under normal circumstances VLCFA are present in very small amounts in most tissues except brain where they are important components of myelin, while ULCFA are normally found only in tissues such as retina (Aveldaño and Sprecher, 1987), brain (Poulos et al., 1986b), and spermatozoa (Poulos et al., 1986a). However, in peroxisome-deficient tissues, because of the impairment in oxidation, VLCFA reach significant levels in most tissues while ULCFA levels are elevated only in brain. It is possible that the increased levels of these fatty acids may contribute to the pathophysiology of the various disorders.

ACKNOWLEDGMENTS

This work was supported by grants received from the National Health and Medical Research Council, the Adelaide Children's Hospital Research Foundation, the Channel 7 Children's Medical Research Foundation, and the National Heart Foundation. The authors would like to thank Ms. Peta Knapman for assistance in the preparation of the manuscript.

REFERENCES

Aveldaño M and Sprecher H (1987) Very long-chain (C24 - C36) polyenoic fatty acids of the n-3 and n-6 series in dipolyunsaturated phosphatidylcholines from bovine retina. J Biol Chem 262:1180-1186.

Bligh BG and Dyer WJ (1959) A rapid method of total lipid extraction and purification, Can J Biochem Physiol 37:911-917.

Johnson DW (1990) A synthesis of unsaturated very long-chain fatty acids. Chem Phys Lipids 56:65-71.

Johnson DW, Beckman K, Robinson BS, Fellenberg AF, Poulos A (in press). Monoenoic fatty acids in human brain: Isomer identification and distribution. Lipids.

Lazo O, Contreras M, Bhushan A, Stanley W, Singh I (1989) Adrenoleukodystrophy: Impaired oxidation of fatty acids due to peroxisomal lignoceroyl-CoA ligase deficiency. Arch Biochem Biophys 270:722-728.

Naughton JM (1981) Supply of polyenoic fatty acids to the mammalian brain: The ease of conversion of the short chain fatty acids to their longer chain polyunsaturated metabolites in liver, brain, placenta and blood. Int J Biochem 13:21-32.

Poulos A, Sharp P, Johnson DW (1989) Polyenoic very long-chain fatty acids in peroxisomal disease. Biochemical discrimination of Zellweger's syndrome from other phenotypes. Neurology 39:44-47.

Poulos A, Sharp P, Johnson DW, Easton C (1988) The occurrence of polyenoic very long chain fatty acids with greater than 32 carbon atoms in molecular species of phosphatidylcholine in normal and peroxisome-deficient (Zellweger's syndrome) brain. Biochem J 253:645-650.

Poulos A, Sharp P, Johnson D, White IG, Fellenberg A (1986a) The occurrence of polyenoic fatty acids with greater than 22 carbon atoms in mammalian spermatozoa. Biochem J 240:891-895.

Poulos A, Sharp P, Singh H, Johnson D, Fellenberg AJ, Pollard AC (1986b) Detection of a homologous series of C26 - C38 polyenoic fatty acids in the brains of patients without peroxisomes (Zellweger's syndrome). Biochem J 235:607-610.

Poulos A, Singh H, Paton B, Sharp P, Derwas N (1986c) Accumulation and defective ß-oxidation of very-long-chain fatty acids in Zellweger's syndrome, adrenoleukodystrophy, and Refsum's disease variants. Clin Genet 29:397-408.

Robinson BS, Johnson DW, Poulos A (1990a) Metabolism of hexacosatetraenoic acid (C26:4, n-6) in immature rat brain. Biochem J 267:561-564.

Robinson BS, Johnson DW, Poulos A (1990b) Unique molecular species of phosphatidylcholine containing very-long-chain (C24-C38) polyenoic fatty acids in rat brain. Biochem J 265: 763-767.

Sharp P, Johnson DW, Poulos A (1991) Molecular species of phosphatidylcholine containing very long chain fatty acids in human brain: Enrichment in X-linked adrenoleukodystrophy brain and diseases of peroxisomal biogenesis brain. J Neurochem 56:30-37.

Sharp P, Poulos A, Fellenberg AJ, Johnson D (1987) Structure and distribution of polyenoic very long-chain fatty acids in the brain of peroxisome-deficient patients (Zellweger syndrome). Biochem J 248:61-67.

Singh H, Usher S, Johnson DW, Poulos A (1990) A comparative study of straight chain and branched chain fatty acid oxidation in skin fibroblasts from patients with peroxisomal disorders. J Lipid Res 31:217-225.

Street JM, Johnson DW, Singh H, Poulos A (1989) Metabolism of saturated and polyunsaturated fatty acids by normal and Zellweger syndrome skin fibroblasts. Biochem J 260:647-655.

Wanders RJA, Heymans HSA, Schutgens RB, Barth PG, van den Bosch H, Tager JM (1988a) Peroxisomal disorders in neurology. J Neurol Sci 88:1-39.

Wanders RJA, van Roermund CWT, van Wijland MJA, Schutgens RBH, Heikoop J, van den Bosch H, Schram AW, Tager JM (1987) Peroxisomal fatty acid ß-oxidation in relation to the accumulation of very long-chain fatty acids in cultured skin fibroblasts from patients with Zellweger syndrome and other peroxisomal disorders. J Clin Invest 80:1778-1783.

Wanders RJA, van Roermund CWT, van Wijland MJA, Schutgens RBH, van den Bosch H, Schram AW, Tager JM (1988b) Direct demonstration that the deficient oxidation of very long-chain fatty acids in X-linked adrenoleukodystrophy is due to an impaired ability of peroxisomes to activate very long-chain fatty acids. Biochem Biophys Res Commun 153:618-624.

LONG CHAIN OMEGA 3 POLYUNSATURATES IN FORMULA-FED

TERM INFANTS

R.A. Gibson, M. Makrides,
K.J. Clark, M.A. Neumann, and D.R. Lines

Department of Paediatrics and Child Health
Flinders Medical Centre
Bedford Park S.A. 5042
Australia

INTRODUCTION

Since linoleic acid (LA, $18:2\omega6$) was reported to be essential for the growth and development of infants (Hansen et al., 1963), it has been common for all infant formulas to contain this $\omega6$ polyunsaturated fatty acid at levels equal to, or above, the range found in human breast milk (Gibson and Kneebone, 1981a,b). Long chain (20 and 22 carbon) polyunsaturated fatty acids (LC.PUFA) that are present in breast milk are not included in infant formulas. Until recently little attention has been paid to either the level of the $\omega3$ LC.PUFA precursor, α-linolenic acid (ALA, $18:3\omega3$) or the ratio of LA to ALA (LA:ALA ratio). Most infant formulas contain less than 1% ALA and have a LA:ALA ratio ranging from 8:1 to more than 20:1 (Gibson and Kneebone 1981a).

It has been established that the erythrocyte membranes of pre-term infants fed formula with a high LA:ALA ratio (>19:1) are depleted in $\omega3$ LC.PUFA relative to those fed breast milk (Carlson et al., 1986). Carlson and co-workers (1987) subsequently demonstrated that this depletion can be corrected by supplementing formula with $\omega3$ LC.PUFA (marine oil). The physiological significance of these observations has been established by Uauy et al. (1990), who demonstrated that a supply of $\omega3$ LC.PUFA in the formula of very low birth-weight neonates was necessary to sustain retinal rod function similar to that found in a human-milk-fed group.

Due to numerous production problems, formula manufacturers have been reluctant to include a source of $\omega3$ LC.PUFA in their products. It seemed logical to determine whether it is possible to maintain $\omega3$ LC.PUFA in erythrocyte membranes at levels seen in breast-fed infants by providing an adequate supply of ALA in the formula. Formulas with a low LA:ALA ratio (about 4:1) were created either by increasing the level of ALA (formula B) or by decreasing the level of LA (formula C) and compared with an average Australian product with a high LA:ALA ratio (19:1; formula A), as well as breast-fed infants.

MATERIAL AND METHODS

Healthy full-term infants were enrolled in the study from birth and randomly allocated to one of three formula test diets. Subjects were maintained on the diets for 10 weeks during which the formulas were the only energy source. These were compared to a group (n=9) of 10-week-old breast-fed infants.

An infant formula base containing 80% of the total required fat (as butter fat) and all other nutrients in full measure was supplied by an infant formula manufacturer (Nestle, Australia). Three formulas with differing LA:ALA ratios were created by complementing with various vegetable oils (Table 1).

At the end of the study, 150-200 μl of whole blood were obtained by heel-prick. After removal of plasma, erythrocytes were washed three times in 0.9% NaCl. Total lipids were extracted by the method of Brockhuyse (1974); following transesterification in 1% H_2SO_4 at 70°C for three hours the resulting methyl esters were separated by gas chromatography (McMurchie et al., 1990).

Table 1. Major fatty acid components of infant formulas A, B, and C and breast milk (Gibson and Kneebone, 1981a)

		weight % of total fatty acid methyl esters ± SD			
	fatty acid	formula A	formula B	formula C	breast milk (range)
saturates					
	8:0	0.58±0.02	0.60±0.01	0.60±0.03	0.04-0.13
	10:0	1.53±0.04	1.57±0.04	1.54±0.02	0.40-1.10
	12:0	2.12±0.03	2.17±0.05	2.13±0.02	2.41-4.07
	14:0	8.67±0.08	8.87±0.21	8.72±0.68	5.09-5.63
	16:0	28.04±0.20	27.71±0.54	27.93±0.14	22.44-24.47
	18:0	11.42±0.07	11.70±0.23	11.87±0.05	8.24-9.20
monoenes					
	16:1	1.49±0.01	1.73±0.01	1.67±0.03	3.79-3.92
	18:1	27.55±0.12	25.15±0.36	37.33±0.24	35.00-37.18
ω6 series					
	18:2	13.80±0.31	12.82±0.63	3.47±0.03	7.82-10.75
	20:3	ND*	ND	ND	0.31-0.49
	20:4	ND	ND	ND	0.40-0.71
	22:4	ND	ND	ND	0.10-0.40
ω3 series					
	18:3	0.72±0.01	3.36±0.27	1.08±0.06	0.41-0.59
	20:5	ND	ND	ND	0.16-0.43
	22:5	ND	ND	ND	0.21-0.35
	22:6	ND	ND	ND	0.32-0.64
LA:ALA		19.2	3.8	3.2	18-19

* Not detected or present at <0.05% total fatty acid methyl esters

Table 2. Effect of feeding formulas with different linoleic acid to α-linolenic acid ratios or breast milk on ω3 fatty acid composition of total lipids from erythrocytes

fatty acid	weight % of total fatty acid methyl esters ± SD			
	formula A (n=10)	formula B (n=11)	formula C (n=8)	breast milk (n=9)
ω3 series				
18:3 (ALA)	0.24±0.04[a]	0.44±0.07[b]	0.37±0.04[c]	0.11±0.08[d]
20:5 (EPA)	0.39±0.07[a]	0.80±0.13[b]	1.28±0.17[c]	0.27±0.12[a]
22:5 (DPA)	2.11±0.39[a]	3.40±0.47[b]	3.45±0.45[b]	1.77±0.31[c]
22:6 (DHA)	3.47±0.46[a]	4.78±0.45[b]	4.48±0.49[b]	6.55±1.23[c]
ω3 LC.PUFA	5.97±0.76[a]	8.98±0.65[b,c]	9.30±0.95[b]	8.59±1.58[c]

Values without a common superscript are significantly different at $p<0.05$

RESULTS

The level of incorporation of ALA into the erythrocyte membranes was a reflection of the ALA content of the infant feeds (Tables 1 and 2). The level of LC.PUFA also varied according to the LA:ALA ratio. For example EPA (eicosapentaenoic acid, 20:5ω3) increased inversely to the LA:ALA ratio of the formula. DPA (docosapentaenoic acid, 22:5ω3) and DHA (docosahexaenoic acid, 22:6 n-3) were also higher in the membranes of erythrocytes from infants fed formulas with low LA:ALA ratios (formulas B and C). Formulas B and C resulted in levels of EPA and DPA in erythrocytes above those seen in breast-fed infants but none of the formulas increased the levels of DHA as high as that seen in breast-fed infants.

DISCUSSION

This study demonstrates that by reducing the LA:ALA ratio of the diet from 19:1 to around 4:1 it is possible to increase the level of all ω3 LC.PUFA in erythrocyte membranes. Decreasing the ratio by increasing the ALA content of the formula fat was as effective as lowering the LA content of the fat as measured by erythrocyte ALA, DPA, and DHA. However the low-LA fat blend (formula C) was more effective than the high-ALA fat blend (formula B) at increasing the level of erythrocyte EPA. It has been previously shown in animal studies that EPA incorporation into tissue membranes is greater when included in low-LA diets (McMurchie et al., 1990; Garg et al., 1990).

None of the formulas tested resulted in erythrocyte DHA levels being maintained to breast-fed levels. Breast milk contains DHA in quantities that are biologically significant (Clandinin et al., 1982). Since the LA:ALA ratio of breast milk is about the same as the fat blend in formula A (19:1, Table 1) it is clear that the higher level of DHA in the membranes of breast-fed infants is due to dietary DHA. It is interesting to note that other studies have demonstrated parity in erythrocyte DHA levels between formula-fed and breast-fed infants following consumption of formulas with LA:ALA ratios higher (8-10:1) than the ones used in the current study (Uauy et al., 1990; Innis et al., 1990). However these studies only lasted for a maximum of 28 days.

Some studies report that premature infants fed diets without ω3 LC.PUFA have compromised visual acuity (Carlson et al., 1989) and electroretinogram responses (Uauy et al., 1990) that can be correlated with erythrocyte DHA levels. While the relationship between ω3 fatty acid status and visual acuity has been found even in infants not receiving DHA and EPA supplementations (Carlson et al., 1989), there are some confusing results reported elsewhere. In the study by Uauy et al. (1990), for example, one of the formula groups receiving a fat blend with an LA:ALA ratio of 7.7:1 and not supplemented with ω3 LC.PUFA showed erythrocyte DHA levels equal to the breast-fed group and yet the electroretinogram parameters of this group failed to achieve levels seen in either the breast-fed or the marine oil-supplemented group. Clearly erythrocyte DHA level is not a perfect prediction of retinal function.

Carlson and Salem (1991) have recently reviewed the apparent need for preformed ω3 LC.PUFA in infant formulas. They point out that the use of marine ω3 LC.PUFA in infant foods has not been fully evaluated with regard to safety. Concerns include undesirable components of fish oil and the possible oversupply of DHA and EPA. Thus there is a real need to examine possible alternative supplies of ω3 LC.PUFA or ways of increasing the conversion of ALA to DHA. Infant formulas containing even lower LA:ALA ratios (1:2, 1:1) and perhaps the addition of vegetable oils containing the immediate metabolite of ALA (stearidonic acid, 18:4ω3) may be useful and allow for metabolic control of DHA production.

REFERENCES

Brockhuyse RM (1974) Improved lipid extraction of erythrocytes. Clin Chim Acta 51:341-343.

Carlson SE and Salem N (1991) Essentiality of ω3 fatty acids in growth and development of infants. World Rev Nutr Diet 66:74-86.

Carlson SE, Cooke RJ, Peeples JM, Werkman SH, Tolley EA (1989) Docosahexaenoate (DHA) and eicosapentaenoate (EPA) status of preterm infants: relationship to visual acuity in n-3 supplemented and unsupplemented infants. Pediatr Res 25:285A.

Carlson SE, Rhodes PG, Ferguson MG (1986) Docosahexaenoic acid status of preterm infants at birth and following feeding with human milk or formula. Am J Clin Nutr 44:798-804.

Carlson SE, Rhodes PG, Rao VS, Goldgar DE (1987) Effect of fish oil supplementation on the n-3 fatty acid content of red blood cell membranes in preterm infants. Pediatr Res 21:507-510.

Clandinin MT, Chappell JE, Hein T (1982) Do low weight infants require nutrition with chain elongation-desaturation products of essential fatty acids? Prog Lipid Res 20:901-904.

Garg ML, Thomson ABR, Clandinin MT (1990) Interactions of saturated n-6 and n-3 polyunsaturated fatty acids to modulate arachidonic acid metabolism. J Lipid Res 31:271-277.

Gibson RA and Kneebone GM (1981a) Fatty acid composition of infant formula. Aust Paediatr J 17:46-53.

Gibson RA and Kneebone GM (1981b) Fatty acid composition of human colostrum and mature breast milk. Am J Clin Nutr 34:252-257.

Hansen AD, Weise HF, Boehsche AN, Haggard ME, Adam DJD, Davis H (1963)

Role of linoleic acid in infant nutrition. Pediatrics 31:171-191.

Innis SM, Foote KD, MacKinnon MT, King DJ (1990) Plasma and red blood cell fatty acids of low birth weight infants fed their mother's expressed breast milk or preterm infant formula. Am J Clin Nutr 51:994-1000.

McMurchie EJ, Rinaldi JA, Barnard SL, Patten GS, Neumann M, McIntosh GH, Abbey M, Gibson RA (1990) Incorporation and effects of dietary eicosapentaenoate (20:5n-3) on plasma and erythrocyte lipids of the marmoset following dietary supplementation with differing levels of linoleic acid. Biochim Biophys Acta 1045:164-173.

Uauy RD, Birch DG, Birch EE, Tyson JE, Hoffman DR (1990) Effect of dietary ω3 fatty acids on retinal function of very low birth weight neonates. Pediatr Res 28:485-492.

SEVERE CHANGES IN POLYUNSATURATED FATTY ACIDS IN THE BRAIN, LIVER, KIDNEY, AND RETINA IN PATIENTS WITH PEROXISOMAL DISORDERS

Manuela Martinez

Unidad de Investigacíon Biomédica
Hospital Universitario Materno-Infantil
Valle de Hebron
Barcelona, Spain

INTRODUCTION

Peroxisomal disorders are a new group of congenital diseases in which peroxisomes are deficient in numbers or functionality. The prototype of these disorders is Zellweger syndrome (ZS), an extremely severe disease involving the central nervous system, liver, and kidneys (cerebro-hepato-renal syndrome). Infants with ZS have profound psychomotor retardation and hypotonia from birth, and very soon become blind and deaf. They usually die during the first year of life; autopsy examination reveals dysmyelination, gliosis, and neuronal heterotopias. Peroxisomes are absent from hepatocytes and proximal renal tubules (Goldfischer et al., 1973). The deficiencies in peroxisomal enzymes cause multiple biochemical abnormalities. In relation to lipid metabolism, there is accumulation of very long-chain fatty acids (Brown et al., 1982) due to a deficient peroxisomal ß-oxidation system (Singh et al., 1984) and a decrease in plasmalogen levels (Heymans et al., 1983) attributable to a defect in dihydroxy-acetone phosphate acyltransferase (Datta et al., 1984).

A somewhat milder, generalized peroxisomal disorder is neonatal adrenoleuko-dystrophy (NALD). The absence of peroxisomes seems to be less complete in the liver of patients with NALD (Vamecq et al., 1986), and abundant peroxisomes have been found in their cultured fibroblasts (Arias et al., 1985). Patients with NALD normally live longer than those with ZS. There are other peroxisomal disorders with loss of multiple peroxisomal functions, such as rhizomelic chondrodysplasia punctata (RCDP). In other patients, peroxisomes are present but they are deficient in some enzyme activity, as is the case in peroxisomal 3-oxoacyl-CoA thiolase deficiency in pseudo-Zellweger syndrome (Schram et al., 1987) and the isolated deficiency of bifunctional enzyme reported by Watkins et al. (1989). A different, although related, disease is X-linked adrenoleukodystrophy (X-ALD). In X-ALD, peroxisomes are apparently normal, but the activation of very long-chain fatty acids (VLCFA) seems to be defective (Wanders et al., 1988). Demyelination has a late onset in these patients. A milder variant of X-ALD is adrenomyeloneuropathy (AMN). In AMN, only the peripheral nervous system is primarily affected.

Recent studies (Martinez, 1989; 1990) have shown that the brain, liver, and kidneys of patients with Zellweger and pseudo-Zellweger syndromes have a profound deficiency of docosahexaenoic acid (DHA, 22:6ω3). It is therefore of interest to assess whether peroxisomal patients with other diagnoses, mainly NALD, have the same DHA deficiency, and whether this deficiency also can be detected in other peroxisomal diseases (such as RCDP, bifunctional enzyme deficiency, and even in late onset disorders, such as X-ALD and AMN). This chapter tries to answer these questions, and also reports, for the first time, some remarkable abnormalities in the polyunsaturated fatty acid composition of the retina in a Zellweger patient.

MATERIALS AND METHODS

The brain study involved 11 patients with Zellweger syndrome (ZS), three with neonatal adrenoleukodystrophy (NALD), one with pseudo-Zellweger syndrome (ps-ZS), one with isolated bifunctional enzyme deficiency (BD), and one with X-linked adrenoleukodystrophy (X-ALD). The values in these patients were compared with those obtained in the normal human brain throughout development (38 cases) and in other related disorders or demyelinating diseases (one patient with adrenomyeloneuropathy (AMN) and one with metachromatic leukodystrophy, MLD). In addition, the liver and kidney of some of the patients were studied, as well as the retina of one of the Zellweger cases and the liver of a patient with rhizomelic chondrodysplasia punctata (RCDP). The values in these tissues were compared with those obtained in the normal developing liver (11 cases), kidney (7 cases), and retina (18 cases). The liver fatty acid composition of a patient with Krabbe's disease was also studied for comparison.

Total lipids and plasmalogens were directly transmethylated from the tissues, according to the procedure of Lepage and Roy (1986), with slight variations previously described (Martinez, 1990). The total fatty acid methyl esters (FAME) and plasmalogen dimethyl acetals (DMA) were separated on a 30-m long, 0.25-mm ID SP-2330 capillary column, installed in a Hewlett-Packard 5890 gas chromatograph. A two-step temperature program was used (140-180°C, at 4°C/min; 180-210°C, at 1.7 °C/min), which perfectly separated the whole spectrum of fatty acids, VLCFA as well as PUFA and plasmalogen DMA (see Fig. 1).

The identity of some peaks of interest was confirmed by mass spectrometry (MS) of picolinyl derivatives (Christie et al., 1986) with a Hewlett-Packard 5970B mass spectrometer. Figure 1 is a gas chromatogram of brain total fatty acid methyl esters from a child with NALD. Note the large 22:5ω6 peak, which is identified by MS in Figure 2 as picolinyl 4,7,10,13,16-docosapentaenoate.

RESULTS

The total amount of PUFA, especially 22:6ω3, increases markedly during normal brain development (Fig. 3), until about 1 year of age, when it reaches a total concentration of 10-11 μmol/g. It is important to point out that, under normal conditions, the concentration of docosahexaenoic acid in the human brain during prenatal development is never less than 3 μmol/g, at least during the second half of gestation. As Figure 3 shows, even a fetus of only 26 weeks gestational age surpasses this figure. In contrast, the levels of brain DHA in Zellweger syndrome are usually well below this value (Fig. 4). When considered as molar percent of total fatty acids (Table 1), the 22:6ω3 level in the normal human brain maintains an even more constant value of about 7% throughout infancy, due to the partial dilution of PUFA by myelin fatty

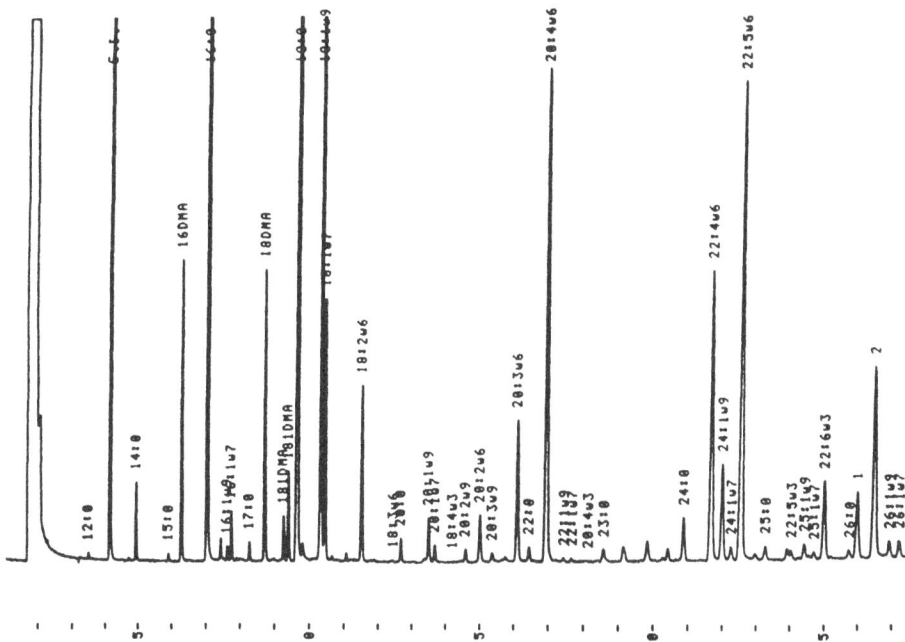

Figure 1. Gas chromatogram of brain total fatty acid methyl esters from a child with NALD.

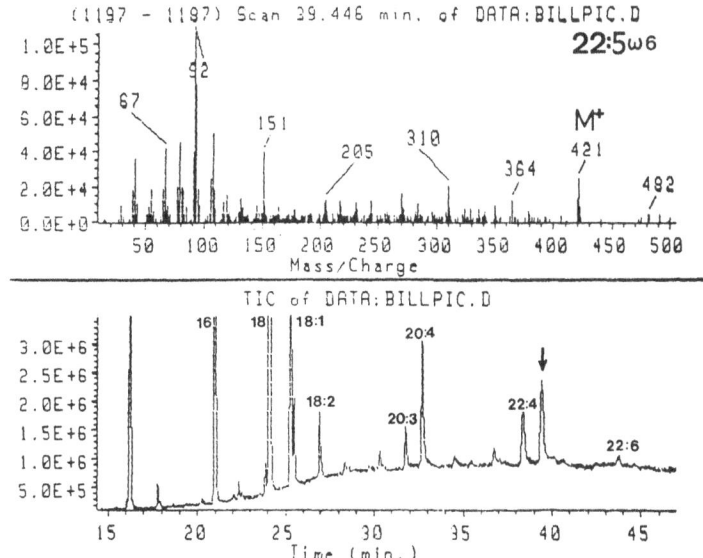

Figure 2. Peaks of interest from Figure 1 identified by mass spectrometry of picolinyl derivatives.

Table 1. Total fatty acid and plasmalogen composition of brain in some peroxisomal disorders and a demyelinating disease, compared with normal controls

	Normal Development Pre-term* (n = 8)	1-9 months (n = 23)	1-8 years (n = 7)	Zellweger 0-1 years (n = 11)	ps-ZS 1 month	NALD 8-31 months (n = 3)	BD 5 months	X-ALD 12 years	AMN 49 years	MLD 17 years
16:O	30.45 ± 1.05	26.89 ± 3.26	17.94 ± 1.07	27.89 ± 4.05	29.51	21.18 ± 1.98	24.46	21.26	19.17	20.79
18:O	22.08 ± 1.07	25.38 ± 1.49	23.95 ± 1.06	25.78 ± 3.97	24.98	26.46 ± 0.10	24.91	24.05	22.61	23.98
18:1ω9	11.35 ± 0.47	12.02 ± 1.13	17.35 ± 1.44	12.80 ± 2.10	11.65	13.16 ± 1.63	12.56	14.84	19.24	15.04
18:1ω7	3.55 ± 0.32	3.60 ± 0.35	4.13 ± 0.42	3.84 ± 1.33	2.55	3.25 ± 0.36	1.84	3.98	3.33	4.00
18:2ω6	0.23 ± 0.03	0.34 ± 0.18	0.61 ± 0.17	2.07 ± 0.68	1.31	1.85 ± 0.32	2.56	1.30	0.83	0.48
20:O	0.10 ± 0.01	0.18 ± 0.07	0.29 ± 0.01	0.36 ± 0.09	0.56	0.37 ± 0.09	0.42	0.41	0.40	0.62
20:1ω9	0.34 ± 0.06	0.33 ± 0.09	0.91 ± 0.27	0.65 ± 0.24	0.60	0.44 ± 0.07	0.44	0.44	1.02	0.40
20:3ω9	0.43 ± 0.11	0.45 ± 0.23	0.23 ± 0.05	0.47 ± 0.30	0.12	0.15 ± 0.04	0.09	0.17	0.40	0.33
20:3ω6	0.58 ± 0.13	1.04 ± 0.43	1.19 ± 0.15	3.40 ± 1.14	3.12	2.41 ± 1.25	2.78	1.04	0.78	0.72
20:4ω6	9.83 ± 0.60	9.04 ± 1.11	7.68 ± 0.87	13.34 ± 2.30	9.51	9.57 ± 2.42	11.04	10.50	8.08	9.65
20:5ω3	0.02 ± 0.01	0.04 ± 0.03	0.03 ± 0.01	0.09 ± 0.07	0.00	0.06 ± 0.07	0.03	ND	0.02	0.02
22:O	0.04 ± 0.03	0.13 ± 0.10	0.35 ± 0.05	0.18 ± 0.08	0.28	0.35 ± 0.11	0.29	0.22	0.28	0.52
22:4ω6	5.06 ± 0.61	5.34 ± 0.58	5.39 ± 0.49	4.63 ± 1.37	6.95	6.55 ± 1.51	6.84	5.28	3.95	4.02
22:5ω6	2.86 ± 0.45	2.33 ± 0.40	1.14 ± 0.28	0.60 ± 0.25	1.42	5.19 ± 4.48	1.01	3.56	1.57	1.65
22:5ω3	0.12 ± 0.02	0.17 ± 0.08	0.20 ± 0.07	0.24 ± 0.09	0.29	0.31 ± 0.27	0.50	0.23	0.19	0.19
22:6ω3	6.50 ± 0.48	6.98 ± 0.97	7.24 ± 0.58	2.30 ± 0.80	3.77	1.30 ± 0.49	5.26	7.41	9.56	9.17
24:O	0.03 ± 0.02	0.29 ± 0.33	1.52 ± 0.30	0.12 ± 0.06	0.28	0.81 ± 0.52	0.33	0.31	0.64	1.37
24:1ω9	0.07 ± 0.03	0.38 ± 0.39	3.38 ± 0.90	0.36 ± 0.20	0.33	1.51 ± 0.70	0.65	0.95	2.12	1.66
26:O	ND	0.03 ± 0.04	0.20 ± 0.06	0.02 ± 0.01	0.03	0.21 ± 0.11	0.08	0.10	0.17	0.18
26:1ω9	ND	0.03 ± 0.05	0.51 ± 0.15	0.05 ± 0.05	0.01	0.27 ± 0.12	0.09	0.14	0.50	0.29
TFA	55828 ± 6199	86349 ± 12107	135926 ± 12086	83162 ± 13168	83374	106260 ± 14707	88483	88620	115986	98099
TP	3114 ± 540	6592 ± 2027	15424 ± 1570	910 ± 724	6428	6923 ± 4335	5671	7333	9941	7523
16DMA	1580 ± 200	2536 ± 567	5297 ± 714	500 ± 334	3353	3291 ± 2334	2397	1889	2376	1801
18DMA	1233 ± 359	3251 ± 1010	5339 ± 315	369 ± 332	2730	2498 ± 1208	2878	3650	4123	3922
18:1ω9 DMA	120 ± 15	327 ± 208	1685 ± 428	24 ± 19	110	332 ± 253	147	824	1830	733
18:1ω7 DMA	181 ± 18	479 ± 307	3103 ± 979	74 ± 29	235	803 ± 681	249	970	1812	1067

*Gestational ages range from 26 to 36 weeks. Values for individual fatty acids are mole percent of the total, which includes some minor fatty acids not reported here. Total fatty acids (TFA) and plasmalogens (total, TP, and individual) are given in nmol/g wet weight of brain tissue. Values for grouped data are means ± SD. ps-ZS, pseudo-Zellweger syndrome. NALD, neonatal adrenoleukodystrophy. BD, isolated bifunctional enzyme deficiency. X-ALD, X-linked adrenoleukodystrophy. AMN, adrenomyeloneuropathy. MLD, metachromatic leukodystrophy. ND, nondetectable.

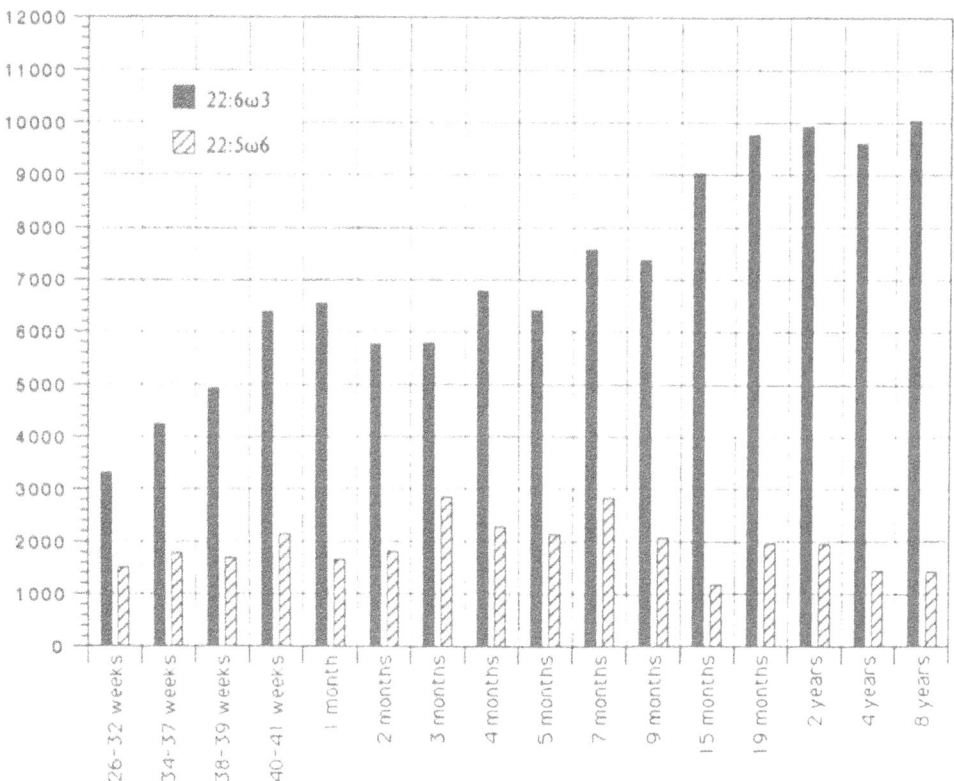

Figure 3. Products of Δ4-desaturation in the normal, developing human brain.

acids in older infants, which masks the increase in DHA. In adulthood the proportion of docosahexaenoate reaches about 9% of total fatty acids. The other product of Δ4-desaturation, 22:5ω6, shows a wider variation during development and a tendency to decrease after about 6 months of age. However, the 22:5ω6 concentration does not normally descend below 1 μmol/g.

Figure 4 shows the important alterations of the two fatty acids, 22:6ω3 and 22:5ω6, found in the brain of peroxisomal patients. In confirmation of previous findings (Martinez, 1989; 1990), all the Zellweger patients had very low brain levels of both 22:6ω3 and 22:5ω6. This was also true when molar percentages were considered (Table 1). All three infants with NALD showed a severe decrease in brain concentrations of docosahexaenoate. However, in contrast to the Zellweger patients, two of the NALD infants had a very important increase in brain 22:5ω6 (the identity of this fatty acid was confirmed by MS, as shown in Fig. 2). It is interesting to point out that, within each pathological condition, the patients represented in the graph are ordered according to age at death. That means that the NALD patients who lived longer were the ones with the higher levels of brain 22:5ω6.

The patient with bifunctional enzyme deficiency showed a much milder, albeit significant, decrease in the brain level of 22:6ω3, and normal or slightly decreased 22:5ω6. The X-ALD patient had also a deficiency in brain 22:6ω3, not so important as the Zellweger patients but quite significant for the patient's age. Like some of the NALD children, the X-ALD patient had an increase in 22:5ω6, less striking but perhaps still statistically significant. In contrast, the AMN patient had a PUFA profile entirely within normal limits.

A useful index for evaluating the ω3 deficiency in brains of peroxisomal patients was the 22:6ω3/22:4ω6 ratio (Fig. 5), which normally remains constant within a narrow range throughout development. Fig. 5 shows that all early onset peroxisomal patients had 22:6ω3/22:4ω6 ratios well below normal. In contrast, the X-ALD patient was within normal limits. This was due mainly to the loss of adrenic acid with demyelination, which compensated for the decrease in docosahexaenoic acid. For the same reason, the two patients with normal 22:6ω3 (AMN and MLD) had an increase in the 22:6ω3/22:4ω6

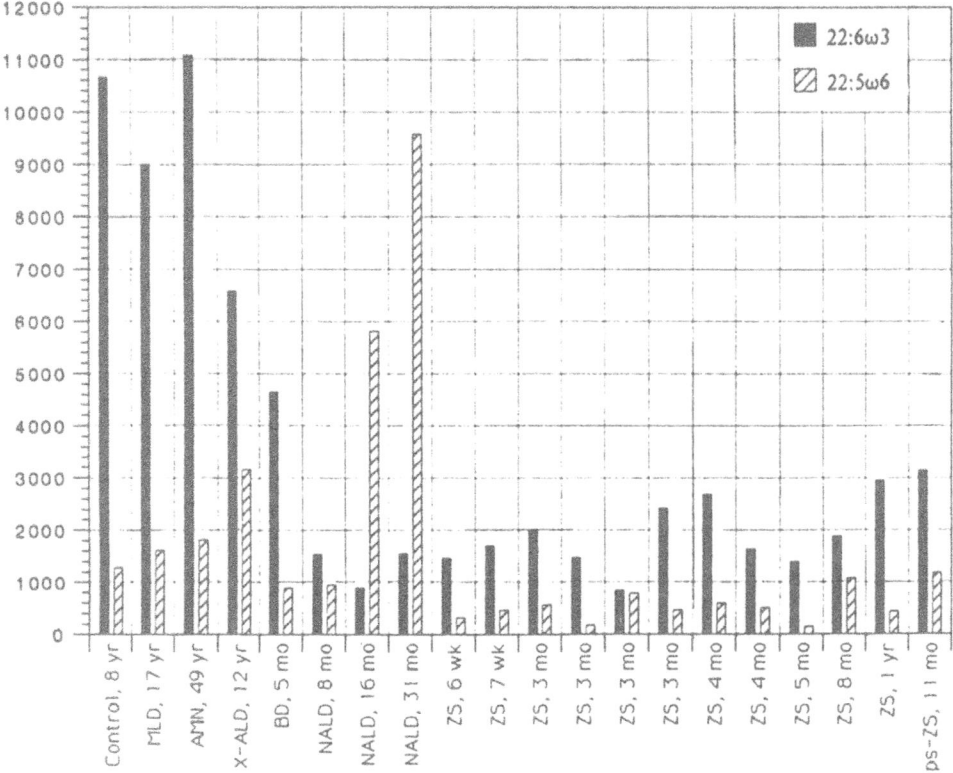

Figure 4. Products of Δ4-desaturation in patients with peroxisomal disorders and controls.

index. Other PUFA abnormalities found in the peroxisomal patients were significant increases in ω6 precursors (Table 1), including the parent fatty acid 18:2ω6 (linoleic acid), 20:3ω6 and, in the case of Zellweger syndrome, 20:4ω6 (arachidonic acid).

The PUFA alterations in the peroxisomal liver and kidney were much more variable than those in the brain. This was partly due to lack of homogeneity of the tissues, especially the kidney, but also to the fact that these tissues are much more strongly influenced by nutritional factors than brain. A different involvement of Δ4-desaturation in different tissues cannot be excluded. In any case, when compared with normal values for a given age, DHA concentration was constantly and significantly decreased in the

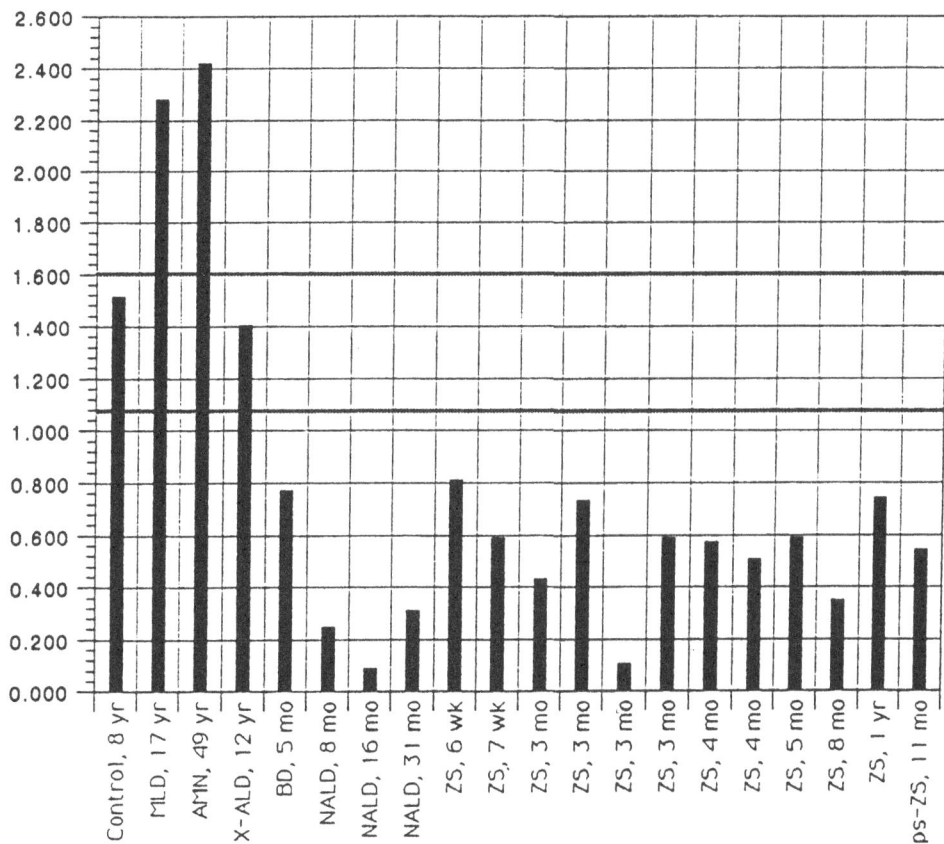

Figure 5. The 22:6ω3/22:4ω6 ratio is the more constant PUFA index throughout prenatal and postnatal brain development. The two horizontal lines indicate the higher and lower limits of this ratio during normal brain development. Note that all peroxisomal patients are well below the lower limit.

liver of all the Zellweger cases, being even lower than the minimum found in a 9-month-old, profoundly undernourished infant (Fig. 6). In the peroxisomal kidney (Fig. 7), DHA was also decreased, and the ratio of 22:6ω3 to 26:0 was more clearly diminished than in any other tissue (Martinez, 1989), due to the important increase in VLCFA in the Zellweger kidney. The other product of Δ4-desaturation, 22:5ω6, was also decreased in all cases in the Zellweger liver, but less so in the Zellweger kidney. The only liver sample of a patient with NALD that could be studied showed a significant increase in 22:5ω6. In contrast, liver from RCDP and AMN patients had low levels of 22:5ω6.

The retina of a Zellweger patient showed extremely low 22:6ω3 values (Fig. 8), even lower than a premature infant of only 24 weeks gestational age. The levels of 22:5ω6 and 22:5ω3 were also low, but 22:4ω6 was within normal limits and 20:4ω6 was somewhat high for the age (Table 2). As Fig. 9 shows, during development the ratio of docosahexaenoate to arachidonate (22:6ω3/20:4ω6) increases steadily in the normal retina, and malnutrition can cause a significant reduction in this index (open circles in Fig. 9), especially when malnutrition occurs early in development (Martinez et al., 1988). However, the reduction in the retinal 22:6ω3/20:4ω6 ratio caused by Zellweger

Table 2. Total fatty acid and plasmalogen composition of the human retina during normal development in two malnourished children and in a patient with a peroxisomal disorder, compared with normal controls

	Prenatal* (n = 9)	Controls			Maln.	Mucovisc.	Zellweger
		0-1 months (n = 5)	19 months	49 years	4 months	9 months	8 months
16:O	26.06 ± 0.83	25.18 ± 0.99	22.27	21.42	23.26	20.88	27.15
18:O	20.99 ± 0.56	21.65 ± 0.99	24.06	24.06	21.77	22.95	19.99
18:1ω9	13.01 ± 0.32	13.00 ± 0.71	12.62	9.71	13.87	12.63	21.64
18:1ω7	3.07 ± 0.53	3.32 ± 0.28	3.64	3.31	3.35	4.49	2.57
18:2ω6	0.98 ± 0.53	1.42 ± 0.69	1.20	1.44	2.58	2.41	4.02
20:O	0.40 ± 0.08	0.53 ± 0.08	0.87	0.68	0.70	0.96	1.00
20:1ω9	0.44 ± 0.10	0.25 ± 0.08	0.37	0.30	0.38	0.47	0.88
20:3ω9	0.42 ± 0.18	0.12 ± 0.07	0.06	0.08	0.14	0.28	0.43
20:3ω6	0.79 ± 0.33	1.21 ± 0.18	1.53	2.17	1.89	1.90	2.14
20:4ω6	13.27 ± 0.57	12.23 ± 0.78	9.90	9.56	12.82	9.88	12.16
20:5ω3	0.09 ± 0.08	0.09 ± 0.06	0.25	0.14	ND	0.08	ND
22:O	0.40 ± 0.07	0.39 ± 0.05	0.63	0.51	0.42	0.55	0.43
22:4ω6	2.46 ± 0.26	2.75 ± 0.26	1.87	1.77	3.56	2.24	1.79
22:5ω6	1.36 ± 0.30	0.83 ± 0.15	0.46	0.87	2.52	2.64	0.21
22:5ω3	0.31 ± 0.17	0.55 ± 0.12	1.22	1.32	0.45	0.76	0.18
22:6ω3	9.15 ± 1.27	11.95 ± 0.25	15.65	19.70	8.07	12.72	0.89
24:O	0.21 ± 0.16	0.21 ± 0.07	0.31	0.37	0.31	0.46	0.49
24:1ω9	0.36 ± 0.15	0.35 ± 0.13	0.23	0.19	0.71	0.41	0.91
26:O	ND	ND	ND	ND	ND	ND	0.20
26:1ω9	ND	ND	ND	ND	ND	ND	0.15
TFA	39786 ± 5635	46396 ± 4075	40096	45981	45281	48297	45736
TP	2762 ± 256	3075 ± 490	2253	2080	2846	2519	136
16DMA	1075 ± 83	1082 ± 169	722	584	917	703	58
18DMA	1343 ± 242	1753 ± 284	1285	1350	1822	1334	61
18:1ω9 DMA	163 ± 29	123 ± 29	98	64	173	177	17
18:1ω7 DMA	181 ± 35	117 ± 33	148	82	134	305	ND

*Gestational ages range from 26 to 36 weeks. Values for individual fatty acids are mole percent of the total, which includes some minor fatty acids not reported here. Total fatty acids (TFA) and plasmalogens (total, TP, and individual) are given in nmol/g wet weight of retinal tissue. Values for grouped data are means ± SD. Maln., malnourished. Mucovisc., mucoviscidosis. ND, nondetectable.

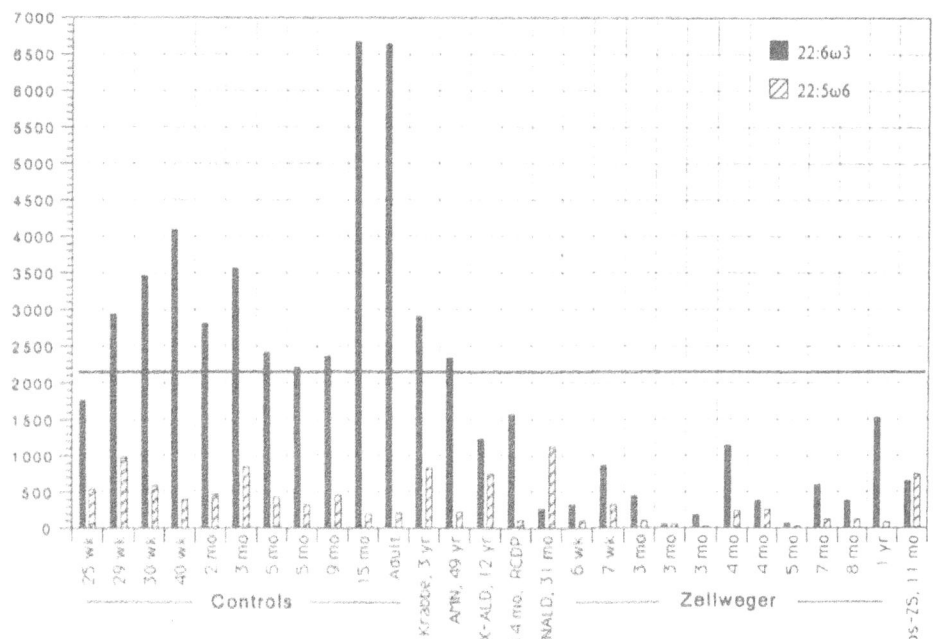

Figure 6. Products of Δ4 desaturation in the liver of patients with peroxisomal disorders and controls. The horizontal line marks the lower limit of DHA (22:6ω3) concentration in the developing human liver, even under poor nutritional conditions. Notice that all peroxisomal disorders, including X-ALD and RCDP, have very low levels of liver DHA.

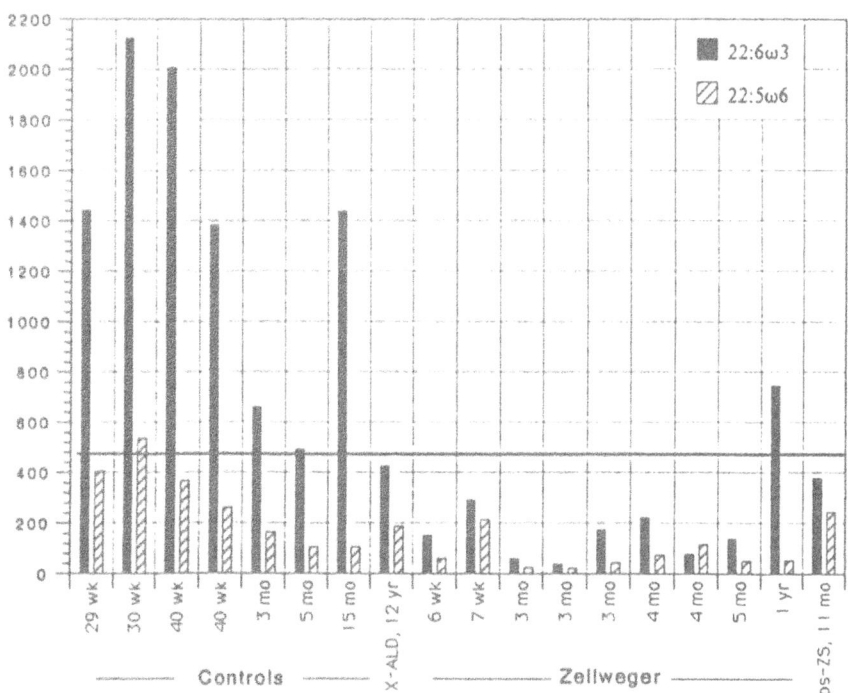

Figure 7. Products of Δ4 desaturation in the kidney of patients with peroxisomal disorders and controls. As in Figure 6, the horizontal line indicates the lowest limit of DHA (22:6ω3) concentration in postnatal life. Because of tissue heterogeneity, the values are even more variable than in the liver. However, all the Zellweger patients but one have DHA levels clearly lower than normal.

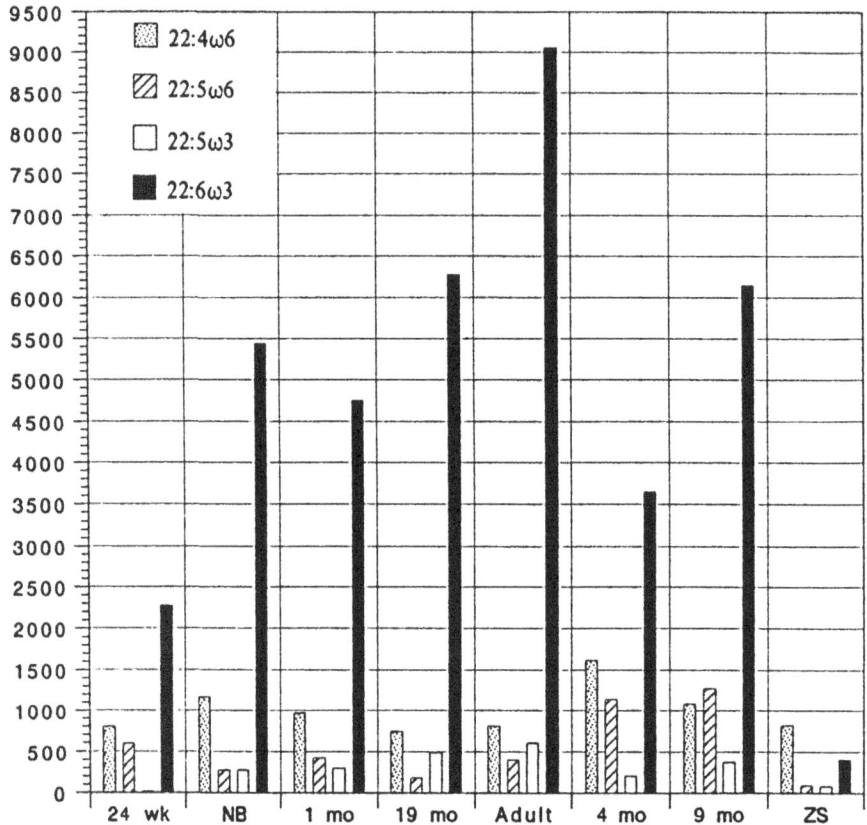

Figure 8. 22-carbon-atom PUFA in the human retina.

syndrome is incomparably more severe. As in other tissues, plasmalogen levels were low in the Zellweger retina, and it was interesting to find the characteristic increase in 26:0 and 26:1 (Brown et al., 1982) in a fundamentally polyunsaturated tissue such as the retina (Table 2).

DISCUSSION

Present findings confirm the existence of a very important deficiency of docosahexaenoic acid in peroxisomal disorders. Of special interest is the extremely low value of 22:6ω3 found in the Zellweger retina. Such low 22:6ω3 levels could, by themselves, explain the neurological and visual defects in peroxisomal patients.

The other product of Δ4-desaturation, 22:5ω6, seems to be constantly decreased in Zellweger syndrome, even in the retina. On the other hand, in neonatal adrenoleukodystrophy—and to a lesser extent in X-ALD—22:5ω6 can be more or less significantly increased, mainly in the brain, as is known to happen in cases of nutritional ω3 deficiency (Galli et al., 1971). It is, therefore, tempting to speculate that 22:5ω6 increases

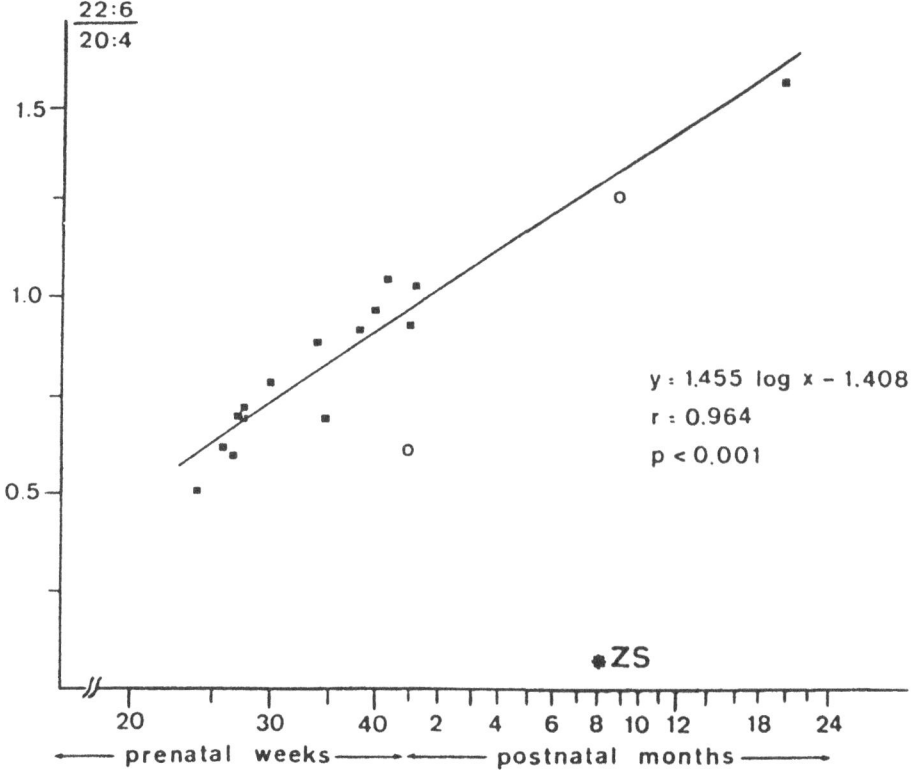

Figure 9. Although the age logarithmic scale indicates gestational weeks and postnatal months, the independent variable in the formula (x) refers to post-conceptual weeks. Filled circles = normal, well-nourished infants; empty circles = malnourished infants; asterisk = Zellweger syndrome (ZS).

to compensate for the $22:6\omega3$ deficiency in neuronal membranes. If this is so, a somewhat less severe clinical picture could be expected from cases with higher $22:5\omega6$ levels compared to those with lower $22:5\omega6$ levels, a hypothesis that seems to be consistent with the longer survival of the NALD patients with high levels of $22:5\omega6$, and with the extreme severity of Zellweger syndrome (very low $22:5\omega6$ levels), compared to other peroxisomal disorders.

Although more cases must be studied before definitive conclusions can be drawn, it seems that in late-onset disorders, such as X-ALD, the decrease in $22:6\omega3$ is much less important than in early onset diseases and is within the range that may be found in nutritional $\omega3$ deficiencies. It may be that the poor PUFA intake of X-ALD patients, who are fed special diets virtually devoid of essential fatty acids, plays a role in their $22:6\omega3$ deficiency. Similarly, the $22:5\omega6$ increase is more moderate than in some cases of NALD, and could be influenced by nutritional $\omega3$ deprivation. The normal PUFA composition of the AMN brain is consistent with the lack of CNS symptoms in these patients.

It remains to be proven whether the $\Delta4$-desaturase system is defective in peroxisomal disorders and, if so, whether the present data suggest a differential activity of the system for ω3 and ω6 fatty acids. Whatever the mechanism, however, the most important fact is that there is a docosahexaenoic acid deficiency of an extraordinary magnitude in early onset peroxisomal disorders—a magnitude not found even in the most extreme cases of ω3 deprivation. It may well be that the changes in peroxisomal membranes are secondary to this ω3 deficiency. Such possibilities open new avenues for research in the field of peroxisomal disorders, including the need to test the biological effects of DHA treatment in these patients.

ACKNOWLEDGMENTS

I thank Dr. H.W. Moser (The Kennedy Institute, Baltimore, USA), Dr. R.J.A Wanders (University Hospital of Amsterdam, The Netherlands), and Dr. I. Lorente (Children's Hospital of Sabadell, Barcelona, Spain) for kindly providing the peroxisomal tissues reported in this chapter. Mass spectrometry of fatty acids was accomplished with funds from the Fondo the Investigaciones Sanitarias de la Seguridad Social, Spain (FIS 89/1623).

REFERENCES

Arias JA, Moser AB, Goldfischer SL (1985) Ultrastructural and cytochemical demonstration of peroxisomes in cultured fibroblasts from patients with peroxisomal deficiency disorders. J Cell Biol 100:1789.

Brown FR, McAdams AJ, Cumins JW, Konkol R, Singh I, Moser AB, Moser HW (1982) Cerebro-hepato-renal (Zellweger) syndrome and neonatal adrenoleukodystrophy: Similarities in phenotype and accumulation of very long-chain fatty acids. Johns Hopkins Med J 151:344.

Christie WW, Brechany EY, Johnson SB, Holman RT (1986) A comparison of pyrrolidide and picolinyl ester derivatives for the identification of fatty acids in natural samples by gas chromatography-mass spectrometry. Lipids 21:657.

Datta NS, Wilson GN, Hajra AK (1984) Deficiency of enzymes catalyzing the biosynthesis of glycerol-ether lipids in Zellweger syndrome. N Engl J Med 311:1080.

Galli C, Trzeciak HT, Paoletti R (1971) Effects of dietary fatty acids on the fatty acid composition of brain ethanolamine phosphoglyceride: reciprocal replacement of n-6 and n-3 polyunsaturated fatty acids. Biochim Biophys Acta 248:449.

Goldfischer S, Moore CL, Johnson AB, Spiro AJ, Vatsmis MP, Wisniewski HK, Ritch RH, Norton WT, Rain I, Gartner LM (1973) Peroxisomal and mitochondrial defects in the cerebro-hepato-renal syndrome. Science 182:62.

Heymans HSA, Schutgens RBH, Tan R, van den Bosch H, Borst P (1983) Severe plasmalogen deficiency in tissues of infants without peroxisomes (Zellweger syndrome). Nature 306:69.

Lepage G and Roy CC (1986) Direct transesterification of all classes of lipids in a one-step reaction. J Lipid Res 27:114.

Martinez M (1989) Polyunsaturated fatty acid changes suggesting a new enzymatic defect in Zellweger syndrome. Lipids 24:261.

Martinez M (1990) Severe deficiency of docosahexaenoic acid in peroxisomal disorders: A defect of Δ4-desaturation? Neurology 40:1292.

Martinez M, Ballabriga A, Gil-Gibernau JJ (1988) Lipids of the developing human retina. I. Total fatty acids, plasmalogens, and fatty acid composition of ethanol-amine and choline phosphoglycerides. J Neurosci Res 20:484.

Schram AW, Goldfinger S, van Roermund CWT, Brouwer-Kelder EM, Collins J, Hashimoto T, Heyman HSA, van den Bosch H, Schutgens RBH, Tager JM, Wanders RJA (1987) Human peroxisomal 3-oxoacyl-coenzyme A thiolase deficiency. Proc Natl Acad Sci USA 84: 2494.

Singh I, Moser AB, Goldfischer S, Moser HW (1984) Lignoceric acid is oxidized in the peroxisome: implications for the Zellweger cerebro-hepato-renal syndrome and adrenoleukodystrophy. Proc Natl Acad Sci USA 81:4203.

Vamecq J, Draye JP, van Hoof F, Misson JP, Evrard P, Verellen G, Eyssen HJ, van Eldere J, Schutgens RBH, Wanders RJ-A, Roels F, Goldfischer SL (1986) Multiple peroxisomal enzymatic deficiency disorders. A comparative biochemical and morphological study of Zellweger cerebro-hepato-renal syndrome and neonatal adrenoleukodystrophy. Am J Pathol 125:524.

Wanders RJA, van Roermund CWT, van Wijland MJA, Schutgens RBH, van den Bosch H, Tager JM (1988) Direct demonstration that the deficient oxidation of very long-chain fatty acids in X-linked adrenoleukodystrophy is due to an impaired ability of peroxisomes to activate very long-chain fatty acids. Biochem Biophys Res Commun 153:618.

Watkins PA, Chen WW, Harris CJ, Hoefler G, Hoefler S, Blake DC Jr, Balfe A, Kelley RI, Moser AB, Beard ME, Moser HW (1989) Peroxisomal bifunctional enzyme deficiency. J Clin Invest 83:771.

DEGRADATION OF PHOSPHOLIPIDS AND PROTEIN KINASE C

ACTIVATION FOR THE CONTROL OF NEURONAL FUNCTIONS

Tetsutaro Shinomura, Hiroyuki Mishima,
Shinji Matsushima, Yoshinori Asaoka,
Kimihisa Yoshida, Masahiro Oka,
and Yasutomi Nishizuka

Department of Biochemistry
Kobe University School of Medicine
Kobe 650, Japan

Biosignal Research Center
Kobe University
Kobe 657, Japan

INTRODUCTION

Protein kinase C (PKC) has been postulated to play key roles in several neuronal processes such as transmitter release, ion channel modulation, receptor downregulation, cross-talk with other signaling systems, neuronal development and regeneration, and long-term potentiation (LTP). The hydrolysis of inositol phospholipids catalyzed by phospholipase C was initially thought to be the sole mechanism for the activation of PKC. Subsequent studies, however, have shown that the receptor-mediated hydrolysis of phosphatidylcholine (PC) may also be involved in the transmembrane signaling (for a review, see Exton, 1990). In addition, in synergy with diacylglycerol (DG), unsaturated free fatty acids (FFAs) such as arachidonic, oleic, linoleic, linolenic, and docosahexaenoic acids dramatically activate some members of the PKC family at the basal level of Ca^{2+} concentration. It is plausible that phospholipase C, phospholipase A_2 and possibly phospholipase D as well are all involved in the activation of PKC. Presumably, this enzyme activation is integrated into the signal-induced membrane phospholipid degradation cascade, prolonging the activation of PKC. The sustained activation of this enzyme appears to be of importance for the long-term cellular responses such as neuronal plasticity and gene activation. This article will briefly discuss a possible link between signal-induced degradation of various membrane phospholipids and PKC activation, and the implications of such a linkage in the control of neuronal functions.

Neurobiology of Essential Fatty Acids
Edited by N.G. Bazan *et al.*, Plenum Press, New York, 1992

Table 1. PKC subspecies structurally identified in mammalian tissues

	Amino acid residues	Molecular weight	Tissue expression
α	672	76,799	Universal
βI	671	76,790	Some tissues
βII	673	76,933	Many tissues
γ	697	78,366	Brain only
δ	673	77,517	Many tissues ?
ϵ	737	83,474	Brain only ?
ζ	592	67,740	Brain, liver, etc. ?
$\eta(l)$	683	77,972	Lung, skin, heart

MOLECULAR HETEROGENEITY AND DIFFERENTIAL TISSUE EXPRESSION

PKC is most prominent in the brain in terms of both quantity and variation. Molecular cloning and enzymological analysis have revealed the existence of multiple subspecies of PKC. Initially, four cDNA clones that encode the α-, βI-, βII- and γ-subspecies were found. Subsequently, another group of cDNA clones, encoding at least four further subspecies having the δ-, ϵ-, ζ- and $\eta(l)$-sequence, have been isolated. These subspecies are all composed of a single polypeptide chain, with the group of α-, βI-, βII- and γ-subspecies each having four conserved (C_1-C_4) and five variable (V_1-V_5) regions. The second group of δ-, ϵ-, ζ- and $\eta(l)$-subspecies lack the region C_2 (for a review, see Nishizuka, 1988; also see Osada et al., 1990; Bacher et al., 1991). All PKC subspecies so far identified possess very similar molecular sizes (Table 1).

Using a combination of immunohistochemical, biochemical, and *in situ* hybridization procedures, the relative activity and individual patterns of expression of multiple PKC subspecies in several tissues and cell types have been examined extensively and clarified in detail (for a review, see Nishizuka, 1988). PKC with the γ-sequence is expressed only in specific cells of the central nervous system as described below. On the other hand, PKCs with the βI- and βII-sequence are expressed in the brain, as well as in other tissues, in different ratios. In contrast, PKC with the α-sequence is widely distributed in many tissues and cell types. In general, one cell type co-expresses more than one subspecies of PKC. These subspecies apparently show a distinct intracellular location. At present, the distribution and biochemical properties of the enzymes encoded by the δ-, ϵ-, ζ- and $\eta(l)$-sequence have been clarified only partially.

SIGNAL ROUTES FOR PKC ACTIVATION

Although the hydrolysis of inositol phospholipids was initially thought to be only one mechanism leading to the activation of PKC, there are several additional routes to provide the DG that is needed for enzyme activation as shown in Figure 1 (for a review, see Farago and Nishizuka, 1990). For instance, PC may also be hydrolyzed to produce DG at a relatively later phase of cellular responses, particularly responses to long-acting signals such as some growth factors (for a review, see Exton, 1990). In addition, both voltage-dependent and receptor-mediated Ca^{2+}-gate opening may cause

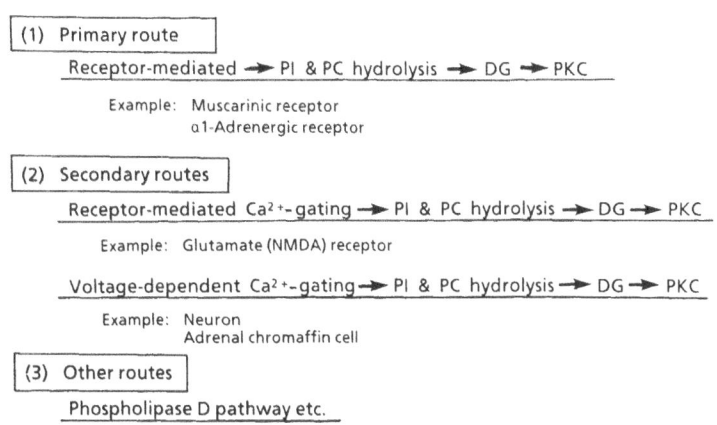

(1) Primary route

Receptor-mediated → PI & PC hydrolysis → DG → PKC

    Example: Muscarinic receptor
             α1-Adrenergic receptor

(2) Secondary routes

Receptor-mediated Ca^{2+}-gating → PI & PC hydrolysis → DG → PKC

    Example: Glutamate (NMDA) receptor

Voltage-dependent Ca^{2+}-gating → PI & PC hydrolysis → DG → PKC

    Example: Neuron
             Adrenal chromaffin cell

(3) Other routes

Phospholipase D pathway etc.

Figure 1. Signal routes for protein kinase C activation. The detailed explanation is given in the text. PI, phosphatidylinositol and its mono- and bisphosphate; NMDA, N-methyl-D-aspartate.

phospholipid breakdown due to the activation of Ca^{2+}-dependent phospholipase C, and probably phospholipase A_2 and phospholipase D as well. The exact biochemical basis for this phospholipid degradation cascade, however, has not been fully understood. As discussed below, unsaturated FFAs are also involved in the activation of PKC. It is thus possible that the signal routes leading to the activation of PKC may greatly vary with cell types, extracellular signals, and perhaps with the time after stimulation.

SYNERGISTIC ACTION OF UNSATURATED FFA AND DG

The α-, ß-, and γ-subspecies of PKC isolated from mammalian tissues require absolutely phosphatidylserine (PS) and Ca^{2+} for their actions, and DG greatly increases an apparent affinity of the enzyme for Ca^{2+}, thereby rendering it active at the micromolar range of Ca^{2+} concentration (Kishimoto et al, 1980; Nishizuka, 1988). The experiments given in Figure 2 confirm this early observation, and show that, in the presence of DG and PS, the addition of arachidonic acid further increases the affinity of PKC for Ca^{2+}, and greatly enhances the reaction velocity, particularly at nearly basal levels of Ca^{2+} concentration. The synergistic action of arachidonic acid and DG was most remarkable for the α- and β-subspecies, which are expressed in many tissues and cell types. The γ-subspecies was activated slightly by arachidonic acid, in the absence of DG and PS, at relatively higher concentrations of Ca^{2+} (Sekiguchi et al., 1987). The γ-subspecies is expressed only in the central nervous tissue after birth (Hashimoto et al., 1988).

The synergistic effect of FFA and DG is observed also with many other naturally occurring cis-unsaturated FFAs such as oleic, linoleic, linolenic, and docosahexaenoic acids, as given in Figure 3. It is worth noting that docosahexaenoic acid is most abundant in the phospholipids in the central nervous system and retina, notably during early postnatal development and synaptogenesis (Scott and Bazan, 1989). Palmitic and stearic acids are inactive. Elaidic acid is also inactive for all PKC subspecies tested.

The results described above were obtained with calf thymus H1 histone as a model substrate. The synergistic action of unsaturated FFAs and DG can be observed with various other model phosphate acceptor proteins such as myelin basic protein as well as with physiological substrate proteins including growth-associated protein-43 (GAP-43,

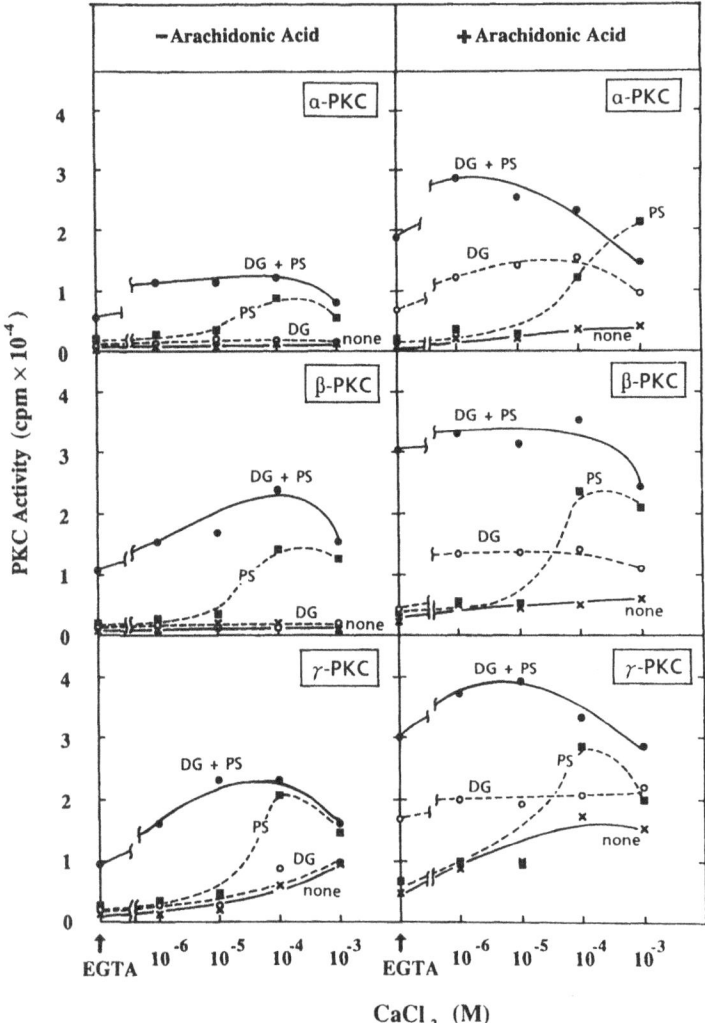

Figure 2. Activation of protein kinase C by PS, DG, and arachidonic acid at various concentrations of Ca^{2+}. Each PKC subspecies was assayed with H1 histone as a model substrate under the conditions described by Shinomura et al. (1991) except that various concentrations of $CaCl_2$ were added in the absence (left panel) or presence (right panel) of 50 μM arachidonic acid. EGTA (5 mM) instead of $CaCl_2$ was added to the reaction mixture where indicated by an arrow.

F1 protein, B50 protein, neuromodulin, see below), although the kinetics of this synergistic action vary slightly with the phosphate acceptor protein used. It may be important to note that, in all substrates so far examined, several PKC subspecies exhibit nearly full activity at basal levels of Ca^{2+} concentration. It is possible that PKC could maintain its activity for a prolonged period of time, if both DG and unsaturated FFAs are available.

IMPLICATIONS FOR CONTROL OF SYNAPTIC PROCESSES

The activation of PKC has been proposed to play a key role in the development of LTP in the hippocampus (for reviews, see Alkon and Rasmussen, 1988; Gispen, 1988;

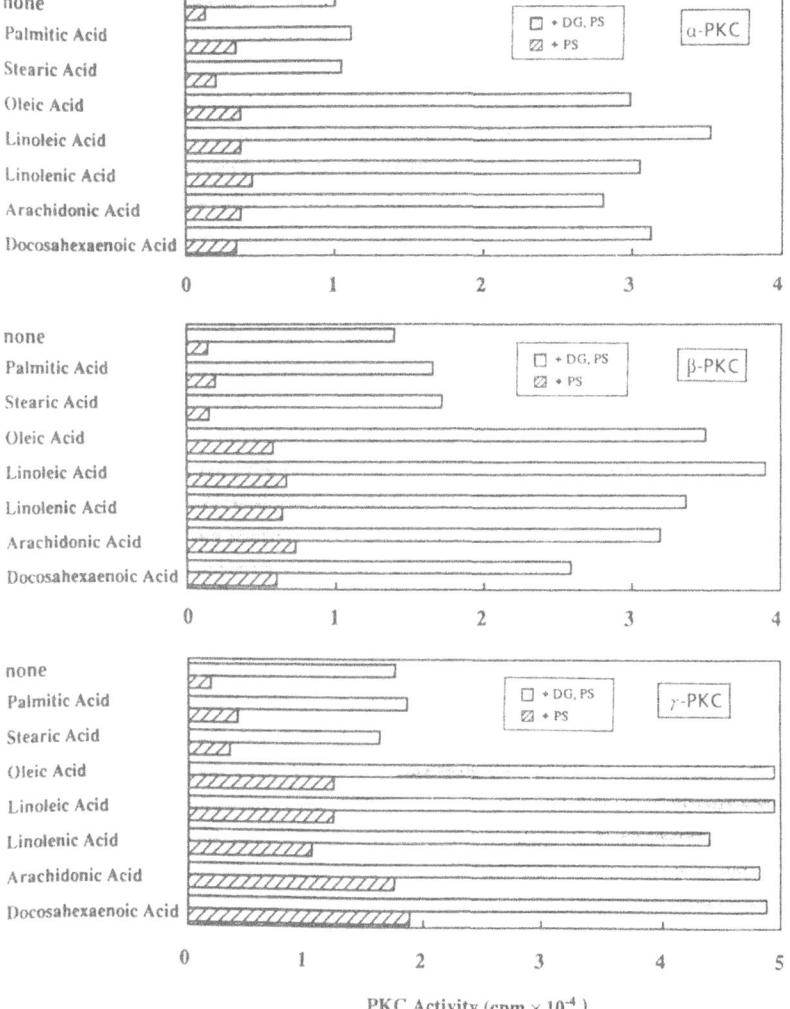

Figure 3. Activation of protein kinase C by various FFAs in the presence of PS, DG, and Ca^{2+}. Each subspecies of PKC was assayed under the standard conditions with H1 histone as a model substrate as described by Shinomura et al. (1991) except that various fatty acids (50 μM each) were added as indicated. Gray bars, assayed in the presence of PS and DG; hatched bars, assayed in the presence of PS.

Linden and Routtenberg, 1989; Tsien et al., 1990). Immunocytochemical analysis has shown that, in the adult rat hippocampus, the γ-subspecies is expressed only in the post- but not in the presynaptic region, as illustrated in Figure 4 (Kose et al., 1990). Consistent with this observation, synaptosomes isolated from the hippocampus region contain the α- and β-subspecies but not the γ-subspecies, although the whole tissue contains a large quantity of the γ-subspecies due to its presence in the pyramidal cells, as shown in Figure 5 (Shearman et al., 1991). The α- and β-subspecies are expressed in both pre- and postsynaptic regions. This figure also shows that synaptosomes isolated from the cerebral tissue contain all of the three subspecies.

GAP-43 is one of the major target proteins of PKC, and is tightly associated with the membrane of presynaptic nerve endings. This protein has been proposed to play

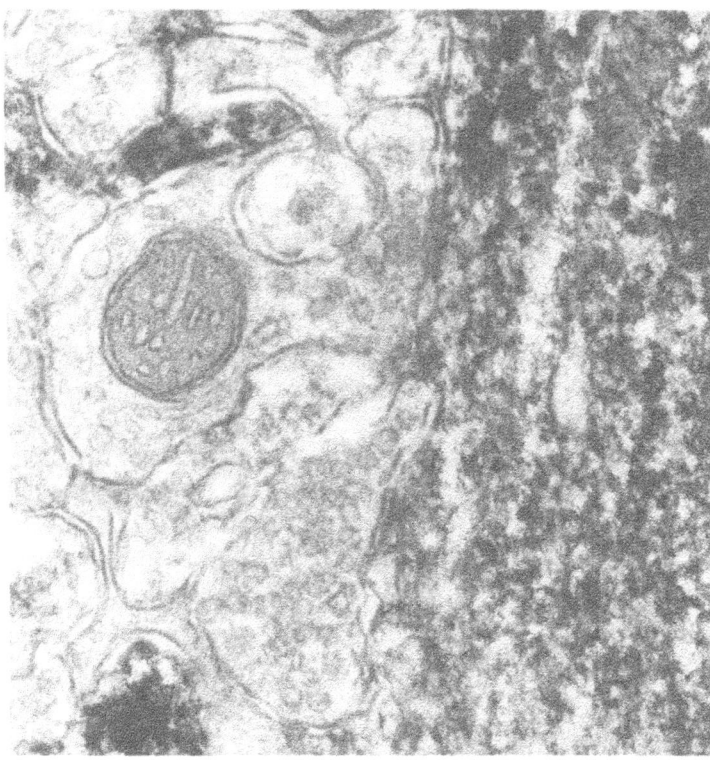

Figure 4. Electron micrograph of pyramidal cell in CA1 region of adult rat hippocampus, showing γ-PKC immunoreactivity. The nerve ending region (left) terminating at the pyramidal cell body (right) does not appear to contain the γ-subspecies. Photograph taken and provided by A. Kose, A. Ito, N. Saito, and C. Tanaka. (For details, see Kose et al., 1990).

various roles in the control of diverse synaptic processes, including not only immediate responses such as transmitter release but also long-term responses such as LTP, synaptogenesis, and regeneration (for a review, see Coggins and Zwiers, 1991). Hypothetical roles of PKC in the pre- and postsynaptic regions of the hippocampus are schematically given in Figure 6.

It has been postulated that unsaturated FFAs may be involved in the maintenance of LTP probably through the activation of PKC (Linden and Routtenberg, 1989). Arachidonic acid has been proposed to be a retrograde messenger for this process from the postsynaptic cell to the presynaptic terminal (Bliss et al., 1989). Although the biochemical mechanism of LTP and the role of PKC in synaptic transmission have not yet been well clarified, the synergistic action of unsaturated FFAs and DG on PKC activation reported herein suggests a potential role for FFAs in the sustained activation of enzyme, particularly the α- and β-subspecies which are present abundantly both in the presynaptic terminal and in the postsynaptic cell. It has been reported, in fact, that in the hippocampus arachidonic acid is released by the activation of phospholipase A_2 upon stimulation (Felder et al., 1990). Docosahexaenoic acid is known to be esterified in large amounts in the phospholipids of the central nervous system, especially during early postnatal development and synaptogenesis (Scott and Bazan, 1989). This fatty acid very actively synergizes with DG to enhance PKC activation, and by itself can activate the γ-subspecies to some extent in the absence of DG. This suggests that docosahexaenoic acid may have some functions in the neuronal processes.

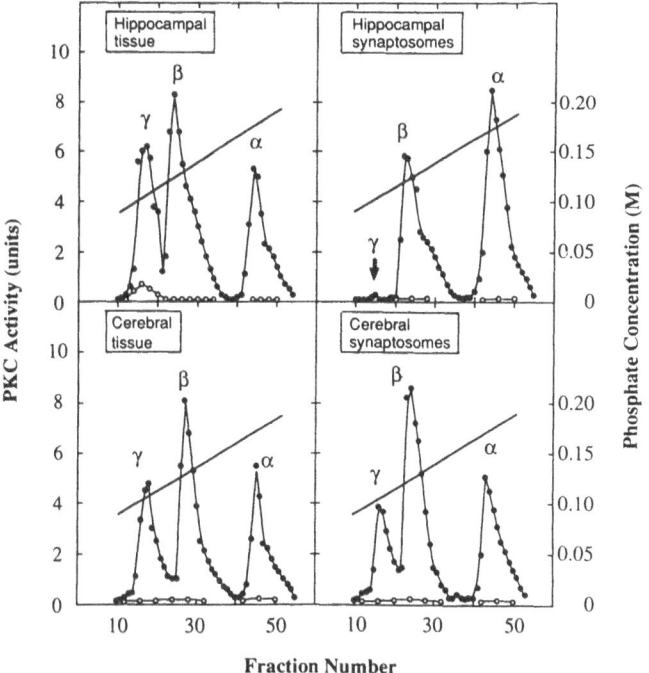

Figure 5. Tissue and synaptosomal expression pattern of protein kinase C subspecies from adult rat hippocampal and cerebral cortical regions. PKC was extracted and subjected to hydroxyapatite column chromatography. Detailed experimental conditions are described elsewhere (Shearman et al., 1991).

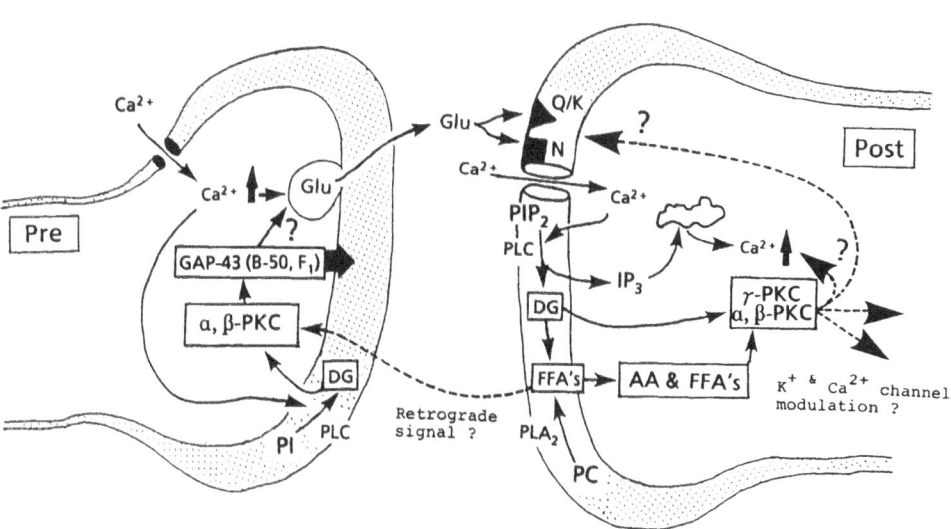

Figure 6. Schematic representation of localization and possible mode of activation of protein kinase C subspecies in the pre- and postsynaptic regions of adult rat hippocampus. Glu, glutamic acid; N, N-methyl-D-aspartate receptor; Q/K, quisqualate/kainate receptor; PLC, phospholipase C; PLA₂, phospholipase A₂.

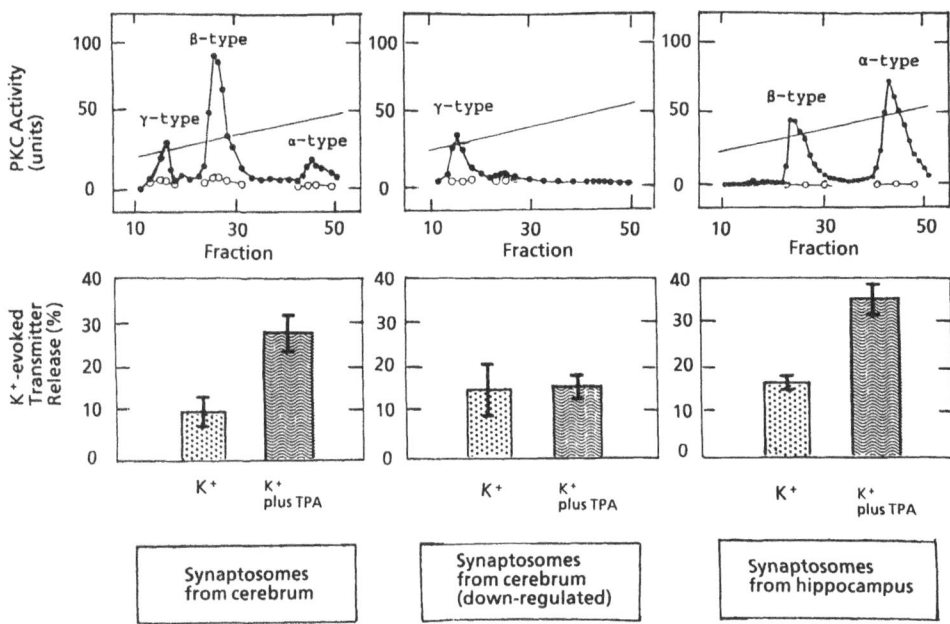

Figure 7. Phorbol ester-enhanced K^+-evoked radioactive noradrenalin release from various synaptosomes. The synaptosomes prepared from the rat cerebral cortex were exposed briefly to phorbol ester to deplete the α- and β-subspecies. Various synaptosomes were loaded with [^{3}H]noradrenalin, and challenged with 41 mM KCl in the presence of phorbol ester or dimethyl-sulfoxide. The detailed experimental conditions are described elsewhere (Oda et al., 1991). TPA, 12-O-tetradecanoylphorbol-13-acetate.

Another role of PKC in neuronal tissues is in the enhancement of evoked transmitter release. The experiments summarized in Figure 7 were designed to show that the α- or β-subspecies but not the γ-subspecies might be involved in the potentiation of the release reaction from the nerve endings. Synaptosomes that contain the PKC subspecies in different proportions can be prepared from various brain regions. It has been shown that pretreatment with phorbol ester under appropriate conditions depletes rapidly only the α- and β-subspecies, with the activity of the γ-subspecies being retained (Oda et al., 1991). Under these conditions, the phorbol ester enhancement of evoked noradrenalin release is diminished. As noted above, synaptosomes prepared from the hippocampus of adult rats have barely detectable levels of the γ-subspecies, whereas those from the cerebral cortex express all three PKC subspecies. In both populations of synaptosomes, phorbol ester enhances K^+-evoked noradrenalin release. These results suggest that the γ-subspecies does not play a role in the release reaction. However, it is possible that the synaptosomes that take up and release the radioactive noradrenalin are only a small portion of the total synaptosomal pool. Although it is presently difficult to identify the precise role of the individual PKC subspecies in synaptic processes, it is attractive to surmise that each member of the PKC family may have one or more very specialized functions.

IMPLICATION FOR CONTROL OF GENE ACTIVATION

With T cells as a model system, it is possible to show that sustained activation of PKC is necessary for gene activation, eventually leading to cell growth and differentiation. Under physiological conditions, DG is metabolized very rapidly by the action of either DG-kinase or DG-lipase. The experiment given in Figure 8 shows that a membrane-permeable synthetic DG, dioctanoylglycerol, even added repeatedly to T cells, disappears very quickly—within several minutes up to 1 hr—depending on the cell density. Under the same conditions, phorbol ester is hardly metabolizable and remains in the membrane for a long period of time irrespective of the cell density. It has been known for some time that, in synergy with ionomycin, a single dose of phorbol ester but not dioctanoylglycerol can induce T-cell activation as measured by interleukin-2 (IL-2) receptor (α) expression as well as by radioactive thymidine uptake into DNA. The results summarized in Figure 9 confirm our previous observation (Berry et al., 1990) that multiple doses of dioctanoylglycerol mimic a single dose of phorbol ester in the induction of T-cell activation. Increasing the duration of PKC activation by treating the cells with multiple doses of this membrane-permeable DG potentiates the levels of IL-2 receptor (α) expression. Similarly, thymidine incorporation into the resting T cell can be greatly enhanced by multiple doses of dioctanoylglycerol, as with a single dose of phorbol ester. Downregulation of PKC does not take place during the entire period of this experiment, although both multiple doses of dioctanoylglycerol and a single dose of phorbol ester cause a similar redistribution of PKC subspecies within the cell (Berry et al., 1989). It is plausible, therefore, that prolonged PKC activation, as well as elevation of the intracellular Ca^{2+} concentration, is required for gene expression leading to cell proliferation.

On the other hand, it is known that an antigenic signal provided by anti-Ti/CD3 antibodies causes only a transient rise in Ca^{2+} and DG within the T cell. The hypothesis that several accessory signals function, at least partially, by prolonging PKC activation, is supported by several recent studies (for a review, see Berry and Nishizuka,

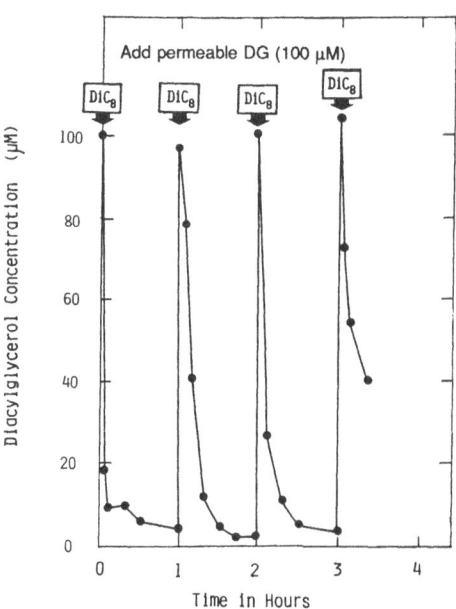

Figure 8. Rapid disappearance of exogenously added synthetic dioctanoylglycerol in human peripheral T lymphocytes. Dioctanoylglycerol was added to purified T cells (10^7 cells/ml) repeatedly, each time at a concentration of 100 μM. At various times, diacylglycerol (DG) was extracted by chloroform/methanol from the entire incubation mixture, and the amount was determined enzymatically with diacylglycerol kinase which was partially purified from rat brain. Detailed experimental conditions are given elsewhere. DiC$_8$, dioctanoylglycerol.

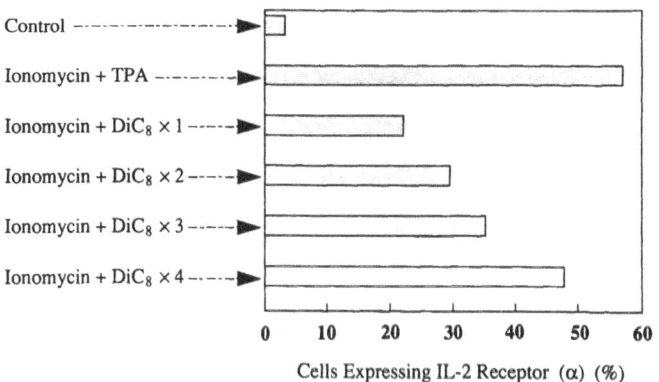

Figure 9. Effect of increasing frequencies of multiple additions of dioctanoylglycerol on IL-2 receptor (α) expression. Accessory cell-depleted T cells (5×10^5 cells/ml) from human venous blood were cultured in RPMI containing 5% autologous serum. Cells were stimulated with either dioctanoylglycerol or phorbol ester in the presence of ionomycin. Dioctanoylglycerol was added repeatedly (1-4 times, each time 25 μM) in the presence of 0.25 μM ionomycin as indicated. A single dose of phorbol ester (10 nM) and ionomycin (1 μM) was also added where specified. Cells were harvested 16 hr after the first stimulation, and IL-2 receptor (α) expression was determined by the direct immunofluorescence staining method. Fluorescence was measured using a flow cytometer. Detailed experimental conditions are given elsewhere. DiC$_8$, dioctanoylglycerol; TPA, 12-O-tetradecanoylphorbol-13-acetate.

1990). Although it is still premature to discuss the precise relationship between the accessory signals and the sustained activation of PKC, it is attractive to investigate whether the accessory signals generate second messengers, such as DG and unsaturated FFAs, that sustain PKC activation for a sufficient period of time to cause T-cell activation, as shown schematically in Figure 10. Szamel et al. (1989) have observed that linoleic and arachidonic acids may potentiate the IL-2 synthesis which is initiated by membrane-permeable DG and ionomycin.

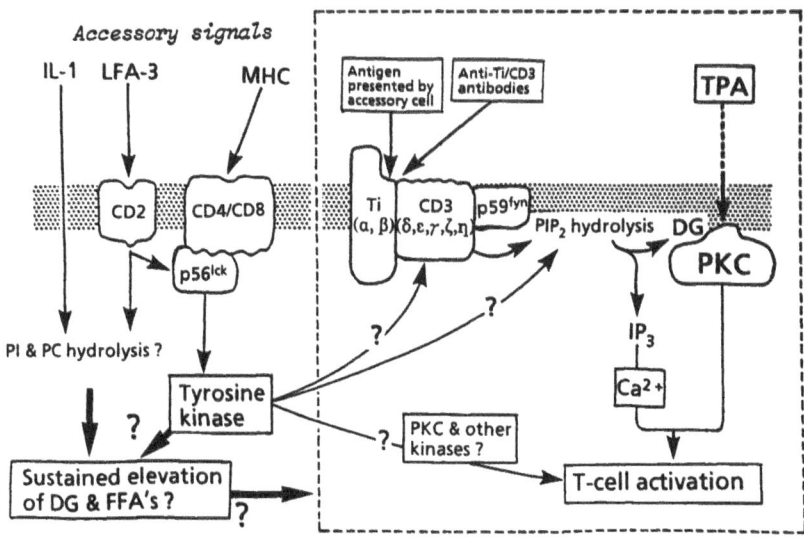

Figure 10. A hypothetical mechanism of T-cell activation. Detailed explanations are given elsewhere (Berry and Nishizuka, 1990).

CONCLUSION

Obviously, the intracellular event involved in the signal-induced cellular response, either short-term or long-term, is the result of interactions among a number of signal transduction pathways. One of these pathways is PKC activation, which is apparently involved in many physiological processes. We once thought of receptor-mediated hydrolysis of inositol phospholipids as the sole route to produce DG that can activate PKC. Subsequently, however, PC was shown to be broken down to produce DG at a relatively later phase of cellular response (for a review, see Exton, 1990). Indeed, phospholipase C that is reactive with PC has been described (Besterman et al., 1986). In recent years, attention has been paid to the receptor-mediated activation of phospholipase A_2 (Axelrod et al., 1988) as well as phospholipase D (Exton, 1990). The synergistic action of DG and unsaturated FFAs described in this article seems to suggest that the signal-induced release of unsaturated FFAs from several phospholipids may take part in prolonging the activation of PKC even when Ca^{2+} concentrations remain at the basal level. Perhaps the activation of PKC is an integral part of the signal-induced degradation cascade of various membrane phospholipids that is catalyzed by phospholipase C, phospholipase A_2, and phospholipase D, as schematically illustrated in Figure 11. Analysis of interactions among various pathways to activate these phospholipases may provide a clue to understand further the mechanism of cellular responses, particularly those that occur over the longer term.

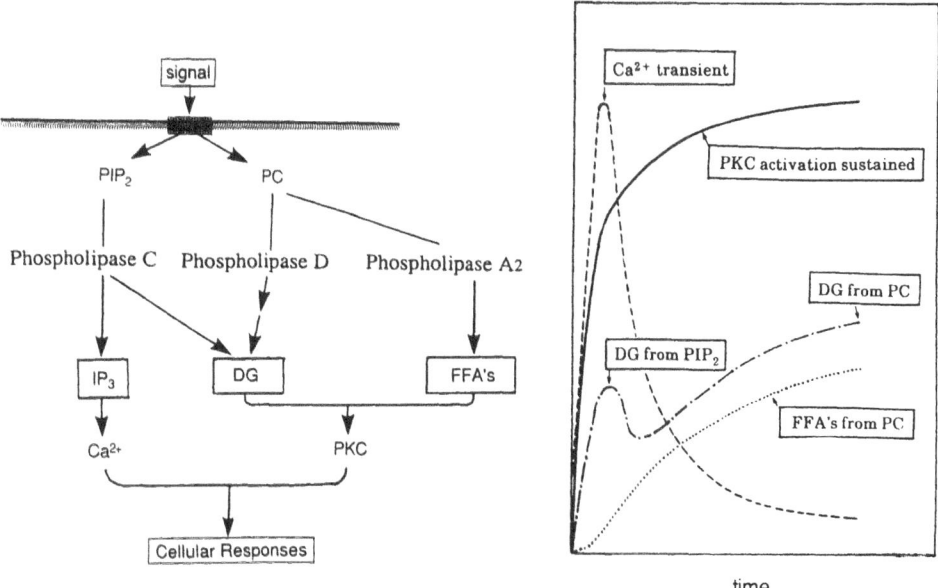

Figure 11. Schematic representation of cellular responses induced by signal-induced degradation cascade of membrane phospholipids.

ACKNOWLEDGMENTS

The skillful secretarial assistance of Mrs. S. Nishiyama, Miss Y. Kimura, and Miss Y. Yamaguchi is cordially acknowledged. This work was supported in part by research grants from the Special Research Fund of the Ministry of Education, Science and Culture, Japan; the Muscular Dystrophy Association, U.S.; the Juvenile Diabetes Foundation International, U.S.; Yamanouchi Foundation for Research on Metabolic Disorders; Sankyo Foundation of Life Science; Merck Sharp & Dohme Research Laboratories; Biotechnology Laboratories of Takeda Chemical Industries; and New Lead Research Laboratories of Sankyo Company.

REFERENCES

Alkon DL and Rasmussen H (1988) A spatial-temporal model of cell activation. Science 239:998-1005.

Axelrod J, Burch RM, Jelsema CL (1988) Receptor-mediated activation of phospholipase A_2 via GTP-binding proteins: Arachidonic acid and its metabolites as second messengers. Trends Neurosci 11:117-123.

Bacher N, Zisman Y, Berent E, Livneh E (1991) Isolation and characterization of PKC-L, a new member of the protein kinase C-related gene family specifically expressed in lung, skin, and heart. Mol Cell Biol 11:126-133.

Berry N and Nishizuka Y (1990) Protein kinase C and T cell activation. Eur J Biochem 189:205-214.

Berry N, Ase K, Kikkawa U, Kishimoto A, Nishizuka Y (1989) Human T cell activation by phorbol esters and diacylglycerol analogues. J Immunol 143:1407-1413.

Berry N, Ase K, Kishimoto A, Nishizuka Y (1990) Activation of resting human T cells requires prolonged stimulation of protein kinase C. Proc Natl Acad Sci USA 87:2294-2298.

Besterman JM, Duronio V, Cuatrecasas P (1986) Rapid formation of diacylglycerol from phosphatidylcholine: A pathway for generation of a second messenger. Proc Natl Acad Sci USA 83:6785-6789.

Bliss TVP, Clements MP, Errington ML, Lynch MA, Williams JH (1989) Presynaptic mechanisms underlying the maintenance of long-term potentiation in the hippocampus. In: Brain signal transduction and memory (Ito M and Nishizuka Y, eds) p 185-196. Tokyo: Academic Press.

Coggins PJ and Zwiers H (1991) B-50 (GAP-43): Biochemistry and functional neurochemistry of a neuron-specific phosphoprotein. J Neurochem 56:1095-1106.

Exton JH (1990) Signaling through phosphatidylcholine breakdown. J Biol Chem 265:1-4.

Farago A and Nishizuka Y (1990) Protein kinase C in transmembrane signalling. FEBS Lett 268:350-354.

Felder CC, Kanterman RY, Ma AL, Axelrod J (1990) Serotonin stimulates phospholipase A_2 and the release of arachidonic acid in hippocampal neurons by a type 2 serotonin receptor that is independent of inositolphospholipid hydrolysis. Proc Natl Acad Sci USA 87:2187-2191.

Gispen WH (1988) Transmembrane signal transduction and ACTH-induced excessive grooming in the rat. Ann NY Acad Sci 525:141-149.

Hashimoto T, Ase K, Sawamura S, Kikkawa U, Saito N, Tanaka C, Nishizuka Y (1988) Postnatal development of a brain-specific subspecies of protein kinase C in rat. J Neurosci 8:1678-1683.

Kishimoto A, Takai Y, Mori T, Kikkawa U, Nishizuka Y (1980) Activation of calcium and phospholipid-dependent protein kinase by diacylglycerol, its possible relation to phosphatidylinositol turnover. J Biol Chem 255:2273-2276.

Kose A, Ito A, Saito N, Tanaka C (1990) Electron microscopic localization of γ- and βII-subspecies of protein kinase C in rat hippocampus. Brain Res 518:209-217.

Linden DJ and Routtenberg A (1989) The role of protein kinase C in long-term potentiation: A testable model. Brain Res Rev 14:279-296.

Nishizuka Y (1988) The molecular heterogeneity of protein kinase C and its implications for cellular regulation. Nature 334:661-665.

Oda T, Shearman MS, Nishizuka Y (1991) Synaptosomal protein kinase C subspecies: B. Down-regulation promoted by phorbol ester and its effect on evoked norepinephrine release. J Neurochem 56:1263-1269.

Osada S, Mizuno K, Saido TC, Akita Y, Suzuki K, Kuroki T, Ohno S (1990) A Phorbol ester receptor/protein kinase, nPKCη, a new member of the protein kinase C family predominantly expressed in lung and skin. J Biol Chem 265:-22434-22440.

Scott BL and Bazan NG (1989) Membrane docosahexaenoate is supplied to the developing brain and retina by the liver. Proc Natl Acad Sci USA 86:2903-2907.

Sekiguchi K, Tsukuda M, Ogita K, Kikkawa U, Nishizuka Y (1987) Three distinct forms of rat brain protein kinase C: Differential response to unsaturated fatty acids. Biochem Biophys Res Commun 145:797-802.

Shearman MS, Shinomura T, Oda T, Nishizuka Y (1991) Synaptosomal protein kinase C subspecies: A. Dynamic changes in the hippocampus and cerebellar cortex concomitant with synaptogenesis. J Neurochem 56:1255-1262.

Shinomura T, Asaoka Y, Oka M, Yoshida K, Nishizuka Y (1991) Synergistic action of diacylglycerol and unsaturated fatty acid for protein kinase C activation: Its possible implications. Proc Natl Acad Sci USA 88:5149-5153.

Szamel M, Rehermann B, Krebs B, Kurrle R, Resch K (1989) Activation signals in human lymphocytes. J Immunol 143:2806-2813.

Tsien RW, Schulman H, Malinow R (1990) Peptide inhibitors of PKC and CaMK block induction but not expression of long-term potentiation. Adv Second Messenger Phosphoprotein Res 24:101-107.

DISPOSITION KINETICS OF PHOSPHOLIPID LIPOSOMES

Pietro Palatini

Department of Pharmacology
University of Padova
Largo E. Meneghetti 2
35131 Padova
Italy

ABSTRACT

This article reviews the disposition of intravenously injected phospholipid liposomes and discusses the problems related to its kinetic modeling. The processes responsible for the plasma clearance of liposomes are examined in detail and it is shown that mechanisms other than reversible distribution to the extravascular space are, as a rule, responsible for the biphasic plasma clearance patterns that are typically observed following bolus intravenous injection of liposomes. Accordingly, a one-compartment open model is generally sufficient to describe the disposition kinetics of phospholipid vesicles.

Two factors may be responsible for the observation of a biphasic decline of plasma liposome concentration. The first factor is the presence of different liposomal species with different kinetic behaviors. Kinetically distinct vesicles are present in preparations of liposomes that are heterogeneous in size, since the larger vesicles are cleared at a faster rate than the smaller ones. Different liposomal species may also originate in the plasma as a result of: i) fusion between phospholipid vesicles with generation of larger liposomal structures; and ii) interaction with high-density lipoproteins (HDL) with consequent production of either liposomes that have acquired apoproteins or lipoprotein particles enriched in phospholipids. Both these species are cleared by specific mechanisms at rates different from that of the original vesicle. The second factor is a time-dependent decrease in clearance due to progressive saturation of the retention capacity of the cells that take up liposomes.

A convex concentration-time decay curve has also been reported. This decay pattern is consistent with a concentration (dose)-dependent elimination. As this observation relates to only one type of liposome (small unilamellar vesicles composed of sphingomyelin and cholesterol), its relevance to the disposition of liposomes of different size and composition remains to be established.

INTRODUCTION

Interest in phospholipid liposomes stems from both their therapeutic potential as drug delivery systems and their intrinsic pharmacological activity (Bruni and Palatini, 1983). Successful exploitation of either property is strictly dependent on an adequate knowledge of their behavior and fate in the body. Since selective drug delivery to specific tissues is the main application of liposomes, efforts have been principally concentrated on "site-directed targeting," i.e., selective uptake of liposomes by the target cells. As a result of these studies, a detailed picture of the tissue distribution of systemically administered liposomes has emerged (Gregoriadis, 1988a). However, the kinetics of the disposition processes are still far from being completely characterized. In contrast to more common drugs, which form true solutions in body fluids, phospholipid liposomes are colloidal suspensions. Accordingly, they are cleared from the blood by mechanisms different from those governing the clearance of substances dispersed at the molecular level. This has prevented the use of the pharmacokinetic models currently applied to the analysis of the disposition kinetics of drugs. Although, as will be shown below, some kinetic models have been proposed to describe the disposition of certain types of liposomes, a model of general applicability is still lacking and a quantitative analysis of liposome disposition is hardly feasible. The development of a general kinetic model has been hindered by uncertainty about the mechanisms of cell-liposome interaction actually operating *in vivo*. In addition, qualitatively different types of cell-liposome interactions may take place depending on the chemical nature of the phospholipid polar head. For example, both receptor- and non-receptor-mediated endocytotic processes have been observed (Margolis, 1988: Nishikawa et al., 1990).

The purpose of this paper is not to provide a comprehensive review of the present knowledge of liposome pharmacokinetics. The methodological problems of such pharmacokinetic investigations (use of membrane-bound or entrapped hydrophilic markers, marker retention in relation to liposome stability, use of biochemical or spectroscopic techniques for monitoring the marker) have been treated in detail in recent reviews (Hwang, 1987; Senior, 1987; Gregoriadis, 1988b) and will not be discussed here. In this paper, an attempt will be made to provide an interpretation of the plasma decay curves of intravenously administered liposomes in the light of the known mechanisms of interaction of liposomes with blood and tissue constituents. A short introductory overview of such interactions will, therefore, follow. The mechanisms of cell-liposome interaction have been reviewed in detail by Poste (1980) and, more recently, by Margolis (1988) in relation to the physical state of the phospholipid vesicles. The interaction with plasma lipoproteins is the subject of a recent review by Williams and Tall (1988).

INTERACTION OF LIPOSOMES WITH CELLS
AND PLASMA LIPOPROTEINS

Fusion

Since phospholipid vesicles can readily fuse with each other under proper conditions, fusion was initially postulated to be the principal mechanism of interaction of liposomes with cells (Bruni and Palatini, 1983). Fusion results in incorporation of the vesicle lipids into the cell membrane and release of the entrapped material into the cytoplasm. Early experiments showing a uniform distribution of liposome-encapsulated fluorescent probes into the cytoplasm appeared to lend support to the idea of cell-liposome fusion (Weinstein et al., 1977). However, subsequent observations prompted a reinterpretation of these experiments, since it was shown that such an intracellular distribution of the liposome-entrapped probe can also result from endocytosis or simple adhesion of the lipid vesicle to the cell and subsequent permeabilization of the liposome (van Renswoude and Hoekstra, 1981; Straubinger et al., 1983). There is now a growing recognition that fusion between phospholipid liposomes and the plasma membrane of cells is a rare event under physiological conditions (Margolis, 1988). The sialic acid residues of cell surface glycolipids probably represent the main obstacle to the juxtapositional fusion of phospholipid vesicles with cell membranes. However, exceptions concerning some specialized types of cells may be possible. For example, a 30% fusion of liposomes bound to adrenal chromaffin cells has been reported (Lelkes and Friedman, 1985).

Endocytosis

The term endocytosis is used here to designate any process in which regions of the plasma membrane fuse and detach to form intracellular vesicles. These processes include phagocytosis, receptor-mediated endocytosis, and pinocytosis (also referred to as non-receptor-mediated or fluid-phase endocytosis). Although the mechanisms underlying these three processes are fundamentally different (Darnell et al., 1990), the fate of the internalized liposomes is the same irrespective of the type of process involved, namely degradation by the lysosomal enzymes after fusion of the endocytic vacuole with lysosomes or incorporation into the plasma membrane. Phagocytosis of phospholipid liposomes by the cells of the mononuclear phagocyte system (MPS) is a well-established phenomenon (Senior, 1987; Allen, 1988; Gregoriadis, 1988b). This system consists of the circulating macrophages, the fixed macrophages of liver (Kupffer cells), and those of spleen and bone marrow. Liposomes, similarly to any foreign particulate matter, are rapidly removed from the circulation mainly by the Kupffer cells and, to a lesser extent, by the other macrophage populations. The MPS represents, therefore, a major obstacle to an effective targeting of liposomes to other cell types. The rate of uptake of liposomes by the MPS depends on liposome size, physical state, composition, and dosage. Small unilamellar vesicles (SUV) are taken up much more slowly than are multilamellar (MLV) or large unilamellar vesicles (LUV). In addition, the less fluid the liposome bilayer, the slower the rate of uptake by the MPS (Allen, 1988; Gregoriadis, 1988b). An extremely rapid uptake is observed with certain phospholipids, such as phosphatidylserine, the polar head of which is specifically recognized by the cells of the MPS (Nishikawa et al., 1990; Allen et al., 1988; Palatini et al., 1991). As the liposome dosage increases, the mechanisms of liposome uptake by the MPS become progressively saturated and, consequently, liposome clearance decreases (Senior, 1987; Gregoriadis, 1988b). This aspect will be treated in more detail below.

Internalization of phospholipid liposomes is also operated by non-phagocytic cells such as fibroblasts, cells of the kidney, lymphocytes, hepatocytes, etc. (Margolis, 1988; Poste, 1980; Mietto et al., 1989). This non-receptor-mediated endocytosis appears to be strictly dependent on the size of liposomes, the optimal size being 50-100 nm (roughly corresponding to that of coated pits), whereas vesicles larger than 400 nm are not apt to be endocytosed (Machy and Leserman, 1983; Heath et al., 1985).

The observation of saturable cell-liposome binding isotherms (Blumenthal et al., 1977) and the finding that trypsin treatment prevents such binding (Mietto et al., 1989) have been interpreted as indicating the presence of receptors on the cell surface mediating liposome endocytosis as well as other types of cell-liposome interactions (Margolis, 1988). However, definitive evidence in favor of receptor-mediated endocytosis by non-phagocytic cells of vesicles composed exclusively of phospholipids is still lacking. Convincing evidence of a receptor-mediated endocytosis by liver parenchymal cells has been obtained for liposomes that have acquired apoproteins from circulating

HDL. A receptor-mediated internalization that involves not endocytosis of whole vesicles but translocation of single phospholipid molecules is operated by the transporter for aminophospholipids. This transporter is a Mg^{2+}- and ATP-dependent protein present in the erythrocyte membrane, which selectively translocates phosphatidylserine and phosphatidylethanolamine monomers to the inner leaflet of the membrane bilayer (Daleke and Huestis, 1985).

Lipid and Protein Exchange Between Cells and Liposomes

Reversible transfer of phospholipid and cholesterol molecules between liposomes and cells has long been known (Bruckdorfer and Graham, 1976). The transferred lipids originate exclusively from the outer monolayer of the liposomes. Various lipid transfer proteins present in human plasma can promote this lipid exchange (Tollefson et al., 1985). Transfer of plasma membrane proteins from the cell surface to the interacting liposomes may also occur.

Adsorption of Liposomes to the Cell Surface

All phospholipid vesicles readily adhere to the cell surface. The subsequent events depend on the physical state and size of the liposomes. As far as fluid liposomes are concerned, it appears that only the larger ones (MLV and LUV) can remain adsorbed to the cell surface and survive extensive washing (Poste, 1980). In contrast, SUV seem to lose their integrity upon adhesion. Only liposomal ghosts or fragments may remain adsorbed and the entrapped substances are leaked out (Margolis et al., 1988). Solid liposomes tend to be more strongly adsorbed to cell surfaces and retain their integrity irrespective of their size (Blumenthal et al., 1977; Margolis, 1988).

Modification of Liposome Permeability and Size

Upon incubation with cells in either serum-free (van Renswoude and Hoekstra, 1981; Szoka et al., 1979) or serum-containing (Allen and Cleland, 1980) media, liposomes become more permeable and leakage of entrapped substances is observed. Decreasing the fluidity of the liposomes (e.g., by the addition of cholesterol) decreases the cell-induced liposome leakage (Allen and Cleland, 1980). It has also been shown that SUV can be transformed into larger lipid aggregates upon interaction with cells. This applies to both solid and fluid vesicles, but it appears that only a small fraction of SUV (less than 5%) undergoes such an enlargement (Margolis et al., 1982).

Interaction of Liposomes with Plasma Lipoproteins

Phospholipid vesicles can interact with, and become destabilized by, plasma lipoproteins. This destabilization is the result of the insertion of apoproteins into the liposomal phospholipid bilayer and the exchange of lipidic material between lipoproteins and liposomes (Senior, 1987; Williams and Tall, 1988). The ensuing modifications of the liposomes may have important consequences for the kinetics of both the entrapped substances that are released from the vesicle, and the phospholipid component that, upon acquisition of apoproteins or transfer to the lipoprotein particle, is eliminated by the specific mechanisms governing the clearance of the lipoproteins (Williams et al., 1986). Although interactions have been observed with all types of plasma lipoproteins (VLDL, LDL, and HDL) (Williams and Tall, 1988), the lipoproteins mainly responsible for liposome destabilization are HDL (Scherphof et al., 1978). Depending on the relative concentrations of HDL and liposomes and the lipid composition of the latter, three types of reaction products can be generated:

i) liposomes retaining their characteristic structure, despite the acquisition of a small amount of apoproteins; ii) discoidal complexes of phospholipids encircled by apoproteins, resulting from complete disruption of liposomes; and iii) modified lipoproteins which are enriched in phospholipids, but depleted of cholesterol and proteins.

Integrity of the vesicular structure is in part preserved when liposomes are administered at very high doses (200-800 mg kg^{-1}) that exceed the disruptive capacity of HDL apoproteins (Williams and Scanu, 1986). Stabilization of the liposomal structure can be obtained by inclusion of unesterified cholesterol that, by decreasing the fluidity of the lipid bilayer, renders it less easily penetrated by apoproteins (Guo et al., 1980). By the same token, sphingomyelin liposomes, which are in a solid state at body temperature, are not disrupted by lipoproteins (Allen, 1981). Resistance to disruption depends also on the diameter of the vesicle; large liposomes are less readily penetrated by apoproteins than are SUV and exhibit, therefore, a greater stability (Scherphof and Morselt, 1984). As mentioned above, liposomes that have acquired apoproteins are able to interact with hepatic apoprotein receptors and, provided they gain access to the extravascular space, they can be taken up by liver parenchymal cells (Williams et al., 1986).

When cholesterol is not included in liposomes (Damen et al., 1981) or the concentration of the liposomal phospholipid is low with respect to that of HDL (Bienvenue et al., 1985), a unidirectional phospholipid transfer from liposomes to HDL takes place. *In vitro*, discoidal complexes are formed, resembling the discoidal particles that are believed to be the nascent form of HDL (Guo et al., 1980). *In vivo*, formation of spherical particles is observed, without any discoidal intermediate (Tall, 1980). This may be due either to a very rapid conversion of disks into spheres by lecithin: cholesterol acyltransferase, as is the case with nascent HDL, or to an insertion of the liposomal phospholipid into preexisting HDL particles. The latter possibility is suggested by the observation that HDL actually take up phospholipids from chylomicrons and VLDL *in vivo* during lipolysis of these low density particles (Patsch et al., 1978; Redgrave and Small, 1979). Whatever the underlying mechanism, phospholipid transfer from liposomes to HDL leads to the formation of HDL with altered phospholipid composition and an increased ratio of phospholipid to apoprotein (the phospholipid content of HDL can be as much as doubled by liposomes) (Williams and Scanu, 1986). By contrast, unesterified cholesterol, which is present on the surface of HDL, can be almost totally lost upon interaction with cholesterol-poor liposomes (Williams and Scanu, 1986).

EXTRAVASATION OF LIPOSOMES: ANATOMICAL AND PHYSIOLOGICAL CONSIDERATIONS

Phospholipid liposomes are unable to leave the vascular space by passing across the cells of the capillary endothelium. The process of transcytosis, whereby various physiological macromolecules cross the endothelial barrier, does not apply to phospholipid vesicles. This process entails migration of the endocytic vacuole formed at the lumenal surface of the endothelial cell to the ablumenal membrane and externalization of the vacuole content on the extravascular space by exocytosis. However, the size of the endothelial cells (about 70 nm in diameter) does not permit accommodation of particles larger than about 50 nm (Poste, 1983; Poznansky and Juliano, 1984). In addition, as pointed out above, endocytosed liposomes do not bypass the lysosomal system, but undergo intracellular degradation or incorporation into the membrane. Egress of liposomes from the circulation is, therefore, dependent on the presence of pores, i.e., spaces between cells. On the basis of the structure of endothelium and basal lamina and

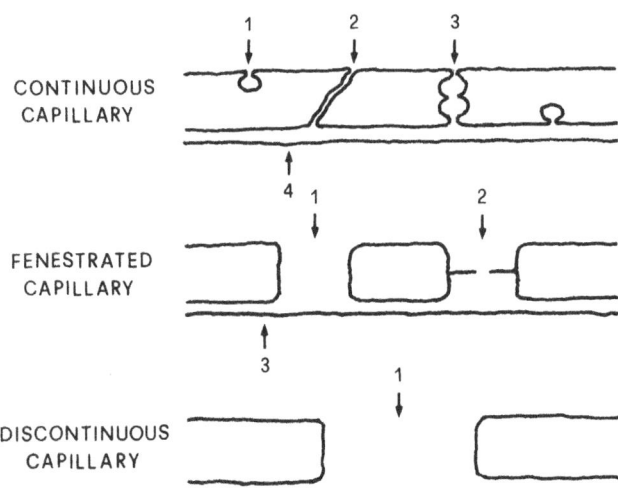

Figure 1. Schematic representation of the possible transfer pathways for colloidal particles and macromolecules across continuous, fenestrated, and discontinuous capillaries. Only the distinctive features of each type of capillary are shown. Continuous capillary: 1) pinocytotic vesicle; 2) intercellular space; 3) transendothelial channel; 4) basal lamina. Fenestrated capillary: 1) open fenestra; 2) diaphragmed fenestra; 3) basal lamina. Discontinuous capillary: 1) intercellular gap. (Modified from Taylor and Granger, 1984).

the width of their intercellular spaces, blood capillaries can be classified into three main groups: continuous, fenestrated, and discontinuous (sinusoids). The possible transfer pathways for macromolecules and particulate materials across these three types of capillaries are schematically represented in Figure 1, which summarizes the results of morphological studies (Taylor and Granger, 1984). A macromolecule or particle can cross the endothelium of continuous capillaries i) by entering into pinocytotic vesicles (average diameter 50 nm); ii) by passing through intercellular spaces, which measure 2-6 nm in width; and/or iii) by passing through transendothelial channels. These channels are believed to originate from one or more vesicles opening simultaneously on either side of the endothelium. Their internal diameter is 10-40 nm. Stomatal diaphragms may reduce the diameter to 5-11 nm.

Diaphragmed or open *fenestrae* are the distinctive features of fenestrated capillaries. *Fenestrae* are circular openings with a diameter of 40-60 nm. Except in kidney capillaries, *fenestrae* are for the most part provided with a diaphragm, the actual porosity of which is unknown. The basement membrane is freely permeable to particles with diameters between 5 and 11 nm, although no structural discontinuity is discernible. This membrane does, however, reduce the rate of penetration of larger particles.

The discontinuous capillary, which is found almost exclusively in liver, spleen, and bone marrow, is characterized by the absence of a basement membrane and the presence of very large endothelial gaps with diameters ranging between 100 and 1,000 nm.

The results of the permeability measurements for particles and macromolecules are, however, at variance with those of the morphological studies (Taylor and Granger, 1984). According to the former studies, two types of pores appear to exist in each type of capillary, a small- and a large-pore population, the structural correlates of which are not readily apparent. This discrepancy may be the result of mechanical damage and fixation artifacts in electron microscopic studies (Moghimi et al., 1990). The diameter

Table 1. Functional pores of the blood capillaries and their putative structural correlates

Capillary	Functional pores (diameter in nm)	Structural correlates
Continuous capillary	Small (13.4-16)	Diaphragmed transendothelial channels
	Large (40-56)	Transendothelial channels without diaphragm
Fenestrated capillary	Small (9.2-10.6)	Diaphragmed *fenestrae*?
	Large (36-50)	Open *fenestrae*
Discontinuous capillary	Small (18)	?
	Large (66)	Intercellular gaps

Based on the data of Taylor and Granger, 1984.

of the functional pores and their putative structural correlates are given in Table 1. Although the transendothelial channels are reasonable correlates of the functional pores of continuous capillaries, the significance of this transient pathway is uncertain. Macromolecular transfer through such channels has been shown to be negligible, when compared to transfer across *fenestrae* or the endothelial gaps of discontinuous capillaries (Taylor and Granger, 1984). The most striking inconsistency between physiological and ultrastructural data is observed in relation to discontinuous capillaries, where functional pores appear much narrower than the endothelial gaps evidenced by morphological studies (diameter 100-1,000 nm). Such a difference cannot be the result of artifacts of the fixation techniques but is likely due to the presence of a barrier beyond the sinusoidal wall. The interstitial matrix in the space of Disse may constitute such a barrier (Taylor and Granger, 1984).

The above considerations lead to the prediction that even the smallest SUV (about 19 nm) (Hwang, 1987) are too large to pass across continuous capillaries, since intercellular gaps are too narrow (2-6 nm), whereas pinocytotic vesicles and transendothelial channels do not provide an efficient transport system for liposomes (see above). The same can be anticipated for fenestrated capillaries since, with the exception of the kidney, *fenestrae* are generally occluded by diaphragms and the basal lamina hinders the diffusion of particles with a diameter larger than about 12 nm (Taylor and Granger, 1984). Thus, extravasation of liposomes appears a realistic possibility only for SUV in those tissues that contain sinusoidal capillaries. MLV and LUV, which have diameters generally exceeding 100 nm, are not predicted to cross discontinuous capillaries either.

Experimental data are in good agreement with these predictions. Roerdink et al. (1981) have shown that MLV with a mean diameter of 120 nm are taken up intact *in vivo* only by Kupffer cells, not by liver parenchymal cells. However, the phospholipid component of the injected vesicles has also been found in association with the parenchymal tissue. As the uptake by liver parenchyma requires extravasation, whereas the uptake by Kupffer cells does not, such observations demonstrate that liposomes of this size cannot pass out of the liver sinusoids. However, liposomal phospholipids can be transported to the extravascular space by HDL, following incorporation into these lipoproteins in the plasma compartment. That this is indeed the case is shown by the observation that the fates of the entrapped material and the lipidic component are the same when injected liposomes are resistant to HDL (Gotfredsen et al., 1983). Analogous results have been obtained by other authors. Rahman et al. (1982) observed uptake of MLV (500 nm in diameter) almost exclusively by Kupffer cells. By contrast, SUV (60-80 nm) were found to be taken up efficiently also by liver parenchymal cells. Gotfredsen et al. (1983) observed significant uptake by the parenchymal cells only when the diameter of the injected liposome population was less than 100 nm (mean diameter 70 nm). They also observed a lag phase (about 30 min) for the uptake by parenchymal cells, whereas uptake by Kupffer cells was virtually immediate. No lag phase was observed *in vitro*. This suggests that the delayed association with parenchymal cells *in vivo* was not due to a slower uptake process but to the time required for extravasation. Subsequent studies by Spanjer et al. (1986) confirmed these results and showed that, depending on the phospholipid composition, up to 96% of SUV taken up by the liver may be associated with parenchymal cells, i.e., may escape phagocytosis by the MPS and cross the capillary barrier. At variance with the sinusoids, both continuous and fenestrated capillaries have been shown to be impermeable to either small or large liposomes even in conditions such as inflammation, in which blood vessel permeability is significantly increased (Poste, 1983). The liposome extravasation observed in the lungs does not constitute an exception to this since it is due to phagocytosis by blood monocytes which then migrate into the alveoli (Poste, 1983).

DISPOSITION KINETICS

A biphasic plasma decay curve is quite often observed following intravenous injection of phospholipid liposomes, with a faster initial decay followed by a slower one. For substances that form true solutions in body fluids, such a plasma concentration profile is the result of two simultaneous processes: reversible distribution to the extravascular space (the main cause of the rapid initial decay) and elimination. Accordingly, concentration-time data are analyzed on the basis of a multicompartment open model (Gibaldi and Perrier, 1982).

As noted above, large liposomes do not distribute outside the plasma compartment. Even the observation that SUV can pass out of the vascular space does not constitute *per se* evidence that a reversible distribution takes place to the interstitial fluids and the lymphatic system. If egress from plasma is unidirectional, then it is an elimination process (irreversible loss of the substance from the site of measurement) that cannot be responsible for a biphasic decay (Rowland and Tozer, 1989). This question has been addressed by Hwang et al. (1982) by injecting into mice SUV prepared from sphingomyelin and cholesterol with a mean diameter of 18.7 nm. These liposomes are most suitable for detecting a possible reversible distribution, since they are small enough to pass out of the discontinuous capillaries and remain intact in the body fluids for a long time. A biphasic decay was observed with such SUV and the extrapolated apparent volume of distribution was found to be 1.28 times larger than the volume occupied by the erythrocytes. This constitutes evidence that small liposomes may equilibrate in a volume somewhat larger than the vascular space. However, considering that the extrapolation method overestimates the apparent volume of distribution (Gibaldi and Perrier, 1982), it may be concluded that even with the smallest of the liposomes the extravascular space that equilibrates with the plasma compartment is very limited. From the data of Hwang et al. (1982), it can be calculated that the ratio of the volume of the central compartment (V_c) to the extrapolated volume of distribution (V_d) is 0.78. The value of the V_c/V_d ratio, the limits of which are 0 to 1, is indicative of the multicompartmental character of a disposition kinetics. The smaller the numerical value of

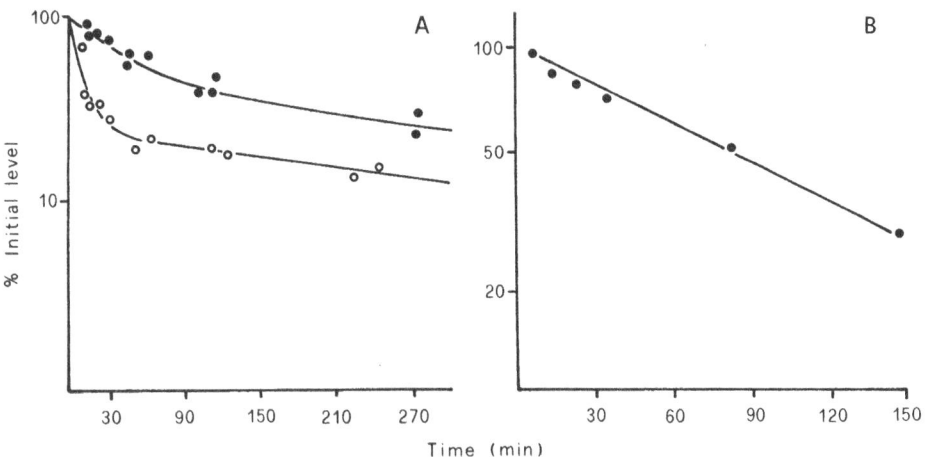

Figure 2. Plasma decay of phospholipid liposomes as a function of vesicle size. A) Unilamellar (filled circles) and multilamellar (open circles) liposomes heterogeneous in size. B) Unilamellar liposomes homogeneous in size. Data obtained after bolus intravenous administration to rats of liposomes prepared from phosphatidylcholine/cholesterol in a 2/1 molar ratio. (Reproduced with permission from Juliano and Stamp, 1985.)

V_c/V_d, the greater the multicompartmental character. Conversely, as V_c/V_d approaches 1, the multicompartment model collapses to the one-compartment model. From a practical point of view, the one-compartment model approximation has been shown to be completely satisfactory when $V_c/V_d = 0.78$ (Wagner, 1983).

The above considerations have made it evident that factors other than reversible distribution must be generally responsible for the biphasic plasma concentration decline of intravenously injected liposomes. Juliano and Stamp (1975), by showing that small vesicles are cleared at a slower rate than large vesicles, were the first to demonstrate that a biphasic plasma decay may result from heterogeneity in size of the injected liposomes. Accordingly, a biexponential clearance pattern is observed with hetero-geneous vesicle populations, whereas a monoexponential decay is found with liposome samples homogeneous in size (Fig. 2). These results were later confirmed by other authors (Sharma et al., 1977). Different decay rates for small and large liposomes are, however, observed only with liposomes that are stable in the plasma. Certain SUV, such as those containing phosphatidylserine, have a strong tendency to coalesce and soon give rise to larger vesicles in the plasma. As a consequence, the same decay rate is observed irrespective of the size of the administered liposomes (Juliano and Stamp, 1975). As discussed above, the size of SUV may also increase upon interaction with cells. Thus, the size of circulating liposomes may in general be different from that of the administered liposome preparation. This "uncertainty principle," as defined by Margolis (1988), should always be taken into consideration when interpreting the results obtained after administration of liposomes liable to modifications in the plasma.

Gregoriadis and Neerunjun (1974) observed that the biphasic decay pattern of intravenously injected MLV is converted into a linear one upon increasing the liposome dose (Fig. 3). These authors suggested that the biphasic decline may be due to the presence of two elimination pathways: a faster one, due to a saturable uptake by the MPS, and a slower one, due to uptake by the liver parenchymal cells. When the reten-tion capacity of the MPS becomes saturated, then only the slower elimination pathway

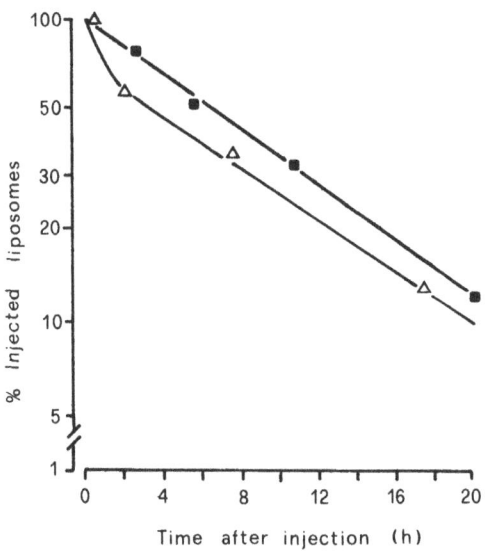

Figure 3. Plasma decay of MLV as a function of dose. Positive MLV (phosphatidylcholine/cholesterol/stearylamine in a 7/2/1 molar ratio) were injected intravenously into rats. Open triangles, about 27 mg lipid kg^{-1}; Filled squares, about 105 mg lipid kg^{-1}. (Modified from the data of Figures 1 and 3.)

remains operative and the decay rate decreases. If, on the other hand, the storing capacity of the MPS is rapidly exhausted by administration of very large doses, then only the slower decay phase is detected. It was later shown that MLV cannot pass out of the endothelial barrier. It is, therefore, unlikely that the slow decay phase observed by Gregoriadis and Neerunjun (1974) was due to liposome uptake by the liver parenchymal cells. More likely, the biphasic decay pattern was the result of the heterogeneous size distribution of the injected MLV. As noted above, it is not necessary to postulate the existence of two distinct elimination pathways to explain the different elimination rates of large and small liposomes, since they are taken up at different rates by the MPS. However, the possibility that a biphasic plasma decay of liposomes may result from a time-dependent saturation of the storing capacity of the MPS cannot be ruled out. Conversion of a biexponential decline into a linear one with an increase in the liposome dose has been confirmed by other authors using more homogeneous vesicle preparations (Kao and Juliano 1981; Souhami et al., 1981). Still more convincing support for this idea has recently come from the demonstration of a time-dependent decrease in spleen and liver clearances following continuous intravenous infusion of MLV 300-350 nm in diameter.

Although the presence of two parallel elimination pathways is an unlikely possibility for MLV, it is, however, theoretically possible for SUV that can be extravasated. Beaumier et al. (1983) observed a downwardly curved plasma decay (Fig. 4) after intravenous administration of SUV (mean diameter 18.7 nm). This decay pattern is typical of substances eliminated either exclusively through a capacity-limited (Michaelis-Menten) pathway or through parallel first-order and capacity-limited pathways (Kume et al., 1991). On the basis of these and other considerations, Beaumier et al. (1983) proposed that the elimination kinetics of SUV are best described by a model involving two parallel pathways: a capacity-limited one due to uptake by the Kupffer cells, and a linear one due to pinocytosis by the liver parenchymal cells.

It has been shown with various types of "inert" colloids that uptake by the MPS is promoted by the interaction of the colloidal particles with certain plasma factors known

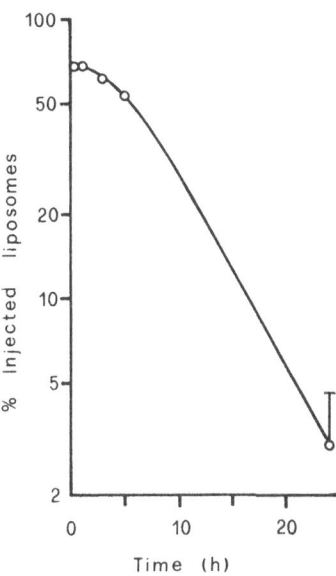

Figure 4. Time course of liposome plasma concentration after intravenous injection into mice of SUV prepared from sphingomyelin and cholesterol in a 2/1 molar ratio. The average dose was 6 mg lipid kg^{-1}. (Reproduced with permission from Hwang, 1987.)

as opsonins. It has also been observed that the rate of colloid clearance can be reduced by injection of large doses that deplete the stores of circulating opsonins (see Kao and Juliano, 1981; Senior et al., 1985; Senior, 1987; Moghimi et al., 1990 and references therein). One such opsonic factor for phospholipid liposomes is α_2-macroglobulin. This protein renders neutral liposomes negatively charged and accelerates their uptake by the MPS (Senior et al., 1985). Opsonic factors specific for hepatic or splenic phagocytic cells have also been identified. Neither liver- nor spleen-specific opsonins have affinity for liposomes made up of sphingomyelin or saturated phosphatidylcholines. This may explain the long half-life of "rigid" liposomes (Moghimi and Patel, 1989). In principle, a dose-dependent decrease in clearance may also be the result of a depletion of the opsonic factors required for the uptake by the MPS. However, Kao and Juliano (1981) were able to show that the slower plasma decay that follows the administration of large liposome doses results from a saturation of the uptake capacity of the MPS rather than from a depletion of plasma opsonins.

The foregoing observations have shown that the liposome disposition kinetics may exhibit saturability in three respects: i) A time-dependent saturation of the retention capacity of the cells of the MPS. Obviously, this saturation is also dose-dependent, since the higher the dose, or the infusion rate, the faster the saturation. ii) A dose-dependent saturation of the uptake mechanism. iii) A dose-dependent saturation of the blood components with which liposomes interact. This is generally observed at doses higher than those necessary for saturation of either cell uptake mechanisms or retention capacities. For example, the disruptive capacity of HDL can be exceeded only at doses ranging between 200 and 800 mg kg^{-1} (Williams and Tall, 1988). Depletion of opsonins has not been observed at doses (300-400 mg kg^{-1}) that largely saturate the MPS retention capacity (Kao and Juliano, 1981).

An additional mechanism that may be responsible for a biphasic plasma decay of phospholipid liposomes has recently emerged from the analysis of the kinetic behavior of multilamellar phosphatidylserine (PS) liposomes administered intravenously to rats

(Palatini et al., 1991). Liposomes prepared from this phospholipid have particular characteristics. The electrostatic repulsion between the negatively charged polar heads of PS results in an increased spacing between the phospholipid molecules of the vesicle bilayer. This has important consequences for the uptake by the MPS, since it has been shown that a loose phospholipid packing favors the interaction with the cells of the MPS (Allen et al., 1988; Mietto et al., 1989), although an excessive surface charge density may have an opposite effect (Mietto et al., 1989). More importantly, the polar head of this phospholipid (the phosphorylserine group) is specifically recognized by the MPS (Allen et al., 1988). Because of their bilayer expansion, PS-containing liposomes are also extremely reactive towards HDL (Bienvenue et al., 1985). Upon interaction with HDL, PS vesicles are destroyed and the lipid is transferred to the lipoprotein particles, which become larger than the native ones. The phospholipid incorporated into HDL is then removed from the plasma along with its lipoprotein carrier. Overall, the disposition of multilamellar PS liposomes is due to three processes: metabolic degradation in the plasma, uptake by the MPS, and incorporation into HDL with subsequent elimination of the HDL-bound PS. A model envisaging these processes is shown in Figure 5A. This model gives rise to an equation

$$C = C_0 \left(\frac{K - k_1 - k_2}{K - k_2} e^{-Kt} + \frac{k_1}{K - k_2} e^{-k_2 t} \right)$$

which predicts a biphasic decay. The initial phase is characterized by a rate constant, $K = k_1 + k_3 + k_4$, reflecting the plasma clearance of intact PS liposomes through all elimination pathways, whereas the terminal phase reflects the clearance of HDL-bound PS.

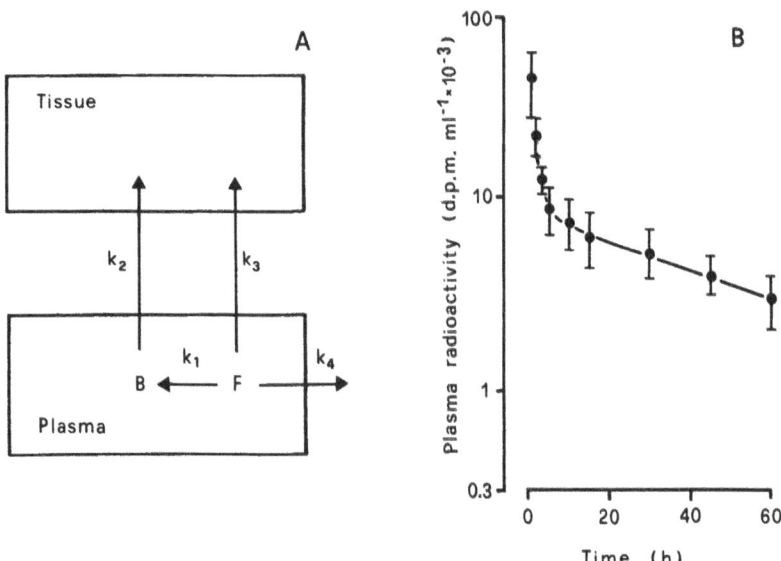

Figure 5. A) Pharmacokinetic model for PS disposition. F and B designate free PS (intact liposomes) and HDL-bound PS, respectively. All rate processes represented are assumed to be first order; k_1 to k_3 are the apparent rate constants for incorporation of PS into HDL, elimination from plasma of HDL-bound PS, and tissue uptake of PS liposomes, respectively; k_4 is the sum of all apparent rate constants for the formation of PS metabolites in plasma. B) PS plasma decay curve after bolus intravenous administration to rats of multilamellar PS liposomes (mean diameter 140 nm) at a dose of 2 mg kg^{-1}. (Reproduced with permission from Palatini et al., 1991.)

As shown in Figure 5B, the plasma PS concentration undergoes an extremely rapid initial decline (t½ = 0.85 min) that accounts for the clearance of more than 90% of the injected dose. This is consistent with the rapid nature of the processes of PS metabolism, uptake by the MPS, and incorporation into HDL. During the terminal linear phase, the PS concentration decreases with a half-time (40 min) very similar to that of the phospholipid component of rat HDL (47 min) (Stein and Stein, 1966). Thus, the slower phase is most likely due to the elimination of PS incorporated into such lipoproteins. It should be noted that the biexponential decline observed with PS liposomes is due to formation in the plasma of two species with different elimination rates. Thus, such a situation is not dissimilar to the administration of a preparation of liposomes that are heterogeneous in size. In either case a multiexponential plasma decay results from the presence of kinetically distinct species that are measured as a single entity. The scheme shown in Figure 5 may constitute a plausible model for the disposition of all those phospholipid liposomes that do not survive HDL-induced disintegration, provided they are not administered at saturating doses. This model implies that a membrane-associated marker be used, since the measurement of an entrapped marker would not permit the detection of the terminal decay phase. It is, on the other hand, known that the latter method can be satisfactorily applied only to those liposomes that are stable in the plasma (Gregoriadis, 1988b).

In conclusion, the observations summarized in this paper have made it apparent that until more general concepts of liposome clearance mechanisms are evolved, a pharmacokinetic model encompassing all the aspects of liposome disposition is likely to prove exceedingly complex. Although a one-compartment model may be satisfactory from a practical point of view, various assumptions of non-linear kinetics are to be built into the model. These assumption are:

i) Time- and dose-dependent decrease in clearance due to saturation of the retention capacity of the MPS. This decrease in clearance has been convincingly demonstrated only for large liposomes, although saturation of the MPS can be obtained also with SUV (Allen, 1988; Kao and Juliano, 1981).

ii) Dose-dependent decrease in clearance due to saturation of the uptake mechanism. This capacity-limited elimination has been observed only with very small SUV composed of sphingomyelin and cholesterol (Beaumier et al., 1983; Hwang, 1987). Accordingly, it has been proposed that the Michaelis constant for SUV uptake by the MPS is much lower than that for large liposomes (Hwang, 1987). This is equivalent to saying that Kupffer cells exhibit a greater affinity for SUV than for MLV or LUV. This difference in affinity may, however, be only apparent, since the liposomal concentration is expressed as moles of phospholipid, not as vesicle number, per unit of volume. At equal phospholipid concentrations, the number of SUV is far higher than that of MLV or SUV. This may explain why the endocytotic processes can be much more readily saturated with SUV.

iii) Saturable interactions with plasma factors such as lipoproteins and opsonins. This type of saturation may in practice be neglected, since it occurs at extremely high dosages. However, such interactions introduce a further complicating factor as they can give rise to different liposomal species that are cleared at different rates.

The above assumptions, which have not yet been verified for all types of liposomes, are likely to render the computation of the pharmacokinetic parameters very complex (Gibaldi and Perrier, 1982). In the absence of specific information on the behavior *in vivo* of a given liposomal preparation, a model-independent analysis appears safer at the present time. It should, however, be remembered also that model-independent parameters such as clearance, mean residence time, or half-time of the terminal phase are meaningless unless the amount of liposomal lipid and mean vesicle diameter are specified. Whatever the type of analysis, care should be exercised in the interpretation of the parameters when the administered vesicles are heterogeneous in size or liable to modifications in the plasma. In these instances, either average parameters or parameters that refer to a species different from the injected one are obtained. For example, with PS liposomes (Palatini et al., 1991) the mean residence time, calculated as AUMC/AUC, was an average parameter that did not represent the time necessary for elimination of 63.2% of the administered dose, as is usually assumed (Gibaldi and Perrier, 1982). Rather, it turned out to be the time required for more than 90% of the dose to be eliminated. In addition, the half-time of the terminal phase was not the elimination half-time of the injected vesicles, but the half-life of PS incorporated into HDL.

REFERENCES

Allen TM (1981) A study of phospholipid interactions between high density lipoproteins and small unilamellar vesicles. Biochim Biophys Acta 640:385.

Allen TM (1988) Interaction of liposomes and other drug carriers with the mononuclear phagocyte system. In: Liposomes as drug carriers (Gregoriadis G, ed.) p 37. New York: John Wiley.

Allen TM and Cleland LC (1980) Serum-induced leakage of liposome contents. Biochim Biophys Acta 597:418.

Allen TM, Williamson P, Schlegel RA (1988) Phosphatidylserine as a determinant of reticuloendothelial recognition of liposome models of the erythrocyte surface. Proc Natl Acad Sci USA 85:8067.

Beaumier PL, Hwang KJ, Slattery JT (1983) Effect of liposome dose on the elimination of small unilamellar sphingomyelin/cholesterol vesicles from the circulation. Res Commun Chem Pathol Pharmacol 39:227.

Bienvenue A, Vidal M, Sainte-Marie J, Philippot J (1985) Kinetics of phospholipid transfer between liposomes (neutral or negatively charged) and high-density lipoproteins: A spin-label study of early events. Biochim Biophys Acta 835:557.

Blumenthal R, Weinstein JN, Sharrow SD, Henkart P (1977) Liposome-lymphocyte interaction: Saturable sites for transfer and intracellular release of liposome content. Proc Natl Acad Sci USA 74:5603.

Bruckdorfer KR and Graham JM (1976) The exchange of cholesterol and phospholipids between cell membranes and lipoproteins. In: Biological membranes, Vol. 3 (Chapman D, Wallach DFH, eds) p 103. London: Academic Press.

Bruni A and Palatini P (1983) Biological and pharmacological properties of phospholipids. Prog Med Chem 19:111.

Daleke DL and Huestis WH (1985) Incorporation and translocation of aminophospholipids in human erythrocytes. Biochemistry 24:5406.

Damen J, Regts H, Scherphof G (1981) Transfer and exchange of phospholipids between small unilamellar liposomes and rat plasma high density lipoproteins. Dependence on cholesterol content and phospholipid composition. Biochim Biophys Acta 665:538.

Darnell J, Lodish H, Baltimore D (1990) Molecular cell biology. p 555. New York: Scientific American Books.

Gibaldi M and Perrier D (1982) Pharmacokinetics. p 45. New York: Marcel Dekker.

Gotfredsen CF, van Berkel TJC, Krujt JK, Goethals A (1983) Cellular localization of stable solid liposomes in the liver of rats. Biochem Pharmacol 32:3389.

Gregoriadis G, ed. (1988a) Liposomes as drug carriers. New York: J Wiley.

Gregoriadis G (1988b) Fate of injected liposomes: Observations on entrapped solute retention, vesicle clearance and tissue distribution in vivo. In: Liposomes as drug carriers (Gregoriadis G, ed) p 3. New York: John Wiley.

Gregoriadis G and Neerunjun DE (1974) Control of the rate of hepatic uptake and catabolism of liposome-entrapped proteins injected into rats. Possible therapeutic applications. Eur J Biochem 47:179.

Guo LSS, Hamilton RL, Goerke S, Weinstein JN, Havel RJ (1980) Interaction of unilamellar liposomes with serum lipoproteins and apolipoproteins. J Lipid Res 21:993.

Heath TD, Lopez NG, Papahadjopoulos D (1985) The effects of liposome size and surface charge on liposome-mediated delivery of methotrexate-aspartate to cells *in vitro*. Biochim Biophys Acta 820:74.

Hwang KJ (1987) Liposome pharmacokinetics. In: Liposomes. From biophysics to therapeutics (Ostro MJ, ed) p 109. New York: Marcel Dekker.

Hwang KJ, Luk K-F S, Beaumier P (1982) Volume of distribution and transcapillary passage of small unilamellar vesicles. Life Sci 31:949.

Juliano RL and Stamp D (1975) The effect of particle size and charge on the clearance rates of liposomes and liposome encapsulated drugs. Biochem Biophys Res Commun 63:651.

Kao YJ and Juliano RL (1981) Interactions of liposomes with the reticuloendothelial system. Effects of reticuloendothelial blockade on the clearance of large unilamellar vesicles. Biochim Biophys Acta 677:453.

Kume Y, Maeda F, Harashima H, Kiwada H (1991) Saturable, non-Michaelis-Menten uptake of liposomes by the reticuloendothelial system. J Pharm Pharmacol 43:162.

Lelkes PI and Friedman JE (1985) Interaction of French-press liposomes with isolated bovine adrenal chromaffin cells: Characterization of the cell-liposome interactions. J Biol Chem 260:1796.

Machy P and Leserman LD (1983) Small liposomes are better than large liposomes for specific drug delivery *in vitro*. Biochim Biophys Acta 730:313.

Margolis LB (1988) Cell interactions with solid and fluid liposomes *in vitro*: lessons for 'liposomologists' and cell biologists. In: Liposomes as drug carriers (Gregoriadis G, eds) p 75. New York: J Wiley.

Margolis LB, Victorov AV, Bergelson LD (1982) Lipid-cell interactions. A novel mechanism of transfer of liposome-entrapped substances into cells. Biochim Biophys Acta 720:259.

Moghimi SM and Patel HM (1989) Serum opsonins and phagocytosis of saturated and unsaturated phospholipid liposomes. Biochim Biophys Acta 984:384.

Moghimi SM, Illum L, Davis SS (1990) Physiological and physicochemical considerations in targeting of colloids and drug carriers to the bone marrow. CRC Crit Rev Therapeutic Drug Carrier Systems 7:187.

Mietto L, Boarato E, Toffano G, Bruni A (1989) Internalization of phosphatidylserine by adherent and non-adherent rat mononuclear cells. Biochim Biophys Acta 1013:1.

Nishikawa K, Arai H, Inoue K (1990) Scavenger receptor-mediated uptake and metabolism of lipid vesicles containing acidic phospholipids by mouse peritoneal macrophages. J Biol Chem 265:5226.

Palatini P, Viola G, Bigon E, Menegus AM, Bruni A (1991) Pharmacokinetic characterization of phosphatidylserine liposomes in the rat. Br J Pharmacol 102:345.

Patsch JR, Gotto AM, Olivercrona T, Eisenberg S (1978) Formation of high density lipoprotein-like particles during lipolysis of very low density lipoproteins *in vitro*. Proc Natl Acad Sci USA 75:4519.

Poste G (1980) The interaction of lipid vesicles (liposomes) with cultured cells and their use as carriers for drugs and macromolecules. In: Liposomes in biological systems (Gregoriadis G, Allison AC, eds) p 101. New York: John Wiley.

Poste G (1983) Liposome targeting *in vivo*: Problems and opportunities. Biol Cell 47:19.

Poznansky MJ and Juliano RL (1984) Biological approaches to the controlled delivery of drugs. Pharmacol Rev 36:277.

Rahman YE, Cerny EA, Patel KR, Lau EH, Wright BJ (1982) Differential uptake of liposomes varying in size and lipid composition by parenchymal and Kupffer cells of mouse liver. Life Sci 31:2061.

Redgrave TG and Small DM (1979) Quantitation of the transfer of surface phospholipid of chylomicrons to the high density lipoprotein fraction during the catabolism of chylomicrons in the rat. J Clin Invest 64:162.

Roerdink F, Dijkstra J, Hartman G, Bolscher B, Scherphof B (1981) The involvement of parenchymal, Kupffer and endothelial liver cells in the hepatic uptake of intravenously injected liposomes. Biochim Biophys Acta 677:79.

Rowland M and Tozer TN (1989) Clinical pharmacokinetics. p 13. Philadelphia: Lea and Febiger.

Scherphof G and Morselt H (1984) On the size-dependent disintegration of small unilamellar phosphatidylcholine vesicles in rat plasma. Evidence of complete loss of vesicle structure. Biochem J 221:423.

Scherphof G, Roerdink F, Waite M, Parks J (1978) Disintegration of phosphatidylcholine liposomes in plasma as a result of interaction with high-density lipoproteins. Biochim Biophys Acta 542:296.

Senior JH (1987) Fate and behavior of liposomes *in vivo*: A review of controlling factors. CRC Crit Rev Therapeutic Drug Carrier Systems 3:123.

Senior JH, Crawley JCW, Gregoriadis G (1985) Tissue distribution of liposomes exhibiting long half-lives in the circulation after intravenous injection. Biochim Biophys Acta 839:1.

Sharma P, Tyrrel DA, Ryman BE (1977) Some properties of liposomes of different sizes. Biochem Soc Trans 5:1146.

Souhami RL, Patel HM, Ryman BE (1981) The effect of reticuloendothelial blockade on the blood clearance and tissue distribution of liposomes. Biochim Biophys Acta 674:354.

Spanjer HH, van Galen M, Roerdink FH, Regts J, Scherphof GL (1986) Intrahepatic distribution of small unilamellar liposomes as a function of liposomal lipid composition. Biochim Biophys Acta 863:224.

Stein Y and Stein O (1966) Metabolism of labelled lysolecithin, lysophosphatidylethanolamine and lecithin in the rat. Biochim Biophys Acta 116:95.

Straubinger RM, Hong K, Friend DS, Papahadjopoulos D (1983) Endocytosis of liposomes and intracellular fate of encapsulated molecules: Encounter with low pH compartment after internalization in coated vesicles. Cell 32:1067.

Szoka FC, Jacobson K, Papahadjopoulos D (1979) The use of aqueous space markers to determine the mechanisms of interaction between phospholipids vesicles and cells. Biochim Biophys Acta 551:295.

Tall AR (1980) Studies on the transfer of phosphatidylcholine from unilamellar vesicles into plasma high density lipoproteins in the rat. J Lipid Res 21:354.

Taylor A and Granger DN (1984) Exchange of macromolecules across the microcirculations. In: Handbook of physiology. Vol IV (Renkin ME, Michel CC, eds) p 467. Bethesda MD: American Physiological Society.

Tollefson JH, Faust R, Albers JJ, Chait A (1985) Secretion of a lipid transfer protein by human monocyte-derived macrophages. J Biol Chem 260:5887.

van Renswoude J and Hoekstra D (1981) Cell-induced leakage of liposome contents. Biochemistry 20:540.

Wagner JG (1983) Significance of ratios of different volumes of distribution in pharmacokinetics. Biopharm Drug Dispos 4:263.

Weinstein JN, Yoshikami S, Henkart P, Blumental R, Hagins WA (1977) Liposome-cell interaction: Transfer and intracellular release of a trapped fluorescent marker. Science 195:489.

Williams KJ and Scanu AM (1986) Uptake of endogenous cholesterol by a synthetic lipoprotein. Biochim Biophys Acta 875:183.

Williams KJ and Tall AR (1988) Interaction of liposomes with lipoproteins: Relevance to drug delivery systems and to the treatment of atherosclerosis. In: Liposomes as drug carriers (Gregoriadis G, ed) p 93. New York: John Wiley.

Williams KJ, Tall AR, Tabas I, Blum C (1986) Recognition of vesicular lipoproteins by the apolipoprotein B,E receptor of cultured fibroblasts. J Lipid Res 27:892.

BEHAVIORAL AND MORPHO-FUNCTIONAL CORRELATES OF BRAIN

AGING: A PRECLINICAL STUDY WITH PHOSPHATIDYLSERINE

Maria Grazia Nunzi, Diego Guidolin,
Lucia Petrelli, Patrizia Polato,
and Adriano Zanotti

Fidia Research Laboratories
Via Ponte della Fabbrica 3/A
35031 Abano Terme, Italy

INTRODUCTION

In the aging brain, changes in membrane lipid composition or content have been related to reduction of membrane fluidity, loss of enzymatic activities, and decreased efficiency of signal transduction and transport mechanisms (Schroeder, 1984). This has led to the assumption that administration of endogenously occurring phospholipids may preserve both structural and functional integrity of central nervous system membranes and prevent, or restore, neuronal deficits that occur with aging or age-related neurodegenerative disorders. Furthermore, exogenous phospholipids may participate in phospholipid metabolism, yielding biologically active intermediates in response to physiopathological phenomena (Toffano and Bruni, 1980; Bruni, 1988).

Among the characteristics of normal and pathologic aging is a decrease in cholinergic and monoaminergic neurotransmission. Furthermore, in age-related neuropathological conditions, certain populations of neurons, e.g., the cholinergic neurons of the basal forebrain, undergo degenerative changes (Whitehouse et al., 1982), and the extent of synaptic loss in the cerebral cortex correlates with the degree of cognitive impairment (De Kosky and Scheff, 1990). Spatial memory, a form of episodic memory severely affected in human dementia, is also impaired in aged rodents. Although many brain areas have been implicated in spatial memory processes, recent studies have focused on the hippocampal formation and, in particular, on the cholinergic septo-hippocampal and glutamatergic entorhinodentate pathways (Fisher et al., 1989; Geinisman et al., 1986).

Phosphatidylserine (PS) constitutes the major acidic phospholipid in the brain where, in addition to its role in Na^+-K^+-ATPase activity (Palatini et al., 1972), it provides a membrane cofactor for full activation of protein kinase C (PKC) (Castagna et al., 1982). Furthermore, PS extracted and purified from bovine brain has been shown to affect neurotransmission and behavioral performance in experimental animals. PS administration stimulates brain catecholaminergic turnover (Toffano et al., 1978) and attenuates the impaired release of acetylcholine (ACh) in the cerebral cortex of aged rats (Pedata et al., 1985). Treatments with PS increase learning and memory functions in aged rodents (Corwin et al., 1985; Drago et al., 1981) and prevent the age-associated decay in avoidance behavior (Zanotti et al., 1987).

In this chapter we report that dysfunctions in both systems underlie spatial memory impairments in aged rats and that oral PS administration restores both spatial memory and related neuroanatomical substrates affected by the aging process.

MATERIALS AND METHODS

Male Sprague-Dawley rats (Charles River, Italy) were maintained on a 12-hr light: 12-hr dark cycle, with free access to water and food (Standard Diet No. 4RF18, Italiana Mangimi, Milano). An aqueous suspension of PS replaced normal drinking water. The concentration of the phospholipid was adjusted throughout the course of the experiment to ensure an average daily intake of 50 mg/kg of PS per rat (Nunzi et al., 1987). In the behavioral study, PS administration started 1 week after screening and lasted until the end of behavioral testing.

Young-adult (5 month old) and aged (21-24 month old) rats were tested in the Morris water maze for spatial reference memory. The swim path and latency to reach the hidden platform were automatically recorded. Aged rats were selected as impaired when their mean escape latencies were above the 99% confidence limit of the young-adult group. The remaining rats constituted the old nonimpaired group. Rats were trained in two blocks of four trials each day of each test week (7th and 12th weeks). After the last trial of test week 7, the platform was removed and rats were given a single "spatial probe" trial for evaluation of searching behavior.

After behavioral testing, animals were perfused and the brains processed for either immunocytochemistry or electron microscopy. Vibratome sections through the septal complex were immunostained using monoclonal antibodies to either choline acetyltransferase (ChAT) or nerve growth factor receptor (NGFR) and the peroxidase-antiperoxidase method. Morphometric analysis was carried out with a computerized image analysis system. Counting of axospinous synapses was performed in the middle molecular layer of the dentate gyrus according to the unbiased serial section technique (Cruz-Orive, 1980).

RESULTS

In the Morris water maze, mean escape latencies during the screening test indicated that a subpopulation of aged rats was impaired in the acquisition of the spatial task. Mean escape latencies of young-adult, aged nonimpaired, aged impaired, and aged impaired PS-treated rats are shown in Figure 1. The performances of old impaired control rats did not change at either retesting, compared to the screening test, and continued to be significantly different from that of both young-adult and aged nonimpaired animals. With PS treatment, however, aged impaired rats showed improved spatial behavior at both retesting times, as shown by the significant decrease in escape latencies when compared to the screening period. The ability of rats to use spatial cues to locate the platform in the pool was evaluated in the "spatial probe" trial. Old impaired control rats did not show any spatial bias for the target quadrant, suggesting that higher escape latencies in this group were due to impaired ability to use spatial information in locating the hidden platform. In PS-treated rats the searching behavior was focused on the prior platform site, suggesting an improved retention of spatial information.

In order to relate PS effects on spatial behavior with morphological parameters, we evaluated the effects of PS administration on hippocampal synaptic plasticity and morpho-

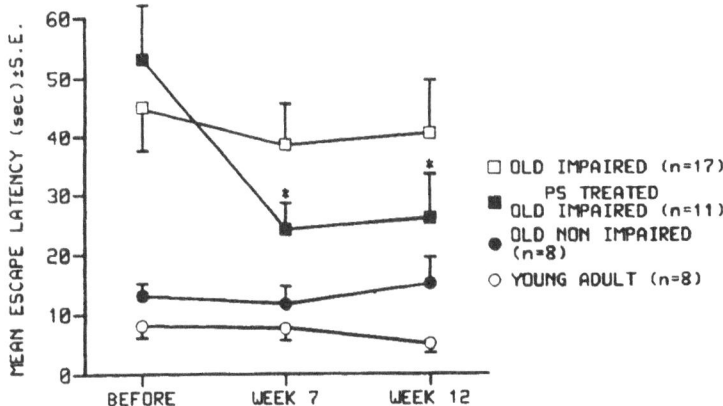

Figure 1. Morris water maze test. Effect of chronic PS treatment on the performance of old impaired rats. * p<0.01 Dunnett's test

functional properties of basal forebrain cholinergic neurons in the behaviorally characterized rats. A quantitative analysis of axospinous synapses was carried out in the middle molecular layer of the dentate gyrus, which represents the main terminal field of entorhinal afferents to the hippocampal formation. Aged impaired and nonimpaired rats showed a statistically significant decrement in the number of axospinous synapses on granule cell dendrites, relative to young-adult animals. However, a decreased incidence of perforated synapses, i.e., synapses with discontinuous postsynaptic density, was observed only in aged memory-deficient rats. In these animals, PS treatment restored the number of axospinous synapses and the percentage of perforated ones to values similar to those of young-adult animals.

Quantitative estimation of morpho-functional properties of the cholinergic neuronal population in the medial septum and diagonal band indicated that degenerative changes occurred in aged rats with spatial memory deficits. In particular, in aged impaired animals the number of ChAT-positive cell bodies was markedly reduced relative to young-adult and aged nonimpaired rats (p<.05; Tukey's test). Similarly, NGFR immunolabeling in the nuclei of both areas was significantly decreased in the memory impaired group, compared to both young-adult and aged nonimpaired animals. However, in aged impaired rats treated with PS, changes in cholinergic cell numbers and NGFR immunoreactivity in these areas were not significant (Fig. 2).

DISCUSSION

One of the most consistent findings in rodent studies is that aged rats are impaired in tasks that young animals solve using spatial information (Wallace et al., 1980; Gage et al., 1984). Chronic oral administration of PS improved spatial behavior of aged memory-impaired rats, further extending previous observations indicating amelioration by PS of memory deficits in aged rats tested in a eight-arm radial maze (Bartus and Dean, 1985). Taken together, these data confirm the positive effect of PS administration on the performance of aged animals in learning/memory tasks.

Electrophysiological studies on the hippocampal formation of aged rats indicate reduced synaptic efficacy (De Toledo-Morrel and Morrel, 1985), possibly as a conse-

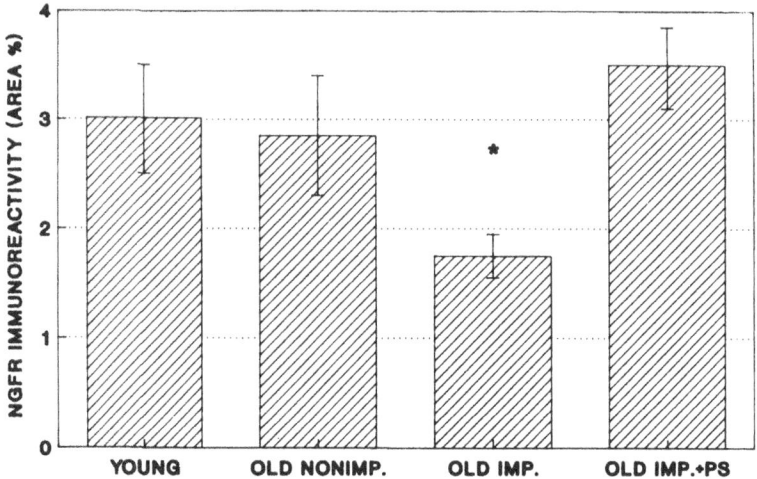

Figure 2. Effect of chronic PS treatment on NGFR immunolabeling in the diagonal band of old impaired rats. * p<0.05; Tukey's test.

quence of the age-dependent loss of synaptic contacts and changes in synaptic morphology in this brain area (Bertoni-Freddari et al., 1986). In particular, the impaired spatial performance of aged rats in a radial maze has been related to selective loss of perforated axospinous synapses in the entorhinodentate pathway (Geinisman et al., 1986). Restoration by PS of the numerical density of perforated synapses in the dentate gyrus of aged memory-impaired rats might, therefore, represent *per se* a structural correlate of the improvement in spatial memory deficits following oral PS administration.

Degenerative changes in cholinergic neurons of the basal forebrain in aged rats also correlate with spatial memory impairments (Fischer et al., 1989; Koh et al., 1989). In the central nervous system, NGF acts by inducing ChAT activity (Vantini et al., 1989). In addition, NGF levels have been reported to be reduced in the hippocampus of aged rats (Larkfors et al., 1987). Age-induced loss of NGFR in cholinergic neurons projecting to the hippocampus may further impair retrograde trophic support by NGF, thus promoting degenerative changes in cholinergic neurons and impairment of cognitive functions. In this context, re-establishment of NGFR density by PS might be causally related to increased ChAT levels in cholinergic neurons, as suggested by restoration of ChAT-positive cell numbers in basal forebrain nuclei of aged impaired, PS-treated animals. These effects on cholinergic metabolic properties could, in turn, be related to improvement by PS of ACh release and synthesis in cortical innervation areas of aged rats (Vannucchi and Pepeu, 1987; Casamenti et al., 1991).

Whether the morpho-functional effects exerted by PS administration depend upon direct or indirect actions of the phospholipid on brain function is at present under investigation. Although most of the PS given by the oral route is metabolized in the intestinal tract, a small fraction reaches the systemic circulation, suggesting the possibility of effects of PS *per se* or its metabolites either directly in the CNS or, alternatively, via modulation of immune-endocrine functions. Another hypothesis that warrants further investigation is that orally administered PS, through activation of PKC in mucosal intestinal cells (see Bruni et al., in this Symposium), modulates transport of ions and/or nutrients affecting brain function. Stimulation of various transport processes has been reported to result from PKC activation (Kikkawa and Nishizuka, 1986; Homma et al., 1990).

In summary, degeneration of basal forebrain cholinergic neurons and loss of synapses correlate with cognitive changes during aging in both rats and man, suggesting a common pathological process in both species. Long-term oral PS administration restores biochemical properties of cholinergic neurons in the septo-hippocampal system, enhances hippocampal synaptic plasticity and improves cognitive functions in aged memory-deficient rats. Since PS treatment prevents or improves biological and behavioral deficits inherent to the aging process in experimental animals, this phospholipid may find application as a therapeutic agent for age-associated memory dysfunctions (Delwaide et al., 1986; Crook et al., 1991).

REFERENCES

Bartus RT and Dean RL (1985) Developing and utilizing animal models in the search for an effective treatment for age-related memory disturbances. In: Normal aging, Alzheimer's disease and senile dementia (Gottfries CG, ed) pp 288-321. Brussels: University of Brussels Press.

Bertoni-Freddari C, Giuli C, Pieri C, Paci D (1986) Quantitative investigation of the morphological plasticity of synaptic junctions in rat dentate gyrus during aging. Brain Res 366:187.

Bruni A (1988) Autacoids from membrane phospholipids. Pharmacol Res Commun 20:529.

Casamenti F, Scali C, Pepeu G (1991) Phosphatidylserine reverses the age-dependent decrease in cortical acetylcholine release: a microdialysis study. Eur J Pharmacol 194:11.

Castagna M, Takai Y, Kaibuchi K, Sano K, Kikkawa V, Nishizuka Y (1982) Direct activation of calcium-activated, phospholipid-dependent protein kinase by tumor-promoting phorbol esters. J Biol Chem 257:7847.

Corwin J, Dean III RL, Bartus RT, Rotrosen J, Watkins DL (1985) Behavioral effects of phosphatidylserine in the aged Fischer 344 rat: amelioration of passive avoidance deficits without changes in psychomotor task performance. Neurobiol Aging 6:11.

Crook TH, Tinkleberg J, Yesavage J, Petrie W, Nunzi MG, Massari DC (1991) Effects of phosphatidylserine in age-associated memory impairment. Neurology 41:644.

Cruz-Orive LH (1980) On the estimation of particle number. J Microsc 120:15.

De Kosky ST and Scheff SW (1990) Synapse loss in frontal cortex biopsies in Alzheimer's disease: Correlation with cognitive severity. Ann Neurol 27:457.

Delwaide PJ, Hurlet A, Hambourg AM, Klieff M (1986) Double-blind randomized study of phosphatidylserine in senile demented patients. Acta Neurol Scand 73:136.

De Toledo-Morrel L and Morrel F (1985) Electrophysiological markers of aging and memory loss in rats. Ann NY Acad Sci 444:296.

Drago F, Canonico PL, Scapagnini U (1981) Behavioral effects of phosphatidyl-serine in aged rats. Neurobiol Aging 2:209.

Fischer W, Gage FH, Bjorklund A (1989) Degenerative changes in forebrain cholinergic nuclei correlate with cognitive impairments in aged rats. Eur J Neurosci 1:34.

Gage FH, Bjorklund A, Stenevi U, Dunnett SS, Kelly PAT (1984)Intrahippocampal septal grafts ameliorate learning impairments in aged rats. Science 225:533.

Geinisman Y, De Toledo-Morrell L, Morrell F (1986) Loss of perforated synapses in the dentate gyrus: morphological substrate of memory deficit in aged rats. Proc Natl Acad Sci USA 83:3027.

Homma T, Burns KD, Harris RC (1990) Agonist stimulation of $Na^+/K^+/Cl$-cotransport in rat glomerular mesangial cells. J Biol Chem 265:17613.

Kikkawa U and Nishizuka Y (1986) The role of protein kinase C in transmembrane signalling. Annu Rev Cell Biol 2:149.

Koh S, Chang P, Collier TJ, Loy R (1989) Loss of NGF receptor immunoreactivity in basal forebrain neurons of aged rats: Correlation with spatial memory impairment. Brain Res 498:404.

Larkfors L, Ebendal T, Whittemore SR, Persson H, Hoffer B, Olson L (1987) Decreased level of nerve growth factor (NGF) and its messenger RNA in the aged rat brain. Mol Brain Res 3:55.

Nunzi MG, Milan F, Guidolin D, Toffano G (1987) Dendritic spine loss in hippocampus of aged rats. Effect of brain phosphatidylserine administration. Neurobiol Aging 8:501.

Palatini P, Dabboni-Sala F, Bruni A (1972) Reactivation of a phospholipid-depleted sodium potassium-stimulated ATP-ase. Biophys Acta 288:413.

Pedata F, Giovannelli L, Spignoli G, Giovannini MG, Pepeu G (1985) Phosphatidylserine increases acetylcholine release from cortical slices in aged rats. Neurobiol Aging 6:337.

Schroeder F (1984) Role of membrane lipid asymmetry in aging. Neurobiol Aging 5:323.

Toffano G and Bruni A (1980) Pharmacological properties of phospholipids liposomes. Pharmacol Res Commun, 12:408.

Toffano G, Leon A, Mazzari S, Teolato S, Orlando P (1978) Modification of noradrenergic hypothalamic system in rats injected with phosphatidylserine liposomes. Life Sci 23:1093.

Vannucchi MG and Pepeu G, (1987) Effect of phosphatidylserine on acetylcholine release and content in cortical slices from aging rats. Neurobiol Aging 8:403.

Vantini G, Schiavo N, Di Martino A, Polato P, Triban C, Callegaro L, Toffano G, Leon A (1989) Evidence for a physiological role of nerve growth factor in the central nervous system of neonatal rats. Neuron 3:267.

Wallace JE, Kranter EE, Campbell EA (1980) Animal models of declining memory in the aged. Short-term and spatial memory in aged rats. J Gerontol 35:255.

Whitehouse PJ, Price DL, Struble RG, Clark AW, Coyle JT, De Long MR (1982) Alzheimer's disease and senile dementia: Loss of neurons in the basal forebrain. Science 215:1237.

Zanotti A, Rubini R, Calderini G, Toffano G (1987) Pharmacological properties of phosphatidylserine: effects on memory function. In: Nutrients and brain function (Essman WB, ed) pp 95. Karger: Basel.

RECEPTOR COUPLING TO PHOSPHOINOSITIDE SIGNALS

P. Kurian, L. J. Chandler,
R. Patel, and F. T. Crews

University of Florida College of Medicine
Department of Pharmacology
Gainesville, FL 32610, USA

A number of neurotransmitters have been shown to activate phospholipase C (PLC) through a G protein and/or Ca^{2+} pathway resulting in hydrolysis of membrane phosphatidylinositol 4,5-bisphosphate [PtdIns(4,5)P_2], leading to the formation of inositol 1,4,5-trisphosphate [Ins(1,4,5)P_3] and diacylglycerol (DAG). Growth factors stimulate phospholipase $C\gamma1$ (PLCγ) and phosphatidylinositol 3-kinase (PtdIns 3-kinase) by tyrosine phosphorylation of these enzymes. Ins(1,4,5)P_3 mobilizes intracellular calcium from a non-mitochondrial, ATP-dependent pool. Ins(1,4,5)P_3 can be phosphorylated to inositol 1,3,4,5-tetrakisphosphate [Ins(1,3,4,5)P_4] by inositol polyphosphate 3-kinase (Bansal and Majerus, 1990). Ins(1,3,4,5)P_4 modulates cytoplasmic Ca^{2+} levels by mobilizing Ca^{2+} or sequestering Ca^{2+} into storage pools mobilized by Ins(1,4,5)P_3 (Gawler et al., 1990; Hill et al., 1988). Ins(1,4,5)P_3 and Ins(1,3,4,5)P_4 are further catabolized by a variety of inositol phosphate-specific phosphatases (Fig. 1). PtdIns 3-kinase, which phosphorylates phosphoinositides at the 3-position, has been implicated in the transduction of mitogenic signals. The products of PtdIns 3-kinase are the novel phosphoinositides, phosphatidylinositol 3-phosphate, phosphatidylinositol 3,4-bisphosphate, and phosphatidylinositol 3,4,5-trisphosphate, which are not hydrolyzed by PLC (Ruderman et al., 1990).

NEUROTRANSMITTER COUPLING TO PtdIns(4,5)P_2 HYDROLYSIS

The antidepressant agent lithium has proven useful for studying agonist-stimulated phosphoinositide hydrolysis *in vitro* due to its ability to inhibit the phosphatases that metabolize inositol monophosphates. In the presence of lithium, there is a buildup of inositol phosphates, particularly the inositol monophosphates, which can be measured when the inositol phospholipids are prelabeled with [³H]inositol. Although the biochemical mechanism of action of lithium in relation to its therapeutic effects is not known, it may relate to inhibition of inositol phosphate metabolism and depletion of inositol for reincorporation into inositol phospholipids (Berridge et al., 1989). Lithium has also been reported to attenuate muscarinic receptor-linked Ins(1,3,4,5)P_4 formation in brain (Whitworth and Kendall, 1988). Therefore, phosphoinositide hydrolysis experiments carried out in the presence of lithium may not provide an accurate measure of inositol polyphosphate formation. This may be especially true in relation to $Ins(1,3,4,5)P_4$ formation. To alleviate the possible complications introduced by the use of lithium, we developed a method for measuring phosphoinositide hydrolysis that does not require the presence of lithium. Our results showed that inositol phosphate formation in the absence of lithium is greatly increased when additional inositol is included in the incubation medium in concentrations normally present in cerebral spinal fluid (Fig. 2). Also, inositol had a much larger potentiating effect on carbachol-stimulated $Ins(1,3,4,5)P_4$ formation than on $InsP_1$, $InsP_2$, or $InsP_3$ formation.

HPLC separation of $[^3H]InsP_3$ formed during stimulation indicated that two isomers were present, $Ins(1,4,5)P_3$, the active isomer for release of intracellular calcium, and $Ins(1,3,4)P_3$, an inactive $Ins(1,3,4,5)P_4$ metabolite (Fig. 3). Approximately equal amounts of the two $[^3H]InsP_3$ metabolites were found after 5 min of carbachol stimulation, indicating that a significant fraction of $Ins(1,4,5)P_3$ is phosphorylated by the $Ins(1,4,5)P_3$ 3-kinase to form $Ins(1,3,4,5)P_4$ and is subsequently dephosphorylated by

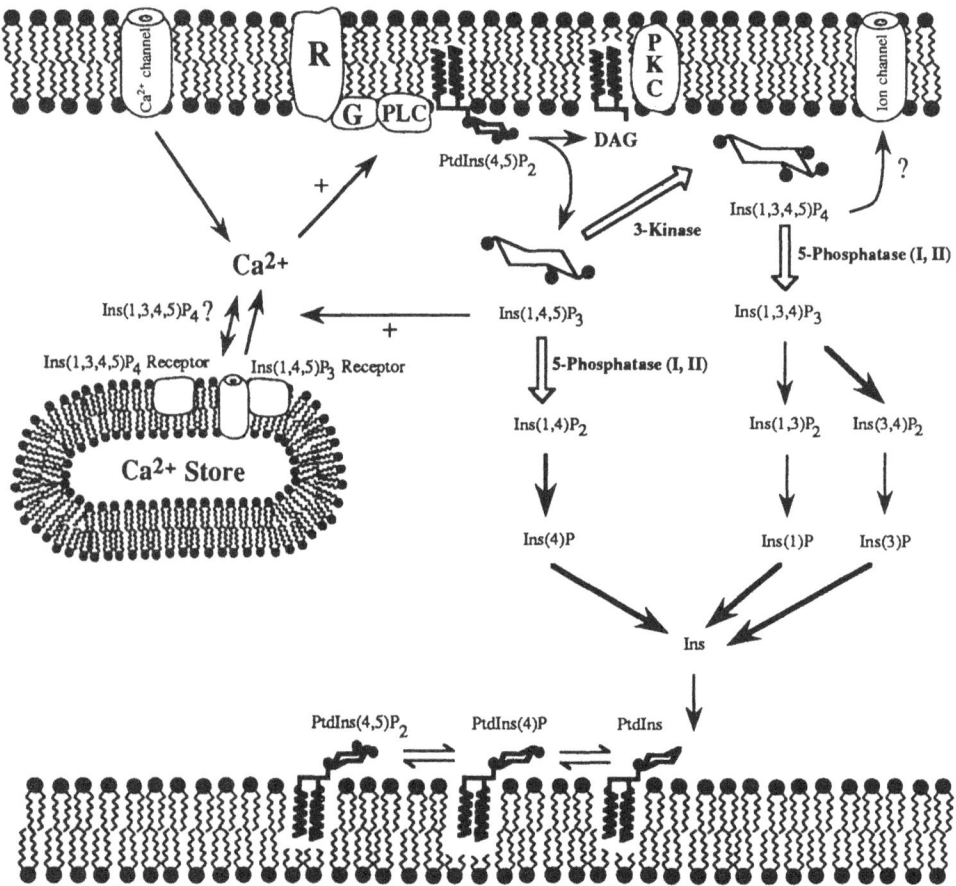

Figure 1. Schematic diagram of neurotransmitter-stimulated inositol polyphosphate cascade. Receptor (R) acts through guanine nucleotide binding protein (G) to stimulate phospholipase C (PLC). Increased intracellular Ca^{2+} through membrane channels or release from intracellular stores may activate PLC causing $PtdIns(4,5)P_2$ hydrolysis generating $Ins(1,4,5)P_3$ and DAG. $Ins(1,4,5)P_3$ releases Ca^{2+} from intracellular stores and/or is converted to $Ins(1,3,4,5)P_4$, which may have actions on both intracellular Ca^{2+} stores and/or membrane channels.

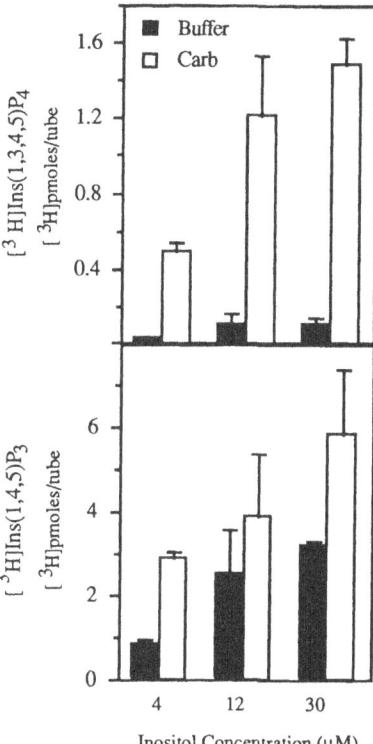

Figure 2. Stimulation of phosphoinositide hydrolysis in the absence of lithium. Cerebral cortical slices were labeled with various concentrations of inositol for 60 min. Slices were stimulated with carbachol for 5 min and the various inositol phosphates separated as described by Chandler et al., 1991.

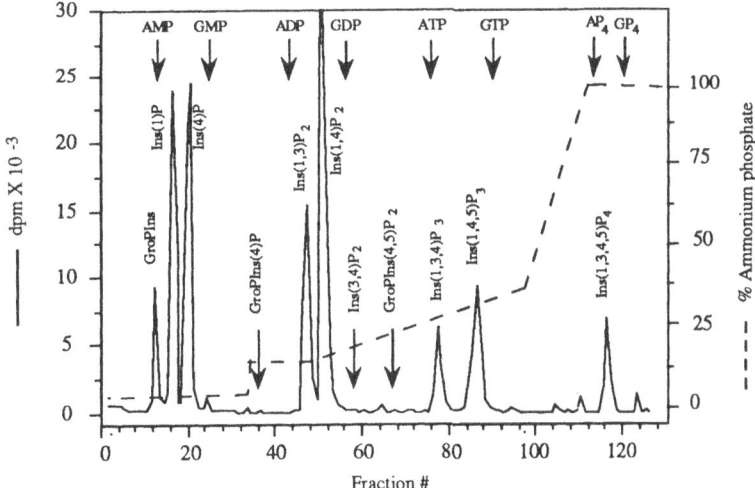

Figure 3. Chromatographic separation of inositol polyphosphates by HPLC. Cerebral cortical slices were labeled with 12 μM [³H]inositol for 60 min and stimulated with carbachol for 5 min. The various [³H]inositol phosphates were extracted and spiked with the indicated nucleotides and separated by SAX HPLC.

a 5-phosphatase to Ins(1,3,4)P$_3$. Comparisons of the effects of various agonists on [^{3}H]Ins(1,3,4,5)P$_4$ and [^{3}H]InsP$_3$ formation in the absence of lithium indicate that only carbachol, a cholinergic agonist, and quisqualate, a glutamatergic agonist, significantly increase [^{3}H]InsP$_3$ and [^{3}H]Ins(1,3,4,5)P$_4$ levels. The lack of response of the other agonists may be due to different mechanisms of coupling to phosphoinositide hydrolysis as has been suggested previously, particularly for norepinephrine (Chandler and Crews, 1989; Crews et al., 1988; Pontzer and Crews, 1990).

The excitatory neurotransmitters acetylcholine and glutamate are involved in neuronal plasticity, which is thought to be an essential component of learning and memory. The loss of cognitive ability in Alzheimer's disease and in age-associated memory impairment has been suggested to be secondary to a loss of central nervous system cholinergic transmission. Drugs that specifically disrupt cholinergic transmission have profound effects on learning and memory. A loss of cholinergic neurons clearly occurs early in the course of Alzheimer's disease when memory loss is the only prominent symptom. In studies using experimental models of Alzheimer's disease, lesioning cholinergic neurons also disrupts the ability of animals to learn. Glutamate has also been implicated in memory processes. Drugs that block glutamate receptors, particularly the N-methyl-D-aspartate (NMDA) receptor subtype, can produce cognitive deficits. An *in vitro* model of synaptic plasticity, long-term potentiation (LTP), is thought to be mediated in part through NMDA receptors. These studies suggest that both cholinergic and glutamatergic signals play an important role in memory processes and cognitive function.

We have investigated receptor transduction systems activated by acetylcholine and glutamate in order to determine which ones may be relevant to synaptic plasticity associated with learning and memory. A variety of receptors have been shown to stimulate the hydrolysis of phosphoinositides to produce inositol polyphosphate and diacylglycerol second messengers. Early studies of PtdIns(4,5)P$_2$ hydrolysis in brain slices in the presence of lithium showed that norepinephrine, carbachol, serotonin, glutamate, calcium ionophores, and potassium-induced depolarization could all stimulate PtdIns(4,5)P$_2$ hydrolysis (Gonzales and Crews, 1984; Gonzales et al., 1985). The ability of calcium ionophores and depolarization to stimulate PtdIns(4,5)P$_2$ hydrolysis in brain slices led to the suggestion that calcium influx can activate a phospholipase responsible for PtdIns(4,5)P$_2$ hydrolysis. *In vitro* experiments with membranes prepared from rat cerebral cortex showed that GTP as well as calcium can stimulate phosphoinositide hydrolysis (Gonzales and Crews, 1988). Furthermore, nanomolar changes in intracellular calcium levels cause parallel changes in PtdIns(4,5)P$_2$ hydrolysis (Chandler and Crews, 1990a,b). These studies and others indicate that both receptor- and depolarization-induced calcium influx leads to phosphoinositide hydrolysis (Fig. 4). In membrane experiments polyphosphoinositides are the preferred substrates for PLC. Taken together, these findings suggest that phosphoinositide-specific phospholipase(s) coupled to PtdIns(4,5)P$_2$ hydrolysis may be activated by either receptor activation of a GTP binding protein or stimulation of calcium flux.

Although the stimulation of Ins(1,3,4,5)P$_4$ formation produces a large signal with carbachol and glutamate, other receptors which are not coupled to Ins(1,3,4,5)P$_4$ formation do not appear to produce significant stimulation in the absence of lithium. We have previously proposed that at least a portion of the norepinephrine signal is mediated by calcium influx (Chandler and Crews, 1989). Thus, these finding suggest that receptors are differentially coupled to phosphoinositide hydrolysis by differences in efficacy and mechanism of coupling.

To further investigate differences in coupling of receptors to phosphoinositide hydrolysis, membrane preparations of rat cerebral cortex were incubated with [^{3}H]phosphoinositides in the presence of various compounds. Consistent with previous studies,

GTPγS and calcium were observed to stimulate PtdIns(4,5)P$_2$ hydrolysis. However, in the presence of GTPγS, only carbachol further increased PtdIns(4,5)P$_2$ breakdown. Glutamate, quisqualate, and norepinephrine had little or no effect (Fig. 4). These findings suggest that muscarinic receptors activated by carbachol are coupled to phospholipase C activation through a guanine nucleotide binding protein (G protein). Since both G$_o$ and G$_q$ have been reported to activate PLC (Smrcka et al., 1991), glutamate and quisqualate may couple to PLC activation through G proteins not active in our membrane preparation. Studies have shown that G$_o$, but not G$_q$, is sensitive to inhibition by pertussis toxin. Furthermore, glutamate-stimulated PtdIns(4,5)P$_2$ hydrolysis is sensitive to pertussis toxin (Sugiyama et al., 1987; Nicoletti et al., 1988) whereas muscarinic receptor-stimulated PtdIns(4,5)P$_2$ hydrolysis is not. Muscarinic receptors may thus be linked to PtdIns(4,5)P$_2$ hydrolysis through G$_q$ while glutamate receptors may be linked to PtdIns(4,5)P$_2$ hydrolysis through both G$_o$ and calcium flux, depending on the subtype of glutamate receptor. The differential coupling of muscarinic receptors and glutamate receptors may have functional implications for the formation of second messengers by allowing specific interactions between receptors. Muscarinic and glutamate receptors—but not norepinephrine, angiotensin II (AngII), 5-hydroxytryptamine (serotonin), or [8-arginine]vasopressin (AVP)—stimulate the formulation of Ins(1,3,4,5)P$_4$. In this regard, muscarinic and glutamate receptors appear to be uniquely coupled to PtdIns(4,5)P$_2$ hydrolysis and Ins(1,3,4,5)P$_4$ formation, which may be important for memory and learning. Recently muscarinic cholinergic-regulated PLC (PLCB) has been purified and was found to be stimulated by GTPγS and carbachol in rabbit cortical membranes (Carter et al., 1990).

A comparison of carbachol-stimulated InsP$_3$ and Ins(1,3,4,5)P$_4$ formation with neuronal firing rates in hippocampal slices suggests an important function for Ins(1,3,4,5)P$_4$ in the regulation of electrophysiological responses. After the preparations were exposed to low concentrations of carbachol, we observed the well-documented increase in firing rate. However, with higher concentrations of carbachol, cell firing actually decreased. This abatement of firing occurred only at concentrations of carbachol that also increased Ins(1,3,4,5)P$_4$ formation. Pharmacological studies suggest that abatement of firing is linked to activation of M1 muscarinic receptors and subsequent hydrolysis of PtdIns(4,5)P$_2$ (Pontzer and Crews, 1990). In preliminary intracellular studies, abatement of firing was associated with a large, sustained membrane depolarization, which may be sufficient to change other voltage-dependent processes, including removing the Mg^{2+} block from NMDA receptors.

The interaction of cholinergic and glutamatergic receptors may occur through the formation of Ins(1,3,4,5)P$_4$, which appears to be particularly increased by these transmitters. Ins(1,3,4,5)P$_4$ has been shown to release intracellular Ca^{2+} (Gawler et al., 1990), to initiate refilling of intracellular Ca^{2+} stores (Hill et al., 1988), and to open a cation channel in plasma membranes (Irvine and Moor, 1986). Muscarinic-stimulated Ins(1,3,4,5)P$_4$ formation may produce a large, sustained depolarization through these or other mechanisms. The implications of a large, sustained depolarization caused by muscarinic receptors are manifold. One receptor thought to be involved in neuronal excitation leading to plasticity as well as excitotoxicity is the NMDA receptor.

Activation of the NMDA receptor opens a nonspecific cation channel, allowing a large calcium influx. Depolarization and calcium influx appear to be important for long-term potentiation, a model of synaptic plasticity, whereas excessive NMDA-mediated calcium influx may also lead to neurotoxicity. However, NMDA-gated cation channels are blocked by Mg^{2+} at normal neuronal levels of polarization. We hypothesize that a muscarinic cholinergic-stimulated increase in Ins(1,3,4,5)P$_4$ induces a depolarization that removes the Mg^{2+} block. This allows a much greater NMDA-mediated Ca^{2+} flux to occur and thereby links glutamate and acetylcholine in the production of neuronal plasticity and excitotoxicity.

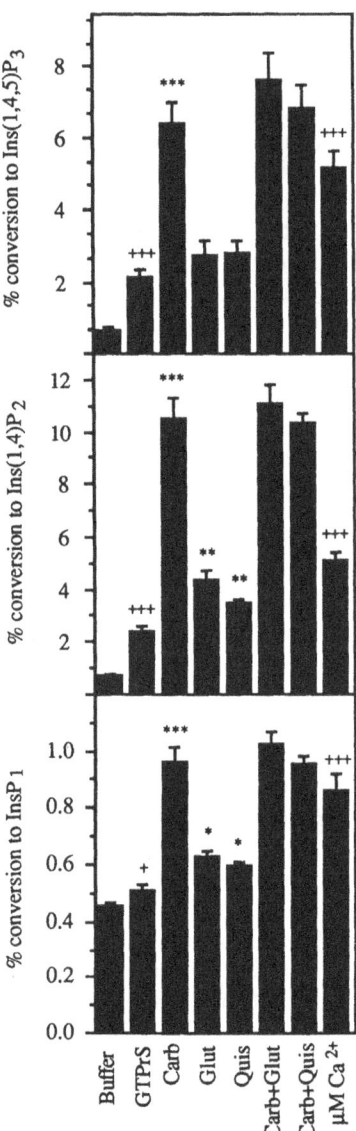

Figure 4. Stimulation of phosphoinositide hydrolysis in isolated membranes from rat cerebral cortex. The various agonists were added in the presence of 1 μM GTPγS: 1 mM carbachol (Carb), 1 mM glutamate (Glut), 100 μM quisqualic acid (Quis). Asterisks or pluses indicate significant difference from GTPγS or buffer value.

EFFECTS OF AGING AND ETHANOL ON PtdIns(4,5)P$_2$ HYDROLYSIS

Since aging and ethanol abuse are associated with various types of human dementia and reduction in cognitive ability, we investigated the effects of ethanol and changes during aging on certain components of the phosphoinositide cascade. Since there is evidence suggesting an increase in the cytosolic calcium concentration during aging, we investigated the effects of muscarinic and other agonists on receptor-stimulated inositol polyphosphate formation in young (7 month old) and senescent (29 month old) rat cerebral cortical and hippocampal slices. To investigate the effects of aging on agonist-stimulated $InsP_3$ and $Ins(1,3,4,5)P_4$ formation, young (6-7 month old) and old (26-28 month old) Fischer-344 rat cerebral cortical or hippocampal slices were incubated with [^{3}H]inositol (20 μCi, 12 μM) for 60 min and challenged with various agonists. Carbachol and quisqualate stimulated $InsP_3$ and $Ins(1,3,4,5)P_4$ formation in young and old rat cerebral cortical slices. In old rat cerebral cortical slices, carbachol-stimulated $Ins(1,3,4,5)P_4$ formation was reduced by 44%. Angiotensin II-stimulated $InsP_3$ was increased 219% in old rat cortex. Carbachol and quisqualate stimulated $Ins(1,3,4,5)P_4$ formation in young and old rat hippocampal slices. There was no influence of aging either on the basal level or on the maximal response to carbachol or quisqualate in hippocampal slices. Data suggest diminished muscarinic-stimulated $Ins(1,3,4,5)P_4$ in senescent rat cortex but not in hippocampus. Furthermore, [^{32}P]$Ins(1,3,4,5)P_4$-specific binding in cerebellum was less in old rats than in young animals (Fig. 5).

Ethanol increases intracellular Ca^{2+} levels in PC12 cells. It has also been suggested that ethanol may activate PLC in hepatocytes (Hoek et al., 1990). We have investigated effects of ethanol (100 mM) on carbachol-stimulated [^{3}H]phosphoinositide hydrolysis by rat cerebral cortical membranes. Carbachol increased the polyphosphoinositide hydrolysis, leading to the formation of $Ins(1,4,5)P_3$ and $Ins(1,4)P_2$. Ethanol neither increased inositol polyphosphate formation in the presence of GTPγS nor decreased carbachol-stimulated inositol polyphosphate formation (Fig. 6). We conclude that ethanol does not affect G protein-coupled PLC in rat cerebral cortex. In other studies we found that ethanol (500 mM) significantly accelerated the metabolism of inositol polyphosphates in rat brain homogenates (Chandler et al., 1991). Pharmacologically relevant concentrations of ethanol (30-100 mM) do not affect the enzymes involved in $Ins(1,3,4,5)P_4$ catabolism.

GROWTH FACTORS AND PHOSPHOINOSITIDES

Normal cell proliferation and differentiation are regulated by the action of growth factors on cell-surface receptors possessing tyrosine kinase activity. Binding of growth

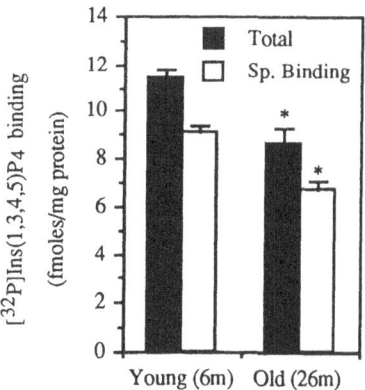

Figure 5. Decreased $Ins(1,3,4,5)P_4$ specific binding in senescent rat cerebellar homogenates. Cerebella from young (6 month old) and old (26 month old) Fisher-344 rats were homogenized in eight volumes (11 mg protein/ml) of ice-cold assay buffer (25 mM sodium acetate, 25 mM KH_2PO_4, 2 mM EDTA, and 1 mM dithiothreitol, pH 5.0). Each assay contained 0.3 pM [5-^{32}P]$Ins(1,3,4,5)P_4$ (20 000 dpm) and 550 μg protein in a final volume 160 μl. $Ins(1,3,4,5)P_4$ was used at 10 μM to define the nonspecific binding component. Shown are the means $\pm$ SEM of three experiments each performed in duplicate. Asterisk indicates significant difference from corresponding value in young rats (*P < 0.05; Student's t test).

factors such as insulin, platelet-derived growth factor (PDGF), and epidermal growth factor (EGF) to their receptors induces receptor oligomerization and autophosphorylation of the clustered receptors (Ullrich and Schlessinger 1987). Autophosphorylation on receptor tyrosine residues in turn induces a conformational change in the phosphorylation of cellular substrates in the signal transduction pathway for the growth factor (Ruderman et al., 1990). In addition to transmembrane receptor tyrosine kinases, there are cytoplasmic tyrosine kinases, including the v-*src*, v-*abl*, and v-*fes/fpm* oncogene products (Hunter and Cooper, 1985). For both receptor and cytoplasmic tyrosine kinases, tyrosine kinase activity is essential for bioactivity such as mitogenesis. This has been shown experimentally in that site-specific mutagenesis of certain receptor tyrosine kinases, including removal of the autophosphorylation site and inactivation of the ATP binding site on the kinase domain, inhibit transmission of the growth factor signal into the cell (Chou et al., 1987). Experimental evidence establishing a strong linkage between receptor and cytoplasmic tyrosine kinases and their ability to influence cell proliferation is supplied by studies in which structural modifications of these tyrosyl kinases induce the constitutive expression of a signal in their mitogenic pathways, thus evoking their oncogenic potential (Hunter and Cooper, 1985).

Transmission of the mitogenic signal through activation of these tyrosine kinases may occur through the generation of second messengers and phosphorylation of proteins that are capable of influencing gene expression, cellular metabolism, cytoskeletal remodeling, and mitogenesis. While the cellular substrates in the signal transduction pathways of growth factors and proto-oncogene products have not been fully identified, the functional commonality of their ability to phosphorylate tyrosine residues suggests that different growth factor receptors may use common cellular substrates. One possible mechanism through which tyrosine kinases may transmit their signal intracellularly is via polyphosphoinositide metabolism. Polyphosphoinositides and their metabolites have been shown to be crucial as signal-transducing elements in response to ligand stimulation of a number of receptors including α_1-adrenergic, muscarinic cholinergic, vasopressin, bombesin, EGF, and PDGF receptors (Fain, 1990). Recently it was discovered that the activity of PtdIns 3-kinase, a novel enzyme in the

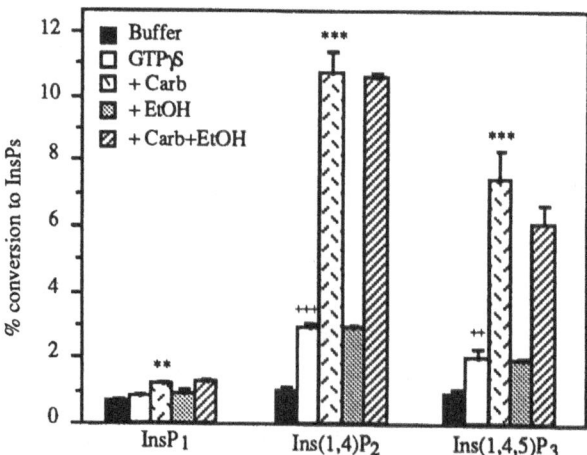

Figure 6. Carbachol stimulation of phosphoinositide hydrolysis is unaffected by ethanol in isolated membranes from rat cerebral cortex. Ethanol (EtOH, 100 mM) was added in the presence or absence of 1 mM carbachol (Carb). Asterisks or pluses indicate significant difference from 1 μM GTPγS or buffer value (**/++ P < 0.01, ***/+++ P < 0.001). Results are mean ± SEM of eight determinations in two separate experiments.

phosphoinositide cascade (Whitman et al. 1988), as well as that of phosphoinositide-specific PLCγ, is increased by tyrosine phosphorylation of these enzymes on tyrosine residues in response to growth factor stimulation (Nishibe et al., 1990). This has generated a lot of excitement as to the mechanisms of growth factor action at the subcellular level.

In addition to PtdIns 3-kinase and PLCγ, a number of other cytoplasmic proteins appear to associate with growth factor-activated receptors. Among these are $p21^{ras}$, GTPase activating protein (GAP), and $p74^{raf}$ (Anderson et al., 1990). These proteins have diverse enzymatic functions and demonstrate no sequence homology in their catalytic domains. However, they have the ability to physically associate with tyrosine kinases through Src homology (SH) domains 2 and 3 (Koch et al., 1991). SH2 and SH3 domains are noncatalytic regions approximately 100 and 45 amino acid residues in length, respectively, and are evolutionarily conserved among a number of receptor tyrosine kinase substrates including PtdIns 3-kinase, PLCγ, GAP, and Src tyrosine kinases (Margolis et al., 1990). In non-receptor tyrosine kinases, similar homologous regions have been identified as $p60^{c\text{-}src}$ and $p98^{c\text{-}fps}$. It is hypothesized that tyrosyl phosphorylation of receptor tyrosine kinase substrates may induce SH2- and SH3-mediated formation of heterooligomeric complexes localized to the plasma membrane. These complexes and/or their enzymatic products in turn may serve as a critical coupler between growth factor stimulation and signal transduction of activated receptor tyrosine kinases (Anderson et al., 1990).

There is experimental evidence to support the hypothesis that SH2 and SH3 domains mediate aggregation of tyrosine kinase substrates to growth factor-activated receptors as a mechanism of activation. For example, insulin- and IGF-1-induced mitogenesis of SH-SY5Y human neuroblastoma cells shows elevated levels of PtdIns 3-kinase activity, compared to levels in unstimulated cells in antiphosphotyrosine and antireceptor immunoprecipitates (Fig. 7). Thus, the ability of PtdIns 3-kinase (Fig. 8) and PLCγ to be immunoprecipitated by antiphosphotyrosine antibody in a concentration-dependent manner and to be coimmunoprecipitated by antigrowth factor receptor antibody suggests a physical association between the ligand-activated receptor and the enzyme, which is mediated by SH2 and SH3 domains.

Tyrosine kinase substrates such as PtdIns 3-kinase and PLCγ and their effects on phosphoinositide metabolism may constitute the primary mitogenic signal for a number of growth factors and certain proto-oncogene products. However, our understanding of the role of the phosphoinositide cycle in mediating the growth factor response is incomplete. For example, while two enzymatic products, $Ins(1,4,5)P_3$ and DAG, are formed as a result of PLCγ-mediated hydrolysis of $PtdIns(4,5)P_2$ and while their functions as regulators of Ca^{2+} flux from the endoplasmic reticulum and in protein kinase C activation, respectively, are well established (Fain, 1990), the functions of the products of PtdIns 3-kinase activity—i.e., $PtdIns(3)P$, $PtdIns(3,4)P_2$, and $PtdIns(3,4,5)P_3$— have not yet been elucidated. PtdIns 3-kinase products possessing a monoester phosphate group on the D-3 position of the inositol ring have been shown to be resistant to PLC-mediated hydrolysis and to be metabolized by only phospho-monoesterases (Serunian et al., 1990). Thus, it has been suggested that in contrast to the classical phosphoinositide pathway, the phospholipid products of PtdIns 3-kinase themselves may be metabolically active rather than their hydrolytic products (Downes and Macphee, 1990). One possible function of these phosphoinositides may involve their ability to reorganize the cytoskeleton for mitosis through their ability to dissociate gelsolin bound to actin, thereby generating free ends for actin polymerization (Cunningham et al., 1991). Recently it has been reported that activation of PtdIns 3-kinase can also occur through a pertussis toxin-sensitive G protein and directly by GTPγS (Kucera and Rittenhouse, 1990), in addition to tyrosyl phosphorylation. This is particularly exciting because all previous mechanisms of activation for PLC isozymes

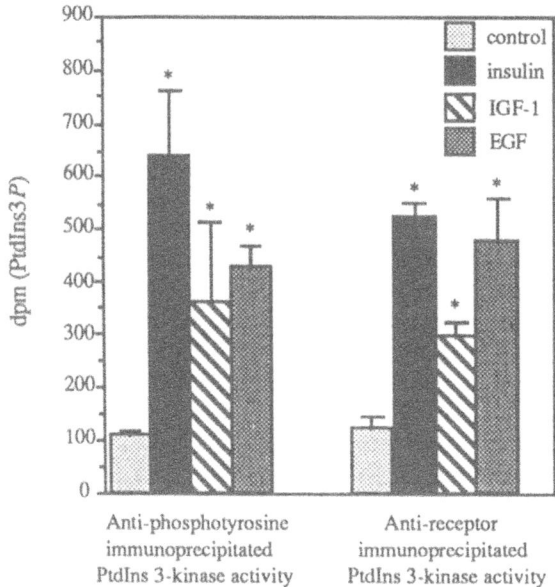

Figure 7. Antiphosphotyrosine and antireceptor antibody immunoprecipitated PtdIns 3-kinase activity from growth factor-stimulated SH-SY5Y human neuroblastoma cells. After a 10-min incubation with 1 μM insulin, IGF-1, or EGF at 37°C, cells were lysed for 30 min at 4°C. Lysates were immunoprecipitated with either Sepharose-conjugated antiphosphotyrosine antibody for 2 hr or by anti-insulin, anti-IGF-1, or anti-EGF receptor antibody for 90 min followed by a 45-min incubation with Protein G PLUS/Protein A-Agarose at 4°C. Washed immunoprecipitates were then incubated with PtdIns and 15 μCi [γ^{32}P]ATP at a final concentration of 0.2 mg/ml for another 30 min. The phosphoinositides were extracted and separated by TLC. The structure of PtdIns(3)P was determined by SAX HPLC (data not shown). *Significantly different from control (p<0.05).

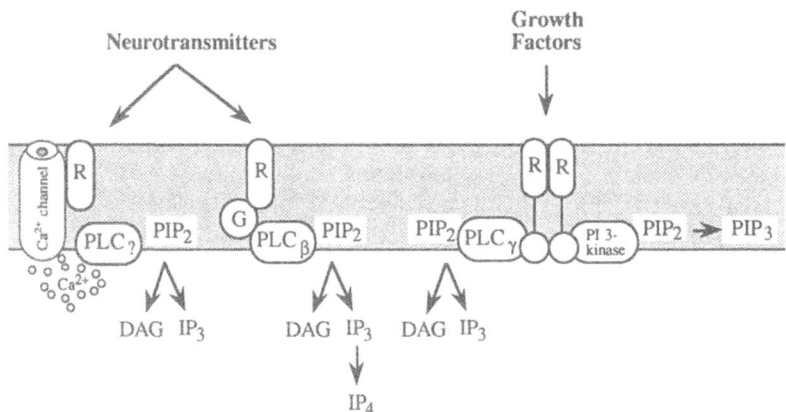

Figure 8. Schematic model depicting separate pathways for activating phospholipase C (PLC) coupled to PtdIns(4,5)P$_2$ [PIP$_2$] hydrolysis and PtdIns 3-kinase. The neurotransmitter receptors (R) are coupled to PLCβ through a guanine nucleotide binding protein (G) and stimulate the formation of Ins(1,4,5)P$_3$ [IP$_3$] and Ins(1,3,4,5)P$_4$ [IP$_4$]. Another pathway for activating PLC is through stimulation of calcium influx which can then directly activate PLC$_?$. This pathway appears to form only Ins(1,4,5)P$_3$. There is physical association between growth factor-activated receptor and enzyme, mediated by SH2 and SH3 domains. The activities of PtdIns 3-kinase and PLCγ are increased by tyrosine phosphorylation in response to growth factor stimulation, leading to the formation of PtdIns(3,4,5)P$_3$ and Ins(1,4,5)P$_3$, respectively. PIP$_3$, PtdIns(3,4,5)P$_3$; IP$_3$, Ins(1,4,5)P$_3$.

have been through receptors with seven transmembrane domains and not through receptor tyrosine kinases.

In summary, $PtdIns(4,5)P_2$ hydrolysis occurs through a number of mechanisms. Certain receptors appear to act mainly through G proteins whereas other receptors may predominantly stimulate hydrolysis by mobilizing calcium. Although many receptors increase levels of $Ins(1,4,5)P_3$, glutamate and cholinergic muscarinic receptors stimulate the formation of large amounts of $Ins(1,3,4,5)P_4$. $Ins(1,3,4,5)P_4$ has specific binding sites in brain and appears to act as a second messenger which mediates signals different from those mediated by $Ins(1,4,5)P_3$. It is hypothesized that muscarinic receptors modulate NMDA receptors by causing a depolarization sufficient to remove the Mg^{2+} block, thereby allowing the neuronal plasticity essential for memory and learning. PtdIns 3-kinase and phosphoinositide-specific PLCγ activity is increased by tyrosine phosphorylation of these enzymes on tyrosine residues in response to growth factor stimulation. There is physical association between the ligand-activated receptor and the enzyme, mediated by SH2 and SH3 domains.

ACKNOWLEDGMENTS

This research was supported by US Public Health Service grants AA06069 and AG06660 from the National Institutes of Health, Bethesda, Maryland.

REFERENCES

Anderson D, Koch CA, Grey L, Ellis C, Moran MF, Pawson T (1990) Binding of SH2 domains of phospholipase Cγ1, GAP, and Src to activated growth factor receptors. Science 250:979-981.

Bansal VS and Majerus PW (1990) Phosphatidylinositol derived precursors and signals. Annu Rev Cell Biol 6:41-67.

Berridge MJ, Downes CP, Hanley MR (1989) Neural and developmental actions of lithium: A unifying hypothesis. Cell 59:411-419.

Carter HR, Wallace MA, Fain JN (1990) Purification and characterization of $PLC_{\beta m}$, a muscarinic cholinergic regulated phospholipase C from rabbit brain membrane. Biochim. Biophys. Acta 1054:119-128.

Chandler LJ and Crews FT (1989) Calcium stimulated versus G-protein mediated phosphoinositide hydrolysis in brain. In: Neurochemical aspects of phospholipid metabolism (Freysz L, Hawthorne JN, Toffano G, eds) pp 135-140. Padova, Italy: Liviana Press.

Chandler LJ and Crews FT (1990a) Calcium versus G-protein mediated phosphoinositide hydrolysis in rat cerebral cortical synaptoneurosomes. J Neurochem 55:1022-1030.

Chandler LJ and Crews FT (1990b) Calcium versus G-protein activated phosphoinositide hydrolysis in synaptoneurosomes from young and old rats. Ann NY Acad Sci 568:187-192.

Chandler LJ, Kurian P, Crews FT (1991) Effects of ethanol on inositol 1,3,4,5-tetrakisphosphate metabolism by rat brain homogenates. Alcoholism: Clinical and Experimental Research 15:136-140.

Chou CK, Dull TJ, Russel DS, Gherzi R, Lebwobl D, Ullrich A, Rosen OM (1987) Human insulin receptors mutated at the ATP-binding site lack protein tyrosine kinase activity and fail to mediate postreceptor effects of insulin. J Biol Chem 262:1842-1847.

Crews FT, Gonzales RA, Raulli R, McElhaney R, Pontzer N, Raizada MK (1988)

Interaction of calcium with receptor stimulated phosphoinositide hydrolysis in brain and liver. Ann NY Acad Sci 522:88-95.

Cunningham CC, Stossel TP, Kwiatkowski DJ (1991) Enhanced mobility in NIH 3T3 fibroblasts that overexpress gelsolin. Science 251:1233-1236.

Downes CP and Macphee CH (1990) *Myo*-inositol metabolites as cellular signals. Eur J Biochem 193:1-18.

Fain JN (1990) Regulation of phosphoinositide-specific phospholipase C. Biochim Biophys Acta 1053:81-88.

Gawler DJ, Potter BVL, Nahorski SR (1990) Inositol 1,3,4,5-tetrakisphosphate induced release of intracellular Ca^{2+} in SH-SY5Y neuroblastoma cells. Biochem J 272:519-524.

Gonzales RA and Crews FT (1984) Characterization of the cholinergic stimulation of phosphoinositide hydrolysis in rat brain slices. J Neurosci 4:3120-3127.

Gonzales RA and Crews FT (1988) Guanine nucleotide and calcium-stimulated inositol phospholipid hydrolysis in brain membranes. J Neurochem 50:1522-1528.

Gonzales RA, Feldstein JB, Crews FT, Raizada MK (1985) Receptor mediated inositide hydrolysis is a neuronal response: Comparison of primary neuronal and glial cultures. Brain Research 345:350-355.

Hill TM, Dean NM, Boynton AL (1988) Inositol 1,3,4,5-tetrakisphosphate induces Ca^{2+} sequestration in rat liver cells. Science 242:1176-1178.

Hoek JB, Taraschi TF, Higashi K, Rubin E, Thomas AP (1990) Phospholipase C activation by ethanol in rat hepatocytes is unaffected by chronic ethanol feeding. Biochem J 272:59-64.

Hunter T and Cooper JA (1985) Protein-tyrosine kinases. Annu Rev Biochem 54:897-930.

Irvine RF and Moor RM (1986) Micro-injection of inositol 1,3,4,5-tetrakisphosphate activates sea urchin eggs by a mechanism dependent on external Ca^{2+}. Biochem J 240:917-920.

Koch CA, Anderson D, Moran MF, Ellis C, Pawson T (1991) SH2 and SH3 domains: Elements that control interactions of cytoplasmic signaling proteins. Science 252:668674.

Kucera GL and Rittenhouse SE (1990) Human platelets from 3-phosphorylated phosphoinositides in response to α-thrombin, U46619 or GTPγS. J Biol Chem 265:5345-5348.

Margolis B, Li N, Koch A, Mahammadi M, Hurwitz DR, Zilberstein A, Ullrich A, Pawson T, Schlessinger J (1990) The tyrosine phosphorylated carboxy terminus of the EGF receptor is a binding site for GAP and PLC-γ. EMBO J 9:4375-4380.

Nicoletti F, Wroblewski JT, Fadda E, Costa E (1988) Pertussis toxin inhibits signal transduction at a specific metabolotropic glutamate receptor in primary cultures of cerebellar granule cells. Neuropharmacology 27:551-556.

Nishibe S, Wahl MI, Hernandez-Sotomayer SM, Tonks NK, Rhee SG, Carpenter G (1990) Increase of the catalytic activity of phospholipase C-γ1 by tyrosine phosphorylation. Science 250:1253-1256.

Pontzer NJ and Crews FT (1990) Desensitization of muscarinic stimulated hippocampal cell firing is related to PI hydrolysis and inhibited by lithium. J Pharmacol Exp Ther 253:921-929.

Ruderman NB, Kapeller R, White MF, Cantley LC (1990) Activation of phosphatidylinositol 3-kinase by insulin. Proc Natl Acad Sci USA 87:1411-1415.

Smrcka AV, Hepler JR, Brown KO, Sternweis PC (1991) Regulation of polyphosphoinositide-specific phospholipase C activity by purified G_q. Science 251:804-807.

Serunian LA, Haber MT, Fukui T, Kim JW, Rhee SG, Lowenstein JM, Cantley LC (1990) Polyphosphoinositides produced by phosphatidylinositol 3-kinase are poor substrates for phospholipase C from rat liver and bovine brain. J Biol Chem 264:17809-17815.

Sugiyama H, Ito I, Hirono C (1987) A new type of glutamate receptor linked to inositol phospholipid metabolism. Nature 325:531-533.

Ullrich A and Schlessinger J (1990) Signal transduction by receptors with tyrosine kinase activity. Cell 61:203-212.

Whitman M, Downes CP, Keeler M, Keeler T, Cantley LC (1988) Type I phosphatidylinositol kinase makes a novel inositol phospholipid, phosphatidylinositol 3-phosphate. Nature 332:644-646.

Whitworth P and Kendall DA (1988) Lithium selectively inhibits muscarinic receptor stimulated inositol tetrakisphosphate accumulation in mouse cerebral cortex slices. J Neurochem 51:258-265.

DIACYLGLYCEROL COMPOSITION AND METABOLISM

IN PERIPHERAL NERVE

J. Eichberg[1] and X. Zhu[2]

[1]Department of Biochemical and Biophysical Sciences
University of Houston
Houston, TX 77204 USA

[2]Department of Cell Biology
Baylor College of Medicine
Houston, TX 77030 USA

ABSTRACT

The content and molecular species composition of 1,2-diacylglycerol (DAG) in rat sciatic nerve was determined and compared with the molecular species profiles for glycerophospholipid classes in order to gain information concerning the metabolic pathways of DAG formation. The level of DAG in freshly dissected epineurium-free nerve (44 ± 2 pmol/mg wet weight) was 10-40% of that in other tissues and cultured cells. The predominant DAG molecular species were 18:0/20:4 (30%) and 16:0/18:1 (17%). In comparison with phospholipid molecular species patterns, DAG was characterized by a substantial but lower proportion of the 18:0/20:4 species than was found in phosphoinositides, and a significant fraction of saturated species such as those found in phosphatidylcholine. In nerve from diabetic rats, both the content and arachidonoyl-containing molecular species of DAG were reduced. These species were also decreased in individual glycerophospholipids, except for phosphatidylinositol. The distribution of molecular species in phosphatidic acid (PA) did not resemble that of any other phospholipid. A large rise in DAG content occurred when nerve was incubated *in vitro*. Molecular species analysis indicated that phosphoinositides were the main source, especially during the initial period. This process was virtually abolished in a Ca^{2+}-free medium and probably reflects a response to tissue injury. Evidence was obtained for the isomerization of DAG to 1,3-diacylglycerol during incubation. PA content and molecular species composition of incubated nerve did not change. However, inclusion of propranolol, a PA phosphatase inhibitor, caused a 40% accumulation of PA within 10 min, suggesting that formation of this phospholipid is continuous. These findings support the conclusion that DAG is principally derived from phosphoinositides by phospholipase C hydrolysis, but a minor fraction could be derived from phosphatidylcholine either by the action of phospholipase C or via phospholipase D and PA phosphatase. The metabolic origins of PA appear to be diverse.

Neurobiology of Essential Fatty Acids
Edited by N.G. Bazan *et al.*, Plenum Press, New York, 1992

INTRODUCTION

The importance of 1,2-diacylglycerol (DAG) as a glycerophospholipid-derived second messenger is now widely appreciated. Numerous reports using either radioisotopes or mass measurements have documented that DAG is liberated in response to a variety of agonists in a number of tissues and cells. Among the pathways for receptor-linked DAG production (Fig. 1), the best known is the phospholipase C-catalyzed degradation of the phosphoinositides (Rana and Hokin, 1990). Recent findings have shown that DAG can also originate from the action of this enzyme on phosphatidylcholine and phosphatidylethanolamine. In addition, the presence of phospholipase D in animal tissues has been established and a pathway for DAG generation from phosphatidylcholine by the sequential actions of this enzyme and phosphatidic acid phosphatase has been clearly demonstrated. Finally, it is possible that DAG levels may be affected as a secondary consequence of hormone-activated sphingomyelin breakdown, i.e., production of DAG concomitant with the reaction of the released ceramide with phosphatidylcholine to transfer the phosphocholine head group and re-synthesize sphingomyelin (Dennis et al., 1991).

The principal function for DAG generated as a result of receptor activation is considered to be as a natural activator of several isoforms of protein kinase C. However, the neutral lipid can also be converted by diacylglycerol kinase to phosphatidic acid (PA), which accumulating evidence suggests has a variety of biological activities associated with Ca^{2+} homeostasis, modulation of signal transduction, and cell growth and proliferation (Loffelholz 1989; Billah and Anthes 1990; Exton, 1990).

Recently, work in our laboratory has documented the presence of muscarinic cholinergic receptors coupled to phosphoinositide breakdown in sciatic nerve and has also shown that nerve possesses phospholipase D activity that is stimulated by phorbol ester (Day et al., 1991; Eggen and Eichberg, unpublished data). We also found that the level of DAG is reduced in sciatic nerve from streptozotocin-induced diabetic rats and that this decrease is largely confined to arachidonoyl-containing molecular species (Zhu and Eichberg, 1990a). In light of these findings, it seemed worthwhile to examine the metabolic origin and fate of DAG in nerve from normal animals. The approach we have adopted is to analyze DAG content and its molecular species profile and to compare these with the molecular species composition of glycerophospholipids in fresh-

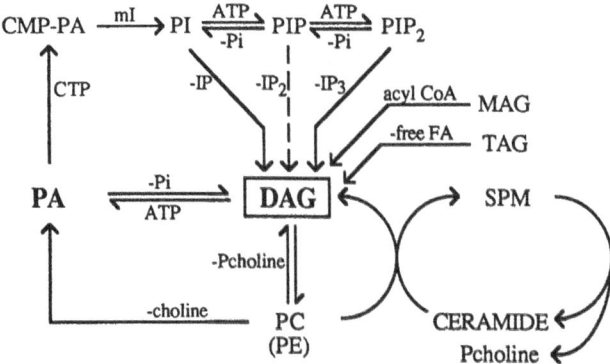

Figure 1. Pathways of DAG metabolism. PA: phosphatidic acid; PC: phosphatidylcholine; PE: phosphatidylethanolamine; CMP-PA: phosphatidyl-CMP (CDP-diacylglycerol); PI: phosphatidylinositol; PIP: phosphatidylinositol-4-phosphate; PIP_2: phosphatidylinositol-4,5-bisphosphate; SPM: sphingomyelin; MAG: monoacylglycerol; TAG: triacylglycerol; Pcholine: phosphocholine; mI: myo-inositol.

ly dissected sciatic nerve as well as in nerve incubated *in vitro* under a variety of conditions. The utilization of molecular species analysis to identify the precursor(s) of DAG and to examine its metabolism has been employed by several other investigators (Kennerly, 1987; Pessin and Raben, 1989; Lee and Hajra, 1991).

DAG CONTENT AND MOLECULAR SPECIES IN SCIATIC NERVE

In intact sciatic nerve, individual nerve fibers are surrounded by the endoneurial connective tissue layer, whereas bundles of fibers are arranged in fascicles and ensheathed by the perineurium. The entire nerve is covered by another connective tissue sheath, the epineurium, in which are embedded clusters of adipocytes. To measure DAG levels accurately, we found it necessary to purify DAG from nerve lipid extracts by thin layer chromatography (TLC) before assaying it by conversion to phosphatidic acid using [^{32}P]-ATP and a 1,2-diolein standard (Wright et al., 1988; Zhu and Eichberg, 1990a). Upon measurement of DAG in nerve frozen *in situ*, very different levels were present in proximal and distal regions (121±8 and 365±75 pmols/mg wet weight, respectively; means ± SEM for three rats each region). This disparity could be correlated with the presence of an adherent fat pad in the distal portion of the nerve which contained about 500 pmol DAG/mg wet weight and also with a high level of triacylglycerol in this region. A more relevant measure of DAG in the nerve itself was obtained after removing the epineurium from freshly dissected tissue (i.e., de-sheathing). This procedure yielded a DAG content of 44±2 pmol/mg wet weight (10 rats), with no difference between proximal and distal regions. This concentration is approximately 10-40% of the DAG content reported in several tissues and cultured cell lines (Preiss et al., 1986; Wright et al., 1988; Sunako et al., 1990).

For molecular species analysis, purified DAG was converted to the benzoate derivative and DAG benzoates were in turn purified by TLC. Molecular species were separated by HPLC using a reverse phase C18 column and a linear gradient of acetonitrile/isopropanol (Zhu and Eichberg, 1990a). Individual peaks were detected by UV absorption at 230 nm and calculation of areas was performed with the aid of a 1,2-dilaurin internal standard. Molecular species were identified: i) by their retention time in comparison with individual DAG standards, and ii) by recovery of individual DAG benzoate peaks, transmethylation, and identification of the fatty acid methyl esters produced by gas chromatography.

More than 20 peaks were obtained when derivatized DAG was subjected to HPLC separation. The major DAG molecular species were 18:0/20:4 (30%) and 16:0/18:1 (17%) (Fig. 2). Small amounts of several other polyunsaturated species were present, including 16:0/20:4 (3%), 18:0/22:6 + 18:1/18:2 (5%), and 16:0/18:2 (4%). Other identifiable species containing only saturated or monounsaturated fatty acids accounted for 21%. For comparison, the molecular species distributions of DAG in the fat pad attached to the distal portion of nerve and in the epineurium were determined. The profiles were strikingly different from that of the de-sheathed nerve. The principal molecular species in the adherent fat were: 16:0/22:6 + 18:2/18:2 (16%); 18:0/22:6 + 18:1/18:2 (34%); 16:0/18:2 (13%); 16:0/18:1 + 18:0/18:2 (12.5%) and 16:0/16:0 (14%). Less than 1% of DAG was accounted for by the 18:0/20:4 species. The corresponding proportions of these molecular species in the epineurium were 10%, 23%, 12%, 17%, and 2%; in addition, appreciable amounts of 18:0/20:4 (10%) and 18:1/18:1 (12%) were also present. It is likely that DAG composition in the fat pad and to a lesser extent in the epineurium is closely related to the metabolism of triacylglycerol in nerve and associated structures, in which the most abundant fatty acids are palmitic (24%), oleic (33%), and linoleic (29%) acids (Lin et al., 1985).

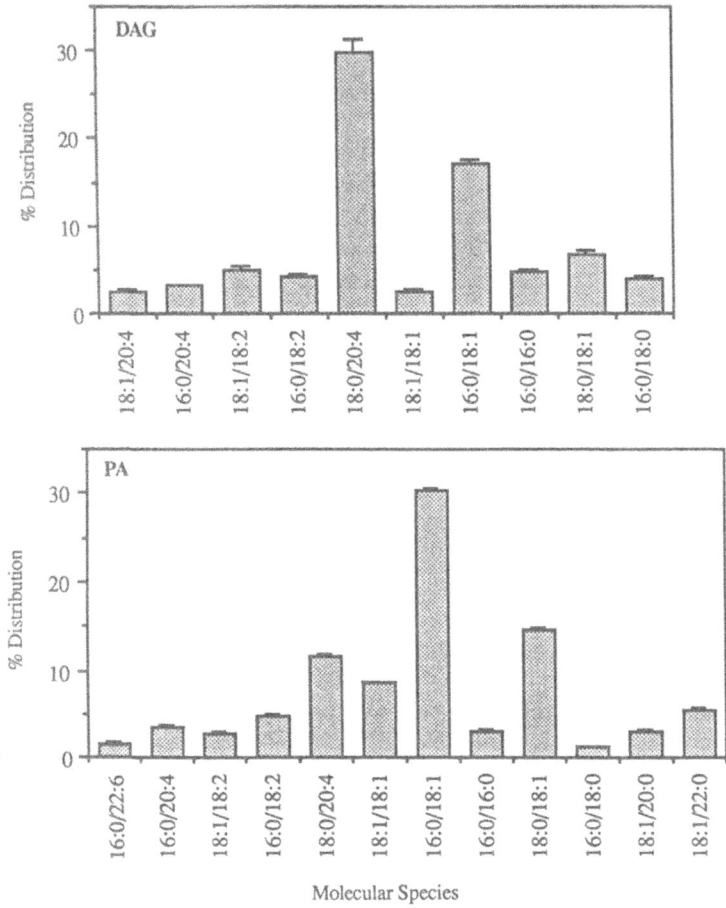

Figure 2. Percent distribution of principal molecular species in sciatic nerve of DAG and PA. Results are the means ± SEM for nerves from 10 rats for DAG and from 13 rats for PA. Some peaks eluted by HPLC contained more than one molecular species as indicated in parentheses. These were: 16:0/22:6 (18:2/18:2); 18:1/20:4 (18:1/22:5, 14:0/14:0); 18:1/18:2 (18:0/22:6); 16:0/18:2 (14:0/18:1, 18:1/22:4); 16:0/18:1 (18:0/18:2). The first listed species makes up more than 50% of each peak in total nerve diacylglycerophospholipids.

MOLECULAR SPECIES OF NERVE GLYCEROPHOSPHOLIPIDS

To reveal the possible sources of DAG in peripheral nerve, analysis of glycerophospholipid molecular species was undertaken. This was accomplished by two-dimensional TLC separation of de-sheathed nerve lipids followed by elution and phospholipase C hydrolysis of individual phospholipid classes (Mavis et al., 1972). When the liberated DAG was converted to the benzoate derivative and purified by TLC, one band was obtained, except in the case of ethanolamine-containing phospholipids which yielded discrete bands for DAG benzoates and alkenylacylglycerol benzoates. Polyphosphoinositides and PA were separated by one dimensional TLC, recovered, and subjected to hydrolysis with alkaline phosphatase (Ehle et al., 1985). DAG thereby released from PA

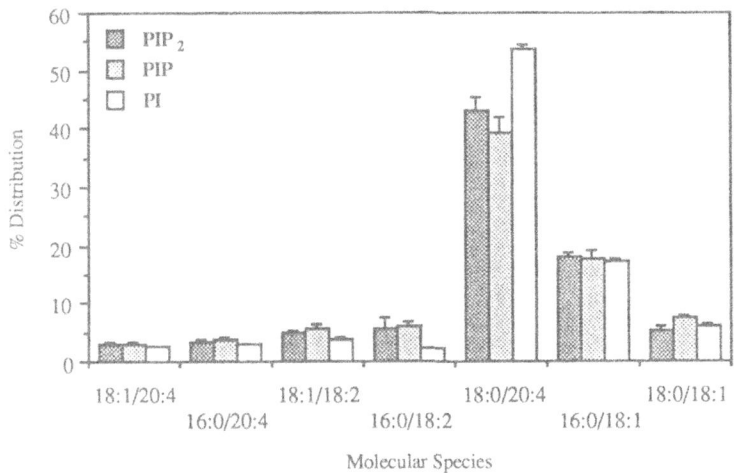

Figure 3. Percent distribution of principal phosphoinositide molecular species. Results are means ± SEM for nerves from four to eight rats.

was derivatized and analyzed, whereas PI produced from the polyphosphoinositides was treated with phospholipase C to obtain DAG.

Upon HPLC separation, DAG benzoates from each phospholipid class exhibited a distinct molecular species distribution. Not surprisingly, the pattern for all three phosphoinositides was very similar and was characterized by an abundance of the 18:0/20:4 species (Fig. 3). The distribution profile was strongly reminiscent of that previously reported for brain (Holub et al., 1970), including the slightly greater proportion of 18:0/20:4 in PI than in the polyphosphoinositides. In contrast, the principal phosphatidylcholine molecular species was 16:0/18:1 (52%) and substantial amounts of 16:0/16:0 and 18:0/18:1 were also present (Fig. 4). Interestingly, the ratio of 18:0/20:4 to 16:0/20:4, the two principal arachidonoyl-containing species, was about 18 to 1 in phosphoinositides, whereas it was 0.4 to 1 in phosphatidylcholine. This marked difference may prove useful as an indicator of the origins of DAG and PA (see below).

Phosphatidylethanolamine was characterized by an abundance of oleoyl-containing molecular species (Fig. 4) and this was also true for ethanolamine plasmalogen (data not shown). The diacyl lipid also contained nearly 10% of a species migrating in the position for 18:0/22:6 + 18:1/18:2. The pattern for phosphatidylserine was distinctive in that nearly 40% of its molecular species composition was accounted for by 18:0/18:1 and further, significant quantities of species containing 20 through 24 carbon fatty acids were present (Fig. 4).

Comparison of the molecular species profile for DAG with those of individual phospholipid classes reveals the DAG profile to be closely similar to the pattern for phosphoinositides, consistent with DAG production primarily by phospholipase C hydrolysis of these phospholipids. However, there are indications that a fraction of DAG arises from one or more other phospholipids. One indication is that the proportion of the 18:0/20:4 species was less in DAG than in phosphoinositides; a second is that the ratio of 18:0/20:4 to 16:0/20:4 DAG species was between 9 and 10, intermediate between PI and PC; and a third is that significant quantities of 16:0/16:0 and 16:0/18:0 were present in DAG but were scarcely detectable in phosphoinositides.

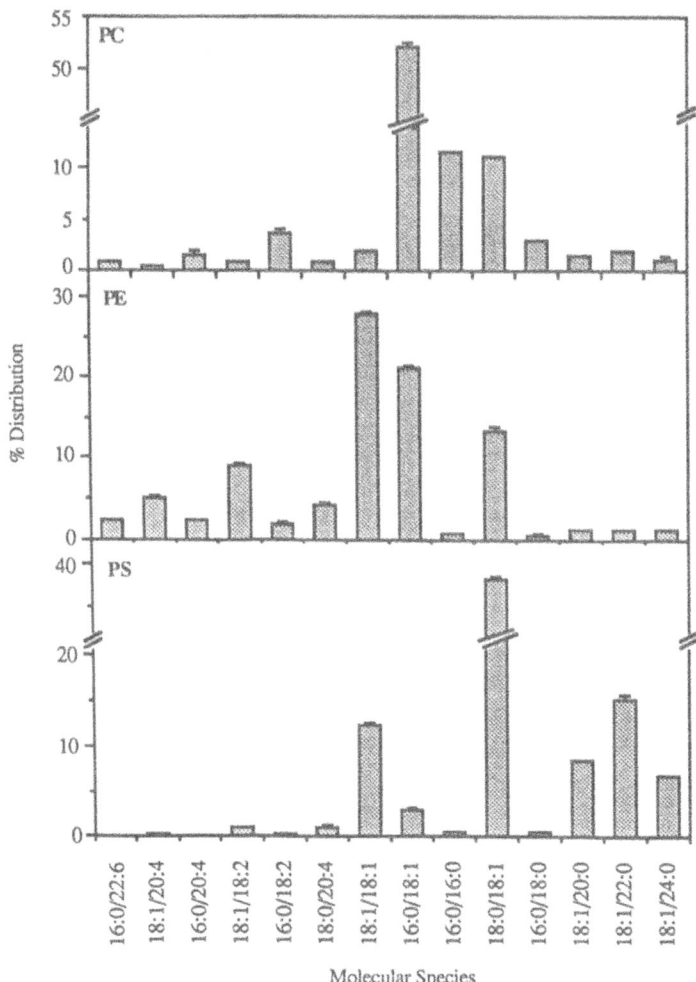

Figure 4. Percent distribution of principal molecular species in phosphatidylcholine (PC), phosphatidyl-ethanolamine (PE), and phosphatidylserine (PS). Results are means ± SEM for nerves from seven or eight rats.

The molecular species composition of PA did not resemble that of any other phospholipid (Fig. 2). The major species were 16:0/18:1 (30%), 18:0/18:1 (14%), and 18:0/20:4 (12%) and the ratio of 18:0/20:4 to 16:0/20:4 species was 3.6. This pattern could reflect the diverse sources of PA which can include not only its formation via phospholipase D from phosphatidylcholine and from phosphoinositides by the combined actions of phospholipase C and diacylglycerol kinase, but also an unknown contribution as a result of *de novo* synthesis from glycerophosphate.

A few analyses of the phospholipid molecular species in epineurium and fat pad were performed. In general, the molecular species profiles in the epineurium resembled those found in de-sheathed nerve, whereas phospholipids associated with the fat pad contained a greater proportion of linoleoyl- and arachidonoyl-containing molecular species.

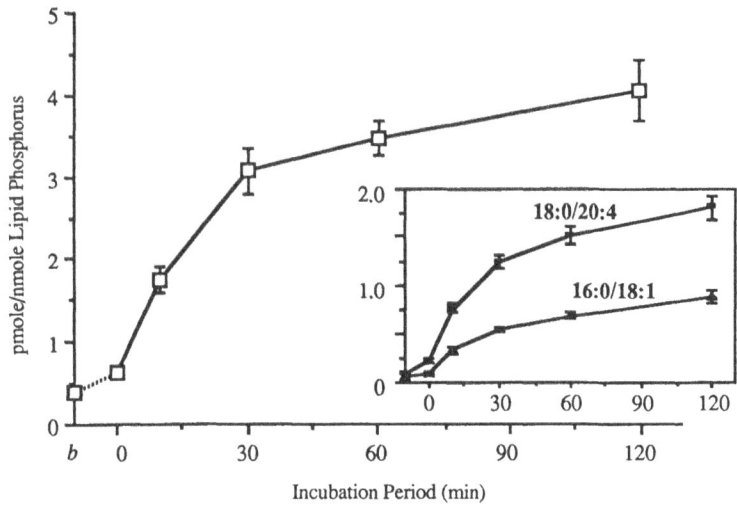

Figure 5. Time course of total DAG generation during nerve incubation. Sciatic nerves were cut into segments and incubated at 37°C for the times indicated. DAG was purified and its mass was determined according to Zhu and Eichberg, 1990a. Results are means ± SEM for nerves from five rats. The value at *b* indicates DAG content in freshly dissected nerve. (De-sheathed nerve contains 115 nmoles lipid P/mg wet weight.) Content at zero time was measured in nerves that had been cut into segments and kept in either ice cold saline or KRBG for 10 min prior to incubation. Inset: Time course of 18:0/20:4 and 16:0/18:1 DAG generation in the same samples.

EFFECT OF NERVE INCUBATION ON DAG CONTENT AND MOLECULAR SPECIES

In view of our findings that nerve possesses phospholipase C coupled to muscarinic receptors and also contains phospholipase D, studies were initiated to examine the factors that influence DAG production in sciatic nerve incubated *in vitro*. De-sheathed nerve was cut into segments while being maintained at 0°C in ice-cold Krebs-Ringer bicarbonate medium containing 5 mM glucose (KRBG). During this procedure, the content of DAG increased 25-50% above that in uncut nerve, indicating that sectioning alone enhanced DAG generation (Fig. 5). Subsequent incubation of the segments in KRBG at 37°C elicited a much larger, time-dependent rise in DAG content. After 10 min, DAG levels were nearly three times higher than the levels in sectioned nerve; this value increased 5 to 6 times after 30 min and continued to rise slowly for the next 90 min (Fig. 5).

The levels of all DAG molecular species rose during incubation but at different rates. The fastest rising species were 18:0/20:4 and 16:0/18:1, which increased about 20 and 17 times respectively from the basal level in unsectioned nerve after 2 hr (Fig. 5, inset). Other species rose more gradually over this time period; for example, 18:0/18:1 increased 12-fold, linoleoyl and docosahexaenoyl-containing species 5-6 fold, and the saturated species (16:0/16:0 and 16:0/18:0) no more than 3-fold. Calculation reveals that the molecular species pattern of DAG liberated during nerve sectioning as well as during the first 10 min of incubation contained 49% 18:0/20:4 and hence more closely resembles the profile of phosphoinositides than does the distribution in DAG from freshly dissected nerve. Taken together, these results strongly suggest that the preponderance of DAG released during incubation arises from phosphoinositides and that this process is especially prominent at early times. However, the increase in all DAG molecular species with time indicates that other phospholipids also undergo breakdown to yield DAG.

Incubation conditions were modified in order to investigate what factors would affect DAG liberation. There was no difference in DAG generation in the presence of 0, 5, or 30 mM glucose. Neither the inclusion of 10 mM Li^+, which might be expected to deplete the supply of myo-inositol used for phosphoinositide synthesis (Zhu and Eichberg, 1990b), nor the addition of 3 mM myo-inositol, which could enhance phosphoinositide formation, had any effect. Carbamylcholine (1 mM), which has been shown to increase inositol phosphate release in nerve (Day et al., 1991), did not measurably increase DAG levels, but any increase would be small and hence would almost certainly be masked by the large quantity of DAG produced in the absence of agonist. Other agents without effect included the PA phosphatase inhibitor, propranolol (Pappu and Hauser, 1983), the DAG kinase inhibitor, R59949 (de Chaffoy de Courcelles et al., 1989), and the DAG lipase inhibitor, RG80267 (Sutherland and Amin, 1982).

The accumulation of DAG was greatly diminished when nerve segments were incubated in isotonic saline and was completely abolished in KRBG from which Ca^{2+} was omitted and to which 1 mM EGTA was added. Indeed, under the latter conditions the proportion of 18:0/20:4 DAG was reduced by more than 50% from that in freshly dissected nerve, suggesting metabolism of preexisting neutral lipid had taken place. These observations allow us to consider that DAG produced in sectioned and incubated nerve is derived from phospholipid precursors in damaged portions of the tissue to which Ca^{2+} in the medium has access, thereby activating phospholipase C. Consequently, removal of the cation effectively inhibits this process. Thus, it may be possible to define incubation conditions in which the release of DAG from traumatized tissue can be minimized so that studies of receptor-mediated cell signaling responses involving DAG can be conducted.

EVIDENCE FOR THE FORMATION OF 1,3-DIACYLGLYCEROLS DURING NERVE INCUBATION

During a 2-hr incubation of nerve, a very small unidentified peak that eluted just

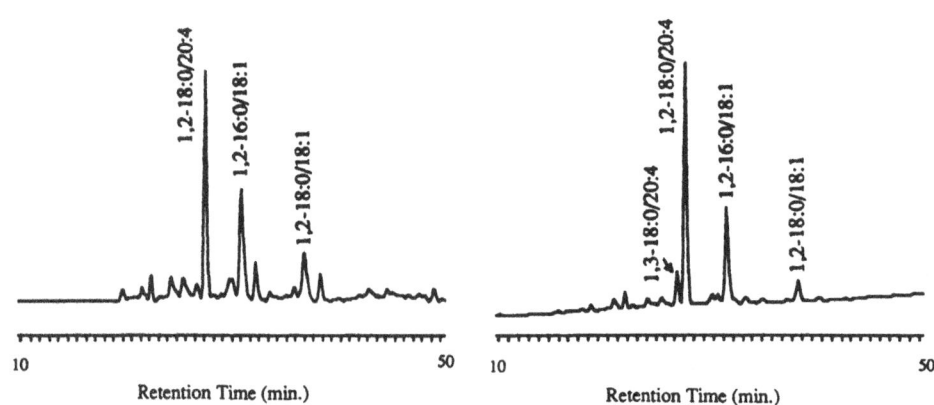

Figure 6. Formation of 1,3-diacylglycerol during nerve incubation. HPLC separation of DAG molecular species from 2 freshly dissected nerves (left panel) and after a 2-hr incubation of segments from one-half of a sciatic nerve (right panel). To display equal heights for the 18:0/20:4 peaks as shown, the absolute scale of the right panel was reduced 7-fold.

before 18:0/20:4 DAG increased nearly 100 times and accounted for 23% of the combined peaks (Fig. 6). This molecular species was tentatively identified as 1,3-18:0/20:4 diacylglycerol as follows: DAG was isolated following incubation and converted to PA by reaction with *E. coli* DAG kinase (Preiss et al., 1986), for which 1,3-diacylglycerol is not a substrate. Separation of the residual DAG molecular species showed that all of them had virtually disappeared except for the unknown. When the remaining peak was recovered and subjected to fatty acid analysis by methylation and gas chromatography, two major peaks, identified as stearoyl and arachidonoyl methyl esters, were present.

The formation of 1,3-diacylglycerol isomers from DAG by acyl migration during chemical manipulations is well known and suitable precautions must be taken to avoid this artifact (Serdarevich, 1967; Kodali et al., 1990). However, as far as we are aware, extensive isomerization during tissue incubation has not been previously noted. Kodali et al. (1990) found that when DAG was incubated in aqueous solution with egg phosphatidylcholine, approximately 6% was converted to the 1,3-isomer after 2 hr. The greater isomerization we observed during this time could have been due to a component in the incubation medium which enhanced the rate of this chemical process or conceivably an enzymatic process could be in part involved. Curiously, the addition of propranolol (0.6 mM) to the incubation medium significantly increased the rate of 1,3-diacylglycerol formation.

ALTERATIONS IN PHOSPHATIDIC ACID LEVELS DURING NERVE INCUBATION

The content of PA in freshly dissected nerve was estimated from the sum of its molecular species to be at least 200 pmol/mg wet weight or about five times the level of DAG. Incubation of nerve for up to 60 min failed to affect either the amount of PA present or its molecular species composition. The effect of several agents added to the incubation medium was examined (Fig. 7). After 10 min incubation, carbachol, which

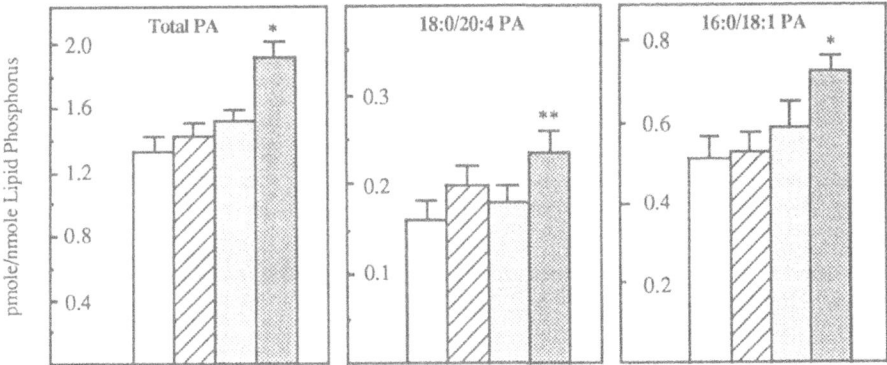

Figure 7. Effects of carbachol, 4-ß-phorbol dibutyrate, and propranolol on levels of total nerve PA and its major molecular species. Nerve segments were incubated in KRBG for 10 min either in the absence of drug or in the presence 1 mM carbachol, 1 μM phorbol dibutyrate, or 0.6 mM propranolol. PA was purified and its molecular species were separated. Total PA was calculated as the sum of its major molecular species which represented about 75% of all PA. Data are means ± SEM for four groups of nerve segments pooled from two experiments. *p<0.01, **p<0.05 different from drug-free controls as determined by one way ANOVA followed by a Tukey-Kramer test.

might be expected to elevate the DAG level and hence enhance PA synthesis, did not measurably change the amount of this phospholipid. 4-ß-Phorbol dibutyrate, which stimulates phospholipase D in nerve (Eggen and Eichberg, unpublished data), showed a tendency to increase PA content, but the effect did not attain statistical significance. However, propranolol increased the content of PA by 40% without appreciably altering its molecular species composition. The combination of propranolol and phorbol ester produced a small additional increment. The rise in PA content evoked by propranolol was most likely brought about by the inhibition of PA phosphatase and, therefore, this result implies that this phospholipid is being continuously generated in nerve. The source of this newly synthesized PA remains to be unequivocally identified. However DAG is unlikely to be the principal precursor because if it were, synthesis of the estimated additional 80 pmol per mg wet weight of PA elicited by propranolol should have resulted in a significant decline in DAG accumulation and none was observed. A second route would be *de novo* synthesis of PA by utilization of the free fatty acid pool in nerve, in which the major components are palmitic, stearic, and oleic acids (Yao et al., 1981). A third source might be via phospholipase D hydrolysis of phosphatidyl-choline. Although the molecular species profile of PA does not match that of phospha-tidylcholine, it is possible that a small pool of the latter phospholipid with a molecular species pattern distinct from the bulk of phosphatidylcholine might be selectively degraded to PA.

ALTERATIONS OF DAG AND GLYCEROPHOSPHOLIPID MOLECULAR SPECIES IN NERVE FROM DIABETIC RATS

Evaluation of DAG composition and metabolism in nerve from diabetic animals became of interest to us in view of the findings that i) reduced nerve Na^+-K^+-ATPase activity in experimental diabetic neuropathy can be restored by inclusion of either phorbol ester or DAG in the assay medium (Lattimer et al., 1989), and ii) the continued turnover of a discrete pool of PI which appears to be necessary for the maintenance of normal enzyme activity may be depressed in diabetic nerve because of the depletion of a critical compartment of myo-inositol in the tissue needed for PI synthesis (Zhu and Eichberg, 1990b; Simmons and Winegrad, 1991). Diminished phosphoinositide turnover could be expected to lower the DAG content in diabetic nerve and hence possibly reduce activation of a protein kinase C-dependent step involved in regulation of Na^+-K^+-ATPase.

We measured DAG content and molecular species distribution in rats injected with streptozotocin and sacrificed 8-12 weeks later. The content of DAG in de-sheathed nerve from diabetic animals was decreased 23% on a tissue wet weight basis as compared to age-matched controls (Zhu and Eichberg, 1990a). Furthermore, the molecular species of DAG from diabetic nerve showed a 29% decrease in the level of 18:0/20:4 DAG and an even sharper fall in 16:0/20:4 DAG and 18:1/20:4 DAG (49% and 47% respectively). There was also an indication that the proportion of linoleoyl-containing species was elevated, but this change was not clear cut because several of these species co-elute with others (e.g., 18:1/18:2 and 18:0/22:6) upon HPLC and hence are not obtained pure. These findings are consistent with the decrease in arachidonic acid and the increase in linoleic acid in the lipids of several tissues from diabetic animals (Holman et al., 1983; Lin et al., 1985).

Molecular species analysis of phospholipids from diabetic animals revealed that phosphatidylcholine, phosphatidylethanolamine, and PA lose up to 50% of all arachidonoyl-containing species but very little change occurs in PI. Except for phosphatidylcholine, the decrease in 16:0/20:4 was greater than that of 18:0/20:4. The depletion of arachidonic acid in diabetic tissues may be explained by the reported marked deficiency in Δ6 desaturase and partial loss of Δ5 desaturase activities which are necessary for its synthesis (Eck et al., 1979; Poisson, 1985; Dang et al., 1989). In this connection, it is noteworthy that when diabetic rats are fed a diet supplemented with γ-linolenic acid, the immediate precursor of arachidonic acid, the characteristic reduction in peripheral nerve conduction velocity is prevented (Julu, 1988; Tomlinson et al., 1989 Cameron et al., 1991). Moreover, γ-linoleic acid administration has also been reported to be beneficial for human diabetic neuropathy (Jamal and Carmichael, 1990).

CONCLUSIONS

The analysis and comparison of molecular species distribution for individual glycerophospholipid classes and DAG in freshly dissected nerve clearly demonstrates the existence of distinctive patterns for each. While these data cannot unambiguously establish the precursor(s) or metabolic pathways for DAG formation, they can provide valuable clues. A reasonable conclusion is that nerve DAG is derived principally by phospholipase C hydrolysis of phosphoinositides, but that a portion is generated by degradation of other phospholipids, mainly phosphatidylcholine. Thus, the molecular species profile for DAG resembles the distribution in phosphoinositides, yet contains saturated species such as 16:0/16:0 and 16:0/18:0, which are virtually absent in PI but are relatively abundant in phosphatidylcholine. Moreover, the ratio of 16:0/20:4 DAG to 18:0/20:4 DAG is intermediate between that for phosphoinositides and phosphatidylcholine. Finally, the finding that in diabetic nerve, 16:0/20:4 DAG is much more depressed than 18:0/20:4 DAG, coupled with the observations that PI exhibits virtually no loss of these molecular species and that PC shows a nearly 50% decrease in both species, can be understood if most DAG is assumed to arise through breakdown of these two phospholipids. However, additional contributions from phosphatidylethanolamine and phosphatidylserine, which could serve as major sources of 18:1/18:1 DAG, via de novo synthesis by acylation of monoacylglycerol (which might furnish some of the fully saturated molecular species) or through dephosphorylation of PA, cannot be discounted.

The rapid generation of DAG during nerve incubation is almost certainly a reflection of tissue injury which exposes phosphoinositides to the action of a phospholipase C that is activated by Ca^{2+}. Although this phenomenon has thus far hindered studies of the physiological responses of nerve phospholipid metabolism to receptor activation by agonists, the likelihood that it can be controlled should make these studies feasible in the future.

Lastly, the molecular species profile of PA indicates that it originates from diverse sources. At present, available information from experiments using propranolol suggests that a combination of de novo synthesis and phospholipase D-catalyzed hydrolysis of phosphatidylcholine may be the major pathways involved. However, it can also be calculated from molecular species data obtained from freshly isolated nerve that phosphoinositides could serve as the precursor of about 25% of PA through phospholipase C degradation followed by phosphorylation of the DAG produced. It is to be hoped that further investigations of changes in molecular species distribution in DAG and glycerophospholipids, together with experiments employing radioisotopes, will increase our knowledge of DAG metabolism in normal nerve and in pathological states such as diabetes.

ACKNOWLEDGMENTS

This work was supported by NIH grant DK30577.

REFERENCES

Billah MM and Anthes JC (1990) The regulation and cellular functions of phosphatidylcholine hydrolysis. Biochem J 269:281-291.

Cameron NE, Cotter MA, Robertson S (1991) Essential fatty acid supplementation: Effects on peripheral nerve and skeletal muscle function and capillarization in streptozocin-induced diabetic rats. Diabetes 40:532-539.

de Chaffoy de Courcelles D, Roevens P, Van Beller H, Kennis L, Somers Y, De Clerck F (1989) The role of endogenously formed diacylglycerol in the propagation and termination of platelet activation. J Biol Chem 264:3274-3285.

Dang AQ, Kemp K, Faas FH, Carter WJ (1989) Effects of dietary fats on fatty acid composition and Δ5 desaturase in normal and diabetic rats. Lipids 24: 882-889.

Day NS, Berti-Mattera LN, Eichberg J (1991) Muscarinic cholinergic receptor-mediated phosphoinositide metabolism in rat sciatic nerve. J Neurochem 56:1905-1913.

Dennis EA, Rhee SG, Billah MM, Hannun YA (1991) Role of phospholipases in generating lipid second messengers in signal transduction. FASEB J 5:2068-2077.

Eck MG, Wynn JO, Carter WJ, Faas FH (1979) Fatty acid desaturation in experimental diabetes mellitus. Diabetes 28:479-485.

Ehle H, Mueller E, Horn A (1985) Alkaline phosphatase of calf intestine hydrolyzes phospholipids. FEBS Lett 183:413-416.

Exton JH (1990) Signaling through phosphatidylcholine breakdown. J Biol Chem 265: 1-4.

Holman RT, Johnson SB, Gerrard JM, Mauer SM, Kupcho-Sandberg S, Brown DM (1983) Arachidonic acid deficiency in streptozotocin-induced diabetes. Proc Nat Acad Sci USA 80:2375-2379.

Holub BJ, Kuksis A, Thompson W (1970) Molecular species of mono- di- and triphosphoinositides of bovine brain. J Lipid Res 11:558-564.

Jamal GA and Carmichael H (1990) The effect of gamma-linolenic acid on human diabetic neuropathy: A double-blind placebo-controlled trial. Diabetic Med 7:319-323.

Julu POO (1988) Essential fatty acids prevent slowed nerve conduction in streptozotocin diabetic rats. J Diabetic Complications 2:185-188.

Kennerly DA (1987) Diacylglycerol metabolism in mast cells. J Biol Chem 262:16305-16313.

Kodali DR, Tercyak A, Fahey DA, Small DM (1990) Acyl migration in 1,2-dipalmitoyl-sn-glycerol. Chem Phys Lipids 52:163-170.

Lattimer SA, Sima AAF, Greene DA (1989) In vivo correction of Na$^+$,K$^+$-ATPase activity in diabetic nerve by protein kinase C agonists. Am J Physiol 256: E264-E269.

Lee C and Hajra AK (1991) Molecular species of diacylglycerols and phosphoglycerides and the postmortem changes in the molecular species of diacylglycerols in rat brains. J Neurochem 56:370-379.

Lin C-J, Peterson RG, Eichberg J (1985) The fatty acid composition of glycerolipids in nerve, brain and other tissues of the streptozotocin diabetic rat. Neurochem Res 10:1453-1465.

Loffelholz K (1989) Receptor regulation of choline phospholipid hydrolysis. Biochem Pharmacol 38:1543-1549.

Mavis RD, Bell RM, Vagelos PR (1972) Effect of phospholipase C hydrolysis of membrane phospholipids on membranous enzymes. J Biol Chem 247:2835-2841.

Pappu AS and Hauser G (1983) Propranolol-induced inhibition of rat brain cytosolic phosphatidate phosphohydrolase. Neurochem Res 8:1565-1575.

Pessin MS and Raben DM (1989) Molecular species of 1,2-diglyceride stimulated by α-thrombin in culture fibroblasts. J Biol Chem 264:8729-8738.

Poisson JP (1985) Comparative *in vivo* and *in vitro* study of the influence of experimental diabetes on rat liver linoleic acid Δ6 and Δ5 desaturation. Enzyme 34:1-4.

Preiss J, Loomis CR, Bishop WR, Stein R, Niedel JE, Bell RM (1986) Quantitative measurement of *sn*-1,2-diacylglycerols present in platelets, hepatocytes and *ras* and *sis*-transformed normal rat kidney cells. J Biol Chem 258:764-769.

Rana RS and Hokin LE (1990) Role of phosphoinositides in transmembrane signaling. Physiol Rev 70:115-163.

Serdarevich B (1967) Glyceride isomerization in lipid chemistry. J Am Oil Chem Soc 44:381-393.

Simmons DA and Winegrad AI (1991) Elevated extracellular glucose inhibits an adenosine-(Na^+, K^+)-ATPase regulatory system in rabbit aortic wall. Diabetologia 34:157-163.

Sunako M, Kawahara Y, Hirata K, Tsuda T, Yokoyama M, Fukazaki H, Takai Y (1990) Mass analysis of 1,2-diacylglycerol in cultured rabbit vascular smooth muscle cells. Hypertension 15:84-88.

Sutherland CA and Amin D (1982) Relative activities of rat and dog platelet phospholipase A_2 and diglyceride lipase: Selective inhibition of diglyceride lipase by RHC80267. J Biol Chem 257:14006-14010.

Tomlinson DR, Robinson JP, Compton AM, Keen P (1989) Essential fatty acid treatment: Effects on nerve conduction, polyol pathway and axonal transport in streptozotocin diabetic rats. Diabetologia 32:655-659.

Wright TM, Rangan LA, Shin HS, Raben DM (1988) Kinetic analysis of 1,2-diacylglycerol mass levels in cultured fibroblasts. J Biol Chem 263:9374-9380.

Yao JK, Dyck PJ, Van Loon JA, Moyer TP (1981) Free fatty acid composition of human and rat peripheral nerve. J Neurochem 36:1211-1218.

Zhu X and Eichberg J (1990a) 1,2-Diacylglycerol content and its arachodonyl-containing molecular species are reduced in sciatic nerve from streptozotocin-induced diabetic rats. J Neurochem 55:1087-1090.

Zhu X and Eichberg J (1990b) A pool of myo-inositol needed for phosphatidyl-inositol synthesis is depleted in diabetic nerve. Proc Nat Acad Sci USA 87:9818-9822.

CONTRIBUTORS

Koji Abe, Department of Neurology, Tohoku University School of Medicine, Sendai, Japan

Mayumi Abe, Center for Neurobiology and Behavior, Howard Hughes Medical Institute, Columbia University, New York, NY 10037, USA

B.W. Agranoff, Mental Health Research Institute, Departments of Psychiatry, Biological Chemistry and Pharmacology, Neuroscience Laboratory Building, University of Michigan, Ann Arbor, MI 48104-1687, USA

G. Aleppo, Institute of Pharmacology, University of Catania, School of Medicine, Catania, Italy

R.E. Anderson, Departments of Ophthalmology and Biochemistry, Baylor College of Medicine, Houston, TX 77030, USA

Masashi Aoki, Department of Neurology, Tohoku University School of Medicine, Sendai, Japan

Tsutomu Araki, Department of Neurology, Tohoku University School of Medicine, Sendai, Japan

E. Aronica, Institute of Pharmacology, University of Catania, School of Medicine, Catania, Italy

Yoshinori Asaoka, Department of Biochemistry, Kobe University School of Medicine, Kobe 650, Japan; and Biosignal Research Center, Kobe University, Kobe, 657, Japan

Marta I. Aveldaño, Instituto de Investigaciones Bioquímicas, Universidad Nacional del Sur-CONICET, 8000 Bahía Blanca, Argentina

Nicolas G. Bazan, LSU Eye Center and Neuroscience Center, Medical Center School of Medicine in New Orleans, Louisiana State University, 2020 Gravier Street, Suite B, New Orleans, LA 70112, USA

K. Beckman, Department of Chemical Pathology, Adelaide Medical Centre for Women and Children, North Adelaide, South Australia 5006

R. Bell, Departments of Biochemistry and Medicine, Duke University Medical Center, Durham, NC 27710, USA

Dale L. Birkle, Department of Pharmacology and Toxicology, West Virginia University, Health Science Center North, Morgantown, WV 26506, USA

Michelle Bonneil, INSERM Unité 130, 18 avenue Mozart, 13009 Marseille, France

Jean-Marie Bourre, INSERM Unité 26, Hôpital Fernand Widal, 75475 Paris cedex 10, France

Alessandro Bruni, Department of Pharmacology, University of Padova, Padova, Italy

V. Bruno, Institute of Pharmacology, University of Catania, School of Medicine, Catania, Italy

D. Burns, Departments of Biochemistry and Medicine, Duke University Medical Center, Durham, NC 27710, USA

Zenobia Byczko, Department of Physiology and Biophysics, Dalhousie University, Halifax, Nova Scotia, Canada B3H 4H7

P.L. Canonico, Chair of Pharmacology, University of Pavia, School of Dentistry, Pavia, Italy

R.O. Carlson, Mental Health Research Institute, Departments of Psychiatry, Biological Chemistry and Pharmacology, Neuroscience Laboratory Building, University of Michigan, Ann Arbor, MI 48104-1687, USA

P. Cassutti, Center of Neuropharmacology & Institute of Pharmacological Sciences, University of Milan, Italy

L.J. Chandler, University of Florida College of Medicine, Department of Pharmacology, Gainesville, FL 32610, USA

Jean Chaudière, INSERM Unité 26, Hôpital Fernand Widal, 75475 Paris cedex 10, France

Tak-Ming Chiu, Neurology Department, Harvard Medical School, and Neurology Department & Brain Imaging Center, McLean Hospital, Belmont, MA 02178, USA

M.T. Clandinin, Nutrition and Metabolism Research Group, Department of Foods & Nutrition and Department of Medicine, 533 Newton Research Building, University of Alberta, Edmonton, Alberta, Canada T6G 2C2

K.J. Clark, Department of Paediatrics and Child Health, Flinders Medical Centre, Bedford Park S.A. 5042, Australia

Michel Clément, INSERM Unité 26, Hôpital Fernand Widal, 75475 Paris cedex 10, France

A. Copani, Institute of Pharmacology, University of Catania, School of Medicine, Catania, Italy

Michael A. Crawford, Institute of Brain Chemistry and Human Nutrition, Hackney Hospital, Homerton High Street, London ES 6BE, UK

F.T. Crews, University of Florida College of Medicine, Department of Pharmacology, Gainesville, FL 32610, USA

V. D'Agata, Institute of Pharmacology, University of Catania, School of Medicine, Catania, Italy

D.S. Damron, Dept. Biological Sciences, Kent State University, Kent, OH 44242, USA

Raymond A. Deems, Department of Chemistry, University of California, San Diego, La Jolla, California 92093, USA

Edward A. Dennis, Department of Chemistry, University of California, San Diego, La Jolla, California 92093, USA

R.V. Dorman, Dept. Biological Sciences, Kent State University, Kent, OH 44242, USA

Odile Dumont, INSERM Unité 26, Hôpital Fernand Widal, 75475 Paris cedex 10, France

Georges Durand, INRA, CNRZ Jouy-en-Josas, France

J. Eichberg, Department of Biochemical and Biophysical Sciences, University of Houston, Houston, TX 77204, USA

Akhlaq A. Farooqui, Department of Medical Biochemistry, 1645 Neil Ave., Rm 471, The Ohio State University, Columbus, OH 43210, USA

Steven J. Feinmark, Department of Pharmacology, Howard Hughes Medical Institute, Columbia University, New York, NY 10037, USA

S.K. Fisher, Mental Health Research Institute, Departments of Psychiatry, Biological Chemistry and Pharmacology, Neuroscience Laboratory Building, University of Michigan, Ann Arbor, MI 48104-1687, USA

E.J. Freeman, Dept. Biological Sciences, Kent State University, Kent, OH 44242, USA

R. Galimberti, Center of Neuropharmacology & Institute of Pharmacological Sciences, University of Milan, Italy

C. Galli, Institute of Pharmacological Sciences, University of Milan, Via Balzaretti, 9, 20133, Milan, Italy

R.A. Gibson, Department of Paediatrics and Child Health, Flinders Medical Centre, Bedford Park S.A. 5042, Australia

William C. Gordon, LSU Eye Center and Neuroscience Center, Louisiana State University Medical Center School of Medicine in New Orleans, New Orleans, LA 70112, USA

Diego Guidolin, Fidia Research Laboratories, Via Ponte della Fabbrica 3/A, 35031 Abano Terme, Italy

A.K. Hajra, Mental Health Research Institute, Departments of Psychiatry, Biological Chemistry and Pharmacology, Neuroscience Laboratory Building, University of Michigan, Ann Arbor, MI 48104-1687, USA

T.F.R. Hamm, Dept. Biological Sciences, Kent State University, Kent, OH 44242, USA

Y. Hannun, Departments of Biochemistry and Medicine, Duke University Medical Center, Durham, NC 27710, USA

Shaul Harel, Department of Neurobiology, Weizmann Institute of Science, Rehovot, Israel

K. Hargreaves, Neurological Research Unit, Department of Surgery, 146 Stuart Street, LaSalle Building, Queen's University, Kingston, Ontario, Canada K7L 3N6

Yutaka Hirashima, Department of Medical Biochemistry, 1645 Neil Ave., Rm 471, The Ohio State University, Columbus, OH 43210, USA

Lloyd A. Horrocks, Department of Medical Biochemistry, 1645 Neil Ave., Rm 471, The Ohio State University, Columbus, OH 43210, USA

Keizo Inoue, Faculty of Pharmaceutical Sciences, University of Tokyo, Hongo, Bunkyo-ku, Tokyo 113, Japan

D.W. Johnson, Department of Chemical Pathology, Adelaide Medical Centre for Women and Children, North Adelaide, South Australia 5006

Kenichiro Katsura, The Laboratory for Experimental Brain Research, Department of Neurobiology, Experimental Research Center, University Hospital S-221, 85 Lund, Sweden

Jun-ichi Kawagoe, Department of Neurology, Tohoku University School of Medicine, Sendai, Japan

Kyuya Kogure, Department of Neurology, Tohoku University School of Medicine, Sendai, Japan

C.A. Koutz, Department of Ophthalmology, Baylor College of Medicine, Houston, TX 77030, USA

Ichiro Kudo, Faculty of Pharmaceutical Sciences, University of Tokyo, Hongo, Bunkyo-ku, Tokyo 113, Japan

Baruch Kunievsky, Department of Neurobiology, Weizmann Institute of Science, Rehovot, Israel

P. Kurian, University of Florida College of Medicine, Department of Pharmacology, Gainesville, FL 32610, USA

Huguette Lafont, INSERM Unité 130, 18 avenue Mozart, 13009 Marseille, France

Jerzy W. Lazarewicz, Fidia-Georgetown Institute for the Neurosciences, Georgetown University School of Medicine, Washington DC, USA and Medical Research Centre, Polish Academy of Sciences, Warsaw, Poland

P. Lecchi, Center of Neuropharmacology & Institute of Pharmacological Sciences, University of Milan, Italy

C. Lee, Mental Health Research Institute, Departments of Psychiatry, Biological Chemistry and Pharmacology, Neuroscience Laboratory Building, University of Michigan, Ann Arbor, MI 48104-1687, USA

Xiang-Duan Li, Institute of Clinical Pharmacology, Shanghai Medical University, Shanghai, 200032, China

D.R. Lines, Department of Paediatrics and Child Health, Flinders Medical Centre, Bedford Park S.A. 5042, Australia

M. Makrides, Department of Paediatrics and Child Health, Flinders Medical Centre, Bedford Park S.A. 5042, Australia

F. Marangoni, Institute of Pharmacological Sciences, University of Milan, Via Balzaretti, 9, 20133, Milan, Italy

F. Märki, Research Department, Pharmaceuticals Division, Ciba-Geigy Ltd., CH-4002 Basel, Switzerland

Manuela Martinez, Unidad de Investigacíon Biomédica, Hospital Universitario Materno-Infantil, Valle de Hebron, Barcelona, Spain

Shinji Matsushima, Department of Biochemistry, Kobe University School of Medicine, Kobe 650, Japan; and Biosignal Research Center, Kobe University, Kobe, 657, Japan

Jorge H. Medina, Instituto de Biología Celular, Universidad de Buenos Aires, Paraguay 2155, 3° piso, Buenos Aires, Argentina

Lucia Mietto, Fidia Research Laboratories, Abano Terme, Italy

Hiroyuki Mishima, Department of Biochemistry, Kobe University School of Medicine, Kobe 650, Japan; and Biosignal Research Center, Kobe University, Kobe, 657, Japan

Makoto Murakami, Faculty of Pharmaceutical Sciences, University of Tokyo, Hongo, Bunkyo-ku, Tokyo 113, Japan

Mary G. Murphy, Department of Physiology and Biophysics, Dalhousie University, Halifax, Nova Scotia, Canada B3H 4H7

Gilles Nalbone, INSERM Unité 130, 18 avenue Mozart, 13009 Marseille, France

M.A. Neumann, Department of Paediatrics and Child Health, Flinders Medical Centre, Bedford Park S.A. 5042, Australia

F. Nicoletti, Institute of Pharmacology, University of Catania, School of Medicine, Catania, Italy

Yasutomi Nishizuka, Department of Biochemistry, Kobe University School of Medicine, Kobe 650, Japan; and Biosignal Research Center, Kobe University, Kobe, 657, Japan

Maria Grazia Nunzi, Fidia Research Laboratories, Via Ponte della Fabbrica 3/A, 35031 Abano Terme, Italy

P.J. O'Brien, Health Research Associates, Rockville, MD 20850, USA

Masahiro Oka, Department of Biochemistry, Kobe University School of Medicine, Kobe 650, Japan; and Biosignal Research Center, Kobe University, Kobe, 657, Japan

T. Okazaki, Departments of Biochemistry and Medicine, Duke University Medical Center, Durham, NC 27710, USA

Paolo Orlando, Servizio Radioisotopi, Universita Cattolica Sacro Cuore, Roma, Italy

Neville N. Osborne, Nuffield Laboratory of Ophthalmology, Oxford University, Walton Street, Oxford OX2 6AW, UK

Pietro Palatini, Department of Pharmacology, University of Padova, Largo E. Meneghetti 2, 35131 Padova, Italy

Gérard Pascal, INRA, CNRZ Jouy-en-Josas, France

R. Patel, University of Florida College of Medicine, Department of Pharmacology, Gainesville, FL 32610, USA

B.C. Paton, Department of Chemical Pathology, Adelaide Medical Centre for Women and Children, North Adelaide, South Australia 5006

Lucia Petrelli, Fidia Research Laboratories, Via Ponte della Fabbrica 3/A, 35031 Abano Terme, Italy

A. Petroni, Institute of Pharmacological Sciences, University of Milan, Via Balzaretti, 9, 20133, Milan, Italy

J. Pfeilschifter, Research Department, Pharmaceuticals Division, Ciba-Geigy Ltd., CH-4002 Basel, Switzerland

Michèle Piciotti, INSERM Unité 26, Hôpital Fernand Widal, 75475 Paris cedex 10, France

A. Pisani, Institute of Pharmacology, University of Catania, School of Medicine, Catania, Italy

Patrizia Polato, Fidia Research Laboratories, Via Ponte della Fabbrica 3/A, 35031 Abano Terme, Italy

A. Poulos, Department of Chemical Pathology, Adelaide Medical Centre for Women and Children, North Adelaide, South Australia 5006

G. Racagni, Center of Neuropharmacology & Institute of Pharmacological Sciences, University of Milan, Italy

B.S. Robinson, Department of Chemical Pathology, Adelaide Medical Centre for Women and Children, North Adelaide, South Australia 5006

Elena B. Rodriguez de Turco, LSU Eye Center and Neuroscience Center, Louisiana State University Medical Center School of Medicine in New Orleans, New Orleans, LA 70112, USA

E. Salinska, Medical Research Centre, Polish Academy of Sciences, Warsaw, Poland

Marek Samochocki, Department of Neurochemistry, Medical Research Centre, Polish Academy of Sciences, 3 Dworkowa str., 00-784 Warsaw, Poland

C. Schalkwijk, Centre for Biomembranes and Lipid Enzymology, Padualaan 8, 3584 CH Utrecht, The Netherlands

James H. Schwartz, Center for Neurobiology and Behavior, Howard Hughes Medical Institute, Columbia University, New York, NY 10037, USA

P. Sharp, Department of Chemical Pathology, Adelaide Medical Centre for Women and Children, North Adelaide, South Australia 5006

Tetsutaro Shinomura, Department of Biochemistry, Kobe University School of Medicine, Kobe 650, Japan; and Biosignal Research Center, Kobe University, Kobe, 657, Japan

Bo K. Siesjö, The Laboratory for Experimental Brain Research, Department of Neurobiology, Experimental Research Center, University Hospital S-221, 85 Lund, Sweden

H. Singh, Department of Chemical Pathology, Adelaide Medical Centre for Women and Children, North Adelaide, South Australia 5006

Douglas J. Steel, Department of Pathology, Howard Hughes Medical Institute, Columbia University, New York, NY 10037, USA

A.M. Stinson, Department of Biochemistry, Baylor College of Medicine, Houston, TX 77030, USA

Marta Stockert, Instituto de Biología Celular, Universidad de Buenos Aires, Paraguay 2155, 3° piso, Buenos Aires, Argentina

Joanna Strosznajder, Department of Neurochemistry, Medical Research Centre, Polish Academy of Sciences, 3 Dworkowa str., 00-784 Warsaw, Poland

E.B. Stubbs, Jr., Mental Health Research Institute, Departments of Psychiatry, Biological Chemistry and Pharmacology, Neuroscience Laboratory Building, University of Michigan, Ann Arbor, MI 48104-1687, USA

M. Suh, Nutrition and Metabolism Research Group, Department of Foods & Nutrition and Department of Medicine, 533 Newton Research Building, University of Alberta, Edmonton, Alberta, Canada T6G 2C2

Grace Y. Sun, Biochemistry Department, University of Missouri School of Medicine, Columbia, MO 65212, USA

Anoopkumar Thekkuveettil, Center for Neurobiology and Behavior, Howard Hughes Medical Institute, Columbia University, New York, NY 10037, USA

C. Tromba, Center of Neuropharmacology & Institute of Pharmacological Sciences, University of Milan, Italy

D. Trotti, Center of Neuropharmacology & Institute of Pharmacological Sciences, University of Milan, Italy

S. Usher, Department of Chemical Pathology, Adelaide Medical Centre for Women and Children, North Adelaide, South Australia 5006

H. van den Bosch, Centre for Biomembranes and Lipid Enzymology, Padualaan 8, 3584 CH Utrecht, The Netherlands

Giampietro Viola, Fidia Research Laboratories, Abano Terme, Italy,

A. Volterra, Center of Neuropharmacology & Institute of Pharmacological Sciences, University of Milan, Italy

R.D. Wiegand, Department of Ophthalmology, Baylor College of Medicine, Houston, TX 77030, USA

Bryan T. Woods, Neurology Department, Harvard Medical School, and Neurology Department and Brain Imaging Center, McLean Hospital, Belmont, MA 02178, USA

J.T. Wroblewski, Fidia-Georgetown Institute for the Neurosciences, Georgetown University School of Medicine, Washington DC, USA

Ephraim Yavin, Department of Neurobiology, Weizmann Institute of Science, Rehovot, Israel

Kimihisa Yoshida, Department of Biochemistry, Kobe University School of Medicine, Kobe 650, Japan; and Biosignal Research Center, Kobe University, Kobe, 657, Japan

Lin Yu, Department of Chemistry, University of California, San Diego, La Jolla, CA 92093, USA

Adriano Zanotti, Fidia Research Laboratories, Via Ponte della Fabbrica 3/A, 35031 Abano Terme, Italy

X. Zhu, Department of Cell Biology, Baylor College of Medicine, Houston, TX 77030, USA

Luis M. Zieher, Instituto de Biología Celular, Universidad de Buenos Aires, Paraguay 2155, 3º piso, Buenos Aires, Argentina

INDEX

Menhaden oil, 223, 227
Metabotropic receptors, 137-139,
    141-143
Microcirculation, 49
MK-801, 74, 79, 80, 82-84, 141, 142
Molecular species, 173, 178-180, 198,
    204, 232, 233, 236, 287-289, 296,
    300, 301, 332, 334, 335, 337,
    414-424
  composition of 1,2-diacylglycerol, 414
Mongolian gerbils, 83, 84, 184
Multilamellar vesicles (MLV), 378, 379,
    382-385, 388
Muscarinic receptors, 76, 77, 252, 258,
    260, 265, 404, 410, 420
Muscarinic acetylcholine receptors, 171

Na^+-K^+-ATPase, 123, 152, 153, 212,
    217, 394, 423
5'-N'ethylcarboxy-aminoadenosine
    (NECA), 91, 97, 98, 100
$[^3H]$-labeled binding, 93
Neuroblastoma cell line, 58
Neurodevelopmental disorder, 308
Neuron
  cell death, 155
  cerebellar, 73, 137-142
  firing rates, 404
  forebrain cholinergic, 396, 398
  glycerophospholipids in, 417
  membranes, 11, 21, 91, 101, 172,
    309, 326
  peripheral, 215, 227, 414, 417,
    424
  plasticity, 137, 362, 403, 405, 410
  toxicity, 137, 138
Neurotoxic damage, 155
Neurotransmission, 74, 81, 156, 183, 184,
    192, 330, 394
N-methyl-D-aspartate (NMDA), 58, 65,
    73-85, 123, 137-143, 147, 149,
    152, 184, 253, 260, 364, 368,
    403
  receptor, 73-75, 77-82, 84, 138,
    139, 141, 149, 155, 403, 404,
    410
    antagonist (MK-801), 74, 79, 80,
    82-84, 141, 142
Nordihydroguaiaretic acid (NDGA),
    122, 130-133
5'-Nucleotidase, 91, 93, 96, 97, 99

Omega-3 fatty acids, 172, 175, 176, 197,

Omega-3 fatty acids (continued)
    200, 202-204, 206-208, 297, 298,
    303
Omega-6 fatty acids, 171-173, 175, 200,
    202-204, 208, 359
Ouabain, 138, 140, 152-154
Oxidative damage, 47

Peroxisomes, 236, 237, 332, 333,
    338, 348
  diseases of, 310, 332, 334, 336,
    338, 348, 351, 353, 356-359
Pertussis toxin, 19, 77, 84, 261,
    404, 409
8-Phenyltheophylline, 92, 95
Phorbol ester, 19, 261, 265, 278,
    280-282, 369-371, 415, 423
  binding regions, 281
Phosphatase, 11, 20, 109, 112,
    224, 277, 403, 414, 415, 417,
    421, 423
Phosphatidic acid (PA), 11, 12, 16,
    19-21, 43, 57, 76, 179, 255, 256,
    277, 300, 414-416, 422
Phosphatidylcholine (PC), 12, 15, 19, 20,
    27, 37, 42, 57, 62, 106, 171, 173,
    198, 200-205, 208, 223, 224,
    232-235, 245, 246, 255, 256, 265,
    277, 300, 332, 334, 335, 362, 384,
    385, 414, 415, 418, 419, 422-424
  biosynthesis, 201, 202, 204, 208
Phosphatidylethanolamine (PE), 12, 27,
    37, 62, 63, 197, 198, 200-205, 208,
    209, 215, 223, 224, 245, 255, 256,
    277, 300, 379, 415, 418, 419, 423,
    424
  methyltransferase, 197, 202-205, 209
Phosphatidylinositol (PI), 12, 14, 15, 30,
    42, 43, 61, 171, 175, 176, 187,
    198, 239, 245, 249, 252, 255, 256,
    260, 269, 270, 272, 276, 277, 284,
    300, 364, 400, 414, 415
  3,4-bisphosphate, 400
  4,5-bisphosphate, 260, 276, 277, 400
  3-kinase, 400
  3-phosphate, 400
  3,4,5-trisphosphate, 400
Phosphatidylserine (PS), 12, 27, 65, 139,
    175, 187, 198, 208, 224, 245, 255,
    256, 281, 282, 318, 326, 327, 364,
    378, 379, 384, 386, 394, 418, 419,
    424
  interaction sites, 281

Protein kinase C (PKC) (continued)
  activation, 19, 20, 139, 143, 271,
    280-282, 365, 366, 368, 394
  regulation, 276, 278
  subspecies, 362-365, 369, 370
    PKC-α, 263-265
    PKC-γ, 261, 263-265
Proton magnetic resonance
    spectroscopy, 268

Quinacrine, 16, 73, 74, 76, 77, 79, 80,
    82, 83, 253-255

Rat
  cerebral cortex, 18, 100, 149, 253,
    369, 403-406, 408
  fetal brain, 317
  glomerular mesangial cells, 1, 6
  hippocampal slices, 73, 79, 80, 406
  sciatic nerve, 414
Receptor
  adenosine, 91, 98, 100, 101
  excitatory amino acid, 74-76, 81,
    84, 141, 149, 184, 322
  -mediated endocytosis, 378
  metabotropic, 137-139, 141-143
  muscarinic (M3), 76, 77, 252, 258, 260,
    265, 404, 410, 420
    cholinergic, 171
  -regulated phospholipases,
    cross talk, 19
  tyrosine kinase, 18, 407, 408, 410
  tyrosine kinase substrate, 408
  tyrosine residues, 407
Refsum's disease, 332, 333
Retina, 58, 77, 115, 118, 206, 208,
    215, 216, 218, 220, 232-239,
    241, 242, 260-265, 286, 287,
    288-292, 296-301, 303-305,
    332, 337, 339, 348, 349, 354,
    355, 357, 364
  degeneration, 291
  development, 206
  function, 290, 345
  lipids, 237, 298, 301
Retinal pigment epithelium (RPE),
    287-289, 292, 296, 297, 301,
    303-305
Retinopathy, 308, 310
Rhizomelic chondrodysplasia punctata,
    333, 348, 349
Rhodopsin, 206, 290, 291, 298,
    302, 303

Rod outer segment (ROS), 206,
    286-291, 298, 302, 303,
    305
  integral proteins, 288
  phospholipids, 286, 288-290

Salmon oil, 224, 225
Scopolamine binding, 171
Shuttle box test, 219
Signal transduction, 11, 19, 21, 35,
    43, 73, 74, 78-81, 84, 85, 171,
    177, 179, 276, 277, 282, 283, 284,
    317, 319, 322, 372, 394, 407, 408,
    415
Site-directed targeting, 377
SK-N-SH neuroblastoma cells, 171-173,
    175-179
Small unilamellar vesicles (SUV), 376,
    378-380, 382-386, 388
Spatial memory, 244, 394-397
  impairments, 395, 397
Spermatozoa, 232, 234, 238-242, 332,
    333, 339
Sphingolipidoses, 282, 283
Sphingolipids, 73, 78, 276, 277, 282-284
  in signal transduction, 277
Sphingomyelin cycle, 276, 282, 283
Sphingosine, 16, 276, 277, 282-284
Src homology (SH) domains, 408
Streptozotocin, 415, 423
Stroke, 43, 49, 50, 122
Subcellular distribution, 16, 159, 236
Substrate specificity, 12, 29
Superoxide dismutase, 149, 152, 153,
    227, 312
Synaptic plasticity, 395, 398, 403, 404
Synaptoneurosomes, 74, 252-258
Synaptosomes, 57, 58, 62, 63, 67,
    74, 121-124, 128, 132, 149-154,
    164, 215, 216, 220, 252, 366,
    369
  membranes, 18, 200, 201, 205, 206,
    326-330

T cells, 369-371
Temporal lobe
  epilepsy, 268
  seizures, 270
Thromboxane, 35, 43, 47, 49, 74, 83, 84,
    116, 151, 317
  B_2, 74, 83, 84, 317
Tissue cultured cells
  astrocytes, 115, 149, 150

The manufacturer's authorised representative in the EU is Springer
Nature Customer Service Centre GmbH, Europaplatz 3, 69115 Heidelberg,
Germany. If you have any concerns regarding our products, please
contact ProductSafety@springernature.com

Printed and bound by CPI Group (UK) Ltd, Croydon, CR0 4YY
23/04/2026
02095629-0012